ENCYCLOPAEDIA OF
NANO EDUCATION

ENCYCLOPAEDIA OF
NANO EDUCATION

James William

ANMOL PUBLICATIONS PVT. LTD.
NEW DELHI - 110 002 (INDIA)

ANMOL PUBLICATIONS PVT. LTD.
H.O.: 4374/4B, Ansari Road, Darya Ganj,
New Delhi-110 002 (India)
Ph.: 23278000, 23261597

B.O.: No. 1015, Ist Main Road, BSK IIIrd Stage
IIIrd Phase, IIIrd Block,
Bangalore - 560 085 (India)
Visit us at: www.anmolpublications.com

Encyclopaedia of Nano Education

First Published, 2006

ISBN 81-261-2976-X

PRINTED IN INDIA

Printed at Mehra Offset Press, Delhi.

Contents

Preface

This book entitled "Encyclopaedia of Nano Education" is one of the existing nanotech institutions, their projects, reports and education courses and research areas. A comprehensive account of nanotech companies, organisation and institutions. An overview of relationship between nanotechnology and nano education is presented herein. The modern global Issues regarding nanoparticles, nanostructured materials and polymer nanotechnology are discussed in detail. The subject areas of nanocomposites and nanostructured materials are covered details about their respective properties and applications. The inter-relationship among today's nanoworld, nano education and the all important process of self-assembly is highlighted. An overview of nanotech products is presented.

There are so many other applications of nano chemistry ranging from mundane to incredibly profound. Other applications include energy storage materials, photovoltaic cells, fuel cells, polymer nanocomposites, super-tough nanocoatings, landmine detectors, and thousands more. Although it is a field that has already seen its share of hype, there is a general consensus that over the long term, nano chemistry is likely to reconfigure everything. Nanostructured materials do not represent a new phenomenon. However, the ability to probe, manipulate, understand and engineer matter at atomic scales has only recently come within our grasp. Developments such as the invention of the Scanning Tunneling Microscope in 1981 have since made nanoscale science a reality. Today, the term nanotechnology is used to describe specific construction and manipulation performed on an atomic and molecular scale. Generally, systems with a maximum size of less than a hundred nanometers are classified as nanotechnology. Nano Education is a completely new production process in a molecular world, opening up possibilities far beyond what we have known before. Nature provides the inspiration for this new technology – from time immemorial, nanomachines have been at work in all organic cells, building our world of plants, animals and people from a store of chemical elements.

Experts world over believe nanotechnology will have a more far-reaching impact than the silicon integrated chip with many more applications than electronics. Nanotechnology is expected to revolutionize certain areas, such as semiconductors, pharmaceuticals and materials. The total societal impact of nanotechnology is expected to be much greater than that of the silicon integrated circuit because it is applicable in many more fields than electronics. Nanotechnology experts believe that because of nano chemistry, we would see more changes in the next 30 years than we saw in the entire last century. This book tends to cover the aforementioned subject areas in their broadest possible terms. All useful supplementary research and reference tools have been included for readers' further investigations in the subject area of Nano chemistry. This book seeks to explain this complex field of study. It also seeks to provide an understanding of what is happening now in the field of nano chemistry, as well as what to expect in future by generations to come.

—Editor

1

Nano-tech Institutions & Nano Education: The Relationship

NanoStrategies is information—in the form of briefings, white papers, and presentations. It is also information alerts, trend tracking, competitive analysis, education services, partnering/mentoring offerings, and general consulting services. NanoStrategies can provide you with insights into how nanotechnologies can affect your business plan, your operations, and your ability to serve your organization's mission.

NANOTECH INSTITUTIONS, ORGANISATIONS AND COMPANIES

IBM

IBM's nanotechnology research aims to devise new atomic- and molecular-scale structures and devices for enhancing information technologies, as well as discover and understand their scientific foundations.

Leading the development of nanotechnology, IBM's scientists have made *numerous breakthroughs* in the study of these nano-scale technologies.

In particular, carbon nanotubes and scanning probes derived from the atomic force microscope — cousin of the scanning tunneling microscope — show particular promise in enabling dramatically improved circuits and data storage devices. Research on nanoparticles leads to applications in biomedicine as well as hard disk drive storage. Photonic bandgap materials — on-chip nanoscale structures the size of a wavelength of light — will manipulate light as optical waveguides, splitters and routers. Research into nanomechanical information storage, such as IBM's Millipede project, continues to increase the possibilities for increased areal storage density.

IBM's research into nano-scale structures that self-assemble may one day obviate the need to "hand-position" atoms. Nanotechnology will allow the design and control of the structure of an object on all length scales, from the atomic to the macroscopic enabling more efficient and vastly less expensive manufacturing processes and providing the hardware foundation for future information technology.

Nanoscale science and technology group

The members of the Nanoscale science and technology group at the IBM T.J. Watson Research Center are:

RSMs:

- Phaedon Avouris (avouris@us.ibm.com) manager
- Joerg Appenzeller (joerga@us.ibm.com)
- Jia Chen (chenjia@us.ibm.com)
- James Hannon (jbhannon@us.ibm.com)
- Yves Martin (ymartin@us.ibm.com)
- James Tsang (jtsang@us.ibm.com)

Postdocs:

- Xiaohui Qiu (xqiu@us.ibm.com)
- Zhihong Chen (zchen@us.ibm.com)
- Yu-Ming Lin (yming@us.ibm.com)
- Vasili Perebeinos (vperebe@us.ibm.com)
- Christian Klinke (cklinke@us.ibm.com)
- Marcus Freitag (mfreitag@us.ibm.com)

Collaborators:

- Prof. Uzi Landman, Georgia Tech (uzi.landman@physics.gatech.edu)
- Dr. Norton D. Lang, IBM Research (lang@watson.ibm.com)
- Dr. Alain Rochefort, CERCA (rochefort@cerca.umontreal.ca)

Nanotechnology Organizations

- Foresight Institute—nonprofit educational organization formed to help prepare society for anticipated advanced technologies.
- Institute of Nanotechnology, The fosters, develops, and promotes all aspects of science and technology.
- NanoBusiness Alliance—industry association founded to advance the emerging business of nanotechnology and microsystems.

- VISION Online—virtual institute committed to training and promotion in the ultra-precision and nanotechnologies.
- National Space Society Molecular Manufacturing Shortcut Group (MMSG)—promotes the development of nanotechnology as a means to facilitate the settlement of space.
- Russian Society of Scanning Probe Microscopy and Nanotechnology—active information interchange, research, and practical applications.
- Nano2Life—setting the basis for a virtual European Nanobiotech Institute, focused on the understanding of the nanoscale interface between biological and non-biological entities

Nanoscale Science Department

Welcome to the website of the *Nanoscale science and technology group* at the IBM T. J. Watson Research Center, Yorktown Heights, New York.

Our group uses scanning tunneling microscopy (STM) and atomic force microscopy (AFM) as structural probes, and, along with electron beam lithography, as tools for the modification of materials at the atomic and nanometer scales and the fabrication and study of nano-electronic devices.

Currenly, we are investigating carbon nanotubes, nanolithography and silicon nanoelectronics

NanoBusiness Alliance

Creating a collective voice for the emerging small tech industry.

Murdock Cautions U.S. Risks Losing Technology Lead

Executive Director Sean Murdock testified before Congress along with Floyd Kvamme, Chair of the President's Advisors on Science and Technology, Matthew Nordan, VP of Research at Lux Research, and Jim O'Connor, VP of the Early Stage Accelerator, Motorola on the question of where the US stands in nanotechnology.

Ford, Boeing, Northwestern Forge Research Alliance

Ford Motor Company has announced the intent to form a strategic alliance with The Boeing Company and Northwestern University to research nanotechnology the first such alliance of its kind in the automotive industry.

Small Times Media Acquired By PennWell

Ardesta, LLC, a leader in the creation of and investment in "Small Technology" companies, announced the acquisition of Small Times Media, LLC by PennWell Corporation, an international publisher of business and trade magazines.

Nano Deemed Best Long-Term Energy Alternative

Breakthroughs in nanotechnology could open up the possibility of moving beyond the United States's current alternatives for energy supply by introducing technologies that are more efficient, inexpensive, and environmentally sound, according to a new science policy study by Rice University.

National Cancer Institute Awards $26.3 Million

The National Cancer Institute (NCI), part of the National Institutes of Health (NIH), has announced the implementation of a major component of its $144.3 million five-year initiative for nanotechnology in cancer research. First year awards totaling $26.3 million will help establish seven Centers of Cancer Nanotechnology Excellence (CCNEs).

NASA and Google Agree on Databse Development

NASA and Google have signed a memorandum of understanding (MOU) that outlines plans for cooperation on a variety of areas, including large-scale data management, massively distributed computing, bio-info-nano convergence, and encouragement of the entrepreneurial space industry. The MOU also highlights plans for Google to develop up to 1 million square feet within the NASA Research Park at Moffett Field.

PCAST Releases First Report On Nanotechnology R&D

The President's Council of Advisors on Science and Technology report: The National Nanotechnology Initiative at Five Years: Assessment and Recommendations of the National Nanotechnology Advisory Panel, is PCAST's first assessment of the Federal Government's nanotechnology research efforts in its role as the National Nanotechnology Advisory Panel. President Bush designated PCAST as the NNAP by Executive Order in July 2004, thereby fulfilling the requirement in the 21st Century Nanotechnology Research and Development Act of 2003.

NanoCommerce And SEMI NanoForum Join Forces

The premier industry and nanotechnology organizations, SEMI, Small Times Media, and the NanoBusiness Alliance are combining their leading conferences into one showcase event NanoCommerce/NanoForum 2005. Attendees will benefit from the integration of the best of both events the commercialization focus of NanoCommerce and the technical business integration emphasis of NanoForum.

Nanotechnology News Online

The World Internet News Service provides industry news gathered from more than 150 reliable sources stationed around the world. Coverage includes breaking business news, technology developments, research updates, public policy issues and important event announcements.

NSET Releases Strategic Plan For National Nanotechnology Initiative

The Nanoscale Science, Engineering, and Technology (NSET) Subcommittee of the National Science and Technology Council's Committee on Technology has released its 2004 Strategic Plan for the Federal R&D program in nanotechnology. This report, which was developed with the support of the National Nanotechnology Coordination Office, updates the original strategic plan of the National Nanotechnology Initiative (NNI) for the next 5 to 10 years.

Local Oxidation

Here, we focus on our recent work on the *local oxidation* of semiconductors and thin metal films under ambient conditions. The capability to oxidize metals and semiconductors with nanometer-scale resolution opens a variety of possibilities to improve current electronic devices such as field-effect transistors and allows the fabrication of novel electronic devices such as single-electron transistors.

This site consists of the following two sections (to access them you may also use the navigation bar on the left):

- *AFM Oxidation:* We use a negatively biased AFM tip in the contact mode to write 10-100 nm thin oxide lines in semiconductors and metals.
- *Current-induced local oxidation (CILO):* We employ high lateral current densities in thin metal films to form 10-50 nm thin oxide barriers

The atomic-force microscope (AFM) image above gives an example where oxide lines of about 20 nm width were used to define the silicon dioxide pattern "IBM NANO" on a silicon wafer.

Nanoscale Science

Nanoscale science began in the IBM's Zurich Research Laboratory with the groundbreaking invention of the scanning tunneling microscope (STM), for which Gerd Binnig and Heinrich Rohrer were awarded the Nobel Prize for Physics in 1986. Equally important was the development of the atomic force microscope (AFM) by Gerd Binnig, Christoph Gerber, and Calvin Quate in the mid-1980s. Work in this research area encompasses novel applications of scanning probe methods and related techniques with which one can manipulate individual atoms and molecules to construct nanometer-sized structures. Scientists' goal is to construct functional devices such as transistors out of nature's fundamental building blocks. Increasing efforts have been dedicated to the development of sensors for chemical and biological interactions, stress, magnetization, etc. in extremely minute amounts of materials as well as into the development of such SPM techniques as microscopy based on chemical interactions, nuclear magnetic resonance or dynamic forces. Research work also includes imaging of nanoscopic magnetic structures with the aim to improve our understanding of ferromagnetism at its limits.

Low-temperature Scanning Tunneling Microscopy and Atomic/molecular Manipulation

We investigate the fundamental properties of single atoms and molecules on solid surfaces. Within this field we are specifically interested in the growth of ultrathin insulating films, the manipulation and buildup of atomic-scale nanostructures, and the mechanical and electrical properties of individual molecules.

Our experiments exploit the extreme versatility and sensitivity of a low-(variable)-temperature STM. Such a machine is not only a nanoanalytical instrument to perform imaging and vibrational/electronic spectroscopy on the atomic scale, but can also be used as an atomic-scale tool to assemble atomic structures, manipulate molecules and their conformations, and induce chemical reactions including the synthesis and dissociation (forming and breaking bonds) of individual molecules.

Overview of basic manipulation processes that can be induced with an STM

- *Lateral manipulation:* The transfer of atoms/molecules along the surface employing for example attractive/repulsive forces between the tip and the adsorbate. » Vertical manipulation: The reversible transfer of atoms/molecules between the surface and the STM tip employing additionally electronic/vibrational excitation of the adsorbate by inelastic tunneling.
- *Desorption:* Similar to vertical manipulation, but desorption of individual adsorbates directly into the surrounding gas phase.
- *Dissociation:* Selective bond breaking within a molecule by means of inelastic tunneling processes.
- *Synthesis:* Selective bond formation between two molecular units employing lateral manipulation followed by electronic/vibrational excitation.
- *Change of intramolecular conformation:* Attractive/repulsive forces between the STM tip and the adsorbate can be applied to change the intramolecular conformation of usually specially designed molecules.

National Centre of Competence in Research Nanoscale Science, University of Basel, Switzerland

Nanoscale Science

"Nanoscale Science" as a National Center of Competence in Research (NCCR) is a long-term interdisciplinary research effort focusing on nanoscale structures and aiming to provide new impact and ideas for the life sciences, for the sustainable use of resources, and for information and communications technologies. Within the NCCR, the University of Basel has the role of a leading house, maintaining a network of university institutes, federal research institutes and industrial partners in which scientists from a wide variety of disciplines work closely together. We show how one can measure the signal from slow jumps of a single molecule between metastable positions using a setup where the molecule is fixed to one lead and one of the coupling strengths is controlled externally. Such a measurement yields information about slow processes deforming the molecule...

The Pacific Northwest National Laboratory (PNNL)

The Pacific Northwest National Laboratory (PNNL), operated by Battelle for the U.S. Department of Energy, is a recognized leader in nanomaterials and nanobiology. Together with our colleagues at other Battelle managed labs, Brookhaven National Laboratory, the National Renewable Energy Laboratory and Oak Ridge National Laboratory, we represent a diverse collaborative team in nanoscience, nanoengineering and nanotechnology. We are significant contributors to the nanorevolution of this Century.

At PNNL we are committed to the discovery of fundamental phenomena and the application of this knowledge to develop commercial products. Our mission is to make revolutionary strides in putting nanotechnology to work for the benefit of humanity. We invite you to browse our site and contact us about your needs. Whatever your interests, nanotechnology has the potential to create impressive improvements.

Committee for the 21st Century: Nanoscience and Nanotechnology—Opportunities and Challenges in California

Advances in nanoscience and nanotechnology will likely have major implications for the health, wealth and peace of society in future decades. An increase in research interest is resulting in the worldwide rapid production of knowledge in this field. In turn, there is a hope that this will lead to dramatic changes in the ways that materials, devices, and systems are understood and created. Some estimates suggest that the market for nanoproducts could lead to a trillion-dollar, multi-industry impact over the next ten to 15 years. However, there are also concerns about nanoscience and nanotechnology: that the demands for new technological products may be outpacing the slower and more deliberate science that creates the technology; that the existing silicon semiconductor industries may not be capable of embracing a new technology; or that a monster technology will be created that will damage society. There is also a possibility that nanoscience and nanotechnology in the future will not live up to the promise glimpsed at the present.

At the request of the Committee for the 21st Century, the California Technology, Trade and Commerce Agency and the Semiconductor Industry Association, CCST prepared a briefing on the opportunities and challenges afforded to California in nanoscience and nanotechnology. At issue is the opportunity for California's research institutions and industries to be leaders in research and development of this field while considering the social, ethical and legal implications of a radically new technology.

The briefing, presented to the Committee on January 20, 2004, concluded that while California is home to several of the world's leading nanotechnology research institutions and also has a healthy industry base, several serious challenges remain, including:

- Loss of Existing Assets — the loss or degradation of any of California's existing high-tech assets — such as our skilled workforce, concentration of venture capital, strong research institutions — would be extremely problematic.
- Scope — though we often think of nanotechnology as a single field, it actually represents a collective advance across several disciplines. Development will certainly not be uniform, and there is the potential that public concern about implausible applications may halt research across the board.
- Transition — existing industries will have to undergo radical transformation to survive such changes as the shift from silicon-based computer chips to carbon nanotube-based chips.
- Intellectual Property (IP) — nanotechnology will lead us to uncharted legal territory (can you patent an atom?) and inefficiencies in the transfer of IP between universities, government and industry could dampen business activity.

The briefing provides recommendations for California's policymakers to support our existing strengths in nanotechnology and prepare the state for the future. If California takes appropriate steps to leverage its existing advantages and to effectively cope with the challenges that may arise, it should be able to maintain its leadership in nanotechnology in the decades to come. Now is the time to lay the foundation for this future.

Electronic Device Failure Analysis Society (EDFAS)

The 2005 EDFAS Board of Directors Election results are in! Thanks to all of you who took the time to participate in the on-line voting. We had four highly qualified and enthusiastic candidates all of whom had expressed their vision and desire to contribute to the future of your society. Unfortunately, only two openings existed for the 2005-2009 term. The new members of the EDFAS BOD will be Matt Thayer from AMD (current incumbent re-elected to full term) and Tracy Myers of LSI Logic (new member). Tracy and Matt have long been involved with EDFAS and with ISTFA and will bring their experience and expertise to bear to advance the goals and objectives of EDFAS. Their seats are effective 1 September 2005 for a four year term. On behalf of the entire society, thanks also to Ingrid DeWolf of IMEC and Philippe Perdu of CNES for their interest and involvement in the society as evidenced by their candidacies. As future openings on the BOD occur, hopefully they will again consider being candidates.

Finally, many thanks to our departing BOD member, Dr. Chuck Hawkins of the University of New Mexico. Chuck has served EDFAS for many years as Education Chair, EDFA Editor, BOD member, and most recently, as Financial Officer. He has also been involved with ISTFA as a tutorial lecturer and paper author/presenter. Chuck's wisdom and counsel will be sorely missed on the BOD.

ICON & CBEN

Nano Coalition Unveils Environmental, Health and Safety Database

The International Council on Nanotechnology (ICON) and Rice University's Center for Biological and Environmental Nanotechnology (CBEN) today launched the world's first online database of scientific findings related to the benefits and risks of nanotechnology. The database can be accessed by clicking here. This environmental health and safety database marks the first effort to integrate the vast and diverse scientific literature on the impacts of nanoparticles, which are tiny pieces of matter with dimensions measuring between 1-100 nanometers and containing between tens and thousands of atoms. The database is the result of the collected efforts of Rice researchers, the chemical industry and the U.S. Department of Energy.

The report of the interagency workshop held in March 2004 to identify and articulate the relationship of nanoscale science and technology to the nation's energy future is now available. Development of new nanoscale materials, as well as the methods to characterize, manipulate, and assemble them, creates an entirely new paradigm for developing new and revolutionary energy technologies. The primary outcome of the workshop was the identification of nine research targets that participants thought should be given high priority in energy-related science and technology in which nanoscience is expected to have the greatest impact. The report contains descriptions of many examples indicative of outcomes and expected progress in each of these research targets.

The report of an interagency workshop convened to solicit input from the research community on the NNI research agenda at the intersection of nanoscale science and technology and biology, with a particular focus on research for biomedical applications of nanotechnology, is now available. "Nanobiotechnology: Report of the National Nanotechnology Initiative Workshop" concludes that nanotechnology is in the process of driving a new wave of medical innovation. The report focuses on topics in which major progress is expected from advanced nanotechnologies, including molecular imaging technologies, quantitative analytical tools, quantitative integration of information, physical models of the cell as a machine, better *ex vivo* tests and improvements in current laboratory techniques, and drug delivery and intelligent therapeutics.

Europlasma

Gas plasma surface treatment is an ideal technology for the surface modification of plastic, textile & film products while enabling hydrophilic and hydrophobic coatings.

Surface treatment & modification

Europlasma offers a wide range of gas plasma surface treatment equipment, from lab scale to industrial application, tailored to the needs of the following markets & applications:

- *Electronics:* PCB cleaning & etching. All types, desoxidation, activation of polyimde film, etc...
- *Automotive:* Pre-treatment before painting, flocking
- *Textile & Filtration:* Hydrophilisation, hydrophobisation of textiles, filters, film.
- *Medical:* Pre-treatment of balloon catheters, syringes, valves before gluing
- *Aeronautic & aerospace:* Coating of insulation products (foams), electronics

In addition to our wide range of gas plasma systems, we also offer to our customers a fully equipped process development lab on an industrial scale and a subcontracting service facility.

NanoStrategies—From Nanotechnology Now

"... an immense technical, business, and social tsunami is coming along shortly ... *companies are already being forced to deal with a blinding rate of technological change. This speed-up will likely increase, and definitely spread."* (our emphasis). Leaping the Abyss: Putting Group Genius to Work Gayle Pergamit and Chris Peterson.

"In not too many decades we should have a manufacturing technology able to: Build products with almost every atom in the right place; Do so inexpensively; Make most arrangements of atoms consistent with physical law. Often called nanotechnology, molecular nanotechnology or molecular manufacturing, it will let us make most products lighter, stronger, smarter, cheaper, cleaner and more precise."

Nanotechnology: It's a Small, Small, Small, Small World Ralph C. Merkle

Who we are: Our consulting team includes top nanotech research scientists, product development specialists, nano-scale manufacturers, tech-focused venture capitalists, business development strategists, management experts, and government liaisons.

What we offer: We are dedicated to supporting the creation of successful nanotech ventures through the applied research phase, product development, start-ups and established firms seeking to develop or invest in nanotechnology.

What you need: Whether you need to: resolve a technical problem, prototype your innovation, research an opportunity, develop a business or marketing plan, raise capital, structure a deal, recruit a management team, or find a strategic partner, we can help. NanoStrategies has the resources, expertise and connections throughout the global nanotech community to bring all the pieces together for your success.

Disruptive Technology: An emerging need among many established businesses is an understanding of how disruptive innovations occurring at the nano-scale will create both crisis and opportunities for their future survival and success. We'll work with your in-house strategic planners to mine the data, study the acceleration trends, and chart out a strategy to take advantage of rather than be overrun by any relevant nano-innovations affecting your industry.

Nano Trends Expertise: Knowing what's on the horizon that may change or eliminate whole industries, change our national and global economy, change legislative and regulatory governance will be critical to any businesses future. These changes will make some companies powerful and influential and diminish others. The ability to adapt and embrace ever-changing technologies relies on careful analysis and strategic vision, innovation and who we partner with in the next few months and years.

Nano Information: NanoStrategies is information—in the form of briefings, white papers, and presentations. It is also information alerts, trend tracking, competitive analysis, education services, partnering/mentoring offerings, and general consulting services. NanoStrategies can provide you with insights into how nanotechnologies can affect your business plan, your operations, and your ability to serve your organization's mission.

Strategic Partnership: Connecting with the right partners who make it their business to know what's going on in all aspects of disruptive technologies may be critical to the future success of your business. A small investment of time and energy now can have a profound impact on where your business, industry or agency ends up five or 10 years from now.

The ASME Nanotechnology Institute

The Nanotechnology Institute of ASME International is dedicated to furthering the art, science and practice of nanotechnology. The Institute is a clearinghouse for ASME activities in nanotechnology and provides interdisciplinary programs and activities to bridge science, engineering, and applications. This site provides a one-stop resource for researchers and practitioners about ASME nanotechnology activities as well as links to other sites. The opinions, findings, and conclusions expressed in the linked web sites are those of the authors and do not necessarily reflect the views of ASME International.

euspen

Who we are

The european society for precision engineering and nanotechnology (euspen) was established in 1999 with funding from the European Community under the 'Competitive and Sustainable Growth' Programme. Now established as a self-supporting, not-for-profit membership organisation, euspen is a body of more than 550 individual members and approximately 90 corporate members from leading industrial organisations, universities and research institutes world-wide.

What we do

euspen links industrialists to researchers across more than 30 countries in an influential and extended network in its mission to:

- Advance the arts, sciences and technology of precision engineering, micro engineering and nanotechnology.
- Promote its dissemination through education and training and
- Facilitate its exploitation by science and industry.

Increasing demands for products delivering enhanced functionality, improved speed, reduced costs and increased reliability, drives an ever greater requirement for increased precision and miniaturisation, today down to the nanoscale. Both development of commercial nanotechnology based products and research into nanoscience rely upon the provision of instruments and tools to fabricate, manipulate, measure and sense at the nanoscale.

The focus of euspen is on the development and exploitation of such precision, micro and nanotechnologies:

- product/technology research and devleopment
- tools and instruments
- associated suppliers; service providers; users and networks.

It aims to promote and nurture core technologies crucial to the further development of the industry.

Assistance is also maintained by the EU through the project VisionOnline. This initiative primarily aids euspen in providing the community with intensive, high calibre and cost effective training in precision engineering, micro engineering and nanotechnology.

Key 'VISION Online' Project Tasks

1. To provide multidisciplinary technical training using traditional and online media techniques. euspen 1-day tutorials are developed into 1-week intensive

training coures around Europe. (1-day euspen tutorials also run prior to all euspen international conferences.) During this project, these modules will be offered in part-time online format.

2. To generate innovative Internet services including electronic 'Nanotechnology' patent library, expertise search facilities and discussion forum.
3. Organise industrial seminars and working groups.
4. Compile portfolio of projects that industry wish research and which is presently hindering industrial growth

Institute of Atomic-Scale Engineering

Featuring some of the inventions of Forrest Bishop

Questions and comments on this material are welcome. (Year of conception in parentheses)

Images: Digital Animations

"Teleportation": Projecting matter beams through space. Capturing and decelerating in a reverse-particle accelerator. Scanning, disassembly and re-assembly of humans and other objects (now called molecular nanotechnology). 7th Grade science paper. (1967)

InterPlanetary Mass Driver (IPMD): Independently invented pellet-stream propulsion. Invented beam-propulsion space transportation network. Hyperbolic beam from high-orbiting Mass Driver intercepts low circular-orbit beam-rider spacecraft. (1981)

Space Cables: Independently invented Orbital Loops. Loop in low Moon orbit launches and lands payloads. (1981)

"South Pole Accelerator": A vertical Mass Driver at the South Pole launches swarms of payloads to the Earth's sphere-of influence, supplies IPMD. (1981)

"The Construction and Utilization of Space-Filling Polyhedra for Active Mesostructures": A multipurpose swarm of connected polyhedra changes shape. Also called Shapeshifter and Active Cell Aggregate (ACA). Posted Dec. 7, 1995 (1992). Download the entire article in PDF format by clicking here.

"Introduction To Shape Shifter"

"A Proposed MNT Active Cell": A design example of a molecular nanotechnology Active Cell. It is used again, with some modifications, in the "Overtool" and "Starseed/ Launcher" studies. (1992-1995). Download the entire article in PDF format by.

"The Overtool: A Proposed Universal Assembler"

A proposed Universal Assembler- a self-replicating machine, which is also capable of manufacturing other products. An ACA with atom manipulators assembles its own components and other structures. (1996)

"A Description of a Universal Assembler": Published in *Proceedings of the IEEE Joint International Symposia on Intelligence and Systems*, ISBN 0-8186-7728-7 Posted Aug 19, 1997.

"The Starseed/Launcher": Appeared in the May, 1996 "prepress" issue of *NanoTechnology* magazine.

A less technical account of the concept.

"The Shape of Things to Come": Appeared in the April, 1996 "prepress" issue of *NanoTechnology* Magazine.

A synopsis of several of the articles above.

"Open Air Space Habitats": A one-gravity, one-atmosphere miniature Ringworld made of carbon nanotubes. Posted Aug 19, 1997 (1997)

"Empirical, Evolutionary Chemistry": Mechano-synthetic genetic-algorithm pathway for novel structures. Interview in *NanoTechnology Magazine*, Dec 2, 1996. Posted Aug 19, 1997

"The Optical Assembler": December 10, 1996 interview with Forrest Bishop on a novel method of atomic assembly using light forces. Lasers position and react individual atoms without using mechanical arms. Posted Mar 16, 1998. (1996)

Cell Rover: A micron-size medical machine made partly of biological structures. (1998)

Rotary Assembler: A simple molecular assembly technique. (1998)

"The Interworld Rapid Transit System (A Transportation Network for the Solar System)": (IRTS)- A variety of energy and matter beam generators in different places exchange with each other and with beam riders. This is the first proposal of its kind. Abstract posted Aug 19, 1997, draft article from 1997 (1967, 1981, 1997)

"Mesoparticle Beam Propulsion": Intermediate size matter beam between particle beam and pellet stream. (Abstract Only) Posted Aug 19, 1997 (1997)

"Some Novel Space Propulsion Systems": Microscale Light Sails (Light-sail swarms without payloads form a mesoparticle matter beam. "Light" can be solar or laser, 1997), Graphene Crossbow, and Mesoparticle Beam Propulsion Presented at the Fifth

Foresight Conference on Molecular Nanotechnology, Nov 6-9 1997, Palo Alto, CA. http://www.foresight.org/Conferences/MNT05/Papers/Bishop Posted Mar 16, 1998.

"Some novel space propulsion systems": Micro-scale Light Sails, Mesoparticle Beam Propulsion, and Ultracold Matter Beam Generators (a new type of particle beam, 1999).

The Institute of Nanotechnology (IoN)

Foundation

The Institute of Nanotechnology (originally the Centre for Nanotechnology founded by Ottilia Saxl in 1994) was one of the world's first nanotechnology information providers, and is now a global leader.

The IoN works closely with governments, universities, researchers, and companies worldwide on developing and promoting all aspects of nanotechnology. It also serves as a key organizer of international scientific events, conferences, and educational courses designed to encourage nanotechnology takeup by industry, as well as stimulating interest in less developed countries.

Current Activities

IoN staff are working on many different projects, but are presently focusing their efforts in a few key areas.

Europe's Premier Information Network

The first is on Nanoforum, an information network that is fully funded by the EU, which reaches out to all of Europe. Nanoforum, www.nanoforum.org serves as a framework for all nanotechnology activities across Europe. It provides the nanocommunity with easy access to specially commissioned up-to-the minute reports, information on new nanotechnology developments, funding and partnership opportunities. Nanoforum partners organize events throughout Europe focussed on industrial applications of nanotechnology, such as smart materials and medical devices. Education and training is also an important area, and the Nanoforum partners organize an annual Summer School, held in a different European location each year. Nanoforum can now justifiably claim to be the premier European nanotechnology knowledge network.

The Club for New Nano and Micro Businesses

From its work with start-ups, the Institute identified a need by new, fragile nano and microtechnology businesses for mentoring, networking and contact with potential customers and investors. It met that need by establishing a club for these companies – the NanoMicroClub (www.nanomicroclub.com). The club gives start-up companies an opportunity to network with each other and the specialists, business advisors, and technology end users, thus helping ensure their survival. After its succesful pilot over the last twelve months in the UK, the aim is to franchise the idea across Euope in the first instance; and later to the Far East.

Online Nano Marketplace

Another project is the online information website and maketplace, www.AZoNano.com (The A to Z of Nanotechnology), which has been designed to meet the needs of industrial end users. The site is free to access, and contains many thousands of articles about new nanotechnologies and their applications – an invaluable tool for companies seeking to identify nanotechnologies relevant to their products or processes.

Technology Transfer

Any queries from industry seeking new technologies are channeled through the Technology Transfer Centre. Many blue chip companies use TTC to identify new nanotechnologies within the research base worldwide that may be relevant to the development of new products. Contact Andy Garland: andy@nano.org.uk

Flagship Conference 2005

With numerous conferences, workshops, and training courses planned throughout the year, the IoN maintains its focus on informing industry and educating the public to ensure responsible development of nanotechnology. Its flagship conference for 2005 is 'Nanotechnology and the Health of the EU Citizen in 2020' (http://www.euronanoforum2005.org/), held in the beautiful city of Edinburgh from 5 to 9 September. Conference speakers include top medical researchers from industry and academia worldwide, and 1000 delegates are expected.

Overseas Links

The Institute's activities are continuing on the European front, but it is extending its reach to new countries and continents in an effort to foster nanotechnology developments and partnerships all over the globe.

With her great interest in collaboration, IoN CEO Ottilia Saxl's aim to advance nanotechnology beyond the Western world has recently led to collaborations with similar organizations in Japan, Korea, Australia and China. "We have developed great relationships with these countries and we're working on closer links with other countries," she says. "There has been a huge investment from the global community in nanotechnology and we are now in the first stages of seeing tangible benefits." The current emphasis is on new biomedical and engineering applications, but she also envisages ecological developments that could provide environmentally-friendly energy for the less-developed countries and allow for more conservational manufacturing practices.

Future Developments

The Institute plans to launch the European Nanotechnology Trade Alliance (ENTA). Many companies have expressed a need for an organization that represents the interests of nanotechnology businesses across Europe. The aim of ENTA will be to act as a bridge between national and regional governments and companies, interface with the public and ensure that member organizations develop new nanotechnologies in a safe and responsible manner.

Research

Since its inception, the Institute has developed very close links with the nanotechnology community. This has enabled it to produce its own unique highly informative reports on nanotechnology activities for Government bodies, major industrial and investment companies worldwide.

Publications

The IoN has produced several important milestone publications:

- 'Opportunities for Industry in the Application of Nanotechnology'
- 'Nanotechnology—The huge opportunity that comes from thinking small'
- 'Nanotechnology Missions to Germany and the East Coast USA'
- 'Nanotechnology—size matters. Building a successful nanotechnology company'. Produced in conjunction with 3i.
- 'What is nanotechnology? CD-ROM
- 'Nanotechnology Mission to the West Coast USA'

Conferences

Conferences organised by the IoN have been described as:

"One of the best conferences I have been to. The conference gave a wonderfully exciting insight into the rapid progress being made in the field today. We are living in a very exciting period."—Ian Pearson, BT Futurist

- "Excellent, very thought provoking"—Martin Bentham, Sara Lee Research
- "An excellent event, very well organised and value for money"—Neil Wilkinson, Skillfast
- "A first class conference full of variety and practical approaches"—Dr. Patrick Ferguson, NEL Technologies Ltd

Recent conferences include:

- 'Advanced Technologies in Crime Prevention and Detection'—November 2004
- 'New Technologies and Smart Textiles for Industry and Fashion'—December 2004

Contracts

Current and previous contracts include:

- 'Benchmarking UK Nanotechnology' – undertaken in conjunction with NPL. The results formed the basis of the John Taylor report 'New Dimensions in Manufacturing—A UK Strategy for Nanotechnology'
- Organisation of Nanotechnology Missions to Germany and the East Coast USA.
- Organisation of Nanotechnology Mission to the West Coast USA to examine the major nanofabrication centres.

To lead the pan-European NanoNetwork of Networks—Nanoforum. The network is aimed at raising awareness of nanotechnology, catalysing transnational research partnerships, developing industrial and postgraduate training in nanotechnology, and informing industry and the public. Partners include the research arm of the French Atomic Energy Authority at Grenoble, CEA-Leti, the Futures Division of the German Association of Engineers VDI-TZ, and coordinator of the Phantoms Nanoelectronics Network—CMP Cientifica.

Establishing the Industrial Nanolandscape in the UK: aimed at identifying the level of current and proposed activities at the nanoscale within UK industry.

European Micro and Nanofabrication Centres as Models for the UK: based on describing the modus operandi of major European Nanotechnology Centres, as possible models for a future UK initiative.

MNT UK Micro and Nanotechnology Tradeshows: a series of tradeshows assembling major manufacturers, sellers and buyers of micro and nanotechnologies, and creating a powerful opportunity for the MNT community to network, find suppliers, customers and new partners.

The Institute has also been instrumental in catalysing UK interest in the EU's Framework 6, working with companies and organisations such as Johnson Matthey, BP, ICI, Unilever, Syngenta, CRL and PERA.

Links with Industry

The Institute is active in forming links between research and industry, and has undertaken specific work for many blue-chip companies in Europe, including 3i, Toshiba, GSK, Unilever, Sharp, Degussa, BNFL and General Electric.

As a membership organisation, current corporate members include Unilever, Degussa, Lot Oriel, Sulzer, Veeco, QinetiQ, INSTM, Toshiba, Merck and ICI.

nanoTEN Nanotechnology Consulting

*nano*TEN is a consulting company specialized in Nanotechnology. Our international team of entrepreneurial scientists is the prime source of Nanotechnology Consulting in the U.S., Latin America, Europe, and Asia Pacific.

*nano*TEN associates have extensive experience in Nanotechnology and are connected to Academic Research, Industry and Government networks in the U.S., Latin America, Europe, and Asia.

Our main business is to bridge the gap between Academic Research, Industry and Finance. For additional information, contact info@nanoten.com.

Why Nanotechnology?

In a 2001 survey covering 1300 industry leaders, the Japan Patent office asked respondents to predict how progress in Nanotech would affect their industry. Over 60% of respondents working in Materials, and over 70% of those working in the Electronics area, answered that manufacturing methods in their industry would have to drastically change, and that further innovation would be needed.

This response shows exceptional expectations for Nanotech, in particular in the Biotech field, where some 90% of respondents felt that Nanotech would lead to greater innovation in their industry. Over half the respondents felt that greater interaction would be necessary between researchers in different fields to achieve this. The number of *Nanotechnology applications influencing our daily life* is increasing steadily.

In a separate survey, nearly 60% of industry leaders felt that technology transfer in Nanotech between the academic world and industry was insufficient, and that the number of Technology Licensing Organizations (TLOs) established in Japan (around 20) was far less than needed, especially when compared to the U.S.

We at *nano*TEN believe that best advice comes directly from leading researchers in the Nanotechnology field, since they can assess best the potential of current and future technologies. In particular, webelieve that the potential of Nanotechnology will by far surpass that of Biotechnology.

Nanovip.com

This is the OLD Nanovip.com' stocks page. The current, up to date stocks page is here.

Important: There is a debate to define which company may be considered a « Nanotechnology » company or not. It seems that a part of our audience expect to find molecular manufacturing companies with self-assembling robots capabilities. Some are considering the "bottom up" technology process of molecular manufacturing as the only "TRUE" nanotechnology. The problem is that as of writing this note, there are no "TRUE" nanotechnology companies according to the above definition. Thus we made the deliberate choice to list on Nanovip.com, companies involved in nanoscale interaction, manipulation, observation and fabrication: the result might not fit anyone's expectation from a "nanotechnology company".

It is easier to define the recent self-called nanotechnology start-ups, but it is more difficult to define already existing companies that are progressively incorporating nanoscale technology into their operations, yet these companies are still, in our opinion, engaged in the path of nanotechnology—a technology interacting with the matter at the nanoscale (atomic / molecular) level.

In this list of public nanotechnology companies, you will find both recent "nanotechnology" start-ups and established companies with various level of involvement in nanoscale technology.

It is up to you to decide if a company's involvement in nanotechnology is enough to match your criteria.

A company with this sign is usually a recent start-up heavily focused on nanotechnology, in contrast with the other older companies only partially involved in nanotechnology.

If you know a nanotechnology company that is not in our list, please contact us and we will add it. You can also contact us to correct an error if any. New: Nanotechnology investing forum ! Discuss about your trading ideas ! This list is NOT an investment recommendation.

Miscellaneous

Merrill Lynch has launched a nanotechnology stocks index. It is traded on the AMEX, under the symbol: NNZ

Punk Ziegel has also launched a nanotechnology stocks index; it can be found here.

AMEX nanotechnology stocks

Biosante Pharmaceuticals, Inc. (Amex: Bpa)—Pharmacutical company developing hormone thearpy products, vaccines and oral delivery. The company also is developing its calcium phosphate nanotechnology (CAP) for novel vaccines, including biodefense vaccines for toxins such as anthrax and ricin, and drug delivery systems.

NASDAQ nanotechnology stocks:

- Accelrys Inc. (Nasdaq: Accl)—Accelrys' computational technologies include industry-leading capabilities for molecular (including nanoscale) graphics and modeling, structure determination and characterization, property prediction, and simulation of chemical and biological behavior.
- Albany Molecular Research, Inc. (Nasdaq: Amri)—Albany Molecular Research, Inc. is a chemistry-based drug discovery and development company, focusing on applications for new small molecule prescription drugs. The company conducts R&D projects and collaborates with many leading pharmaceutical, biotechnology and genomics companies, and is developing new chemistry technology for potential prescription drug applications.
- Altair Nanotechnologies Inc. (Nasdaq: Alti)—Altair owns a proprietary technology for making nanocrystalline materials of unique quality, economically and in large quantities. The company produces closely-sized nanoparticles of titanium dioxide and related ceramic oxide materials and compounds. Applications are found in products such as advanced batteries, fuel cells, thermal spray coatings, specialty paints, catalysts, plastics, cosmetics, solar cells and self-cleaning and sanitizing materials. In line with its cutting-edge position in nanotechnology, Altair is also designing materials to suit the nanoparticle applications of the future.
- American Pharmaceutical Partners, Inc. (Nasdaq: Appx)—American Pharmaceutical Partners, Inc. (APP), a majority owned subsidiary of American BioScience, Inc., is a specialty drug company that develops, manufactures and markets injectable pharmaceutical products. APP manufactures and market ABRAXANE, a proprietary nanoparticle injectable oncology.
- American Superconductor Corporation (Nasdaq: Amsc)—American Superconductor Corporation is a supplier of dynamic reactive power grid stabilization products and the world's principal vendor of high temperature superconductor (HTS) wire and large rotating superconductor machinery.
- Applied Films (Nasdaq: Afco)—Applied Films is a leading provider of technologically advanced thin film deposition equipment and service.
- Arrowhead Research Corporation (Nasdaq sc: Arwr) Arrowhead Research Corporation funds research at universities in pioneering scientific areas, primarily nanotechnology, in return for exclusive rights to commercialize

technologies and associated intellectual property and patents developed as a result of this research.

- Caliper Life Sciences (formerly Caliper Technologies) (Nasdaq: Calp)—Caliper Life Sciences uses its advanced liquid handling and LabChip technologies to create breakthrough tools that accelerate drug discovery and enhance the diagnosis of disease.
- CombiMatrix Corporation (Nasdaq: Cbmx)—CombiMatrix Corporation develops life science research products based on proprietary technologies in the areas of genomics and proteomics. Combimatrix's microarray technology allows for the parallel synthesis of large numbers of nano structured materials.
- DayStar Technologies, Inc. (Nasdaq: Dsti)—DayStar Technologies, Inc. is a renewable energy/photovoltaic (PV) technology development and commercialization business. The company plans to revolutionize solar power components by using nanotechnology to reduce their size and change their shape for a number of applications.
- FEI Company (Nasdaq: Feic)—FEI's focused ion and electron beam technologies provide researchers and manufacturers the ability to modify, fabricate and analyze a wide variety of samples in 3D at nanoscale dimensions.
- Flamel Technologies S.A. (Nasdaq: Flml)—Flamel Technologies S.A. is a drug delivery company that has invented two innovative delivery systems.
- Genus, Inc. (Nasdaq: Ggns)—Atomic Layer Deposition (ALD) is an advanced process for depositing ultra-thin films one atomic layer at a time. Our mission is to delight our customers by providing enabling thin film competitive advantages. We are totally committed to high-productivity and enabling thin film solutions, with best-in-class service and support. We work to provide our customers with new competitive capabilities and manufacturing processes uniquely tailored for their needs.
- Harris & Harris Group, Inc. (Nasdaq: Tiny)—In early 2002, Harris & Harris Group announced that it would concentrate its new business activities exclusively in a single focus: tiny tech, including nanotechnology, microsystems and microelectromechanical systems (MEMS) technology.
- Immunicon Corporation (Nasdaq: Immc)—Immunicon has developed a revolutionary diagnostic platform based on immuno magnetic selection and fluorescence characterization of rare cells in blood. The Company's initial focus is cell-based and molecular diagnostic products for monitoring cancer. The company has patented magnetic nanoparticles called ferrofluids.

- **Introgen Therapeutics, Inc. (Nasdaq: Ingn) Introgen Therapeutics is a biopharmaceutical company focused on the discovery, development and commercialization of molecular therapies for the treatment of cancer and other serious diseases. The company uses nanoparticles to target cancer tumors.**
- **Isonics Corp. (Nasdaq: Ison)—Isonics is a speciality chemical and advanced materials company that develops, commercializes and markets materials which have been sub-atomically engineered to enhance performance.**
- **JMAR Technologies, Inc. (Nasdaq: Jmar)—JMAR is the world's leading developer and manufacturer of collimated laser plasma lithography (CPL)™ sources and systems and X-ray lithography stepper systems for the semiconductor industry.**
- **Komag, Inc. (Nasdaq: Komg)—Komag is an independent supplier of thin-film disks, the primary high-capacity storage medium for digital data. Komag has developped a patterned magnetic recording disks utilizing unique nano imprint lithography.**
- **Kopin HBT, Kopin Corporation (Nasdaq: Kopn)—Kopin Corporation is a developer and provider of small format, high-information content, low cost AMLCDs and microdisplays for personal information devices and Gallium Arsenide wafer-engineered materials for high-performance integrated circuits.**
- **Lumera Corporation (Nasdaq: Lmra)—Lumera develops proprietary polymer materials and products based on these materials for a broad range of applications. We engineer polymer materials at the molecular level: commonly referred to as nanotechnology: to optimize their electrical, optical and/or surface properties.**
- **Magma Design Automation, Inc. (Nasdaq: Lava)—Magma provides leading software for designing highly complex integrated circuits while maximizing Quality of Results with respect to area, timing and power, and at the same time reducing overall design cycles and costs.**
- **Nanogen, Inc. (Nasdaq: Ngen)—Nanogen's mission is to become a leading supplier of molecular diagnostic tests to the medical community and to clinical researchers.**
- **Nanometrics Inc. (Nasdaq: Nano)—Nanometrics, Inc. designs, manufactures and markets advanced thin film, critical dimension, overlay and defect inspection metrology systems.**
- **Nanophase Technologies Corporation (Nasdaq: Nanx)—Nanophase**

Technologies is an industry-leading nanocrystalline materials innovator and manufacturer with an integrated family of nanomaterial technologies.

- Novavax, Inc. (Nasdaq: Nvax)—Novavax, Inc. is a fully integrated specialty pharmaceutical company that is focused on the research, development and commercialization of products utilizing our proprietary drug delivery and vaccine technologies, specifically focusing on the large and growing markets of women's health and infectious diseases.
- NVE corporation (Nasdaq: Nvec)—NVE's enabling technology is called spintronics, which uses an electron's spin rather than its charge to sense, store, or transmit digital data. The technology could yield nothing less than the next generation of microelectronics. MRAM uses spintronics electrons' spin rather than their charge to store data, and is fabricated using nanotechnology.
- Orthovita, Inc. (Nasdaq: Vita)—Orthovita is a biomaterials company with proprietary technologies applied to the development of BioStructures, synthetic biologically active tissue engineering products for restoration of the human skeleton. Fine nano particle composition enhances resorption, facilitating rapid, effective bone remodeling. Our focus is on developing novel products for use in spine surgery and in the repair of osteoporotic fractures.
- Precision Optics Corporation, Inc. (Nasdaq: Poci)—POC specializes in fully integrating optical design, product development and volume production services to support special medical product requirements. The Company is expanding its line of medical and industrial endoscopic products, to develop instruments for single molecule and nanotechnology applications, and for general corporate purposes.
- Research at Intel (Nasdaq: Intc)—Intel's research in Nanodimension computed visual representation.
- Trikon Technologies, Inc. (Nasdaq: Trkn) Trikon Technologies is a leading supplier of equipment in three core technologies in the semiconductor industry – PVD (Physical Vapor Deposition), CVD (Chemical Vapor Deposition) and Plasma Etch. In 2004, the Uk government announce funding award to trikon to develop 300 MM deposition system for nanotechnology.
- Ultratech (Nasdaq: Utek)—Ultratech is a leading provider of tools in nanotechnology markets such as thin film heads, inkjet print heads, optical components, inertial and pressure sensors, and radio frequency (RF) devices.

- Universal Display Corporation (Nasdaq: Panl)—Universal Display Corporation (UDC) is working in the development of innovative organic light emitting device (OLED) technology for future generations of flat panel displays, lasers and light generating devices.
- Pharmacopeia Inc. (Nasdaq: Accl)—Pharmacopeia is in the field of drugs discovery and utilizes its multi-million compound sample collection of novel small molecules and expertise in assay development, high-throughput and ultra high-throughput screening to identify compounds active against drug targets.
- SkyePharma (USA) (Nasdaq: Skye)—Drug delivery company. SkyePharma has performed several inhalation experiments, which have shown that nanotechnology can produce particles in very narrow size distributions, which theoretically allow for targeted delivery in the lung.
- Spire Corporation. (Nasdaq: Spire)—Spire Corporation provides products and services based upon a common technology platform to the solar energy and biomedical industries worldwide. (Notice: the company might get de-listed from the Nasdaq. Also, the ticker of the stock is sometimes recognized as "Spir" instead of "Spire").
- Symyx Technologies, Inc. (Nasdaq: Smmx)—Symyx is the pioneer of high-throughput experimentation for the discovery of materials. We apply our technologies to discover innovative materials for high-value industrial applications in a broad range of industries.
- Veeco Instruments (Nasdaq: Veco)—Veeco's Metrology tools are used to measure at the nanoscale and our Process Equipment tools help create nanoscale devices.
- Westaim Corporation (Nasdaq: Wedx)—The Westaim Corporation creates value for shareholders by partnering with established commercial leaders for rapid market penetration of its technologies and maximum return on investment.

NYSE Nanotechnology Stocks

- 3M Corporate (Nyse: Mmm)—3M is a technology company with leading positions in consumer and office; display and graphics; electronics and telecommunications; health care; industrial; safety, security and protection services; transportation and other businesses. The company is engaged in research and development of nanoscale applications.
- AMCOL International Corporation (Nyse: Aco)—Amcol produces and markets a wide range of specialty minerals used for industrial, environmental

and consumer-related applications. AMCOL's newest venture is in the field of nanocomposite technologies.

- ChevronTexaco Corp: MolecularDiamond Technologies (Nyse (ChevronTexaco): Cvx)—Tiny diamond fragments called higher diamondoids offer untapped potential for advances in the fast-growing, actively funded field of nanotechnology. MolecularDiamond Technologies' goal is to collaborate with leading academic, government, and corporate researchers to facilitate groundbreaking work with diamondoids.
- Dupont Research and Development center (Nyse: Dd)—CR&D is the foundation of our science efforts and has been responsible for most of our major product breakthroughs. CR&D provides both leveraged scientific services to the corporation and long-term research activities.
- General Electric Global Research (Nyse: Ge)—We understand the enormous impact of breakthroughs like atomic scale circuits, nanotube computers and self assembled block copolymers. And that's why we've made long-term investments in these technologies. We're ready to change the world, little by little.
- General Motors Corporation (Nyse: Gm)—General Motors Corp. is a large vehicle manufacturer. GM is expanding its use of new materials, such as nanocomposites, to reduce weight in the vehicle while at the same time providing a quality, recyclable and affordable product.
- Hewlett-Packard's Quantum Science Research (QSR) (Nyse: Hpq)—The QSR group at HP Labs is exploring the limits of the physically possible by focusing on the fabrication of nanometer-scale structures and the measurement and understanding of their properties.
- Keithley Instruments, Inc. (Nyse: Kei)—Keithley Instruments has more than a half-century of experience in pushing out the boundaries of measurement science and has published literature and technical application tutorials that support engineers and scientists developing nanoscale devices and technology.
- Nanoscale science department (IBM research) (Nyse: Ibm)—Our group uses scanning tunneling microscopy (STM) and atomic force microscopy (AFM) as structural probes, and, along with electron beam lithography, as tools for the modification of materials at the atomic and nanometer scales and the fabrication and study of nano electronic devices.
- Philips Research Laboratories (Nyse: Phg)—Nanotechnology provides promising options in several areas that match the business interest of Philips, for example in displays, semiconductor manufacturing and lighting.

- STMicroelectronics (Nyse: Stm)—STMicroelectronics is a global independent semiconductor company and is a leader in developing and delivering semiconductor solutions across the spectrum of microelectronics applications. The company has developed an expertise in nanotechnology.
- Taiwan Semiconductor Manufacturing Co, Ltd. (Nyse: Tsm)—TSMC is committed to leading the dedicated semiconductor foundry business into a new generation of technology and service delivery. The foundy is capable of manufacturing chips at the nanoscale level.

OTC BB Nanotechnology Stocks

- Acrongenomics, Inc. (Otc bb: Agnm)—Acrongenomics Inc. is a publicly trading biotechnology company that specializes in research, development, production, and distribution of solutions for Genomics, Proteomics and Diagnostics, currently focusing on the development of a platform comprised of molecular human diagnostic tools for various diseases.
- Alysium, Corp. (a subsidiary of AuGRID Corporation) (Otc bb: Agrd)—Alysium is a manufacturing and distribution company with three core technologies catering to all tiers of the industrial world. Our expertise covers thermocouples and nanotechnology, as well as the manufacturing and implementation of silicon carbide tools and solutions for the fabrication of wafers.
- Biophan Technologies, Inc. (Otc bb: Biph)—Biophan's mission is to enable biomedical devices to be MRI safe and image compatible. We offer proprietary thin film nanomagnetic particle and carbon composite polymer MRI shielding, RF filtering...
- Cell Robotics International, Inc. (Otc bb: Crii)—Cell Robotics' Laser Microscopy Workstations use photonic forces for manipulation and microdissection of live or fixed specimens in nanospace. The LaserTweezers and LaserScissors are used to capture, incise, excise, and transport microscopic objects without mechanical contact.
- Emergency Filtration Products (Otc bb: emfp)—Air filtration technology. Our manufacturing facility in Taipei is currentlly in full production for our new nanotechnology enhanced NanoMasks™ in preparation for the winter flu season and a possible resurgence of SARS.
- GEMZ Corp. (International Nanotechnology Corp) (Otc bb: Gmzp)—GEMZ Corp. recently incorporated its wholly-owned subsidiary, International Nanotechnology Corp. The Company has as its goal to Develop patented, commercial products using Nanotechnology to harness renewable energy.

- Microfluidicscorp (a division of MFIC Corporation) (Otc bb: Mfic)—Microfluidics, pioneered the Microfluidizer® Material Processors that deliver unparalleled levels of microdispersions, microemulsions, cell disruption, liposomes and injectibles. Through its Microfluidics Division, the Company provides leading equipment, and innovative technology and comprehensive solutions for nanocrystal and other materials processing.
- MIV Therapeutics Inc. (Otc bb: Mivt)—The mission of MIV Therapeutics is to become a recognized world leader in the development of biocompatible coating and therapeutic stent technologies for drug delivery. It uses Nano crystalline structure with drug delivery capability.
- mPhase Technologies Inc. (Otc bb: Xdsl)—mPhase Technologies has entered into a broad sales, marketing and development partnership with Lucent Bell Laboratories for commercializing nano technology creating intelligent power sources.Our mission is to bring this technology to the marketplace, enabling customer products and systems to take advantage of its unique characterics.
- NanoPierce Technologies, Inc. (Otc bb: Npct)—NanoPierce has resolved to systematically develop and market its proprietary connection technology for microelectronics applications.
- Nano Proprietary, Inc. (NNPP) (Otc bb: Nnpp)—A holding company consisting of two high-tech subsidiaries: Applied Nanotech, Inc. (ANI), and Electronic Billboard Technology, Inc. (EBT). These two technical subsidiaries focus on specific technologies, playing an integral part of Nano-Proprietary's vision of providing cutting-edge solutions for the microelectronic and display industries. The Company's focus is Carbon Field Emission (CFE) products, and related electronic display products utilizing thin carbon film, diamond film, and diamond-like carbon film (DLC).
- Pacific Fuel Cell Corp. (Otc bb: Pfce)—The company owns the worldwide rights (excluding Japan) to proprietary PEM (Proton Exchange Membrane) fuel cell technology. Using its extensive international relationships, the company plans to create joint ventures to rapidly commercialize its innovative technology. The company announces Development of PEM Fuel Cell with Carbon Nanotube-Based Electrodes.
- Sono-Tek Corporation (Otc bb: Sotk)—Sono-Tek ultrasonic nozzles provide thinner, more uniform coatings with less liquid than a conventional spray nozzle. Application includes organic coatings in nanotechnology and thin film application.

- US Global Nanospace, Inc. (Otc bb: Usga)—US Global Nanospace, Inc. is a leader in Anti Ballistic material sciences based upon the company's proprietory nanotechnology expertise. USGN is focused on providing state-of-the-art solutions for Homeland Security, Defense, and Aerospace Applications.
- vFinance, Inc. (Otc bb: vfin)—vFinance, Inc. is a rapidly growing financial services company that provides personalized investment banking and brokerage services, through its wholly owned broker dealer subsidiary vFinance Investments, Inc.

Pink Sheet Nanotechnology Stocks

- Cetek Technologies Inc. (Pink Sheet: Ctkh)—The company supplies structural ceramic components, electronic, and customized advanced ceramic products. Cetek Technologies Inc. has secured a license from Motorola for its carbon nanotube triode (CNT) technology, that are to be used for flat panel displays production.
- Cool Chips plc. (Pink Sheet: Colcf)—Cool Chips represent a new development for cooling, refrigeration, and thermal management. One of the first industrial applications of nanotechnology.
- Nanobac Life Sciences, Inc. (Pink Sheet: Nnbp)—Nanobac Life Sciences Inc. is involved in the research of degenerative diseases stemming from Nanobacterial infections. This infection is implicated in the formation of disease-causing calcified plaque in the circulatory system and vital organs.
- Obducat AB (Pink Sheet: Obdcf)—Obducat develops and supplies technologies and processes for production and analysis of micro and nano structures and provide solutions to companies working with information storage, semiconductors, printed circuit boards, and sensors.
- pSivida Limited (Pink Sheet: Psivida)—pSivida Limited is committed to the biomedical nanotechnology sectors. It has as its core focus the development of nano-structured porous silicon (BioSilicon™) for multiple potential applications in human and animal healthcare. (Note January 2004: The company will trade on the german stock market).
- Starpharma Pooled Development Ltd. (Pink Sheet: Sphrf)—Starpharma is creating value from dendrimer-based nanotechnology through.

Australia Nanotechnology Stocks

- Iatia Limited (Asx: Iat)—The company is developing a nanoscale capable imaging technology. IATIA Ltd is focused on the development and marketing of unique imaging products based on its patented Quantitative Phase Imaging (QPI) technology.

Canada Nanotechnology Stocks

- Advectus Life Sciences Inc. (Vancouver: Avx.v)- Advectus Life Sciences Inc. is an emerging life sciences company focused on the development and commercialization of a patented process using nanotechnology for the delivery of approved cancer fighting drugs across the blood-brain barrier for the treatment of brain tumors.
- Nexia Biotechnologies Inc. (Tsx: Nxb)—The company is involved in molecular biology. It produces BioSteel® and biopharmaceuticals. The Company's lead product, BioSteel® (BioSteel®-M, BioSteel®-I), is recombinant spider silk extracted from the milk of transgenic goats which is then spun into fibres.
- Westaim Corporation (Nasdaq and Toronto Stock Exchange: WEDX and WED)—The Westaim Corporation creates value for shareholders by partnering with established commercial leaders for rapid market penetration of its technologies and maximum return on investment.

France Nanotechnology Stocks

- L'Oréal (Cac40: Or)—L'Oréal is at the forefront of cosmetology research, development and production. The company has developed "nanocapsules" as a skin delivery vehicle for some of its cosmetic ingredients.

Netherlands AEX Nanotechnology Stocks

- Philips Research Laboratories (Aex: Phia)—Nanotechnology provides promising options in several areas that match the business interest of Philips, for example in displays, semiconductor manufacturing and lighting.

China Nanotechnology Stocks

- U-Right International Holdings Limited (Hong Kong main board: 627)—The Group is principally engaged in the design, manufacture, distribution and retailing of men and ladies casual wear and invested in Swedish Texcote Group's nano processing technology.

Sweden Nanotechnology Stocks

- Obducat AB (NGM: Obdu b)—Obducat develops and supplies technologies and processes for production and analysis of micro and nano structures and provide solutions to companies working with information storage, semiconductors, printed circuit boards, and sensors.

- Sandvik AB (Stockholm Stock Exchange: Sand)—Sandvik is a high-technology, engineering group with advanced products and a world-leading position within selected areas. The company has developed an Ultra High-Strength Material based on nanotechnology: the Nanoflex (TM).

Switzerland Nanotechnology Stocks

Actelion Ltd & Actelion Pharmaceuticals Ltd (Swx: Atnl)—Actelion combines the best attributes of the biotechnology and pharmaceutical industries, blending biotech innovation, speed and flexibility with pharma disciplines in drug development, regulatory and marketing. The company is involved in molecular biology; this is why it was added to this page.

United Kingdom nanotechnology stocks:

- Pilkington plc (Lse: Pilk) Pilkington is one of the world's largest manufacturers of glass and glazing products for building, automotive and related technical markets. They created the first self-cleaning glass, using a nano—thin layer that breaks down dirt and other organic materials and prevents them from sticking to the glass.
- SkyePharma Plc (Uk Hq) (Lse: Skp)—Drug delivery company. SkyePharma has performed several inhalation experiments, which have shown that nanotechnology can produce particles in very narrow size distributions, which theoretically allow for targeted delivery in the lung.

Nanophase Technologies (NASDAQ: NANX) Nanomatrials Innovation

Nanophase Technologies is an industry-leading nanocrystalline materials innovator and manufacturer with an integrated family of nanomaterial technologies. The entire focus of Nanophase is nanotechnology, with two distinct and patented processes for the preparation and commercial manufacturing of nanopowder metal oxides i.e. Aluminum Oxide, Zinc Oxide, Cerium Oxide, Titanium Dioxide, and several others. Nanophase Technologies Corporation has developed a process, Discrete Particle Encapsulation, to coat the surface of its nanoparticles with a thin polymeric shell that enables compatibility of the particles with a wide variety of fluids, resins and polymers. Also, Nanophase has developed technology to permit the dispersion of its nanoparticles in water and a variety of polar and non-polar organic fluids. This allows Nanophase to supply concentrated, ready-to-use nanoparticle dispersions, eliminating the need for customers to disperse the nanoparticles themselves. Nanophase's family of integrated technologies economically produces nanocrystalline materials, and then nanoengineers those materials to fit a customer need.

An enhanced capability of Nanophase is our ability to process complex multi-component materials. This process has demonstrated the ability to produce homogeneous mixed metal oxide nanopowders where the component materials form solid solutions with well-defined single crystalline phases. Nanocrystalline metal oxides having up to four metallic elements have been successfully produced. Nanophase has the capability to produce a wide variety of single-phase pure and mixed rare earth oxides, as well as pure and mixed transition metal oxides and main group metal oxides. The materials produced by this process have application in ultrafine polishing and chemical-mechanical planarization (CMP), catalysis, fuel cells, electronic materials, and advanced imaging.

All of Nanophase's facilities and practices are certified to the latest ISO 9001:2000 international standards. Where applicable, Nanophase facilities are cGMP (current Good Manufacturing Practices) compliant and the Company is registered as a drug ingredient manufacturer with the FDA. Nanophase's facilities are designed for flexibility and growth, and to maintain concise product/process flow. Since our processes are modular, there is ample room for expansion and fundamental infrastructure exists for a further 100% growth with minimal additional cost. Nanophase's processes are tightly controlled using statistical process control (SPC) under our Lean Six-Sigma Manufacturing discipline and are demonstrably scalable, cost effective and robust.

Nanophase produces engineered nanomaterial products for a variety of diverse markets: personal care, sunscreens, abrasion-resistant applications, environmental catalysts, antimicrobial products, and a variety of ultra-fine polishing applications, including semiconductor wafers, hard disk drives, and optics. Various applications include wood preservation, anti-fouling and anti-microbial coatings, fuel cells, catalytic converters, UV-attenuation coatings, scratch resistant coatings, charge dissipating coatings, deodorant/antiperspirants, depilatory/shaving products, and others. In parallel, new markets and applications are constantly being developed.

The Company markets its products to target markets and customers, including multinational corporations and Fortune 500 companies. As a small company with a strong focus on technology and commercial manufacturing, the Nanophase business model revolves around strategic partnerships. As examples, Nanophase has worldwide mutually exclusive relationships with BASF for the supply of zinc oxide into sunscreen applications and with Rohm and Haas Electronic Materials, formerly Rodel, for ceria dispersions used in CMP (chemical mechanical planarization) semiconductor applications. While products are available as powders, they are also offered as coated materials and/or stable dispersions in aqueous or organic media, providing partners with nanomaterials in readily usable forms which often lead to decreased time to market.

IndiCo.

"IndiaCo is a private equity group that specializes in mobilization of resources for enabling innovation through the lifecycle of an enterprise; IndiaCo has been recognized by various multilateral agencies and is currently supported under the InfoDev initiative of the World Bank." Private Equity IndiaCo Capital provides extensive value-added capital to high growth companies in emerging markets for enhancing technology capabilities and global market expansion. Technology Transfer Nanotechnology Research & Innovation Foundation (IndiaNano) is the Intellectual Property holding arm of IndiaCo Group. Incubation IndiaCo iClusters manages incubation centers for high technology startups under the IndiaCo umbrella.

Nanalyze

Nanotechnology Investment Information for the "Small" Investor

Nanalyze provides investors with a controlled environment to research, analyze, and discuss "true" nanotechnology companies. As a "small" investor who opts not to buy expensive research reports, you have probably experienced the following setbacks while researching potential nanotechnology investments:

- Nowhere can a "small" investor find a consolidated list of nanotechnology companies covered by respected sources in the nanotechnology community
- No active forums are solely devoted to the discussion of nanotechnology companies
- The popular investment message boards do not moderate the quality of posted messages
- No complete listing of "pure" nanotechnology companies
- Lack of regular response to user feedback

Arrowhead Research

Arrowhead Research is a diversified nanotechnology company structured to commercialize products expected to have revolutionary impacts on a variety of industries, including materials, electronics, life sciences, and energy.

There are three strategic components to Arrowhead's business model:

- *Outsourced R&D Program:* Arrowhead identifies patented or patent-pending technologies at universities or government labs and funds additional development of those technologies in exchange for exclusive rights to commercialize the resulting prototypes. Leveraging the resources and infrastructure of these institutions provides Arrowhead with a highly cost-effective development pipeline. Currently, Arrowhead is supporting efforts in stem cell technology, nanomaterials, nanoelectronics, and nanobiotools at the California Institute of Technology and Stanford University.
- *Commercialization Program:* After prototypes have been sufficiently developed in the laboratories, Arrowhead forms or acquires majority-

owned subsidiaries to commercialize the technology and provides the subsidiaries with strategic, managerial, and operational support. At present, Arrowhead owns majority interest in subsidiaries commercializing diverse technologies, including anti-cancer drugs, RNAi therapeutics, and compound semiconductor materials.

- *The Patent Toolbox:* Arrowhead has acquired or exclusively licensed patents and patent applications covering a broad range of nanotechnology. The Company is actively seeking to add to this intellectual property portfolio.

Advance Nanotech

Advance Nanotech Announces Increase in Facility to Portfolio Company Owlstone Nanotech

Advance Nanotech Inc., specializes in the acquisition and commercialization of nanotechnology. Our interests are focused in three nanotechnology areas, namely: electronics, biopharma and materials which we further classify into distinct groups within our *portfolio*. We provide investment to bridge patented innovation with the capital markets. Our development network creates economic and time efficiencies which can advance the development of University research-programs to marketable product. Our process of nanotechnology development and commercialization has a clear aim, to facilitate the introduction of new products, into new markets, thereby creating new expectations. Nanotechnology pushes out the boundaries of science, providing a catalyst for innovation, industry and investment growth.

Cambridge NanoTech Inc.

Layer Atoms

- Turnkey yet client customizable
- Standard models or 3D animated design
- Heated multi-precursor processing
- All sample sizes
- Conformal pinhole-free thin films
- Oxides, nitrides, metals
- Uniform high aspect ratio deposition
- Superior gate dielectrics
- High-k, high breakdown nanolaminates

AAP—American Pharmaceutical Partners

American Pharmaceutical Partners, Inc. (APP) is a specialty pharmaceutical company that fulfills an important need in the medical community as a provider of a broad portfolio of injectable pharmaceutical products, including difficult to manufacture, urgently needed medical products.

Generic Injectable Business

APP's portfolio of more than 150 products ranks the company among the market share leaders in each of its areas of focus—oncology, anti-infective and critical care injectable pharmaceuticals. The company has a deep pipeline of approximately 50 products in various stages of development, which includes a number of product applications pending at the FDA. APP has expanded its product portfolio with the receipt of an industry leading 39 product approvals between 2001 and 2004. As of July 2005, the company continued this momentum with four new generic product approvals from the FDA.

Proprietary Business

The company also is committed to developing a proprietary pharmaceutical program with a strategy to in-license late-stage proprietary products, complete the development work and usher them through the FDA approval process. In early 2005, the FDA approved ABRAXANE™, a proprietary, next-generation taxane for the treatment of breast cancer after failure of combination chemotherapy for metastatic disease or relapse within 6 months of adjuvant chemotherapy. Please visit the Abraxane.com web site for Important Safety Information, Product Information, and Patient Information. The company's proprietary division Abraxis Oncology, markets ABRAXANE™ to leading U.S. oncologists through its national sales force. A comprehensive development program is underway for ABRAXANE™ to expand the indications in breast, lung, ovarian, head and neck, and melanoma cancers. The company believes it has established the only commercial scale protein-engineered nanoparticle manufacturing capacity in the United States.

Advanced Diamond Technologies

Diamond is nature's perfect material. It is the hardest known substance on Earth. It can be polished to be very smooth, and it has a coefficient of friction similar to that of Teflon™. It is chemically inert, naturally repels water and is an outstanding electrical insulator.

Natural diamonds have been prized through the ages for their scarcity and extraordinary beauty. Synthetic diamonds have been used for decades in cutting tools and abrasive applications. But due to the difficulty of integrating diamond with other materials, it is uncommon to find diamond in use today for anything other than cutting applications. But that's about to change.

Diamond is an engineering material.[SM]

Ultrananocrystalline Diamond™ (UNCD™), synthesized in thin film form using a patented process originally developed at Argonne National Laboratory, is known for its ability to be integrated with other materials. Comprised of diamond grains that are 3-5 nm across—a billion-fold smaller than traditional diamond films—UNCD has many desirable properties such as as-deposited smoothness and the ability to be deposited at temperatures compatible with traditional semiconductor manufacturing processes.

Integrating UNCD into other systems opens up new vistas for diamond applications heretofore unimagined. The day has arrived for engineers to consider diamond as an engineering material.

nGimat

*n*Gimat Co. is an intellectual property company and a manufacturer of engineered nanomaterials in the following areas: nanopowders, thin film coatings, and devices. nGimat's CCVD and NanoSpray℠ Processes along with its Nanomiser® Device enable synthesis of nanoparticles and thin films. These processes are easily scalable, thereby enabling low-cost production of engineering materials with controlled composition, size, and morphology.

*n*Gimat™ has expertise in nanomaterial discovery, production, and processing in final products. The Company has over 30 U.S. patents, issued or allowed; many of these have also issued overseas, and over 100 patent applications pending covering its raw materials, processes, equipment, composition of matter, intermediate products and final products.

Applied Films

Applied Films Innovations for the future

Applied Films is a leading provider of technologically advanced thin film deposition equipment and service. Developing and implementing innovative coating solutions we are supplying high-tech markets such as the flat panel display and the solar glass industry.

ONAMI

Oregon Nanoscience and Microtechnologies Institute (ONAMI) "... an unprecedented and powerful collaboration involving Oregon's three public research universities—Oregon State University, Portland State University, University of Oregon; the Pacific Northwest National Laboratory; the state of Oregon; selected researchers from the Oregon Graduate Institute and the Oregon Health and Sciences University School of

Dentistry; and the world-leading "Silicon Forest" high technology industry cluster of Oregon and Southwest Washington."

Alliance for NanoHealth The first collaborative research endeavor aimed solely at bridging the gaps between medicine, biology, materials science, public policy, and nanotechnology. With the principle goal of providing new clinical approaches to saving lives, the Alliance for NanoHealth comprises world-class research institutions and innovative scientists and clinicians from within the Texas Medical Center and greater Houston community. Members include Baylor College of Medicine, M.D. Anderson Cancer Center, Rice University, University of Houston, The University of Texas Health Science Center at Houston, and The University of Texas Medical Branch.

Nanobiotechnology Center (NBTC) The Center brings together experts in their fields from Cornell University, the Wadsworth Center (New York State Health Department in Albany), Princeton University, Oregon Health & Science University, Clark Atlanta University, and Howard University.

Strategic Partnership for Research in Nanotechnology (SPRING): The University of Texas at Austin, Rice University in Houston, The University of Texas at Dallas and The University of Texas at Arlington. "... a partnership to help position the state as a center for education, research and development in the cutting-edge science of nanotechnology."

Use of Carbon Nanotubes in Space NASA, Michigan State University, University of Pennsylvania, Clemson University, Carnegie-Mellon University.

Nanotec NI "Northern Ireland's nanotechnology engine for new wealth creation. Nanotec NI involves the University of Ulster and Queen's University Belfast, together with Invest Northern Ireland, the Department for Employment and Learning, and various Northern Ireland Industries in a strategic partnership."

National Nanotechnology Infrastructure Network (NNIN) "an integrated networked partnership of user facilities, supported by the National Science Foundation, serving the needs of nanoscale science, engineering and technology."

Members of the NNIN include (and are also listed above):

- Cornell Nanoscale Facility at Cornell University
- Stanford Nanofabrication Facility at Stanford University
- Solid State Electronics Laboratory at the University of Michigan
- Microelectronics Research Center at the Georgia Institute of Technology
- Center for Nanotechnology at the University of Washington
- Penn State Nanofabrication Facility at the Pennsylvania State University

- Nanotech at the University of California at Santa Barbara
- Microtechnology Lab at the University of Minnesota
- Nanoscience at the University of New Mexico
- Microelectronics Research Center at University of Texas at Austin
- Center for Imaging and Mesoscale Structures at Harvard University
- Materials Science Research Center of Excellence at Howard University
- Triangle National Lithography Center at NCSU (DUV lithography only) (Affiliate)

Prior to March 2004, was called the National Nanofabrication Users Network (NNUN) a NSF-sponsored partnership of user facilities to provide opportunities for researchers to turn new ideas in nanoscale science and technology into experimental reality.

NanoMEMS Edmonton "A registered public-private partnership, NanoMEMS Edmonton is dedicated to fostering the development and growth of micro—and nano-based commercial enterprise across the region."

NCCR "Nanoscale Science" "a long-term interdisciplinary research effort focusing on nanoscale structures and aiming to provide new impact and ideas for the life sciences, for the sustainable use of resources, and for information and communication technologies." Universität Basel, Institut für Physik

CeNTech "Integrated into the densest network of universities in all over Europe in the German state of North Rhine-Westfalia (NRW), CeNTech provides the ideal environment to direct selected ideas and results of nanotechnological research into technical applications. CeNTech promotes the formation of start-up companies that originate from university research and supports the expansion of companies in the nanotechnology sector. The CeNTech GmbH provides the general conditions for entrepreneurs to further develop their research ideas into marketable products." Münster, Germany

NSF Funded Nanoscale Science and Engineering Centers For Years 2002—2006

White Rose Nanofactory (WRN) WRN consists of universities who are members of The White Rose consortium (www.whiterose.ac.uk) and other regional universities partners in Yorkshire and Humberside.

Institute for Molecular Manufacturing

The Institute for Molecular Manufacturing (IMM) is a nonprofit foundation formed in 1991 to conduct and support research on molecular systems engineering and molecular manufacturing (molecular nanotechnology, or MNT). IMM also promotes guidelines for research and development practices that will minimize risk from accidental misuse or from abuse of molecular nanotechnology.

During the early decades of the 21st century, the advent of practical molecular manufacturing technology will make it possible to fabricate inexpensively almost any structure allowed by the laws of physics. Consequences will include immensely powerful computers, abundant and very high quality consumer goods, and devices able to cure most diseases by repairing the body from the molecular level up.

The adherence to lifecycle environmental design principles reduces considerably the chances of accidental impacts. The risks of intentional abuse are more difficult to manage, and they exist to some extent with all powerful technologies. The latter problem will require more complex social and economic solutions that reduce the incentives for abuse, as well as new technical systems for inhibiting, detecting, and ameliorating abuse.

Recent progress in technologies leading towards molecular manufacturing includes advances with DNA actuators, nanotubes, proteins, and scanning probe techniques.

Molecular nanotechnology research supported by IMM

- IMM announces IMM Freitas Research Fund. $30,000 needed by July 31, 2004.
- In conjunction with IMM, CMP Cientifica is pleased to offer a special price to Senior Associates for the 2003 version of their Nanotechnology Opportunity Report™. Free White Paper also available. Offer expires August 31, 2004.
- IMM announces IMM Molecular Nanotechnology Projects.
- IMM responds to attacks on the feasibility of molecular assemblers.
- IMM announces IMM Prizes in Computational Nanotechnology. Deadlines for 2004 awards will be posted soon.

Molecular machinery parts design work by Dr. Eric Drexler includes a simple pump, a fine motion controller for molecular assembly, and a molecular differential gear. Information on free and Open Source molecular visualization tools. Some design graphics of molecular machine parts, produced by K. Eric Drexler, or by K. Eric Drexler and Ralph C. Merkle, that appear on the Web sites of Foresight Institute and the Institute for Molecular Manufacturing (IMM) are copyrighted by IMM. *It is not necessary to obtain permission to use IMM-copyrighted images for either commercial or non-commercial purposes. Permission to use the images is granted on the condition that credit is given and our copyright notice appears in the publication.* Images should also contain the URL for IMM (www.imm.org).

- IMM funded the writing of the first technical book on nanotechnology,

Nanosystem: Molecular Machinery, Manufacturing, and Computation by Dr. K. Eric Drexler (Wiley Interscience, 1992).

- IMM sponsored the first textbook in the emerging field of nanotechnology applied to medicine, *Nanomedicine* by IMM Research Fellow, Robert Freitas (Landes Bioscience, 1999).
- Work done by Markus Krummenacker on molecular building blocks, that lead to the paper "Steps Toward Molecular Manufacturing" and to the CavityStuffer program, was supported by IMM.

More information on molecular nanotechnology

Now available, the first journal article published on molecular nanotechnology: "Molecular engineering: An approach to the development of general capabilities for molecular manipulation," *Proceedings of the National Academy of Sciences*, September 1981.

IMM is cosponsoring the Foresight Conferences on Molecular Nanotechnology:

- The 1st Conference on Advanced Nanotechnology: Research, Applications, and Policy will be held Oct. 21-24, 2004, Crystal City Marriott Hotel (Washington, DC Area)
- 11th Foresight Conference on Molecular Nanotechnology, October 10-12, 2003; San Francisco Airport Marriott, Burlingame, California.
- 10th Foresight Conference on Molecular Nanotechnology, October 11-13, 2002; Hyatt Hotel, One Bethesda Metro, Bethesda, Maryland.
- Ninth Foresight Conference on Molecular Nanotechnology, November 9-11, 2001; Westin Hotel in Santa Clara, Silicon Valley, California.
- Eighth Foresight Conference on Molecular Nanotechnology, November 3-5, 2000; Bethesda, Maryland.
- Seventh Foresight Conference on Molecular Nanotechnology, October 15-17, 1999; Santa Clara, CA.
- The First EL.B.A.—Foresight Conference on Nanotechnology, April 14-16, 1999; Rome, Italy
- Sixth Foresight Conference on Molecular Nanotechnology, November 12-15, 1998 in Santa Clara, CA.
- Fifth Foresight Conference on Molecular Nanotechnology, November 5-8, 1997, Palo Alto, California.
- Fourth Foresight Conference on Molecular Nanotechnology, November 9-11, 1995, Palo Alto, California. Proceedings were published in the journal Nanotechnology; preprints are available on the main conference page.

ASM Materials Solutions 2000, Oct. 9-12, 2000, America's Center in St. Louis, MO. There were five sessions at this conference (ASM's largest annual event) devoted to nano-materials. One session on Tuesday October 10th, co-sponsored by IMM and organized and co-chaired by IMM President David Forrest, was dedicated to molecular nanotechnology. The purpose, and unifying theme of this session, was to provide an overview of molecular nanotechnology to a portion of the materials community that has been largely insulated from progress in this area. This session was focused toward the view that substantial progress is occurring to develop assembler technology (whether specifically directed to this end or not), and that it has important consequences for all materials systems and the devices made from those materials. Web: http://www.salsgiver.com/people/forrest/asm_solutions_2000.html

ASM Materials Solutions 2001, 5 November 2001, also will include a session on Nanomaterials (From Promise to Production: Perspectives on the Evolution of Nanomaterials) organized and co-chaired by IMM President David Forrest. Web: http://www.salsgiver.com/people/forrest/asm_solutions_2001.html

IMM President David Forrest presented the paper "Molecular Manufacturing Development and Technology Planning," in: Proceedings of the 32nd International SAMPE Technical Conference, Society for the Advancement of Material and Process Engineering, held 5-9 Nov. 2000, Boston, MA. Web: http://www.baverstam.com/nanotech/forrest_SAMPE_2000_Word98.doc

IMM copublishes a newsletter on nanotechnology, *Foresight Update*. The current issue; previous issues are available.

Usenet technical discussion group (moderated) on nanotechnology: sci.nanotech. The sci.nanotech archives are now available at: http://crit.org/critmail/nano_archives.html. See sci.nanotech web site for more information.

Other IMM activities

The Institute for Molecular Manufacturing and the Foresight Institute have proposed guidelines for the development of nanotechnology.

IMM Reports on molecular manufacturing. Most recent additions: *Recent results in SRMS*; *Nanomedicine*.

IMM News and Events.

- IMM personnel have participated as judges in the Feynman Prize in Nanotechnology. Feynman pointed in the direction of nanotechnology in a talk he gave in 1959.
- Senate testimony on nanotechnology has been presented by Eric Drexler. The submitted written testimony and the oral testimony include discussion of environmental applications.

- Dr. Merkle has testified on nanotechnology to the U.S. House of Representatives Committee on Science, Subcommittee on Basic Research, June 22nd, 1999.
- IMM Chairman Neil Jacobstein presented an Aspen Institute Socrates Seminar entitled "Genes, Nanobots, and Megawatts" July 5-8 2001.
- On 23 July 2001, IMM President David Forrest presented an invited talk to ASTM's D-20 Committee on Polymers in Montreal, entitled, "Molecular Manufacturing: Implications for Polymers."
- On 24 September 2001, David Forrest presented an invited seminar on, "Molecular Manufacturing and the National Defense" to the Carderock division of the Naval Surface Warfare Center, in Bethesda, MD.

Helping IMM develop molecular nanotechnology

IMM is a 501(c)3 nonprofit foundation; donations are tax-deductible in the U.S. as permitted by law. Donations at any level are welcome and appreciated.

The IMM Senior Associates Program supports vital research in molecular nanotechnology by providing seed money for new proposed efforts. It enables individuals to play a key role in advancing this technology through their individual or corporate contributions.

By pledging an annual contribution of $250, $500, $1000, or $5000 for five years, Senior Associates join those committed to making a difference in molecular nanotechnology. Activities include special publications, online interaction, and meetings.

A letter to Foresight Institute members

For more information about the Senior Associates Program, or to become a Senior Associate, phone 650-917-1120.

Senior Associate membership donations may be made by VISA, MasterCard, check or international money order valid in the U.S. Make checks payable to "Institute for Molecular Manufacturing"; checks and bank drafts must be in U.S. dollars drawn on a U.S. Bank. Donations with payment by credit card may be faxed, or you may use a secure server for on-line donation.

- This form may be printed and mailed or faxed to IMM.
- Use your VISA or MasterCard to register online on our secure server.

Nanotech RIE3

General Information and Usage

This is a Materials Research Corporation RIE-51 parallel plate, 13.56 MHz system used for etching with fluorine-containing gases (CF_4, SF_6, and CHF_3). The system is used primarily for etching of Si, SiO_2, and Si_3N_4 films. Metals such as tungsten may also be etched. Tool features include: six inch diameter water cooled cathode/substrate platform, pyrex cylinder for plasma confinement and gas flow control, adjustable cathode-anode spacing, fixed DC bias or RF power control and a HeNe laser etch monitor with chart recorder. It is turbo pumped and has no loadlock.

CF_4 / O_2 and SF_6 /O_2 will etch Si, SiO_2 and Si_3N_4 readily since free fluorine is readily liberated in the plasma. The oxygen (up to 40%) initially enhances the fluorine concentration resulting in a higher etch rate. The oxygen also minimizes polymer formation in CF_4/O_2. Too much Oxygen will compete for fluorine available, suppressing the etch rate. Argon can be added to increase the physical component of etching. The highest etch rates are achieved with SF_6 due to the ease of liberating fluorine compared with CF_4. The relative etch rate decreases as one goes from Si to Si_3N_4 to SiO_2. CF_4/H_2 and CHF_3 can be used to selectivity etch SiO_2 over Si and resist due to increased polymer formation from the presence of hydrogen. This polymer layer is thicker on Si and resist than on SiO_2. The trade-off is selectivity versus sidewall profile as the polymer will result in a tapered wall profile. Also, the polymer can be difficult to remove after etching.

The etches have good selectivity to many metals and semiconductors such as Ni, Al, Cr, Ti, GaAs, InP, and GaN. The system generally produces anisotropic etch profiles unless one goes into a purely chemical fluorine etch mode with higher pressure SF_6 processes. The system also has a strong loading effect so that larger substrates and open areas will require more feed gas and higher pressure to compensate. As a result, individual processes need to be characterized.

Equipment Specifications

- Etch gases include: CF_4, CHF_3, SF_6, Ar, O_2, H_2
- Low 1 E -6 ultimate chamber pressure
- 13.56 Mhz excitation freq
- Manual gas control
- Automatic pressure control
- Manual RF tuning network
- Timer circuit for stopping the plasma
- HeNe laser monitoring for etch stop
- Sample size limited to approximately 4 inches
- Masking materials include: Ni, photoresist (limited to low bias/power), Cr, Al.

Contact Information

For additional information regarding the RIE #3 Fluorine-Based System MRC 51", or if you would like to inquire about using the UCSB Nanofabrication Facility, please contact Jack Whaley using the telephone number or e-mail address: Jack Whaley, Phone: (805) 893-8174, E-Mail: whaley@ece.ucsb.edu

MITRE

The MITRE Corporation is a not-for-profit organization chartered to work in the public interest. As a national resource, we apply our expertise in systems engineering, information technology, operational concepts, and enterprise modernization to address our sponsors' critical needs.

MITRE manages three Federally Funded Research and Development Centers (FFRDCs): one for the Department of Defense (known as the DOD Command, Control, Communications and Intelligence FFRDC), one for the Federal Aviation Administration (the Center for Advanced Aviation System Development), and one for the Internal Revenue Service (the Center for Enterprise Modernization). MITRE also has its own independent research and development program that explores new technologies and new uses of technologies to solve our sponsors' problems in the near-term and in the future.

Our History

MITRE was formed in 1958 as a not-for-profit corporation under the leadership of C.W. Halligan. In the beginning, several hundred employees from the Massachusetts Institute of Technology's Lincoln Laboratories came to MITRE to create new technology for the Department of Defense—specifically the Semi-Automated Ground Environment, which used brand new digital computers. The company expanded in 1963 after the Federal Aviation Administration gave the company systems engineering responsibility for the projected National Airspace System. MITRE continued to expand as it took on new challenges for new customers, always focused on serving the public interest. Our work in information technology led to the formation of an FFRDC for the Internal Revenue Service in 2000, which now provides enterprise modernization support to numerous government agencies. Over the years, the company has continued to evolve to meet the changing needs of its sponsors by providing top-notch engineers and scientists experienced in a wide range of technologies.

Staff Statistics

MITRE has 5,700 scientists, engineers and support specialists—65 percent of whom have Masters or Ph.D. degrees. Staff members work on hundreds of different projects across the company, demanding a high level of technical, operational, and domain knowledge.

Facilities and Offices

MITRE has headquarters in Bedford, Massachusetts, and McLean, Virginia, with more than 60 sites around the world, enabling our staff to work closely with our customers.

Areas of Expertise

MITRE works in the broad domains of aviation, defense and intelligence, and enterprise modernization. Our core areas of expertise, listed below, underlie all these domains.

- Acquisition and Systems Analysis
- Communications and Networking
- Information Technology
- Sensors
- Systems Engineering
- Independent Research and Development

MITRE Scientists and Engineers

MITRE scientists and engineers provide our customers with a unique combination of broad operational knowledge and in-depth technical skills through our federally funded research and development centers. MITRE also maintains its own independent research and development program, in which MITRE staff explore new technologies that may provide the answers to the near-term and far-term challenges of our customers. In these research projects, MITRE experts develop concepts and prototypes that may be transferred to government organizations. As the manager of FFRDCs working in the public interest, MITRE also transfers technology to the public whenever appropriate.

Overview

Since its beginnings, MITRE has helped its sponsors acquire large, complex systems—such as those for the military and for aviation—that are technically feasible and cost-effective. MITRE has experts throughout its FFRDCs who focus on this work, such as estimating cost and schedule resources for system development, production, operations, and maintenance of sponsors' projects; developing acquisition strategies and assisting with source selection; performing cost, price, and technical analyses for contract negotiations; and performing financial analyses and identifying process reengineering opportunities.

MITRE also has a technical center of excellence, the Center for Acquisition and Systems Analysis, that provides a broad range of support to the entire corporation in the areas of economic, operational, and systems acquisition analyses. In addition, our

Systems Engineering Process Office provides resources and collects "best practices" to share throughout the company.

Current Work Programs

Some of our projects in this area include:

- Developed the "Program Management Toolkit," a Web-based resource that helps guide program managers through the world of risk assessment and management.
- Developing, instituting, and using metrics to measure the instantaneous, daily, and seasonal performance of the National Airspace System (NAS), providing daily NAS "report cards" to senior Federal Aviation Administration managers.
- Delivered life cycle models tailored to a unique maintenance strategy and risk management tools built around the acquisition strategy and technical characteristics of a sponsor's program.
- Developed a risk management plan for the Navy/Marine Corps Intranet.
- Developing case studies and alternatives for the evolution of the air traffic management system infrastructure, especially for en route and traffic flow management systems—and for the communication, navigation, and surveillance systems.
- Helping the Coast Guard with a plan to modernize and replace its long-range ships, aircraft, and related command and control systems.

Related Information

Articles and News

- Making Army Battle Command Systems Play Together August 2005
- Applying Neuroscience to Hard-to-Solve Problems, *The Edge* Spring 2005
- A New Approach to Analyzing Enterprise Architectures, *The Edge* Fall 2004
- Mapping Among Independently Developed Aviation Information Systems Increases Interoperability, *The Edge* Summer 2004
- Project Alpha Expedites Search for Ideas for a Changing Military May 2004
- Modeling the Airspace Can Improve Aviation Efficiency March 2004
- The Air Force "Practices" for Future Warfare March 2003
- Exploring ADS-B Technology 2001

- How One Group Plows New Ground 2000
- Decision Support Issue, *The Edge* June 1999
- Modeling and Simulation Issue, *The Edge* August 1998

Technical Papers and Presentations

- JFACC Information Management (IM) Capability—2010 July 2005
- Improving the Practice of DoD Architecting with the Architecture Specification Model June 2005
- A Collaboration Evaluation Framework March 2005
- Midway Revisited: Detecting Deception by Analysis of Competing Hypothesis July 2004
- Adaptive Time-Critical Resource Management Using Time/Utility Functions April 2004
- FAA System Security Testing and Evaluation May 2003
- Capturing Behavioral Influences in Synthetic C2: What We've Learned So Far and Where We Need to Go August 2002
- 1-2-3 Invest: Frame of Reference for Public-Sector and Non-Profit Investment Selection Studies October 2001

Websites

- 2005 MITRE Technology Symposium—Investment Strategies and Operational Analysis
- 2004 MITRE Technology Symposium—Investment Strategies and Operational Analysis
- 2003 MITRE Technology Symposium—Investment Strategies
- 2002 MITRE Technology Symposium—Investment Strategies
- Systems Engineering Process Office (SEPO)

2

Nanotech Projects, Reports & Nano Education

Silicon Valley Nano Ventures

Nanotechnology Investing: 2005 A high value guide for Entrepreneurs, Executives, Investors, and Money Managers to learn more about the markets and companies that nanotechnology will enable or disrupt. This comprehensive 240 page report is available for $300! Your purchase includes 2 hours of consulting in your area of interest—a $200 value!

Nano Grind

Project Description

Traditionally, most optical surfaces are produced by grinding followed by lapping (also called grinding in the optics field) and polishing. More recently some lenses were produced by moulding (typically plastic spectacle lenses and lenses for low-cost cameras) but this also required the moulds to be made in the traditional way

Nano Grind aims at producing cheap, smooth and curved optical surfaces (in glass, ceramics or hard metal) using an innovative grinding machine tool with electrolytic in-line dressing (ELID). Nano Grind will give a significant machining cost reduction with equal specs on dimensional accuracy and surface quality. This will be realized by a faster overall production process for direct manufacturing and by offering longer standing moulds for replication. This will result in an important cost reduction for the end-user. Nano Grind will enable production of curved surfaces with average roughness value comparable to polishing. Nano Grind will also enable production of curved surfaces with a very high dimensional accuracy. Combined with the previous objective this will allow high-precision optics with high shape accuracy.

The Market

The basic result of this project will be that products with surfaces of optical quality can be produced without (or with a much shorter) time-consuming polishing phase. In some cases it is not the time reduction effect that dominates the improvement but the fact that the shape-distorting effect of polishing can be abandoned.

The envisaged machine and technology will have the multiple applications.

A first one is the manufacturing of advanced optical components like required for semi-conductor manufacturing. In this field, extreme accuracies, even beyond those of Nano Grind are requested. The project will allow a more efficient manufacturing by eliminating a great part of the polishing effort.

Next to benefit from Nano Grind is the manufacturing of complex lenses (progressives) needed by elderly people. They will become affordable for a larger group of consumers when prizes decrease. And for the same prize, the quality can be improved. Also, the new grinding technology will improve the quality of medical equipment based on advanced optics (aspehericals and free formed) or will decrease the price of existing optical components so that advanced facilities will become available for a less wealthy users group. Another application field are instruments for remote monitoring of environmental characteristics (by satellite or plane) that often make use of sophisticated optics. The availability of the new grinding technology will have a positive influence on the application of more advanced optics in new monitoring programmes. Finally the results of this project will improve the quality of reflectors of automobile spots. This will have a positive influence on the safety conditions of road users under poor light conditions.

Innovations

Current highly accurate ELID processing is only applicable to flat surfaces. Other high precision manufacturing processes are limited to symmetric surfaces or cannot guarantee high shape accuracy. With regard to the state-of-the-art, the main innovation of Nano Grind therefore will be the combination very stiff and accurate positioning elements with an improved ELID process in order to allow producing highly accurate and smooth curved optical surfaces in an efficient way. The process will be applied to various materials covering metals, ceramics and glass.

This requires the development of the following innovative concepts:

- A stiff five-axis positioning system. The envisaged stiffness is about 5 times higher than the current industrial practice.
- A reference measurement system capable of tracing tool and work piece position with nanometer precision
- An industrially reliable ELID grinding technology for curved surfaces. Nano Grind will further refine and develop a novel and exclusive concept of dressing electrode that results in a much more stable process

It should also be mentioned that these innovations will lead to a production process for precision optics that is substantially more eco-efficient than the current industrial practice.

The Project Team

The project team is structured to include the multi-disciplinary competencies necessary for the simultaneous development of a general technology and the push towards the envisaged applications. The partnership consists of 8 partners covering 7 European countries.Katholieke Universiteit Leuven and Cranfield University play the central role of two higher education institutes specialised in precision engineering, bringing in the state-of-the-art competence. Both partners already developed ELID grinding technology in other projects. They have lab set-ups for preliminary experiments available. TNO Industrial Technology, a research organisation specialised in manufacturing processes, and Tekniker, a research organisation specialised in precision engineering also contribute to respectively the development of de ELID process for free form surfaces and to the design and testing of the high precision machine tool.The main exploiting parties are Loh Optikmaschinen, a commercial manufacturer of precision machine tools, Berliner Glas, a commercial manufacturer of precision optics, Centro Ricerche FIAT, being the research centre of a commercial manufacturer of automotive products, and Gravitas 2000 an SME offering design and manufacturing services for high-precision components.

An Application Example

An important field of application is the manufacturing of high precision optical systems for semiconductor lithography, especially for the next generation projection systems. Nano Grind will allow a reduction of production costs and will make fast prototyping possible. The improved ELID process will result in better surface quality and lower subsurface damage.

MAJOR REPORTS ON NANOTECHNOLOGY

Forbes/Wolfe Nanotech Report

Stunning breakthroughs in Nanotechnology are about to transform the future of our economy and make EARLY INVESTORS RICH. And the FIVE PROFIT-TAKERS we own now will collect the lion's share of profits. Don't miss out! Here at FORBES we believe that the secret to profitable investing is to identify the companies and the technologies that will drive future growth and profits. We fully expect our top holdings to DOMINATE this exciting new sector the same way Microsoft, Intel, and Cisco have dominated operating systems, microprocessors, and the Internet. That's why I've made it possible for you to try the *FORBES / Wolfe Nanotech Report* SATISFACTION GUARANTEED. In fact, FORBES' new Nanotech Report is the FIRST and ONLY publication that targets the hottest growth markets of nanotechnology—from molecular computing to self-assembly... from DNA programming to biomedicine.

Now it's your turn to profit. For over eight decades, FORBES has helped investors like you build their wealth light-years ahead of the crowd. Those who have followed our timely advice through our publications and investment advisories have enjoyed many MONEY DOUBLING gains over the past 80 years, while avoiding risk like the plague. But please don't delay. As you'll see, our recommendations are on the verge of big BREAKTHROUGHS. Every day you invest without the help we can bring you is another day you risk being on the sidelines during the next profit wave.

Nano-zilla Strikes Tokyo

Last month, I was lucky enough to attend Japan's Nano Tech 2005 International Conference—the world's largest nanotech gathering held in Tokyo's Big Sight conference tower over three days. Over 30,000 people attended this conference in an effort to learn about new nanotech research and development and catch a glimpse of the latest nano-enabled prototypes from Asia's consumer electronics and chemicals giants.

Will Nanotech Preserve Moore's Law?

"Within 10 years, the entire semiconductor industry will rely on nanotechnology," says Mike Roco, the head of the United States' National Nanotechnology Initiative.

Nanotech Relief for G.I. Joe

As the casualty count rises in Iraq, the safety of our soldiers is paramount in the minds of defense researchers. Nowhere is that more evident than at the Institute for Soldier Nanotechnologies, established at the Massachusetts Institute of Technology in 2002 with a five-year, $50 million grant from the U.S. Army.

Nanotechnologies That Could Change the World

How many technologies have really changed our world over the past five years? E-mail, broadband Internet and cell phones would certainly top the list. These three breakthroughs have spawned industries and companies few of us could have predicted. Experts agree that nanotech will also eventually affect most sectors of business.

Archived Josh Wolfe Webcast:

On Tuesday November 30th, Josh Wolfe, Editor of the Forbes/Wolfe Nanotech Report hosted a one hour live Web-based video discussion, "Nanotechnology: What Every Investor Should Know". Click here to access the archived webcast:

Nanotech Vs. The Green Gang

There could be trouble brewing on the horizon for nanotech. The trouble I am talking about has nothing to do with nanotech's great scientific or commercial promise. It has to do with politics and the media. The "Green Gang" is starting to focus on nanotech research, and governments around the world are listening. There are rumblings that regulations are needed. They say they want to guarantee the safety of the technology and instill confidence in the general public.

2003 Nanotech Product Guide

Skeptics say that it's too early to profit from nanotechnology. We proved them wrong.

Nanotech vs. the Environmentalists

Ask some folks about nanotech's environmental impact and you'll get a cheery story of ending pollution and mopping up the planet's waste.

Top Nanotech Buys For 2005

Last year was a year of mixed blessings for nanotechnology. The year started out on a high note after President George W. Bush signed a $3.7 billion nanotech bill. That sent most nano stocks to even frothier levels. Nanoscience made exceptional progress in 2004, as well.

It's 1993 All Over Again

My team and I recently searched through nearly 6,000 publications over the past eight years and found that the word "nanotechnology" is spreading like wildfire.

Nanotech on the Front Lines

As the U.S. enters into its first full-scale international conflict in more than a decade, military technology has taken front stage.

Beware the Nanobubble

Energy is to an industrial society what oxygen is to the body. Even a few minutes without it and the whole complex organism collapses.

Asian Nanotech Fever Running Hot

Nanotech is white hot in Asia, and the Pacific Rim power players have in their crosshairs the epicenter of nanotech development: American startups and research labs.

Nano Frenzy Hits Wall Street

Did you catch the action in nanotech shares last month? Wall Street is finally tuning into tiny technologies in a big way.

Nanotechnology's Power Brokers

My team at Forbes/Wolfe Nanotech Report surveyed leading investors, scientists, corporate execs and high-ranking government officials to give our subscribers a jump on who the top movers and shakers are in nanotech.

Nano Skepticism

Warren Buffett is fond of saying, "You pay a high price for a cheery consensus" But despite the investor community's budding interest in nanotech, we're nowhere near reaching that cheery consensus.

Lux Research Inc.

The Nanotech Report 2004, a more than 600-page reference study from Lux Research. Lux Research, the leading research and advisory firm focused on nanotechnology, today announced the release of key findings from The Nanotech Report 2004, a more than 600-page reference study. The findings are part of the report's 24-page introduction that can be accessed at no charge from: http://www.luxresearchinc.com/tnr2004

Partial list of The Nanotech Report 2004's key findings:

- Governments, corporations and venture capitalists will spend more than $8.6 billion worldwide on nanotechnology research and development in 2004.
- National and local governments across the world will invest more than $4.6 billion in nanotechnology R&D in 2004. We expect 2004 will be the last year that governments outspend corporations on nanotechnology as activity shifts from basic research to applications development.
- The U.S. government will spend nearly twice as much on nanotechnology this year as it did on the Human Genome Project (HGP) in its peak year. In 2005, the National Nanotechnology Initiative will surpass the HGP on a cumulative basis. The U.S. has now appropriated more than $3.16 billion to fund nanotechnology R&D since 2000 and is proposing $982 million in new funding for FY 2005.
- Established corporations will spend more than $3.8 billion globally on nanotechnology R&D in 2004.
- There is no bubble in nanotech venture capital funding, despite widespread reports to the contrary. VC firms invested just $79 million into nanotechnology companies in the first half of 2004, down from annual totals of $325 million in 2003 and $386 million in 2002. Based upon current fundraising activity, we expect VC investment to total approximately $200 million in 2004.
- Approximately 1,500 total companies worldwide have announced

nanotechnology R&D plans. Eighty percent of them — approximately 1,200 — are start-ups, 670 of which are in the U.S.

- Media coverage of nanotechnology — both positive and negative — is increasingly exponentially. Mentions of the word "nanotechnology" in the popular press rose from 190 in 1995 to 7,316 in 2003; we predict more than 12,000 mentions in 2004.

According to Matthew Nordan, Vice President of Research at Lux Research, "The biggest shift we've seen in the last year is nanotechnology moving out of the lab and onto the production line. Nanotech startups are beginning to make money, with revenue ranges between $10 million to $20 million for those at the top of the ranks. They're partnering with established corporations to develop products in a pattern similar to biotech: Ten of the 30 corporations in the Dow Jones Industrial Average have announced nanotech partnerships."

"Like electricity and the Internet, nanotechnology is a general-purpose technology revolutionizing multiple industries," says Draper Fisher Jurvetson Managing Director Steve Jurvetson. "The Nanotech Report 2004 identifies much of nanotechnology's direct effects and its derivative impact."

The Nanotech Report 2004 is the premier research guide for nanotechnology. The TNR 2004 profiles more than 1,000 companies and features new investment frameworks and strategies, company and academic profiles, patent licensing opportunities and trends, competitive data, interviews with Nobel laureates and industry experts, profiles of the most influential players in nanotechnology, and a technical primer. The report can be purchased by emailing sales@luxresearchinc.com. For more information on The Nanotech Report 2004, visit http://www.luxresearchinc.com/tnr2004

Sizing Nanotechnology's Value Chain—New Report from Lux Research

Sales of products incorporating emerging nanotechnology will rise from less than 0.1% of global manufacturing output today to 15% in 2014, totaling $2.6 trillion. This value will approach the size of the information technology and telecom industries combined and will be 10 times larger than biotechnology revenues, according to a new report from Lux Research entitled "Sizing Nanotechnology's Value Chain." However, sales of basic nanomaterials like carbon nanotubes and quantum dots will total only $13 billion in 2014: Nanotechnology's economic impact will arise from how these fundamental building blocks are used, not from sales of the materials themselves.

The report refutes the popular misconception that nanotechnology is an industry or a sector — it isn't. Instead, nanotechnology is a set of tools and processes for manipulating matter that can be applied to virtually any manufactured good. Rather than envisioning a mythical "nanotechnology market," Lux Research recommends that executives focus on how nanotechnology is being exploited across industry value chains, from basic materials to intermediate products to final goods. The report presents separate forecasts by each value chain stage as well as by sector and region.

"Over the past several years, companies have selectively applied nanoscale innovations to products ranging from the Chevrolet Impala to Merck's anti-emetic drug Emend," explained Matthew Nordan, Vice President of Research at Lux Research. "These initial deployments have proven the value of nanotechnology, setting the stage for an explosion of applications. In 2014, we project that 4% of general manufactured goods, 50% of electronics and IT products, and 16% of goods in healthcare and life sciences by revenue will incorporate emerging nanotechnology." Lux Research predicts that nanotechnology's growth will occur in three phases:

- In the first phase, ending this year, nanotechnology is being incorporated selectively into high-end products. In 2004 revenues from products incorporating emerging nanotechnology will total \$13 billion, \$8.5 billion of which lies in automotive and aerospace applications.
- Through 2009, commercial breakthroughs will unlock markets for nanotechnology innovations, with revenues rising to \$292 billion. Electronics and IT applications will dominate as microprocessors and memory chips built using new nanoscale processes come to market.
- From 2010 onwards, nanotechnology will become commonplace in manufactured goods, with revenues rising to \$2.6 trillion in 2014. Healthcare and life sciences applications will finally become significant in this period as nano-enabled pharmaceuticals and medical devices emerge from lengthy human trials.

The widespread use of nanotechnology in mainstream products will have profound ripple effects. Ten million manufacturing jobs worldwide in 2014 — 11% of total manufacturing jobs — will involve building products that incorporate emerging nanotechnology. Nanotechnology will shift market shares and introduce unconventional competitors: For example, silicon nanowire display printing technologies could cut capital requirements for flat-screen display plants by an order of magnitude, tempting fleet-footed manufacturers like Dell to enter the market. Supply chains will simplify as highly functional materials eliminate steps in manufacturing processes, negatively impacting sub-assembly manufacturers and transportation companies while making value-added taxes more productive for governments than sales taxes.

"Nanotechnology's increasing relevance creates clear mandates for business and government leaders," said Nordan. "Corporations need to develop an explicit nanotechnology strategy — apart from leaders such as DuPont and Praxair, few companies coordinate their nanotechnology activities at all today. Investors should focus on nanotechnology applications in the middle of industry supply chains where profit potential is highest, and consider laying nanotech as a long-term secular trend.

Public sector leaders should focus on fostering nanotechnology demand, not just supply, and establish informed regulation to address health and safety issues."

For "Sizing Nanotechnology's Value Chain," Lux Research built bottom-up, top-down, and evolutionary models of 42 product segments impacted by nanotechnology. The report team populated the models through exhaustive secondary research; interviews with more than 100 executives, researchers, and academics working to commercialize nanotechnology; and integration of macroeconomic data from organizations such as the U.S. Bureau of Economic Analysis and the World Bank. Additionally, Lux Research partnered with UK-based Volterra Consulting — founded by Paul Ormerod, influential economist and author of "Butterfly Economics" — to build advanced evolutionary models that measure how nano-enabled solutions compete for market share with alternatives.

The report is available immediately to clients of Lux Research's Nanotechnology Strategies advisory service. For information on how to become a client, contact Rob Burns, VP of Sales at (646) 723-0708.

Liquid Immersion and Nanoimprint Lithography Will Keep Moore's Law Going for Another Decade

Doomsday scenarios say that Moore's Law — the chip industry's ability to double transistor density every 18 months — is coming to an end. But two new nanolithography techniques will be ready in time to sustain the chip industry's performance curve for at least another decade, according to a new report from Lux Research entitled "Optical Lithography's Last Stand," debuted today in a presentation at SEMI NanoForum 2004.

The semiconductor industry is facing a crisis as current processes for building high-performance chips reach their limits. As heat dissipation makes raising the clock speed of chips impractical — Intel canceled plans to increase the clock speed of its Pentium processors beyond 3.8 GHz last month — the only way chip manufacturers can sustain performance is to put more circuits into less space. That requires a new approach to the most critical step in building chips: the process of lithography, which lays down the pattern of desired circuits on a silicon wafer.

"Lithography is the single most expensive part of the semiconductor manufacturing process, accounting for one-third of the cost of each chip today," said Matthew Nordan, Vice President of Research at Lux Research. "It's also the bottleneck that stands to limit the power of microprocessors and the capacity of memory chips. With $3.9 billion in lithography tools sold last year, it's no surprise that 10 different platforms are now competing to take over when today's process runs out of steam. All of these 10 nanolithography platforms focus on patterning circuits smaller than 100 nm in size."

For the report, Lux Research constructed an assessment tool to rank the 10 competing nanolithography platforms on their potential impact and level of development. Lux Research conducted in-depth interviews with experts on each platform to gather data and reviewed the scores with these experts to ensure accuracy. The report finds that:

- 193 nm liquid immersion lithography will be adopted en masse in 2007 and will be used as a stopgap solution before hitting a performance wall in 2011.
- Nanoimprint lithography looks best-positioned as the long-term solution after 2011 because of its technical advantages, escalating amount of research activity, and feasible solutions to technical challenges.
- Two wild-card platforms may outpace nanoimprint if significant technical breakthroughs are made in the next four years: Extreme ultraviolet (EUV), which has been the subject of longstanding research from the likes of IBM and Intel but has not yet proved viable, and little-known X-ray projection from startup Adelphi Technology.
- All other contenders are unlikely to be used for high-volume lithography.

"Semiconductor industry players will need to prepare for a radically new platform," said Will Arora, Senior Associate at Lux Research. "Semiconductor companies should assess these new platforms based on their merits, not on the millions of dollars in sunk cost plowed into some which no longer look compelling. The five nanoimprint toolmakers — EVG, Molecular Imprints, Nanonex, Obducat, and SUSS MicroTec — need to form a consortium to advance their common interests and battle skepticism from major chipmakers. Companies that make reduction masks and lenses should identify new ways to remain relevant should a non-optical tool like nanoimprint, which would decrease the need for their products, be adopted."

The report is available immediately to clients of Lux Research's Nanotechnology Strategies advisory service. For information on how to become a client, contact Rob Burns, Vice President, Sales, at (646) 723-0708.

Lux Research is the world's premier research and advisory firm focusing on the business and economic impact of nanotechnology and related emerging technologies. Lux Research provides continuous advisory services, customized consulting, and reference studies to corporations, start-ups, financial institutions, and public sector organizations. Our founders and our research staff are the most widely recognized nanotechnology visionaries throughout the world. Visit http://www.luxresearchinc.com/ for more information.

Corporate Nanomaterials Buyers Fail to Get What They Pay For

Corporate nanotechnology researchers frequently purchase nanomaterials — nanoscale structures in pure form, like carbon nanotubes, fullerenes, and quantum dots — which are the basic building blocks of nanotechnology. But these buyers frequently fail to get what they pay for, according to a new report from Lux Research entitled "Nanomaterials: Buyer Beware."

"More than 200 companies worldwide sell nanomaterials today. As a group, they have a frighteningly poor track record, " said Matthew Nordan, Vice President of Research at Lux Research. "In our interviews with nanomaterials buyers, we've heard horror story after horror story from companies that haven't gotten the nanomaterials they expected from suppliers. Nanotubes, metal oxide nanoparticles, and fullerenes generate the most complaints while nanoporous materials and dendrimers generate the least."

Among the experiences that Lux Research documents in its report:

- An electronics company bought samples of carbon nanotubes from multiple vendors and found that up to 30% of what was shipped was actually leftover catalyst used to grow the nanotubes — not the nanotubes themselves.
- A specialty chemicals company purchased samples of carbon nanotubes and found that in at least one case, the sample provided by the vendor included no nanotubes at all — just amorphous carbon soot.
- One widely regarded nanomaterials supplier considered to be a pivotal player in its segment cannot actually supply product in commercial quantities, and secretly resells product from its own competitors when asked to do so.
- One buyer of metal-oxide nanoparticles found many vendors eagerly claimed they could synthesize particles of the buyer's desired size dimension. But none could package or ship them in a way such that they didn't agglomerate into bigger particles, rendering them useless for the buyer's application.

Lux Research projects that nanomaterials supply won't improve quickly. Many suppliers are part-time operations that lack robust quality control, suppliers' ever-changing production processes all but ensure that one batch of material differs from the next, and buyers and suppliers disagree over fundamental terminology. Standards from groups like the American National Standards Institute and the International Organization for Standardization could solve the problem, but they will require years of negotiation to develop. All told, it will take three years for competition to weed out ineffective suppliers and for agreed-upon standards to take shape, according to Lux Research.

"In the interim, nanomaterials buyers and suppliers have no choice but to act defensively," Nordan said. "Buyers should use secret shopping to qualify suppliers, requesting price and availability for large quantities — a kilogram for carbon nanotubes, for example — to weed out suppliers that can't actually deliver. In addition, they should clearly describe to suppliers how they plan to use the material to eliminate ambiguity, and enlist a procurement specialist to finalize T's and C's. On the other side of the table, suppliers must obsessively document each shipment using tools like scanning electron microscopy images, as well as educate customers on how to use the nanomaterial without contaminating or destroying it," he added.

The report is available immediately to clients of Lux Research's Nanotechnology Strategies advisory service. For information on how to become a client, contact Rob Burns, Vice President of Sales, at (646)723-0708.

Lux Research is the world's premier research and advisory firm focusing on the business and economic impact of nanotechnology and related emerging technologies. Lux Research provides continuous advisory services, customized consulting, and reference studies to corporations, start-ups, financial institutions, and public sector organizations. Our founders and our research staff are the most widely recognized nanotechnology visionaries throughout the world.

Other Headlines from Lux Research Inc....

- The Nanotech Report 2004, a more than 600-page reference study from Lux Research
- Advisory Services Veteran Joins Leading Nanotechnology Research Firm
- Sizing Nanotechnology's Value Chain—New Report from Lux Research
- Liquid Immersion and Nanoimprint Lithography Will Keep Moore's Law Going for Another Decade
- Corporate Nanomaterials Buyers Fail To Get What They Pay For

More Industry Headlines...

- Materials research consortium works with Malvern Zetasizer Nano systems
- Advance Nanotech Announces Increase in Facility to Portfolio Company Owlstone Nanotech
- Nanosphere Appoints Chief Marketing Officer
- The Nanoethics Group Launches to Research Dangers in Nanotechnology
- pSivida To Acquire Control Delivery Systems—Creates one of the world's first bio-nanotech companies with product revenues

3

Nanotech Educational Courses, Programmes and Outreach

Nanotechnology Educational Links

Colleges, Universities & Institutions in the World

- Foresight: http://www.foresight.org, Foresight Institute's goal is to guide emerging technologies to improve the human condition. Foresight focuses its efforts upon nanotechnology, the coming ability to build materials and products with atomic precision, and upon systems that will enhance knowledge exchange and critical discussion, thus improving public and private policy decisions.
- NANO.DOT, http://nanodot.org/ —A Foresight Institute website—News & Discussion
- Nanotechnology White Paper Library, http://nanobusiness.org/whitepaper.html—The NanoBusiness Association's "Nanotechnology White Paper Library" is a collection of links to more than 500 nanotechnology white papers, presentations and reports. Most are free; some may require registration with their host website.
- CRN, http://www.crnano.org/—Center for Responsible Nanotechnology
- The Institute of Nanotechnology, http://www.nano.org.uk/—The Institute of Nanotechnology has been created to foster, develop and promote all aspects of science and technology in those domains where dimensions and tolerances in the range of 0.1 nm to 100 nm play a critical role
- UCSC Engineering, http://www.cse.ucsc.edu/research/vision.html—University of California, Santa Cruz / Baskin School of Engineering
- Nanotechnology Initiative Research Group, http://www.me.berkeley.edu/

nti/home.html—Department of Mechanical Engineering University of Berkeley

- Stanford Nanotechnology, http://scpd.stanford.edu/SCPD/courses/contentView/nanotechnology/—Stanford University is at the forefront in multidisciplinary research and innovation in nanoscience and nanotechnology. The Stanford Center for Professional Development, in its commitment to extend Stanford research and teaching to students in industry, has developed this portfolio of courses and programs in this exciting new field.
- NSL NanoStructures Laboratory, http://nanoweb.mit.edu/ —The Nanostructures Laboratory (NSL) at MIT develops techniques for fabricating surface structures with feature sizes in the range from nanometers to micrometers, and uses these structures in a variety of research projects. The NSL is closely coupled to the Space Nanotechnology Laboratory (SNL) with which it shares facilities and a variety of joint programs.
- MIT Stanford UC Berkeley Nanotechnology Forum, http://www.quantuminsight.com/MSB/organization.html—The MIT Stanford UC Berkeley Nanotechnology Forum is an all-volunteer organization. It is organized under the MIT Club of Northern California and has affiliations with MIT, Stanford University, and the University of California at Berkeley. The Forum is run by the Chairman and is guided by the Steering Committee.
- The Foresight Institute—Non-profit institute focused on nanotechnology, the coming ability to build materials and products with atomic precision, and systems to aid knowledge exchange and critical discussion, thus improving public and private policy decisions.
- Nanotechnology Now—Comprehensive and informative nanotechnology portal. Provides introduction to nanotechnology, general information, images, interviews, news, events, research, books, glossary and links.
- Applied Computational Research Society—Nonprofit organization fostering interdisciplinary approaches to modeling and simulation in the advancement of computational science and high technology. Focused on nanoscience and MEMS applications.
- The ASME Nanotechnology Institute—American Society of Mechanical Engineers: nanotechnology resources, events, programs, activities, online courses, webcasts, news and links.
- CMP Cientifica—News, networks, conferences and nanotechnology resources in Europe

- Euspen—European Society for Precision Engineering and Nanotechnology—contains society news, conference information, discussion forums and expertise search. Euspen is a not-for-profit company based at Cranfield University in the UK, with members around the world.
- Exploring the Nanoworld—Movies of nano-structured materials: ferrofluids, memory metals, amorphous metals, LEDs, re-entrant foams, self assembly, DNA, Magnetic Resonance Imaging, atomix imaging, Lego models, Giant Magneto Resistance, silicon, carbon nanotubes, polymers, and semiconductor devices. National Science Foundation supported Materials Research Science and Engineering Center on Nanostructured Materials and Interfaces at the University of Wisconsin-Madison.
- The gRobots Project—An open source project to build and run a globally distributed supercomputer for the simulation of evolvable nanotechnology.
- How Stuff Works: How Nanotechnology Will Work—Animated narrative shows how Nanotechnology has the potential to totally change manufacturing, health care and many other areas.
- IBM Almaden STM Molecular Art—Some of the famous images of atoms and molecules made with IBM's scanning tunneling microscope.
- Institute for Atomic-Scale Engineering—Personal page of Forrest Bishop: published and unpublished concept articles.
- Institute of Nanotechnology (UK)—The Institute of Nanotechnology has been created to foster, develop and promote all aspects of science and technology in those domains where dimensions and tolerances in the range of 0.1 nm to 100nm play a critical role.
- Iran Nanotechnology Policy Studies Committee—A committee for studying nanotechnology trends in Iran and the world.
- Minatec—Centre for innovation in micro and nanotechnology. Teaching, research, and industrial development.
- Minatec—Centre for Innovation in Micro and Nanotechnology in Grenoble. Teaching, research, and industrial development.
- MITRE Corp. Nanosystems Group—Reviews current nanotechnology research and people behind it, downloadable copies of several widely-used articles. Nanoelectronics focus.
- Molecular Manufacturing Shortcut Group—Promotes developing nanotechnology as a way to facilitate space exploration and settlement.
- Moletronics and Nanotechnology references—Links to sites with information mainly on molecular electronics.

- Nano Science & Technology Institute—Events, publications and news in nanotechnology, microtechnology and biotechnology.
- Nano Space—Nano technology as it applies now and in the future to Space Exploration and Studies.
- NanoApex—Nanotechnology portal.
- Nanobase—Loyola College, Maryland. National Science Foundation sponsored site presenting up-to-date sources of information on nanotechnology in the areas of major research centers, funding agencies, major reports and books.
- NanoComputer Dream Team—Multidisciplinary project to make Earth's first nanometer supercomputer via Internet.
- Nanodot—News and Discussion of Coming Technologies: Weblog devoted to nanotechnology-related news stories, with attached discussion forum. (based on Slashdot software)
- Nanoforum.org: European Nanotechnology Gateway—Offers background information, news and developments in industry and research and a selection of publications.
- nanogloss—Online dictionary of nanotechnology.
- Nanoguys—Nanotechnology Career Consultants. Recruiting and placement of staff in nanotechnolgy related positions such as Organic, Inorganic, Biological, Project Management, Research Scientist, Engineer, and Lab Technician.
- Nanomagazine—Interviews with people in the Nanotechnology field.
- NanoManipulator: 3rdTech—Commercial version of the NanoManipulator system originally from University of North Carolina, Chapel Hill.
- NanoManipulator: University of North Carolina—The nanoManipulator provides an improved, natural interface to SPMs (STMs and ATMs) by coupling the microscope to a virtual-reality interface giving a virtual telepresence on the surface, scaled by a factor of about a million to one.
- Nanomedicine—Devoted to the medical applications and implications of molecular nanotechnology. Nanomedicine may be defined as the monitoring, repair, construction and control of human biological systems at the molecular level, using engineered nanodevices and nanostructures. Includes the Nanomedicine Art Gallery. (From the Foresight Institute.)
- Nanoscale Science and Engineering—National Science Foundation. Activities, Information & Education, Research & Development, Connections, and NSF & NNI (National Nanotechnology Initiative) Reports.

- NanoScout—Portal with forums, and links sorted by topic and popularity.
- NanoSpace—The Center for NanoSpace Technologies is a Texas-based nonprofit scientific research and education foundation chartered to conceive, establish, and conduct cutting-edge technology, research, and development in the areas of aerospace, education, energy, life sciences, and shipping & transportation.
- NanotechNews—News for nanotechnologists and investors: regular updates, many links to other nanotechnology websites, archives, search function, newswire.
- Nanotechnology—Institute of Physics quarterly journal
- Nanotechnology—Institute of Physics monthly journal for aspects of nanoscale science and technology
- The NanoTechnology Group—A consortium of nano companies, universities and organizations developing a nano science curriculum for K-12.
- Nanotechnology in Manufacturing—What Next? The Coming Revolution In Manufacturing: Autodesk Technology Forum Presentation by John Walker, May 10th, 1990. Somewhat dated but well worth reading.
- Nanotechnology Journals Database—References, and links to, papers about nanotechnology, from selected scientific journals (eg PRB, PRL, JACS, JAP, JPCB).
- Nanotechweb.org—A portal for nanoscience and nanotechnology online resources for academics, industrialists and investors, as well as for laymen interested in nanotechnnology.
- NanoTechWire.com—Provides information for nanotechnology and research.
- Nanotek—Nanotechnology: articles online, news, pictures, cvs
- nanoTEN—Nanotechnology consulting firm. Services include nanotechnology matchmaking, assessment of trends, feasibility studies, materials and products analysis, and presentation assistance.
- nanoTITAN—Nanotechnology Software, Data & Services: nanoTITAN offers Java libraries and applications, a nanodevice database, the nanoML specification, custom simulation and visualization to the nanotechnology and general scientific communities.
- Nanotube Patent Search—Results from a search of nanotube or nanotube related patents at the United States Patent & Trademark Office
- The Nanotube Site—Michigan State University's Library of Links for the Nanotube Research Community

- **NanoWave—An MIT startup, Nanowave offers SPPE (Scanning Probe Position Encoder) technology for picometer-order (subnanometer) resolution over ranges greater than several inches.**
- **National Science Foundation—Partnership in Nanotechnology (US). Recent technical and policy reports, funding opportunities, profiles of projects currently supported, and links to other nanoscience/technology sites.**
- **Phantoms—European Union sponsored nano electronics network. Contains comprehensive information about European nanotechnology, and links to groups and events worldwide.**
- **Polymer Simulation—Computer simulation of a polymer sticking to a surface.**
- **Scanning Tunneling Microscope (STM)—Describes the invention of the topografiner, a precursor instrument, between 1965 and 1971, and also tells of the STM's development.**
- **Scientific American—Nanotechnology articles, some free and some archived.**
- **Self Replicating Systems—NASA and Self-Replicating Systems: Implications for Nanotechnology.**
- **Small Times—Daily articles covering MEMS, nanotechnology, and microsystems, with a business angle.**
- **Stony Brook Buckyballs—A virtual tour of fullerenes in Laszlo Mihaly's laboratory at the Physics Department at SUNY, Stony Brook.**
- **STT: Netherlands Study Center for Technology Trends—Somewhere in this confusing site is good material in English, though it is often labeled in Dutch, and Dutch material is often labeled in English. STT hopes to bring their project to the attention of a wide audience, thereby contributing to the development of nanotechnology.**
- **Technology Review: Nanotechnology and Materials—Articles, discussion and services for nanotech built on MIT's award-winning monthly magazine.**
- **Thermodynamics Research Laboratory—Classical, Quantum, Nanoscale & Statistical Thermodynamics & Mechanics data and property calculation sites.**
- **Virtual Journal of Nanoscale Science & Technology—A weekly multijournal compilation of the latest research on nanoscale systems. Published by the American Institute of Physics and the American Physical Society.**
- **VISION Online—Virtual Institute supported by the EC, committed to promotion and training in the precision engineering and nanotechnologies. Contains discussion forums, training programmes and patent library.**

- Xerox/Zyvex Nanotechnology—Former Xerox Palo Alto Research Center Nanotechnology page: brief introduction to core concepts of molecular nanotechnology (MNT), and links for further reading.
- Yahoo Groups: Nanotechnology—Active nanotechnology discussion group.
- Yashnanotech—Provides marketing and consultancy services for products developed by nanotechnology. Promotes nanotech awareness through training, educational publications, seminars, workshops and conferences.
- Albany NanoTech "... a fully-integrated research, development, prototyping, pilot manufacturing and education resource managing a strategic portfolio of state-of-the-art laboratories, supercomputer and shared-user facilities and an array of research centers located at the University at Albany—SUNY."
- Arizona State University (ASU) Graduate Research Training Program in Biomolecular Nanotechnology
- Arizona State University (ASU) Center for Applied NanoBioscience
- Arizona State University (ASU) Biodesign Institute
- Arizona State University (ASU) Lindsay Lab—Research in Biophysics, Molecular Electronics and Condensed Matter Physics
- Arizona State University (ASU) Center for BioOptical Nanotechnology
- Arizona State University (ASU) Nanostructures Research Group—Center for Solid State Electronics Research
- Asian Institue of Technology Nanotechnology Studies. Bangkok, Thailand
- Australian National University Research School of Physical Sciences and Engineering
- Australian National University Nanotechnology and Mesoscale Physics
- Ball State University Center for Computational Nanoscience
- Baylor College of Medicine Molecular Biology programs
- Beckman Institute Biological Sensors Group (BSG)
- Berkeley Nanotechnology Club (BNC)
- Brigham Young University Cleanroom A class 10 cleanroom located on the BYU campus, used for research in solid state, optoelectronics, photonics, nanotechnology, microfabrication, MEMS, micro-fluidics, etc.
- Boston University College of Engineering Departmental program in Nanotechnology. "... is concerned with the mechanical and electronic properties of nanometer scale systems as well as macroscopic systems that require nanometer tolerances in fabrication, control, and measurement."

- Boston University Laboratory for Nanometer Scale Engineering
- Boston University Mohanty Group. "The central theme of our research is the exploration of quantum mechanical effects in engineered nanoscale structures and devices with a goal to study fundamental physical phenomena."
- Brown University Nano & Micromechanics Laboratory, part of the Mechanics of Solids and Structures Group in the Division of Engineering
- California Institutes for Science and Innovation: California NanoSystems Institute (CNSI) University of California at Los Angeles and University of California at Santa Barbara California Institute for Telecommunications and Information Technology (Cal-(IT)2) University of California at San Diego and UC Irvine Center for Information Technology Research in the Interest of Society (CITRIS) University of California at Berkeley, University of California at Davis, University of California at Merced, University of California at Santa Cruz Institute for Bioengineering, Biotechnology and Quantitative Biomedical Research (QB3) University of California at Berkely, UC San Francisco, UC Santa Cruz
- California Institute of Technology (Caltech) Materials and Process Simulation Center
- Caltech Roukes Group—Explorations with Three-Dimensional Nanostructures in Physics, Engineering and Biology
- Caltech Kavli Nanoscience Institute (KNI)
- Caltech Solid State Device Physics (SSDP) Research Group. Department of Applied Physics.
- Carnegie Mellon University Nanomaterials for Magnetic, Electronic and Optical Applications
- Carnegie Mellon University Matyjaszewski Group—Research—Organic/Inorganic Nanocomposites
- Center for Applied Research and Technology at CMU The Central Michigan University Research Corporation (CMURC) is a not-for-profit organization established to facilitate innovative research and development opportunities between the university and high technology companies.
- Chalmers University of Technology, Sweden Nanoscale Science
- Chalmers University of Technology, Sweden Microtechnology Centre
- Chiang Mai University Nanomaterials Research Unit
- City College of New York New York Center for Biomedical Engineering

- City College of New York Molecular Beam Epitaxy (MBE) Program
- City University of New York IGERT: Nanostructural Materials & Devices (Integrative Graduate Education and Research Traineeship Program)
- City University of Hong Kong Center Of Super-Diamond and Advanced Films (COSDAF)
- Clarkson University Center for Advanced Materials Processing.
- Clemson University COMSET—Center for Optical Materials Science and Engineering Technologies.
- Clemson University Center for Advanced Engineering Fibers and Films (CAEFF).
- Clemson University Rao Group.
- Columbia University Center for Electronic Transport in Molecular Nanostructures NSF Funded "Nanoscale Science and Engineering Centers" Real Video (2:50) Professor James Yardley speaking about Nanotechnology
- Columbia University Materials Research Science and Engineering Center (MRSEC)
- Columbia University Environmental Molecular Science Institute
- Columbia University NANOWEB—Nanotechnology Resources at Columbia University
- Columbia University Adams Group
- Columbia University Brus Group
- Columbia University Herman Group—Laser Diagnostics & Solid State Physics
- Columbia University Nuckolls Laboratory
- Cornell University Nanofabrication Facility. Part of the National Nanotechnology Infrastructure Network (NNIN)
- Cornell University Craighead Research Group. "... research focuses on creating nanoscale devices..."
- Cornell University Alliance for Nanomedical Technologies
- Cornell University Center for Nanoscale Systems NSF Funded "Nanoscale Science and Engineering Centers"
- Cornell University Nanobiotechnology Center
- Cornell University Molecular BioEngineering Lab
- Cornell University Ober Group Research

- Cornell University Center for Materials Research
- Cornell University Laboratory of Atomic and Solid State Physics (LASSP)
- Cornell University McEuen Group—LASSP
- Cornell University Applied Physics—Nanoscience and Nanotechnology
- Cornell University Applied Physics—Condensed Matter, Solid State, and Materials Research
- Cornell University Applied Physics—Optical Physics, Quantum Electronics, and Photonics
- Cranfield University Nanotechnology Group, Advanced Materials Department
- Curtin University of Technology Nanochemistry Research Institute
- Dalhousie University Zwanziger Lab
- Darmstadt University of Technology Institute of Materials Science—Thin Films Division
- Dartmouth Molecular Materials Group
- Delft University of Technology Kavli Institute of Nanoscience Delft. The Kavli Institute of Nanoscience at Delft University of Technology (The Netherlands) consists of six research groups and a nanofacility cleanroom, all aimed at innovative research at the current frontier of science on a nano-scale.
- Delft University of Technology and Leiden University MSc NanoScience programme. Leiden University and the Delft University of Technology (The Netherlands) offer a unique MSc program on NanoScience, aimed at students who are eager to transcend the traditional borders between scientific disciplines.
- Delft University of Technology Quantum Transport—one of the research groups in the department of Applied Physics
- Delft University of Technology Molecular Biophysics Group
- Drexel Nano Materials Group (Gogotsi's Group)
- Duke University Eom Research Group, Thin Films Laboratory
- Duke University Nanoscience Group
- Earnest — Alumni Magazine for CES at Clemson Nanotechnology research inspires industry collaboration
- Ecole Polytechnique Federale de Lausanne Nanomechanics and Tribology

- Eidgenössiche Technische Hochschule (ETH) NANO-PHYSICS group of the laboratory for solid state physics at ETH Zürich
- Emory University Nie Group. Focus on Drug Delivery, Cancer Therapeutics, Molecular Imaging, Molecular Profiling, and In-vitro Diagnostics.
- Florida State University Molecular Expressions Website
- Free University Berlin Physics Department — Atomic Manipulation Site
- George Washington University Wagner Lab. Synthesis, study and application of solid-state inorganic materials with technologically significant magnetic, electrical, optical, electrochemical or catalytic properties.
- Georgia Institute of Technology Microelectronics Research Center (MiRC). Part of the National Nanotechnology Infrastructure Network (NNIN)
- Georgia Institute of Technology Applied Sensors Laboratory
- Georgia Institute of Technology Zhong Lin (ZL) Wang's Nanoscience and Nanotechnology Group
- Georgia Institute of Technology Nanotechnology and Microfluidics
- Georgia Institute of Technology Nanostructure Research Laboratory
- Georgia Institute of Technology Nanostructure Optoelectronics Group
- Georgia Institute of Technology Molecular Design Institute
- Georgia Tech Research Institute GTRI Nanotechnology Laboratory
- Harvard University Nanoscale Science and Engineering Center—Science of Nanoscale Systems and their Device Applications The Nanoscale Science and Engineering Center (NSEC) is a collaboration among Harvard University, the Massachusetts Institute of Technology, the University of California at Santa Barbara and the Museum of Science in Boston with participation by Delft University of Technology (Netherlands), the University of Tokyo (Japan), and Brookhaven National Laboratory, Oak Ridge National Laboratory and Sandia National Laboratory. NSF Funded "Nanoscale Science and Engineering Centers"
- Harvard University Center for Imaging and Mesoscale Systems (CIMS). Part of the National Nanotechnology Infrastructure Network (NNIN)
- Harvard University Materials Science Group
- Harvard University Leiber Group—a Physical Chemistry research group in the Department of Chemistry and Chemical Biology of Harvard University
- Harvard University Hongkun Park—physics and chemistry of nanostructured materials

- Harvard University Westervelt group. Department of Physics and Division of Engineering and Applied Sciences
- Harvard University George M. Whitesides Research Group
- Harvard University Nanopore Group
- Harvard University Center for Molecular Imaging Research
- Harvard University Narayanamurti Group
- Haverford College "...to bring together faculty from biology, chemistry, physics, and math in the pursuit of an interdisciplinary effort to develop new protein-based biomaterials for the field of nanometer-scale electronic and mechanical device fabrication."
- Hebrew University of Jerusalem Center for Nanoscience and Nanotechnology
- Helsinki University of Technology Optics and Molecular Materials
- Helsinki University of Technology Materials Physics Laboratory
- Heriot-Watt University, Edinburgh, UK The Nano-Optics Group
- Hong Kong University of Science and Technology Institute of NanoMaterials and NanoTechnology (INMT)
- Howard University Materials Science Research Center. Part of the National Nanotechnology Infrastructure Network (NNIN)
- Illinois Institute of Technology Center on Nanotechnology and Society (CONAS) "... to catalyze informed, inter-disciplinary research and education on the implications of nanoscale science and technology for ethical, legal, policy, business, and wider social issues, and with a special focus on the human condition."
- Instituto de Carboquímica Group of Carbon Nanostructures and Nanotechnology
- Institutionen för Fysik och Mätteknik (IFM) Department of Physics and Measurement Technology, Biology and Chemistry—Materials Science Division
- Indiana University Advanced Research and Technology Institute
- Institut Für Neue Materialien, University of the Saarland in Saarbrücken (INM) (Germany) Institute for New Materials
- Institute for Solid State and Materials Research, Dresden Germany Materials Science Division
- Interuniversity Microelectronics Center (IMEC) IMEC today is the largest independent microelectronics research and development center in Europe.

- Iowa State University Condensed Matter Physics — Ames Lab
- Iowa State University Condensed Matter Physics — Canfield Research Group
- Iowa State University Surface Engineering and Molecular Assemblies (SEMA) Laboratory
- Institute for Physical High Technology Jena Molecular Nanotechnology Group
- Jawaharlal Nehru Centre for Advanced Scientific Research Carbon Lab. Prof C. N. R. Rao (FRS) Director.
- Joanneum Research—Institute of Nanostructured Materials and Photonics "Our 15 research units make Joanneum Research one of the largest non-university research institutions in Austria." Affliated with international institutions such as the University of California at Berkeley and the Max Planck Institute for Polymer Research in Germany.
- Johns Hopkins University Materials Research Science and Engineering Center (MRSEC). One of 26 MRSECs funded by the National Science Foundation, is composed of scientists at JHU, Brown University, and the National Institute of Standards and Technology (NIST). Research in the Center focuses on nanostructures made from novel materials that exhibit enhanced magneto-electronic properties.
- Johns Hopkins University Dept of Mechanical Engineering
- Kansas State University Dept of Physics, Interdisciplinary Research
- Katholieke Universiteit Leuven Erasmus Mundus Master of Nanoscience and Nanotechnology
- Kaunas University of Technology, LITHUANIA Research Center for Microsystems and Nanotechnology
- Kobe University, Japan Molecular Electronics and Photonics
- Korea Advanced Institute of Science and Technology (KAIST) Superlattice Nanomaterials Laboratory
- Korea University Spintronic Materials Lab., Division of Materials Science and Engineering
- Kyushu University, Japan Division of Nanoelectronics, Dept. of Electronic Device Engineering
- Lehigh University Center for Advanced Materials and Nanotechnology.
- Lehigh University Department of Materials Science and Engineering.

- London Centre for Nanotechnology (LCN) "... designed to act as a focus for current interdisciplinary nanoscale materials and device research." A joint enterprise between University College London and Imperial College.
- Louisiana Tech University Nanosystems Engineering
- Loyola College—Nanobase International Technology Research Institute.
- Ludwig-Maximilians-Universität (LMU), Germany Center for NanoScience (CeNS)
- Ludwig-Maximilians-Universität (LMU), Germany Nanophysics Group
- Lund University The Nanometer Consortium
- Mahidol University Nanotechnology Center. "Building Capability on Molecular Engineering and Devices."
- Marshall University Norton Imaging Group.
- Massachusetts Institute of Technology (MIT)—BioInstrumentation Laboratory Technologies that are currently being developed in the BI Lab include a servo-controlled laser micro-surgical robot, artificial muscle technology based upon conducting polymer biomimetic materials, the Living Chip, a cell-based high throughput screening system for mass-scale discovery of new drugs, and the Nanowalker, a nano-stepping autonomous robotic instrumentation platform.
- FAQ for MIT Institute for Soldier Nanotechnology (ISN)
- MIT Microsystems Technology Laboratories
- MIT Laboratory for Experimental & Computational Micromechanics (LEXCOM)
- MIT Nanostructures Laboratory
- MIT Surfaces and Structures
- MIT NanoMechanical Technology Laboratory
- MIT Space Nanotechnology Laboratory
- Moungi Bawendi's Group MIT "...quantum mechanical behaviors demonstrated by semiconductor and metallic materials..."
- MIT Center for Materials Science and Engineering
- MIT Department of Materials Science and Engineering
- Max Planck Institute for Biophysical Chemistry Department of NanoBiophotonics
- Max Planck Institute for Solid State Research Synthetic Nanostructures Team

- Metroplex Research Consortium for Electronic Devices and Materials (MRCEDM) combines the research and development capabilities of The University of Texas at Arlington, Southern Methodist University, Texas Christian University and The University of North Texas to assist companies that innovate and manufacture advanced technical components and systems.
- Macdiarmid Institute for Advanced Materials and Nanotechnology A New Zealand research organisation concerned with high quality research and research education in materials science and nanotechnology.
- Michigan State University (MSU) Carbon Nanotubes
- MSU—David Tománek "The Nanotube Site"
- MSU—NSF Center on Low-Cost, High-Speed Polymer Composites Processing
- MSU—Composite Materials and Structures Center
- MSU—Advanced Materials Engineering Experiment Station (AMEES) is managed and directed as part of the Composite Materials and Structures Center
- Michigan Technological University Nanotechnology at Michigan Technological University
- McGill — NanoScience & Scanning Probe Microscopy... to apply the tools, techniques and materials developed in the past few years to the emerging new field of Nanoelectronics.
- McGill — NANOSCIENCE AND NANOTECHNOLOGY Course description
- Nagoya University Department of Molecular Design and Engineering
- Nanobiotechnology Center (NBTC) The Nanobiotechnology Center was established in January 2000 as a Science & Technology Center, with core funding from the National Science Foundation. Nanobiotechnology is an emerging area of scientific and technological opportunity that integrates nano/microfabrication and biosystems to the benefit of both. The Nanobiotechnology Center is characterized by its highly interdisciplinary nature and features a close collaboration between life scientists, physical scientists, and engineers. It has a fully integrated education and outreach effort in which all NBTC faculty participate. The Center brings together experts in their fields from Cornell University, the Wadsworth Center (New York State Health Department in Albany), Princeton University, Oregon Health & Science University, Clark Atlanta University, and Howard University. It also involves the active collaboration of K-12 educators, the Sciencenter Museum in Ithaca, NY, and representatives from industry and the government.

- Nanotechnology Institute (NTI) A collaboration led by Ben Franklin Technology Partners of Southeastern Pennsylvania, Drexel University and The University of Pennsylvania. "... to focus on the transfer of discoveries and intellectual knowledge in the area of nanotechnology from universities to industry partners and on the rapid application and commercialization of this technology to stimulate economic growth."
- Nanyang Technological University Precision Engineering & Nanotechnology Centre
- Nanyang Technological University Micro Electro Mechanical Systems Research
- Nanyang Technological University Advanced Material Research Centre
- Nanyang Technological University Microelectronics Centre
- Nara Institute of Science and Technology Graduate School of Material Science
- National Institute of Advanced Industrial Science and Technology (AIST) Nanomaterials Theory Group
- National Microelectronics Research Centre (NMRC) Ireland Nanotechnology Research
- National Taiwan University Center for Nano Science and Technology (NTU-CNST). "... integrated research projects of this program include nanomaterials, nano-devices, NEMS, and Bio-nano."
- National University of Singapore
- New Jersey Institute Of Technology Device & Material Characterization Laboratory
- New Mexico State University Nano Physics
- New York University Center for Advanced Materials and Nanotechnology
- North Carolina Center for Nanoscale Materials (NCCNM) "...research activities in the center are directed towards understanding the fundamental science of nanoscale materials and utilizing their unique properties for commerical applications."
- North Carolina State University Triangle National Lithography Center. Part of the National Nanotechnology Infrastructure Network (NNIN) (a collaboration with the University of North Carolina at Chapel Hill)
- North Carolina State University Analytical Instrumentation Facility
- North Carolina State University Nanoindentation—studying mechanical properties on the atomic scale.

- North Carolina State University Nanoscale Tribology Laboratory.
- North Dakota State University Center for Nanoscale Science and Engineering
- Northeastern University NSF Center for High-rate Nanomanufacturing (CHN)
- Northwestern University Nanoscale Science and Engineering Center (NSEC) for Integrated Nanopatterning and Detection Technologies NSF Funded "Nanoscale Science and Engineering Centers"
- Northwestern University Mirkin Group
- Northwestern University — Institute for Nanotechnology Established as an umbrella organization for the multimillion dollar nanotechnology research efforts at Northwestern University.
- Northwestern University Center for Transportation Nanotechnology.
- Northwestern University MEMS / Nanotechnology.
- Northwestern University Stupp Laboratory
- Northwestern University Nanomechanics
- Northwestern University Mechanics of MEMS and Nanosystems
- Northwestern University Ruoff Group—Nanoscience and Technology Lab
- Northwestern University Center for Nanofabrication and Molecular Self-Assembly
- Northwestern University Professor Horacio Dante Espinosa—Micro and Nanomechanics Lab
- Northwestern University Center for Quantum Devices
- Northumbria University Advanced Materials Research Institute (AMRI). The Institute's technical focus is in the fields of advanced materials, surface engineering, corrosion and wear.
- Norwegian Institute of Technology (NTH) NMNL (Norwegian Micro and Nano Laboratories), the research infrastructure in Norway hosted by SINTEF (The Foundation for Scientific and Industrial Research at the Norwegian Institute of Technology), the Norwegian University of Science and Technology (NTNU), and the University of Oslo (UiO).
- Norwegian Institute of Technology (NTH) NANOCAPS (at SINTEF).
- Norwegian Institute of Technology (NTH) NANOMAT—Materials Science and Nanotechnology at the Norwegian Microtechnology Center (NMC-MRL) (at SINTEF).
- Nottingham University Nanoscience Group Nanometre scale structures

and nanostructured materials play an increasingly pervasive role in a wide range of scientific disciplines ranging from solid state physics through to molecular biology.

- Nottingham Trent University Polymer Engineering Centre
- Oak Ridge National Laboratory (ORNL) Polymer Science Group Summary of ongoing research aimed at simulating nanomachines and aspects of fundamental chemical physics at the molecular scale.
- Ohio University Research Institute to Advance Studies in Nanotechnology
- Ohio University Nanoscale & Quantum Phenomena Institute
- Ohio University Nanospintronics and Nanomagnetics
- Ohio Supercomputer Center Nanotechnology at OSC
- Oklahoma State University NanoNet Probe Microscopy Laboratory
- Oklahoma State University Department of Physics
- Oregon State University College of Engineering
- Oregon State University Microtechnology-Based Energy, Chemical and Biological Systems (MECS)
- Osaka University Protonic NanoMachine Project.
- Oxford University Department of Materials
- Oxford University Begbroke Science Park Institute of Nanotechnology
- Paul Scherrer Institut Laboratory for Micro- and Nanotechnology
- Penn State Atomic-Scale Measurements and Control
- Penn State Center for Molecular Nanofabrication and Devices
- Penn State Nanofabrication Facility. Part of the National Nanotechnology Infrastructure Network (NNIN)
- Penn State Materials Research Institute
- Penn State Allara Research Group
- Penn State Department of Materials Science and Engineering
- PNNL/UW Joint Institute for Nanoscience
- Portland State University Center for Nanoscience and Nanotechnology
- Princeton University Nanostructure Laboratory.
- Princeton University Ceramic Materials Laboratory.
- Purdue Nanotechnology Initiative "...to demonstrate novel techniques for the design and fabrication of nanoelectronic devices by the chemical

manipulation of nanometer (10-9 meters) sized clusters and molecular wires."

- Nanotechnology At Purdue (PDF) A visual guide to the various programs and collaborations at Purdue. "This brochure highlights many of those projects across the categories of nanomaterials, nanodevices, nano/bio interfaces, nanomanufacturing, computational nanotechnology, and nanometrology/characterization."
- Purdue University Discovery Park—Birck Nanotechnology Center & Ultra-Performance Nanotechnology Center (ULTEC).
- Purdue University David Janes Group.
- Purdue University Laboratory for Chemical Nanotechnology (PLCN).
- Purdue University nanoHub The nanoHub is a new initiative designed to promote the application of computational science to nanotechnology. Using PUNCH, a network-computing software infrastructure, researchers and educators can access and operate simulation tools through standard Web browsers. Computational scientists can share tools with colleagues, experimentalists can operate simulations tools without downloading and installing them, and users can access supercomputing resources transparently. The focus is on state-of-the-art, research-grade tools designed to complement commercial simulation packages.
- Purdue University Center for Nanoscale Devices.
- Purdue University Nanoscale Physics.
- Purdue University Cluster-Based Materials Group.
- Queen Mary University of London Molecular and Materials Physics Group.
- Research Institute for Technical Physics and Materials Science Budapest, Hungary.
- Rensselaer Polytechnic Institute Nanotechnology Center.
- Rensselaer Polytechnic Institute Center for Directed Assembly of Nanostructures NSF Funded "Nanoscale Science and Engineering Centers"
- Rensselaer Polytechnic Institute Prof. Ajayan's Carbon Nanomaterials Technology Group.
- Rice University Center for Nanoscale Science and Technology
- Rice University Colvin Group
- Rice University Halas Nanophotonics Group
- Rice University Tour Group

- Rice University Laboratory for Nanophotonics
- Rice University Center for Biological and Environmental Nanotechnology NSF Funded "Nanoscale Science and Engineering Centers"
- Rice University Environmental & Energy Systems Insitute of Rice University (EESI)
- RMIT Royal Melbourne Institute of Technology—Nanotechnology Program
- Rochester Institute of Technology NanoPower Research Labs
- Rochester Institute of Technology Nano- and Micro-Electromechanical Systems, Nano- and Microsystems, Nanoarchitectronics
- Rutgers University Bio Nano Robotics
- Rutgers University Division of Computer & Information Sciences
- Rutgers University Laboratory for Surface Modification (LSM)
- Seoul National University Center for Science in Nanometer Scale
- Sheffield Hallam University Electronics Research Group, Microsystems & Machine Vision Laboratory (MMVL)
- Sinica Nanoscience and Technology Program
- South Dakota School of Mines and Technology Center for Accelerated Applications at the Nanoscale (CAAN).
- Stanford University Nanofabrication Facility—serves academic, industrial, and governmental researchers across the US in areas ranging from of optics, MEMS, biology, and chemistry, as well as traditional electronics device fabrication and process characterization. Part of the National Nanotechnology Infrastructure Network (NNIN)
- Nanoscience and Nanotechnology Online Seminars
- Explore the universe of nanotubes and bucky balls. Expand your vision of a changed world, from the quality of our goods to the quality of our lives. Presenting the latest nanoscience and nanotechnology concepts, the Stanford Engineering and Science Institute will explore the promise of a wide range of exciting new products and applications capable of transforming and redefining many industries. Learn from Stanford faculty and industry experts the potentially broad impacts of nanotechnology for your business. Learn More
- Stanford University Department of Materials Science and Engineering
- Stanford University Center for Molecular and Genetic Medicine
- Stanford University Geballe Laboratory for Advanced Materials (LAM)

- Stanford University Chemical Engineering
- Stanford University Mechanical Engineering
- Stanford University S.C.Zhang Group
- http://buckminster.physics.sunysb.edu/ Department of Physics and Astronomy.... study the fundamental properties of Fullerenes (Carbon-60) and doped fullerenes
- Stockholm University Sol-Gel Group at the Department of Inorganic Chemistry
- http://www.matscieng.sunysb.edu/tsl/nano/ Thermal Spray Lab
- SUNY—University at Buffalo The Center for Advanced Photonic and Electronic Materials (CAPEM)—Laboratory for Spintronics Research in Semiconductors (LSRS)
- SUNY—University at Buffalo Electronic Packaging Laboratory
- SUNY — Molecular Structure Laboratory Department of Chemistry at SUNY Stony Brook.
- Swiss Federal Institute of Technology Zurich (ETH Zurich) FIRST! Center for Micro- and Nanoscience (Frontiers in Research: Space and Time)
- Tata Institute of Fundamental Research (TIFR) Department of Condensed Matter Physics and Materials Science
- Technical University of Denmark NANO@DTU—organizes the broad spectrum of nanotechnology research and education at DTU.
- Technical University of Denmark Nanotechnology Research Group.
- Technical University of Denmark Center for Atomic-scale Materials Physics.
- Technion—Israel Institute of Technology Microelectronics Research Center
- Technische Universitat Berlin Project Nanostructures
- Tel Aviv University TAU Research Institute for Nanoscience and Nanotechnology
- Texas Christian University Atom Optics Lab
- Texas Engineering Experiment Station (TEES) a member of Texas A&M University System
- Texas A&M University Texas Institute of Intelligent Bio-Nano Materials and Structures for Aerospace Vehicles (TiiMS)
- Texas State—Nanomaterials Applications Center—NAC
- Texas Nanotechnology Initiative A Consortium of industry, universities,

government, and venture capitalists whose goal is to position Texas as the Nanotechnology State

- Towson University Nanotechnology Studies using Scanning Probe Microscopy
- Towson University Nanotechnology Laboratory
- Trinity College Dublin Nanomag Research Group
- Trinity College Dublin Centre for Research on Adaptive Nanostructures and Nanodevices (CRANN)
- TRI/Princeton "... TRI has extended its expertise to provide advanced research and education in polymers, fibers, films, human hair, and porous materials."
- Tufts University The Walt Group... nanosensors and nanostructures.
- University at Albany—SUNY Albany NanoTech "is a fully-integrated research, development, prototyping, pilot manufacturing and education resource managing a strategic portfolio of state-of-the-art laboratories, supercomputer and shared-user facilities and an array of research centers located at the University at Albany—SUNY." Formerly "The Center for Advanced Thin Film Technology".
- University at Albany College of Nanoscale Science and Engineering. "... cross-disciplinary curriculum integrates the fundamental science principles of physics, chemistry, computer science, and biology with the cross cutting fields of nanosciences, nanoengineering, nanoeconomics, and nanobiology."
- University at Albany—SUNY Advanced Computer Modeling for Nanosystems and Processes.
- University at Albany—SUNY Center for Advanced Thin Film Technology—Microsystems Integration Laboratory
- Université Catholique de Louvain UCL/CRMN—Centre de Recherche en Dispostifs et Matériaux Electroniques Micro- et Nanoscopiques (CeRMiN). Research Center in Micro and Nanoscopic Materials and Electronic Devices
- University College Cork Nanotechnology Group—Dr. Gareth Redmond Director. "The Group undertakes research aimed at development and innovative exploitation of nanoscale materials and devices within emerging Information and Communication Technology (ICT) application areas."
- University of Aalborg Institute of Physics and Nanotechnology
- University of Aarhus and Aalborg University iNANO—interdisciplinary Nanoscience Center
- University of Akron Center for Molecular Design and Recognition

- University of Alberta Faculty—National Institute for Nanotechnology "... an integrated, multi-disciplinary institution involving researchers in physics, chemistry, engineering, biology, informatics, pharmacy and medicine. Established in 2001, it is operated as a partnership between the National Research Council and the University of Alberta, and is jointly funded by the Government of Canada, the Government of Alberta and the university."
- University of Alberta Buriak Group
- University of Alberta Centre for Nanoscale Physics
- University of Alberta Microsystems Technology Research Institute—NanoFab. "... an open access micro and nano fabrication facility for use by the University of Alberta research community and other groups including Canadian universities and companies involved in microsystem research and related prototyping development."
- University of Arizona Nanomechanics and Mesoscopic Physics
- University of Arizona AFM Lab
- University of Arkansas Semiconductor Fabrication and NanoScale Characterization Facility
- University of Basel National Center of Competence in Research Nanoscale Science
- University of Basel Nano-Electronics (Mesoscopic Physics)
- University of Bath, UK Materials Research Centre
- University of Bath, UK Centre for Electron Optical Studies
- University of Berlin — Project Nanostructures Focuses on the fabrication, characterization and theoretical modelling of quantum wires and quantum dots.
- University of Birmingham I² NanoTech Centre
- University of Birmingham Nanoscale Chemistry
- University of Birmingham Nanoscale Physics Research Laboratory
- University of Birmingham μ-Nano Consortium
- University of California at Berkeley • Davis • Merced • Santa Cruz The Center for Information Technology Research in the Interest of Society (CITRIS). Includes as members/participants California High-Tech Industry & The State of California.
- University of California at Berkeley Stacy Group "... involved in the development of new synthetic routes that lead to the discovery of materials for emerging technologies."

- University of California at Berkeley Berkeley Sensor & Actuator Center—MEMS Reference Database
- University of California at Berkeley Microstructured Materials Group
- University of California at Berkeley Professor Connie Chang-Hasnain Optoelectronics Research Group
- University of California at Berkeley McEuen Group has moved to McEuen Group—LASSP Cornell University—Laboratory of Atomic and Solid State Physics—Department of Physics
- University of California at Berkeley Alivisatos Group
- University of California at Berkeley Zettl Research Group—Condensed Matter Physics—Department of Physics. An experimental solid state physics group.
- University of California at Berkeley, UC San Francisco, UC Santa Cruz Institute for Bioengineering, Biotechnology and Quantitative Biomedical Research (QB3). QB3 is one of four institutes established through the California Institutes for Science and Innovation
- University of California at Davis, NEAT Nanophases in the Environment, Agriculture, and Technology is a multidisciplinary research and education program which links the fundamental physics, chemistry, and engineering of small particles and nanomaterials to several challenging areas of investigation.
- University of California, Irvine Integrated Nanosystems Research Facility (INRF)
- University of California at Los Angeles Heath Group—Metal & Semiconductor nanocrystal research.
- University of California at Los Angeles Nanoelectronics Research Facility
- University of California at Los Angeles Optoelectronics Group, headed by Prof. Eli Yablonovitch.
- University of California at Los Angeles Stoddart Research Group
- University of California at Los Angeles Wudl Group
- University of California, Riverside Center for Nanoscale Science and Engineering
- University of California, Riverside Nanotechnology Educational Outreach
- University of California, Riverside Biomedical Science & Nanotechnology Laboratory

- University of California at Los Angeles and University of California at Santa Barbara California NanoSystems Institute (CNSI). UCLA & UCSB have joined to build the California Nanosystems Institute, which will facilitate a multidisciplinary approach to develop the information, biomedical, and manufacturing technologies that will dominate science and the economy in the 21st century. CNSI is one of four institutes established through the California Institutes for Science and Innovation
- University of California at Santa Barbara Nanofabrication Facility. Part of the National Nanotechnology Infrastructure Network (NNIN)
- University of California at Santa Barbara Nanoscale Physics—Cleland Group
- University of California at Santa Barbara Awschalom Group
- University of California at Santa Barbara Institute for Quantum Engineering, Science and Technology (iQUEST) (formerly known as the Quantum Institute)
- University of California at Santa Barbara Department of Molecular, Cellular and Developmental Biology (MCDB)
- University of California Santa Cruz Quantum Electronic Group
- UCSC Extension General Interest Page (Current nanotechnology course offerings)
- University of California at San Diego Physics Department—Thin Film Lab—Nanoscience Group. Professor Ivan K. Schuller Director.
- University of California at San Diego Materials and Devices Layer at UCSD. Professor Ivan K. Schuller Director.
- University of California at San Diego MURI Integrated Nanostructured Supersensors. Professor Ivan K. Schuller Director.
- University of California at San Diego Silva Research Group
- University of California at San Diego and UC Irvine California Institute for Telecommunications and Information Technology (Cal-(IT)2). Cal-$(IT)^2$ is one of four institutes established through the California Institutes for Science and Innovation initiative proposed in the year 2000 by Governor Gray Davis. Cal-$(IT)^2$, a partnership between UC San Diego and UC Irvine, seeks to ensure that California maintains its leadership in the telecommunications and information technology marketplace.
- University of Cambridge CUED Nanoscale Science Laboratory
- University of Cambridge Centre of Molecular Materials for Photonics and Electronics

- University of Cambridge the Cavendish Laboratory is the Department of Physics
- University of Cambridge The Nanoscience Centre
- University of Cambridge IRC in Nanotechnology—The Interdisciplinary Research Collaboration (IRC) in Nanotechnology is funded by EPSRC, BBSRC, MRC and the MoD as a collaboration between the University of Cambridge, University College London and the University of Bristol.
- University of Canterbury—NEST Nanostructure Engineering Science and Technology Group
- University of Canterbury Cluster Assembled Nanostructures—Dr. Simon Brown, Research Interests
- Université Catholique de Louvain Research Center on Micro and Nanoscopic Electronic Devices and Materials (CERMIN)
- University of Chicago Materials Research Science and Engineering Center
- University of Cincinnati Nanoelectronics Laboratory
- University of Colorado Team Weimer
- University of Connecticut Nanostructured Materials Program
- University of Copenhagen, Denmark Center for Interdisciplinary Studies of Molecular Interactions.
- University of Copenhagen, Denmark Nano-Science Center
- University of Delaware Center for Molecular and Engineering Thermodynamics (CMET)
- University of Delaware Center for Nanomachined Surfaces
- University of Dublin | Trinity College Centre for Research on Adaptive Nanostructures and Nanodevices.
- University of Durham Department of Physics.
- University of Durham Carbon Nanostructures.
- University of Durham Durham Nanoscale Magnetics Group.
- University of Durham Nanotechnology University Innovation Centre (Nanotechnology UIC).
- University of Glasgow Nanoelectronics Research Centre. Provides a focus for diverse research activities within the University of Glasgow linked by a common interest in Nanoelectronics and Nanofabrication.
- University of Glasgow Spectroscopy of Nanostructures
- University of Glasgow Kelvin Nanotechnology

- University of Greenwich NANO-SCIENCE SIMULATION GROUP
- University of Hamburg Nanostructure Physics Group
- University of Hannover, Germany Institute for Solid State Physics—Department of Nanostructures
- University of Houston Texas Center for Superconductivity (and Advanced Materials) at the University of Houston (TCSUH)
- University of Houston Sharma Group—Mechanical Engineering.
- University of Illinois at Chicago Materials Characterization Laboratory
- University of Illinois at Chicago Microfabrication Applications Laboratory (MAL)
- University of Illinois at Chicago Nanoscale structures of asphaltene steric-colloid and asphaltene micelles & vesicles.
- University of Illinois at Urbana-Champaign Center for Nanoscale Science & Technology (CNST)
- University of Illinois at Urbana-Champaign Beckman Institute for Advanced Science and Technology
- University of Illinois at Urbana-Champaign Beckman Institute for Advanced Science and Technology—Molecular and Electronic Nanostructures
- University of Illinois at Urbana-Champaign Micro and Nanotechnology Laboratory
- University of Illinois at Urbana-Champaign Frederick Seitz Materials Research Laboratory (MRL).
- University of Illinois at Urbana-Champaign MASS Group. "...research in all aspects of the emerging technology of Microelectromechanical Systems."
- University of Jyväskylä Nanoscience Center
- University of Kentucky Materials Research Science and Engineering Center (MRSEC) Advanced Carbon Materials Center
- University of Leeds Centre for Self-Organising Molecular Systems
- University of Leeds Centre for Nano-Device Modelling
- University of Leeds Masters Training Package (MTP) in Nanoscale Science and Technology, run jointly by the universities of Leeds and Sheffield.
- University of Leipzig (Universität Leipzig) Semiconductor Physics Group. Quantum dot physics and device applications, and hybrid organic-anorganic nanostructure research.

- University of Liverpool Centre for Nanoscale Science "... material aspects of nanotechnology and nanoparticle research."
- University of Liverpool Chemistry with Nanotechnology—MChem.
- University of London Nano4CE "... a specialist unit operated by the University with the remit of working with industry to develop product forms that utilise nanomaterials."
- University of Louisville Nanoscale Imaging Facility
- University of Louisville Chemical Engineering Department
- University of Louisville Electro-Optics Research Institute
- University of Louisville MicroTechnology Center
- University of Manchester Manchester Centre for Mesoscience & Nanotechnology
- University of Manchester Nanotechnology Research
- University of Maryland Maryland Center for Integrated Nano Science and Engineering (M-CINSE)
- University of Maryland, College Park — Robert H. Smith School of Business Asian Technology Information Program.
- University of Maryland Materials Research Science and Engineering Center (MRSEC)
- University of Maryland Fuhrer Research Group / Nanoelectronics Research Group
- University of Maryland Drew Research Group / Nano-optics
- University of Maryland LPS Nanostructures Team
- University of Maryland The Williams Lab / Surface Physics
- University of Maryland Co-laboratory for Nanoparticle Based Manufacturing & Metrology—NM^2
- University of Massachusetts Amherst MassNanoTech
- University of Massachusetts Amherst Dinsmore Research Group.
- University of Massachusetts Amherst Materials Research Science and Engineering Center (MRSEC).
- University of Massachusetts Amherst Polymer Science & Engineering Department
- University of Massachusetts Lowell Nano Manufacturing.

- University of Michigan Solid State Electronics Laboratory. Part of the National Nanotechnology Infrastructure Network (NNIN)
- University of Michigan Center for Biologic Nanotechnology
- University of Michigan WIMS Engineering Research Center (ERC) an NSF-University-Industry Collaboration
- University of Michigan Glotzer Group—Laboratory for Computational Nanoscience and Soft Matter Simulation
- University of Michigan Materials Science & Engineering
- University of Michigan Kotov Group—Nano Research Laboratory
- University of Minnesota Minnesota Nano Technology Cluster. Part of the National Nanotechnology Infrastructure Network (NNIN)
- University of Minnesota Center for NanoEnergetics Research (CNER). CNER is an Army funded center created in the spring of 2001 and exists at four university sites, with the University of Minnesota as the lead institution.
- University of Minnesota Nanotechnology Coordinating Office
- University of Minnesota Particle Technology Laboratory (PTL)
- University of Minnesota Molecular Nanosciences Alliance for Interdisciplinary Studies and Activities (MONALISA)
- University of Minnesota Norris Research Group
- University of Montreal Pavillon J.-Armand Bombardier: Nanosciences and Nanotechnologies
- University of Muenster Interface Physics Group
- University of Nebraska—Lincoln Center for Materials Research and Analysis (CMRA)
- University of Nebraska—Lincoln Nanomachining/Nanofabrication at the Center for Electro-Optics.
- University of Nebraska—Lincoln Quantum Device Laboratory
- University of Nebraska—Lincoln Condensed Matter/Materials Physics and Nanotechnology.
- University of Nebraska Medical Center Nanomedicine Group
- University of Nevada, Reno Adams Group
- University of Newcastle upon Tyne Centre for Nanoscale Science. A Multidisciplinary research centre linking scientists, engineers, medical

researchers, and clinicians developing micro- and nanodevices and technologies for applications in biology, biotechnology and medicine.

- University of Newcastle upon Tyne Institute for Nanoscale Science and Technology (INSAT)
- University of Newcastle upon Tyne INEX—a microsystems and nanotechnology R&D, commercialisation and manufacturing organisation.
- University of New Mexico Integrative Nanoscience and Microsystems
- University of New Hampshire The Nano Group
- University of New Hampshire Nanotechnology and Society Research Group
- University of New Hampshire Nanostructured Polymers Research Center
- University of New Hampshire Nanostructured Polymers Research Center—Polymer Research Group (PRG).
- University of New Hampshire Nanostructured Polymers Research Center—Polymer Nanoparticle Laboratory (PNL).
- University of New Hampshire Nanostructured Polymers Research Center—Advanced Polymer Laboratory (APL).
- University of New Mexico Nanomaterials and Nanocharacterization. Part of the National Nanotechnology Infrastructure Network (NNIN)
- University of New Mexico Brinker Group.
- University of North Carolina at Chapel Hill North Carolina Center for Nanoscale Materials (NCCNM)
- University of North Carolina at Chapel Hill Nanoscale Science Research Group (NSRG)
- University of North Carolina at Chapel Hill The nanoManipulator
- University of Notre Dame Center for Nano Science and Technology
- University of Notre Dame Nanofabrication Facility
- University of Notre Dame Nano Science and Technology Center
- University of Nottingham Nottingham Nanotechnology Centre (includes 6 specializations: Quantum Nanostructures, Synthesis of Nanoscale Vessels and Storage Devices, Single Molecule Nanoscience, Biofunctional Surfaces and Materials, Nanofabrication and Characterisation, Biological Application of Nanoparticles)
- University of Oregon Jim Hutchison Lab
- University of Oregon Materials Science Institute

- **University of Oregon CAMCOR—Center for Advanced Materials Characterization in Oregon. "... a full-service, comprehensive materials characterization center available to research institutions and private industry."**
- **University of Oregon Oregon Center for Optics**
- **University of Oxford Carbon and Nanotechnology Group**
- **University of Oxford Carbon Nanotube Group**
- **University of Oxford Nanotechnology Group Pages**
- **University of Oxford Centre for Quantum Computation**
- **University of Pennsylvania LRSM / Laboratory for Research on the Structure of Matter**
- **University of Pennsylvania School of Engineering and Applied Science**
- **University of Pennsylvania Nano/Bio Interface Center**
- **University of Pittsburgh Swanson Center for Micro and Nano Systems**
- **University of Queensland (UQ) Nano Optics Group.**
- **University of Queensland Centre for Microscopy and Microanalysis.**
- **University of Queensland Soft Condensed Matter Group.**
- **University of Queensland NanoMaterials Centre.**
- **University of Queensland Nanotechnology & Biomaterials Centre.**
- **University of Queensland ARC Centre for Functional Nanomaterials (In partnership with the University of New South Wales, the Australian National University and the University of Western Sydney).**
- **University of Rochester Institute of Optics—Nano-Optics Group**
- **University of Sheffield Masters Training Package (MTP) in Nanoscale Science and Technology, run jointly by the universities of Leeds and Sheffield.**
- **University of Sheffield Nanorobotics Technologies for Multidimensional Imaging and Manipulation of Nanoobjects**
- **University of Southern California Laboratory for Molecular Robotics**
- **University of Southern California The Koel Group—Surface Science: Understanding Reactions at Nanoparticles and Catalysts**
- **University of Southern California Nanotechnology & Micro Electro-Mechanical Systems Research Area**
- **University of South Carolina NanoCenter**

- University of South Carolina USC NIRT: Philosophical and Social Dimensions of Nanoscale Research
- University of Southhampton Nanoscale Systems Integration Group (NSI).
- University of Sussex Fullerene Research Centre.
- University of Sussex Nanoscience and Nanotechnology @ Sussex.
- University of Sydney Nanostructural Analysis Network Organisation—Major National Research Facility (NANO—MNRF). Australian facility for nanometric analysis of the structure and chemistry of materials in both physical and biological systems. An unincorporated joint venture between university, industry, state and federal governments.
- University of Technology Sydney Institute for Nanoscale Technology
- University of Technology Sydney Institute for Nanoscale Technology Nanohouse Initiative.
- University of Texas at Arlington NanoFAB Research and Teaching Facility
- University of Texas at Austin Microelectronics Research Center (MRC). Part of the National Nanotechnology Infrastructure Network (NNIN)
- University of Texas at Austin Center for Nano- and Molecular Science and Technology
- University of Texas at Austin Science, Technology and Society (STS) "Explore social impacts of rapid scientific and technological change."
- University of Texas at Dallas NanoTech Institute.
- University of Texas at Dallas Center for Quantum Electronics.
- University of Texas at Dallas MEMS Research Group
- University of Tokyo Quantum Microstructure Devices—Sakaki Laboratory
- University of Toronto Energenius Centre for Advanced Nanotechnology (ECAN)
- University of Toronto Electronic Materials Group
- University of Toronto Geoff A. Ozin Group
- University of Toronto Nanoclub aims to promote nanotechnology at the University of Toronto. We are entirely student-run, but receive support from the Department of Materials Science Engineering.
- University of Twente—Netherlands Strategic Research Orientation Nanolink.
- University of Ulster Nanotechnology Research Institute

- University of Vienna Institute of Materials Physics
- University of Virginia MRSEC—Center for Nanoscopic Materials Design
- University of WalloniaNanoWal—Wallonia Network for Nanotechnologies
- University of Warwick — The Centre for Nanotechnology and Microengineering "...a national centre for excellence in nanotechnology and ultraprecision engineering for both the academic community and manufacturing industries."
- University of Washington Center for Nanotechnology
- University of Washington Center for Nanotechnology—Nanotech Student Association. "NSA fosters interdisciplinary research in nanotechnology, provides a forum for exchange of ideas between students from different disciplines, and promotes nanotechnology at the university level."
- University of Washington NanoTech User Facility (NTUF). Part of the National Nanotechnology Infrastructure Network (NNIN)
- University of Washington Center for Applied Microtechnology (CAM)
- University of Washington Department of Materials Science & Engineering
- University of Washington Molecular Bioengineering & Nanotechnology
- University of Waterloo Nanotechnology Engineering Program
- University of Wisconsin, Madison Nanoscale Science and Engineering Center (NSEC)
- University of Wisconsin, Madison Center for NanoTechnology
- University of Wisconsin, Madison Nanowires
- University of Wisconsin, Madison Applied Superconductivity Center
- University of Wisconsin, Madison Center for Plasma-Aided Manufacturing
- University of Wisconsin, Madison Materials Research Science and Engineering Center on Nanostructured Materials and Interfaces (MRSEC)
- University of Wisconsin, Madison Learning how to Fabricate Nanowires
- University of Wisconsin, Madison Nealey Research Group
- University of Wollongong, Australia Fundamental Properties of Semiconductors Research Centre.
- University of Wollongong, Australia ARC Centre for Nanostructured Electromaterials.
- Uppsala University Division of Materials Science at the Ångström Laboratory.
- University of York York-JEOL Centre: a collaboration between the University of York, Yorkshire Forward and JEOL UK Ltd.

- University of York Nanoscale Quantum Simulations for Nanostructures and Advanced Materials—Nanoquanta
- Vanderbilt University Vanderbilt Institute of Nanoscale Science and Engineering (VINSE).
- Virginia Tech Physics Department—Research Areas—Centers & Institutes.
- Virginia Tech Fiber & Electro-Optics Research Center.
- Virginia Tech Center for Modeling and Simulation in Material Science.
- Wake Forest The Center for Nanotechnology.
- Walter Schottky Institute—Technische Universität München Experimental Semiconductor Physics I (E24)—Prof. Gerhard Abstreiter.
- WASEDA University Establishment of Molecular Nano-Engineering by Utilizing Nanostructure Arrays and its Development into Micro-Systems.
- Washington University in St. Louis Buhro Group.
- Wayne State University Brock Group—Inorganic, Solid State, and Materials Chemistry.
- Wayne State University Jaewu Choi's Group: nano Devices & Systems Laboratory.
- Weizmann Institute of Science Prof. Ehud Shapiro—Laboratory for Biological Nanocomputers & The BIOSPI Project.
- Weizmann Institute of Science Department of Condensed Matter Physics.
- Weizmann Institute of Science Department of Materials and Interfaces.
- Western Michigan University Nanotechnology Computation and Research Center (NRCC) "NRCC's unique niche area is "nanobioenvironmental chemistry" which will build on Western's existing research strengths in nanotechnology approaches to environmental chemistry, biotechnology, and interrelated phenomena and problems."
- Widener University Fullerenes At Widener.
- Albany Institute of Nanotechnology and Applied Sciences—Research facility affiliated with the University at Albany engaged in a variety of nanoscience fields, including MEMS. Pictures and facility descriptions, news, events, staff profiles, bibliography.
- Alberta University—Department of Physics, Centre for Nanoscale Physics: ultrafast microscopy, ultrafast spectroscopy, nanocrystals and ion implantation, modelling of dynamics at solid surfaces, superconductivity, phase transitions and biophysics.

- Analytical Instrumentation Facility—North carolina state university, providing studies in instruments electron & scanned probe microscopy.
- Australian National University—Nanotube group: producing boron nitride and carbon nanotubes using a new relatively low temperature milling process.
- Birmingham University—Nanoscale Physics Research Laboratory: surface modification, cluster physics, sensors, scanning probe microscopy, nano-optics, and thin films.
- California University—Department of Physics: Condensed Matter Physics: Zettl Research Group: research into the electronic, magnetic, and structural properties of novel materials, including nanostructures such as telescoping nanotube bearings. (Berkeley)
- Cambridge University—Department of Engineering: Nanoscale Science Laboratory: nanotubes, nanomagnetics, nanowires, spin dependent transport, nanotribology, sensors, scanning probe microscopy, electron beam lithography, and an introduction to nanotechnology.
- Center for Nano and Molecular Science and Technology at UT Austin—Center for Nano- and Molecular Science and Technology (CNM)is a multidisciplinary, collaborative research center in the University of Texas at Austin. The mission of the CNM is to foster education, science, and engineering in nanoscience and nanotechnology at the University of Texas.
- Cranfield University—Nanotechnology Group, Advanced Materials Department: Research, facilities and courses, including the MSc in Microsystems and Nanotechnology, the first nanotechnology degree course in Europe.
- Georgia Tech—Center for Nanoscience and Nanotechnology. Nanostructure synthesis, sensors, MEMS, photonic crystals, microscopy, nano-lithography.
- Illinois University—The Beckman Institute for Advanced Science and Technology: molecular and electronic nanostructures; computational electronics; scanning tunneling microscopy, including lithography and fabrication; semiconductor nanostructures and photonics; synthesizing and characterizing new materials.
- Imperial College—Undertakes programme of one-year interdisciplinary nanomaterials research masters course. Provides funds as well.
- Integrated Microfabrication Lab at BYU—Semiconductor process laboratory, used for research in solid state electronics, optoelectronics, and microfabrication..

- Laboratory for Thin Films Physics—Studies size effects in phase transitions, structure and surface phenomena in condensed films of various materials. Includes research interests, people, publications, photo gallery and links.
- Newcastle upon Tyne University—Centre for Nanoscale Science & Technology: multidisciplinary research centre linking scientists, engineers, medical researchers, and clinicians developing micro and nanodevices and technologies for applications in biology, biotechnology and medicine.
- Northwestern University—Transportation Center: nanotechnology as applied to the transportation sector.
- Northwestern University—Nanoscience and Technology Lab: synthesis and properties of carbon and boron nitride nanotubes and nanofibers, fullerenes, and how materials synthesis at the sub-micron scale can impact the biological sciences.
- Nottingham University—Nanoscience Group: "bottom up" approaches to nanotechnology, scanning probe microscopes, properties of fullerenes and derivatives.
- Purdue University—Nanoscale Physics: research into nanotubes, nanoclusters, molecular wires, break junctions, scanning probe microscopy (STM and AFM), and field-ion microscopy.
- Rensselaer Polytechnic Institute—Advanced Materials & Nanotechnology Research
- Texas University—Malcolm Brown's Nanopage: ongoing research in the structure and synthesis of nanostructures and high resolution transmission electron microscopy and specialized light microscopy.
- Washington University—Center for study and research in nanotechnology, offering cross-discipline graduate courses in the field.
- Washington University (DURINT)—Defense University Research Initiative on Nanotechnology Project—use combinatorial biology techniques to select surface-specific polypeptides, and then use these Genetically Engineered Proteins for Inorganics (GEPIs) for assembling functional nanostructures for optical and electronic applications.
- Würzburg University (Germany)—Experimental Physics II: development and modification of nanostructured materials such as aerogels, nanostructured foams, carbon nanofibers and composites for different fields of applications mainly in energy storage, energy conversion and thermal insulation systems.

NSTI Nanotechnology Course Series

Summary An overview of the current state of the art in nanotechnology will be presented through 6 (2-day) courses taught by expert instructors renowned in their respective fields of expertise. Participants can select individual short courses to tailor the curriculum to their specific needs.

Who Should Attend

This introductory- to intermediate- level course is suitable for: Managers, Practicing Engineers in R&D, Industrial Scientists on a decision-making level, Executives seeking strategic planning insight, Policy Makers with some technical background, and Academic Researchers developing a strong nano program.

Invited Organizations Include:

- NNCO
- IBM
- Nantero
- Molecular Imprints
- General Electric
- Nanophase
- Hybrid Plastics
- Procter and Gamble
- Altair Nano
- Insert Therapeutics

Nanotechnology Instruments Inc.

Nanotechnology RET Programs

- Summer Program for High School Science Teachers at Northwestern's Institute for Nanotechnology. 2-year, ongoing RET program. Director: Prof. Dennis Lehman
- Virgina Tech Nanoscale Interdisciplinary Research Teams Program. Provides workshops for secondary school educators, includes topics on nanogeoscience, scanning probe microscopy, SEM and other relevant nano topics.

Director: Dr. Susan Eriksson

- Cornell Nanobiotechnology Center RET program
- Harvard University Nanoscale Science and Engineering Summer 2003 RET Program
- Univ. of Wisconsin-Madison RET program

Nanotechnology Degree Programs

- Rice University Professional Masters Degree in Nanoscale Physics: An interdisciplinary nanoscience program which offers studies in many areas of nanoscale science, including carbon nanotubes, nanoshells, and nanobiology. Courses include Nanostructures and Nanotechnology, Nano-optics and Nanophotonics, and Characterization and Fabrication at the Nanoscale.

Nanotechnology REU Programs

- Cornell University Nanobiotechnology program
- Cornell University National Nanofabrication Users Network REU Program
- Northwestern's Institute for Nanotechnology
- University of Conneticut—Chemical engineering at the nanoscale
- Harvard University Nanoscale Science and Engineering Center
- Univ. of California, Santa Barbara Nanotech
- Columbia University Research Experiences for Undergraduates in Nanotechnology
- University of South Carolina Nanoscience REU program
- Rice University Center for Biological & Environmental Nanotechnology REU program
- Univ. of Wisconsin, Madison REU

NSF: Summer Programme for Science Teachers

Ongoing professional development for teachers is an important step in the revitalization of science education. Teachers need opportunities to stay abreast of the latest in technological advances, to develop lasting professional relationships, and to discover ways to share the excitement of interdisciplinary science with their students. In answer to this need, the Center offers an innovative summer program for K-12 science teachers and community college faculty.

Program Description (8-weeks): Participants engage in hands-on nanoscience or nanotechnology related research over the first four weeks under the mentorship of center faculty, postdoctoral fellows, and graduate students. Weekly group meetings, laboratory tours, and events broaden the experience.

For the remaining four weeks of the program, participants fine-tune and test ideas based on their research projects and develop them into effective curricular projects that introduce nanotechnology concepts to pre-college students.

Final projects are presented at the closing symposium and a written report is required. Additional funding may be available to support the presentation of teaching units at regional and/or national meetings.

Eligibility and Application Process: Applicants must be U.S. citizens or permanent residents with a bachelor's degree in the physical sciences or engineering and be currently employed as a K-12 science teacher or community college faculty.

You may download an application by clicking the link below, emailing a request to nanotechnology@northwestern.edu, or calling 847/467-4862.

RET Application: Selections are based upon application materials and the availability of a research project compatible with the applicant's scientific and professional interests.

NSF: Research Experience for Teachers (RET)

What is the RET program?

The Research Experience for Teachers (RET) program is funded by the a National Science Foundation. It provides a professional development opportunity to K-12 science teachers by placing them in a research laboratory during the summer. The program that has been established at the University of Wisconsin-Madison's Materials Research Science and Engineering Center (MRSEC) focuses on materials science and science at the nanoscale, the scale of atoms.

Learn more about the MRSEC Mission and Goals

In recent years the RET program has worked with three Madison, WI middle and high school teachers to create modules relating to advanced and nanoscale materials including shape memory alloys and light emitting diodes. The modules also contain information about how scientist know where atoms are located in a material using the methods of x-ray diffraction and scanning probe microscopy.

Wiess School of Natural Sciences: Professional Master's Programme

"Individuals with an educational background in these interdisciplinary areas are very few in number. Rice University's idea of combining a rare and highly demanded technical education with a modest exposure to training in business will produce students who are truly unique, and these students will be highly recruited by industry."—Ken Smith, co-founder of Carbon Nanotechnologies, Inc.

The Nanoscale Physics program prepares students for a career in nanoscience by combining a strong component in quantum theory, which governs the behavior of systems at the nanoscale, with the study of practical nano- and mesoscale devices. This provides the student with the knowledge required to successfully navigate the emerging field of nanoscale science and nanotechnology. In addition, a year-long course in methods of experimental physics is offered to ensure that students obtain the advanced practical skills valuable to the nanotechnology industry.

Rice is a well-established center for nanotechnology, with researchers active in several departments outside the physics core. The *Rice Center for Nanoscale Science and Technology* includes among its faculty Nobel Laureates R. L. Smalley and R. Curl, as well as esteemed faculty from the physics and astronomy, chemistry, electrical and computer engineering, civil and environmental engineering, chemical engineering, bioengineering, computational and applied math, and mechanical engineering and materials science departments. It focuses on interdisciplinary studies in many areas of nanoscale science including carbon nanotubes, nanoshells, and nanobiology.

The *Center for Biological and Environmental Nanoscience (CBEN)* at Rice fosters the development of this field through an integrated set of programs that aim to address the scientific, technological, environmental, human resource, commercialization, and societal barriers that hinder the transition from nanoscience to nanotechnology. The Center's research focuses on investigating and developing nanoscience at the "wet/dry" interface.

Faculty members involved in the Professional M.S in Nanoscale Physics Program include:

- F. Barry Dunning, Track Director (Physics and Astronomy)
- Douglas A. Natelson, Track Advisor (Physics and Astronomy)
- Andrew R. Barron (Chemistry)
- Linda P. Driskill (Cain Project in Engineering and Professional Communication)
- Jason H. Hafner (Physics and Astronomy)
- Amy M. Jaffe (Baker Institute for Public Policy)
- Thomas C. Killian (Physics and Astronomy)
- Kristen M. Kulinowski (Center for Biological and Environmental Nanotechnology)
- Neal F. Lane (Physics and Astronomy; Baker Institute for Public Policy)
- Kathleen S. Matthews (Dean of Natural Sciences)
- Donald H. Ostdiek (Policy Studies Program)
- Frank R. Toffoletto (Physics and Astronomy)

NTUF Short Course & Nanoscale Science and Technology Workshop

Nanotechnology: Where Science Fiction Meets Reality

Do you ever wonder about our future? Do you ever imagine how discoveries could make the impossible a reality? Do you ever wish you had cures for threatening diseases at hand?

Nanotechnology, while not providing a cure for everything, is defined by the length scale when scientists and engineers discover new phenomena. It provides exquisite new tools to engineer novel materials and devices at the nanoscale, and to study biology. A nanometer, one billionth of a meter, is about 10,000 times narrower than a human hair. Major technological revolutions, including the industrial revolution and the dawn of the information era, have revealed how new discoveries can drastically change our lives. There is no doubt that rapid technological transformations require new paradigms of how to educate the next generation of leaders in academia and industry.

By virtue of their interdisciplinary nature, rapid advances in nanoscale science and technology can only thrive in a collaborative environment in which faculty and students from different disciplines discuss ideas, work together, and share their expertise.

The Center for Nanotechnology at the University of Washington was created in 1997 to address these changing realities. It brings together faculty members and students from the Colleges of Arts and Sciences, Engineering, Pharmacy, and the School of Medicine. The Center enjoys major financial support from the University of Washington Initiatives Fund (UIF) and National Science Foundation Integrative Graduate Education and Research Traineeship (NSF-IGERT) program.

Learn more about Nanotechnology at the University of Washington and our programs.

Director Search—The University of Washington seeks an exceptional researcher with strong leadership, organizational, and team-building skills to become the Director for the Center for Nanotechnology.

Nanoparticulates: Syntehsis and Application

Technology Focus

After years of evolutionary research, industry today uses dry technology for manufacture of carbon blacks and simple oxides. It is practically impossible to use these production units for synthesis of functional inorganic nanoparticles without going through the same costly and time-consuming cycle. However, recent major breakthroughs in academia have placed dry synthesis of nanoparticles on a firm scientific basis, allowing synthesis of these materials in appreciable volumes and at competitive prices. This has created renewed interest in dry processes and products.

The focus now shifts to an integrated process development for final product performance rather than particle characteristics. Special emphasis is placed on the degree of particle agglomeration, and its control, as well as on nanoparticles with designed morphology and layered composition.

Course Objectives

The first day will focus on dry synthesis of functional nanoparticles. We will show how it survived the "valley of death" from the laboratory to manufacturing for selected products. Important theories and examples will be given through the discussion of a number of new products and processes using dry technologies.

The second part of the course focuses on the wet synthesis methods and chemistries used to produce polymer particles in the 10-1000 nanometer range. Careful attention is paid to the analytical techniques that allow confident measurement of phase-separated structure within these composite particles and the varied applications of such polymer particles. The objectives are to:

- Introduce the diversity of polymer particle sizes, shapes, and chemical composition with structural control at the nano scale
- Describe how particles can be made from either reactive or non-reactive processing techniques, including emulsion polymerization and self assembly

The third and fourth parts focus on characterization (analysis and verification) and the wide range of application areas of economic importance. The objectives are to:

- Demonstrate how particles can be characterized by modern instrumental techniques
- Discuss the multitude of applications for polymer nanoparticles
- Provide an interactive learning experience through the use of individual and group exercises and case studies

Course Content —Nano Particulates

A scalable, dry technology for synthesis of high purity nanoparticles with closely controlled characteristics will be presented. This is advantageous over classic wet chemistry technologies (sol-gel or precipitation) as it does not use their multiple processing steps (e.g. washing, drying, calcinations etc.) and high volumes of liquid byproducts.

We will start with the manufacture of optical fibers, carbon blacks, filamentary nickel, pigments and fumed silica and present the flame, hot-wall, laser and plasma reactors for synthesis of metal, alloys, and ceramics. The fundamental physical and chemical phenomena that control these processes will be covered.

Principles for designing processes by combining fluid and particle dynamics will be presented. Emphasis is placed on scalable flame reactors that dominate the value and volume of today's manufacture of nanostructured materials. The versatile flame spray pyrolysis process for synthesis of catalysts, ZnO quantum dots for UV-filters, mixed ceramics and others are highlighted. Synthesis of solid nano- or hollow micro-particles of Bi2O3, CeO2 or Al2O3 by combustion spraying of solutions, emulsions or slurries will be shown along with process design criteria. Specific cases for hot-wall synthesis of selected metals and co-production of solar H2 will be presented. Scale-up will be discussed showing how design correlations are developed with reactors of various sizes along with principles for synthesis of hard- or soft-agglomerates. Coating of nanoparticles with oxide or carbon films in flame and hot wall reactors will be discussed.

All through the course practical problems will be solved both individually and in groups.

Daily Schedule

Overview

- Nanoparticles: Origins, significance and applications
- Is it a bubble?
- Health issues
- Cost
- The evolution of today's industry for manufacture of carbon blacks, fumed silica, pigmentary titania, ZnO, filamentary nickel, optical fibers and, most recently, for metallic and ceramic nanoparticles.
- Pros and cons of current processes: Flame, hot-walls, plasmas, lasers.

Fundamentals

- Definitions, Brownian motion and particle diffusion
- Thermophoretic sampling and particle characterization
- Aerosol coagulation in the continuum and free-molecular regimes, self-preserving distributions, agglomeration, fractal-like particles
- Critical, Kelvin or minimum particle size, condensation and nucleation.

Process Design and Operation

- Controlled flame synthesis of nanoparticles
- Chemistry affects particle characteristics
- Reactor design by computational fluid and particle dynamics
- Process scale-up and correlations
- Hard- and soft-agglomerates.

Novel Products and Applications

- Flame-made catalysts for DeNOx removal (V_2O_5/TiO_2), polymer synthesis (TiO_2/SiO_2) and chiral pharmaceuticals (Pt/Al_2O3) manufacture, automotive CeO_2/ZrO_2.
- Mixed ceramics: Stable ZnO Quantum Dots for UV-filters by doping with silica, Dental nanocomposites (non-agglomerated SiO_2) or Translucent Ta_2O_5/SiO_2 in polymer matrices.
- Hot-wall reactors for metal (Bi, Pd, Al, Zn) and non-oxide (AlN, B4C) nanoparticles and even for co-production of solar H_2.
- Coatings: Carbon or ceramic oxide films on titania or silica nanoparticles.
- Solid or hollow particles by combustion of solutions, emulsions or slurries.

NSTI Outreach

NSTI is chartered with the promotion and integration of small technologies through education, technology and business development. Consequently, the NSTI uses its network not only as a private consultancy, but also as an educational and promotional vehicle in the space of Nanotechnology. Listed below are a few of the scientific, educational and community building projects that the NSTI donates its resources towards:

- *Education – Technology Proceeding Posting:* NSTI freely posts all of its nanotechnology publications into the NSTI website for the promotion of scientific and business growth. These papers are downloaded daily by academics, graduate students, business developers and technology companies, resulting in over 200,000 site visits per month.
- *Educational – National Public Radio – Earth & Sky* NSTI endorses and has donated its network for an Earth and Sky radio educational project focused on nanotechnology for support by the National Science Foundation.
- *Economic Development—Massachusetts:* NSTI is a founding member of the Massachusetts Nanotechnology Initiative and has supported the development of state economic development programs in nanotechnology.
- *Science – Biomolecular Motors White Paper:* NSTI prepares a DARPA commissioned overview of Biomolecular Motors with highlights from the special session at Nanotech 2004.
- *Science – Nanotechnology for Cancer Research:* NSTI endorses, has donated its network and is organizing symposia aligned with a North Eastern University program directed at nanotechnology for cancer detection and treatment.

- *Science – Environmental & Health Studies of Nanotechnology:* NSTI is using the Nanotech conference series as a vehicle to scientifically and socially address health and environmental studies related to nanotechnology, in partnership with NNCO, EPA and Dupont.
- *Science – Chemical Imaging Center – Puerto Rico:* NSTI endorses and supports a proposed NSF sponsored Chemical Imaging Center by providing for a series of symposia for Puerto Rican students and researchers embedded into the Nanotech conference series.
- *Interdisciplinary Community Building:* NSTI acts as a focal point for multi-disciplinary and multi-industry nanotechnology activities. Through the NSTI-Nanotech conference series, the NSTI has partnering agreements with a range of single disciplinary societies including:
- American Physical Society (APS)
- IEEE Electron Device Society
- American Society of Mechanical Engineers (ASME) Nanotech Institute
- American Ceramics Society (ACS)
- *Technology and Business Community Development* At the NSTI-Nanotech 2004, NSTI provided free promotion to leading Technology Licensing offices and international initiatives as a benefit to the nanotechnology community. 2004 participating organizations included:
- National Nanotechnology Coordination Office (NNCO)
- National Nanotechnology Initiative (NNI)
- Boston University, Technology Transfer Office, Nanotechnology Iniative, Photonics Center
- University of Massachusetts Amherst
- Harvard University, Office for Technology and Trademark Licensing
- Tufts University Office for Technology Licensing and Industry Collaboration
- Harvard University—NSEC/NNIN/CIMS
- UC Irvine Office of Technology Alliances / Integrated Nanosystems Research Facility
- University of Massachusetts Lowell
- Swiss Nanotechnology
- German Nanotechnology
- Nanotechnology Australia

- Georgia Nanotechnology
- Massachusetts Nanotechnology
- MCNC Research & Development Institute
- Greater Washington Nanotech Alliance
- MESA+ Institute for Nanotechnology
- Museum of Science, Boston

4

Nanotech Education & Research Facilities

Research Programs on Nanotechnology in the World

Scientific breakthroughs combined with recent research programs in the United States, Japan, and Europe, and various initiatives in Australia, Canada, China, Korea, Singapore, and Taiwan highlight the international interest in nanoscale science and technology. Definitions of nanotechnology vary somewhat from country to country. Nanotechnology as defined for the projects reviewed in this chapter arises from the exploitation of the novel and improved physical, chemical, mechanical, and biological properties, phenomena, and processes of systems that are intermediate in size between isolated atoms/molecules and bulk materials, where phenomena length and time scales become comparable to those of the structure. It implies the ability to generate and utilize structures, components, and devices with a size range from about 0.1 nm (atomic and molecular scale) to about 100 nm (or larger in some situations) by control at atomic, molecular, and macromolecular levels. Novel properties occur compared to bulk behavior because of the small structure size and short time scale of various processes. Nanotechnology's size range and particularly its new phenomena set it apart from the technologies referred to as microelectromechanical systems (MEMS) in the United States or microsystems technologies (MST) in Europe.

It is estimated that nanotechnology is presently at a level of development similar to that of computer/information technology in the 1950s. As indicated in the preceding chapters and as evidenced by the WTEC panel's research and observations during the course of this study, the development of nanoscale science and technology is expected by most scientists working in the field to have a broad and fundamental effect on many other technologies. This helps to explain the phenomenal levels of R&D activity worldwide. This chapter presents an overview of most of the significant nanotechnology research programs in the world. Where possible, a general picture is given of the funding levels of the programs, based on site interviews and publications.

TheUnited States

Various U.S. public and private funding agencies; large companies in chemical, computer, pharmaceutical, and other areas; as well as small and medium-size enterprises provide support for precompetitive research programs on nanotechnology. Most of the supported programs are evolving out of disciplinary research programs, and only some are identified as primarily dealing with nanotechnology. U.S. government agencies sponsored basic research in this area at a level estimated at about \$116 million in 1997 (Siegel et al. 1998), as shown in Table 8.1. The National Science Foundation (NSF) has the largest share of the U.S. government investment, with an expenditure of about \$65 million per year, or about 2.4% of its overall research investment in 1997. In 1998 it expanded its research support to functional nanostructures with an initiative in excess of \$13 million.

NSF activities in nanotechnology include research supported by the Advanced Materials and Processing Program; the Ultrafine Particle Engineering initiative dedicated to new concepts and fundamental research to generate nanoparticles at high rates; the National Nanofabrication User Network (NNUN); and Instrument Development for Nano-Science and Engineering (NANO-95) to advance atomic-scale measurements of molecules, clusters, nanoparticles, and nanostructured materials. A current activity is the initiative, Synthesis, Processing, and Utilization of Functional Nanostructures (NSF 98-20 1997).

In the United States, a number of large multinational corporations, small enterprises, and consortia are pursuing nanotechnology-related research and development activities. Dow, DuPont, Eastman Kodak, Hewlett-Packard (HP), Hughes Electronics, Lucent, Motorola, Texas Instruments, Xerox, and other multinationals have established specialized groups in their long-term research laboratories, where the total research expenditure for nanotechnology research is estimated to be comparable to the U.S. government funding. Computer and electronics companies allocate up to half of their long-term research resources to nanotechnology programs. HP spends 50% of long-term (over 5 years) research on nanotechnology (Williams 1998). Small business enterprises, such as Aerochem Research Laboratory, Nanodyne, Michigan Molecular Institute, and Particle Technology, Inc., have generated an innovative competitive environment in various technological areas, including dispersions, coatings, structural materials, filtration, nanoparticle manufacturing processes, and functional nanostructures (sensors, electronic devices, etc.). Small niches in the market as well as support from several U.S. government agencies through the Small Business for Innovative Research (SBIR) program have provided the nuclei for high-tech enterprises. The university-small business technology transfer (STTR) program at NSF is dedicated to nanotechnology in fiscal year 1999. Two semiconductor processing consortia, the Semiconductor Manufacturing and Technology Institute (Sematech) and the Semiconductor Research Corporation (SRC), are developing significant research activities on functional nanostructures on inorganic surfaces. A series of interdisciplinary centers with nanotechnology activities has been established in the last few years at many U.S. universities, creating a growing public research and education infrastructure for this field. Examples of such centers are

- Brown University, Material Research Science and Engineering Center
- Rice University, Richard Smalley's Center for Nanoscale Science and Technology (CNST)
- University of California–Berkeley, nanoelectronics facilities
- University of Illinois at Urbana, the Engineering Research Center on Microelectronics in collaboration with the Beckman Institute, a private foundation
- University of North Carolina
- University of Texas–Austin
- Rensselaer Polytechnic Institute
- University of Washington (focus on nanobiotechnology)
- University of Wisconsin at Madison (focus on nanostructured materials)

NNUN, mentioned above, is an interuniversity effort supported by NSF at five universities: Cornell, Stanford, University of California–Santa Barbara (UCSB), Penn State, and Howard. It has focused on nanoelectronics, optoelectronics, electromechanical systems, and biotechnology. The Center for Quantized Electronic Structures (QUEST) at UCSB is a national facility developing expertise on underlying physics and chemistry aspects. Hundreds of graduate students have completed their education in connection with these centers in the last few years.

Current interest in nanotechnology in the United States is broad-based and generally spread into small groups. The research themes receiving the most attention include

1. metallic and ceramic nanostructured materials with engineered properties
2. molecular manipulation of polymeric macromolecules
3. chemistry self-assembling techniques of "soft" nanostructures
4. thermal spray processing and chemistry-based techniques for nanostructured coatings
5. nanofabrication of electronic products and sensors
6. nanostructured materials for energy-related processes such as catalysts and soft magnets
7. nanomachining
8. miniaturization of spacecraft systems

In addition, neural communication and chip technologies are being investigated for biochemical applications; metrology has been developed for thermal and mechanical properties, magnetism, micromagnetic modeling, and thermodynamics of nanostructures; modeling at the atomistic level has been established as a computational tool; and nanoprobes have been constructed to study material structures and devices with nanometer length scale accuracy and picosecond time resolution. While generation of nanostructures under controlled conditions by building up from atoms and molecules is the most promising approach, materials restructuring and scaling-down approaches will continue. Exploratory research includes tools of quantum control and atom manipulation, computer design of hierarchically structured materials (e.g., Olson 1997), artificially structured molecules, combination of organic and inorganic nanostructures, biomimetics, nanoscale robotics, encoding and utilization of information by biological structures, DNA computing, interacting textiles, and chemical and bioagent detectors.

Commercially viable technologies are already in place in the United States for some ceramic, metallic, and polymeric nanoparticles, nanostructured alloys, colorants and cosmetics, electronic components such as those for media recording, and hard-disk reading, to name a few. The time interval from discovery to technological application varies greatly. For instance, it took several years from the basic research discovery of the giant magnetoresistance (GMR) phenomenon in nanocrystalline materials (Berkowitz et al. 1992) to industry domination by the corresponding technology by 1997. GMR technology has now completely replaced the old technologies for computer disk heads, the critical components in hard disk drives, for which there is a $20+ billion market (Williams 1998). All disk heads currently manufactured by IBM and HP are based on this discovery. In another example, nanolayers with selective optical barriers are used at Kodak in more than 90% of graphics black and white film (Mendel 1997) and for various optical and infrared filters, which constitute a multibilliondollar business. Other current applications of nanotechnology are hard coatings, chemical and biodetectors, drug delivery systems via nanoparticles, chemical-mechanical polishing with nanoparticle slurries in the electronics industry, and advanced laser technology. Several nanoparticle synthesis processes developed their scientific bases decades ago, but most processes are still developing their scientific bases (Roco 1998). Most of the technology base development for nanoparticle work is in an embryonic phase, and industry alone cannot sustain the research effort required for establishing the scientific and technological infrastructure. This is the role of government (e.g., NSF and NIH) and private agency (e.g., Beckman Institute) support for fundamental research.

Nanotechnology research in the United States has been developed in open competition with other research topics within various disciplines. This is one of the reasons that the U.S. research efforts in nanotechnology are relatively fragmented and partially overlapping among disciplines, areas of relevance, and sources of funding. This situation has advantages in establishing competitive paths in the emerging nanotechnology field and in promoting innovative ideas; it also has some disadvantages for developing system applications. An interagency coordinating "Group on Nanotechnology" targets some improvement of the current situation. The group was established in 1997 with participants from twelve government funding/research agencies to enhance communication and develop partnerships among practicing nanoscience professionals.

Canada

Canada's National Research Council supports nanotechnology through the Institute for Microstructural Science, which has the mission to interact with industry and universities to develop the infrastructure for information technology. The main project, the Semiconductor Nanostructure Project, was initiated in 1990. It provides support for fundamental research at a series of universities, including Queen's, Carleton, and Ottawa Universities.

Japan

The term "nanotechnology" is frequently used in Japan specifically to describe the construction of nanostructures on semiconductors/inorganic substrates for future electronic and computer technologies, and to describe the development of equipment for measurement at nanometer level (Sienko 1998). There are, however, Japanese programs in a number of other areas related to nanotechnology in the broader definition used in this report.

Government agencies and large corporations are the main sources of funding for nanotechnology in Japan; small and medium-size companies play only a minor role. Research activities are generally grouped in relatively large industrial, government, and academic laboratories. The three main government organizations sponsoring nanotechnology in Japan are the Ministry of International Trade and Industry (MITI), the Science and Technology Agency (STA), and Monbusho (the Ministry of Education, Science, Sports, and Culture). Funding for nanotechnology research should be viewed in the context of an overall increased level of support for basic research in Japan since 1995 as a result of Japan's Science and Technology Basic Law No. 130 (effective November 15, 1995), even if the law has not been fully implemented. The data presented below are based on information received from Japanese colleagues during the WTEC visit in July 1997. All budgets are for the fiscal year 1996 (1 April 1996 to 31 March 1997) and assume an exchange rate of ¥115 = $1, unless otherwise stated. The first five-year program on ultrafine particles started in 1981 under the Exploratory Research for Advanced Technologies (ERATO) program; an overview of the results of that program was published in 1991 (Uyeda 1991).

It is estimated that the Agency of Industrial Science and Technology (AIST) within MITI had a budget of approximately $60 million per year for nanotechnology in 1996/97 (roughly 2.2% of the AIST budget). The National Institute for Advancement of Interdisciplinary Research (NAIR) hosts three AIST projects:

1. Joint Research Center for Atom Technology (JRCAT), with a ten-year

budget of about $220 million for 1992-2001 ($25 million per year in 1996)

2. Research on Cluster Science program, with about $10 million for the interval 1992-1997
3. Research on Bionic Design program, with $10 million for 1992-1997, about half on nanotechnology

Other efforts supported to various degrees by MITI include the following:

- the Electrotechnical Laboratory in Tsukuba, which allocates about 17% (or $17 million per year) of its efforts on advanced nanotechnology projects
- the Quantum Functional Devices Program, funded at about $64 million for 1991-2001 (about $6.4 million in 1996)
- the Osaka National Research Institute and the National Industrial Research Institute of Nagoya, which each spend in the range of $2.5-3 million per year for nanotechnology
- the Association of Super-Advanced Electronics Technologies (ASET), a relatively new MITI-sponsored consortium with partial interest in nanotechnology; it has similarities with the U.S. Ultra Electronics program of DARPA

It is estimated that STA investment in nanotechnology research was about $35 million in 1996, mainly within five organizations:

1. Institute of Physical and Chemical Research (RIKEN), where nanotechnology is included in the Frontier Materials Research initiative
2. National Research Institute for Metals (NRIM)
3. National Institute for Research in Inorganic Materials (NIRIM)
4. Japan Science and Technology Corporation (JST—formerly called JRDC), which manages the ERATO program, including four nanotechnology-related projects, each with total budgets of $13-18 million for five years:
   - Quantum Wave Project (1988-1993)
   - Atomcraft Project (1989-1994)
   - Electron Wavefront Project (1989-1994)
   - Quantum Fluctuation Project (1993-1998)

Monbusho supports nanotechnology programs at universities and national research institutes, as well as via the Japan Society for Promotion of Science (JSPS). The most active programs are those at Tokyo University, Kyoto University, Tokyo Institute of Technology, Tohoku University, and Osaka University. The Institute of Molecular

Science and the Exploratory Research on Novel Artificial Materials and Substances program promote new research ideas for next-generation industries (5-year university-industry research projects). The "Research for the Future" initiative sponsored by JSPS has a program on "nanostructurally controlled spin-dependent quantum effects" (1996-2001) at Tohoku University. Monbusho's funding contribution to nanotechnology programs is estimated at ~ $25 million.

In total, MITI, STA, and Monbusho allocated ~ $120 million for nanotechnology in 1996.

Large companies also drive nanotechnology research in Japan. Important research efforts are at six institutions: Hitachi (Central R&D Laboratories, where nanotechnology is ~ 25% of long-term research)—Hitachi has ~ $70 billion per year in sales; NEC (Fundamental Research Laboratories, where nanotechnology is estimated to be ~ 50% of the precompetitive research)—NEC has ~ $40 billion per year in sales; NTT (Atsugi Lab); Fujitsu (Quantum Electron Devices Lab); Sony; and Fuji Photo Film Co. An allocation of 10% of sales for research and development is customary in these companies, with ~ 10% of this for long-term research. Some Japanese nanostructured products already have considerable market impact. Nihon Shinku Gijutsu (ULVAC) produces over $4 million per year in sales of particles for electronics, optics, and arts. Also, there are in Japan, at the United States, private consortia making an increased contribution to nanotechnology R&D:

- The Semiconductor Industry Research Institute of Japan (SIRI), established in 1994, focuses on long-term research with partial government funding
- Semiconductor Leading Edge Technologies, Inc. (SELETE), established by ten large Japanese semiconductor companies in 1996, focuses on applied research and development with an estimated budget of $60 million in 1997
- Semiconductor Technology Academic Research Center (STARC) promotes industry-university interactions

Strengthening of the nanotechnology research infrastructure in the last years has been fueled by both the overall increase of government funding for basic research and by larger numbers of academic and industry researchers choosing nanotechnology as their primary field of research. Potential industrial applications provide a strong stimulus. A systems approach has been adopted in most laboratory projects, including multiple characterization methods and processing techniques. A special Japanese research strength is instrumentation development. The university-industry interaction is stimulated by the new MITI projects awarded to universities in the last few years that encourage temporary hiring of research personnel from industry. Other issues currently being addressed are more extensive use of peer review, promoting personnel mobility and intellectual independence, rewarding researchers for patents, promoting interdisciplinary and international interactions, and better use of the physical infrastructure.

China

Nanoscience and nanotechnology have received increased attention in China since the mid-1980s. Approximately 3,000 researchers there now contribute to this field (Bai 1996). The ten-year "Climbing Project on Nanometer Science" (1990-1999) and a series of advanced materials research projects are the core activities. The Chinese Academy of Sciences sponsors relatively large groups, while the China National Science Foundation (CNSF) provides support mainly for individual research projects. Areas of strength are development of nanoprobes and manufacturing processes using nanotubes. The Chinese Physics Society and the Chinese Society of Particuology are societies involved in the dissemination of nanotechnology research.

India

India's main research activities are on nanostructured materials and electronic devices (Sikka 1995). These involve a combination of research institutions (the Central Electronics Engineering Research Institute in Pilani, the Space Application Center in Ahmedabad, and others), funding organizations (the Centre for Development of Materials in Pune, the Indian Institute of Science in Bangalore, and others), and industry.

Taiwan

Taiwan's major nanotechnology research effort is conducted in the area of miniaturization of electronic circuits. The research is conducted in academic institutions and at the Industrial Technology Research Institute. Government funds for fundamental research are channeled via the National Science Council.

South Korea

A national research focus on nanotechnology was established in Korea in 1995. The Electronics and Telecommunications Research Institute (ETRI) in Taejon, Korea's science city, targets advanced technologies for information and computer infrastructures, with a focus on nanotechnology (ATIP 1998). The emphasis is on nanoscale semiconductor devices and particularly on semiconductor quantum nanostructures and device applications (lasers, modulators, switches and logical devices, resonant tunneling devices, self-assembled nanosize dots, single-electron transistors, and quantum wires).

Singapore

Nanotechnology research received a considerable boost in Singapore by the initiation of a national program in this area in 1995.

Australia

The National Research Council (NRC) of Australia has sponsored R&D in nanotechnology since 1993 (ASTC 1993). Research groups work on synthesis of nanoparticles for membranes and catalysts (University of New South Wales), nanofiltration (UNESCO Center for Membrane Science and Technology), and use of nanoparticles in processing minerals for special products (the Advanced Mineral Products Research Center at the University of Melbourne). AWA Electronics in Homebush has the largest industrial research facility for nanoelectronics in Australia.

European Community (EC)

The term "nanotechnology" is frequently defined in Europe as "the direct control of atoms and molecules" for materials and devices. A more specific definition from H. Rohrer (1997) is a "one-to-one relationship between a nano-object or nano-part of an object and another nano-, micro- or bulk object." The nanotechnology field as defined in this WTEC report includes these aspects, with the clarification that only the specific, distinctive properties and phenomena manifesting at length scales between individual atoms/molecules and bulk behavior are considered.

There are a combination of national programs, collaborative European (mostly EC) networks, and large corporations that fund nanotechnology research in Europe. Multinational European programs include the following:

1. The ESPRIT Advanced Research Initiative in Microelectronics and the BRITE/EURAM projects on materials science in the EC are partially dedicated to nanotechnology.
2. The PHANTOMS (Physics and Technology of Mesoscale Systems) program is a network created in 1992 with about 40 members to stimulate nanoelectronics, nanofabrication, optoelectronics, and electronic switching. Its coordinating center is at the IMEC Center for Microelectronics in Leuven, Belgium.
3. The European Science Foundation has sponsored a network since 1995 for Vapor-phase Synthesis and Processing of Nanoparticle Materials (NANO) in order to promote bridges between the aerosol and materials science communities working on nanoparticles. The NANO network includes 18 research centers and is codirected by Duisburg University and Delft University of Technology.
4. The European Consortium on NanoMaterials (ECNM) was formed in 1996, with its coordinating center in Lausanne, Switzerland. This group aims at fundamental research to solve technological problems for nanomaterials and at improved communication between researchers and industry.
5. NEOME (Network for Excellence on Organic Materials for Electronics) has had some programs related to nanotechnology since 1992.
6. The European Society for Precision Engineering and Nanotechnology

(EUSPEN) was designed in 1997 with participation from industry and universities from six EC countries.

7. The Joint Research Center Nanostructured Materials Network, established in 1996, has its center in Ispra, Italy.

It is expected that the European Framework V will introduce additional programs on nanotechnology, particularly by adding a new dimension in nanobiology in the next four-year plan.

Germany

The Federal Ministry of Education, Science, Research, and Technology (BMBF) in Germany provides substantial national support for nanotechnology. The Fraunhofer Institutes, Max Planck Institutes, and several universities have formed centers of excellence in the field. It is estimated that in 1997 BMBF supported programs on nanotechnology with a budget of approximately $50 million per year. Two of the largest upcoming projects are "CESAR," a $50 million science center in Bonn equally sponsored by the state and federal governments with about one-third of its research dedicated to nanoscience, and a new institute for carbon-reinforced materials near Karlsruhe ($4 million over 3 years, 1998-2001). BMBF is establishing five "centers of competence in nanotechnology" in Germany starting in 1998, with topics ranging from molecular architecture to ultraprecision manufacturing.

U.K.

A network program (LINK Nanotechnology Programme) was launched in the United Kingdom in 1988 with an annual budget of about $2 million per year. The Engineering and Physical Sciences Research Council (EPSRC) is funding materials science projects related to nanotechnology with a total value of about $7 million for a five-year interval (1994-1999). About $1 million is specifically earmarked for nanoparticle research. The National Physical Laboratory established a forum called the National Initiative on Nanotechnology (NION) for promoting nanotechnology in universities, industry, and government laboratories.

France

The Centre National de la Récherche Scientifique (CNRS) has developed research programs on nanoparticles and nanostructured materials at about 40 physics laboratories and 20 chemistry laboratories in France. Synthesis methods include molecular beam and cluster deposition, lithography, electrochemistry, soft chemistry, and biosynthesis. Nanotechnology activity has grown within a wide variety of research groups, including ones focused on molecular electronics, large gap semiconductors and nanomagnetism, catalysts, nanofilters, therapy problems, agrochemistry, and even cements for ductile nanoconcretes. It is estimated that CNRS spends about 2% of its budget and dedicates 500 researchers in 60 laboratories (or about $40 million per year) on projects related to nanoscience and nanotechnology. Companies collaborating to research and produce nanomaterials include Thompson, St. Gobain, Rhône Poulenc, Air Liquide, and IEMN. Also, there is the "French Club Nanotechnologie," aimed at promoting interactions in this field in France.

Sweden

The estimated total expenditure for research on nanotechnology in Sweden is ~ $10 million per year. There are four materials research consortia involved in this field: 1. Ångström Consortium in Uppsala, with a budget of ~ $0.8 million per year in 1998 for surface nanocoatings 2. Nanometer Structures Consortium in Lund with a budget of ~ $3.5 million per year partially supported by ESPRIT (~ $1 million per year) 3. Cluster-based and Ultrafine Particle Materials in the University of Uppsala and Royal Institute of Technology with a budget of ~ $0.4 million per year in 1998 4. Brinell Center at the Royal Institute of Technology

Switzerland

There is a Swiss national program on nanotechnology with a special strength in instrumentation. The most advanced research centers are focused on nanoprobes and molecule manipulation on surfaces (IBM Research Laboratory in Zürich), devices and sensors (Paul Scherrer Institute), nanoelectronics (ETH Zürich), and self-assembling on surfaces in patterns determined by the substrate or template (L'École Polytechnique Fédérale de Lausanne). See the site reports on all of these institutions.

The Netherlands

The most active research centers in the Netherlands are the DIMES institute at Delft University of Technology, which receives one-third support from industry, and the Philips Research Institute in Eindhoven, which researches self-assembling monolayers and patterning on metallic and silicon surfaces. The SST Netherlands Study Center for Technology Trends is completing a study on nanotechnology and aims at promoting increased funding and research networking in the Netherlands (ten Wolte 1997).

Finland

The Academy of Finland and the Finnish Technology Development Center began a three-year nanotechnology program in 1997. The program involves sixteen projects with funding of $9 million for a three-year period (1997-1999) for nanobiology, functional nanostructures, nanoelectronics, and other areas. Research on actuators and sensors is the Finnish area of strength.

Belgium and Spain

Since about 1993 both Belgium and Spain have established nanotechnology programs, centers of excellence, and university-industry interactions.

Multinational Efforts

Large multinational companies with significant nanotechnology research activities in Europe include IBM (Zürich), Philips, Siemens, Bayer, and Hitachi. Degussa Co., with headquarters in Germany, is a commercial supplier since 1940 of microparticles and, now, nanoparticles.

Western Europe has a variety of approaches to funding research on nanotechnology. These are discussed in detail in other studies. IPTS (the Science and Technology Forecast Institute) has conducted a study on nanotechnology research in the EC (Malsh 1997). Other European studies published recently include those by VDI (1996), UNIDO (1997), the U.K. Parliamentary Office of Science and Technology (1997), and NANO network (Fissan and Schoonman 1997). The overall expenditure for nanotechnology research within the EC was estimated in 1997 to be over $128 million per year.

Russia and Other FSU Countries

Support for generation of nanoparticles and nanostructured materials has a tradition in Russia and other countries of the former Soviet Union (FSU) dating back to the mid-1970s; before 1990 an important part of this support was connected to defense research. The first public paper concerning the special properties of nanostructures was published in Russia in 1976. In 1979 the Council of the Academy of Sciences created a section on "Ultra- Dispersed Systems." Research strengths are in the areas of preparation processes of nanostructured materials and in several basic scientific aspects. Metallurgical research for special metals, including those with nanocrystalline structures, has received particular attention; research for nanodevices has been relatively less developed. Due to funding limitations, characterization and utilization of nanoparticles and nanostructured materials requiring costly equipment are less advanced than processing.

Russian government funds are allocated mainly for research personnel and less for infrastructure. Funding for nanotechnology is channeled via the Ministry of Science and Technology, the Russian Foundation for Fundamental Research, the Academy of Sciences, the Ministry of Higher Education, and other ministries with specific targets. The Ministry of Higher Education has relatively little research funding. Overall, 2.8% of the civilian budget in Russia in 1997 was planned for allocation to science. There is no centralized program on nanotechnology; however, there are components in specific institutional programs. Currently, about 20% of science research in Russia is funded via international organizations. The significant level of interest in the FSU can be identified by the relatively large participation at a series of Russian conferences on nanotechnology, the first in 1984 (First USSR Conference on Physics and Chemistry of Ultradispersed Systems), a second in 1989, and a third in 1993.

The Ministry of Science and Technology contributes to nanotechnology through several of its specific programs related to solid-state physics, surface science, fullerenes and nanostructures, and particularly "electronic and optical properties of nanostructures." This last program involves a network of scientific centers: the Ioffe Institute in St. Petersburg, Lebedev Institute in Moscow, Moscow State University, Novgograd Institute of Microstructures, Novosibirsk Institute of Semiconductor Physics, and others. This research network has an annual meeting on nanostructures, physics, and technology, and has developed interactions with the PHANTOMS network in the EC. The U.S. Civilian Research and Development Foundation has provided research funds in the FSU for several projects related to nanotechnology, including "Highly Non-Equilibrium States and Processes in Nanomaterials" at the Ioffe Institute (1996-1998).

Russian government and international organizations are the primary research sponsors for nanotechnology in Russia. However, laboratories and companies privatized in the last few years, such as the Delta Research Institute in Moscow, are under development. With a relatively lower base in characterization and advanced computing, the research focus is on advanced processing and continuum modeling. Research strengths are in the fields of physico-chemistry, nanostructured materials, nanoparticle generation and processing methods, and applications for hard materials, purification, and the oil industry, and biologically active systems (Siegel et al. n.d.). There are related programs in Ukraine, Belarus, and Georgia, mostly under the direction of the respective academies of sciences in these countries, that are dedicated to crystalline nanostructures and advanced structural and nanoelectronic materials. Several innovative processes, such as diamond powder production by detonation synthesis at SINTA in Belarus, are not well known abroad.

Nanotechnology in the United States, Japan, and Western Europe is making progress in developing a suitable research infrastructure. The promise of nanotechnology is being realized through the confluence of advances in two fields: (1) scientific discovery that has enabled the atomic, molecular, and supramolecular control of material building blocks, and (2) manufacturing that provides the means to assemble and utilize these tailored building blocks for new processes and devices in a wide variety of applications. Technology programs cannot be developed without strong supporting science programs because of the scale and complexity of the nanosystems. The overlapping of discipline-oriented research with nanotechnology-targeted programs seems appropriate at this point in time. Highly interdisciplinary and multiapplication nanotechnology provides generic approaches that enable advances in other technologies, from dispersions, catalysts, and electronics to biomedicine. Essential trends include the following:

- learning from nature (including templating, self-assembly, multifunctionality)
- building up functional nanostructures from molecules
- convergence of miniaturization and assembly techniques
- novel materials by design
- use of hierarchical/adaptive simulations

A characteristic of discovery in nanotechnology is the potential for revolutionary steps. The question "what if?" is progressively replaced by "at what cost?" The road from basic research to applications may vary from a few months to decades. Research and development is expensive, and the field needs support from related areas. The R&D environment should favor multiapplication and international partnerships. Based on the data for 1996 and 1997 collected during this WTEC study, 1997 government expenditures for nanotechnology research were at similar absolute levels in the United States, Japan, and Western Europe. (Estimated OECD data for 1997 indicated GDPs of \$4.49 billion for Japan, \$7.76 billion for the United States, and \$7.00 billion for Western Europe.) The largest funding opportunities for nanotechnology are provided by NSF in the United States (approximately \$65 million per year for fundamental research), by MITI in Japan (approximately \$50 million per year for fundamental research and development), and by BMBF in Germany (approximately \$50 million per year for fundamental and applied research). Large companies in areas such as dispersions, electronics, multimedia, and bioengineering contribute to research to a larger extent in Japan and the United States than in Europe.

While multinational companies are pursuing nanotechnology research activities in almost all developed countries, the presence of an active group of small and medium-size companies introducing new processes to the market is limited to the United States. In the United States, individual and small-group researchers as well as industrial and national laboratories for specialized topics have established a strong position in synthesis and assembly of nanoscale building blocks and catalysts, and in polymeric and biological approaches to nanostructured materials. The Japanese large-group research institutes, and more recently academic laboratories, have made particular advances in nanodevices and nano-instrumentation. The European "mosaic" provides a diverse combination of university research, networks, and national laboratories with special performance in dispersion and coatings, nanobiotechnology, and nanoprobes.

With a relatively lower base in characterization and computing infrastructure, the research focus in Russia is on physico-chemistry phenomena, advanced processing, and continuum modeling. Interest and economic support, particularly for device-related research, is growing in China, Australia, India, Taiwan, Korea, and Singapore.

The pace of revolutionary discoveries that we are witnessing now in nanotechnology is expected to accelerate in the next decade worldwide. This will have a profound impact on existing and emerging technologies in almost all industry sectors, in conservation of materials and energy, in biomedicine, and in environmental sustainability.

Research Centers

- Institute for Molecular Manufacturing - nonprofit foundation carrying out research aimed at developing molecular manufacturing (molecular nanotechnology).

- Rice University - Center for Nanoscale Science and Technology - devoted to nurturing science and technology at the nanometer scale.
- Cornell NanoScale Science & Technology Facility (CNF) - subjects of research encompass physical sciences, life sciences, and engineering, particularly with inter-disciplinary emphasis.
- Cornell University - Nanobiotechnology Center (NBTC) - devoted to studying the convergence of nano-microfabrication and biosystems.
- University of Washington - Center for Nanotechnology University of Birmingham (United Kingdom) - Nanoscale Physics Research Laboratory - advancing the frontiers of the physics, chemistry, and technology of nanometre-scale structures, devices, and processes.
- University of Newcastle upon Tyne - Centre for Nanoscale Science and Technology - multidisciplinary research centre developing micro- and nano-devices for biomedical applications.
- University of California, Santa Barbara Nanotech - offers expertise in compound semiconductor-based device fabrication providing a full range of processes to the scientific and research communities.
- Center for Nanoscience and Nanotechnology - Georgia Institute of Technology - working to improve the human condition by providing fundamental and applicable knowledge to nanotechnology and its related fields of science.
- University of Michigan - Center for Biologic Nanotechnology - dedicated to the application of nano-molecular architecture to medicine.
- University of Notre Dame Nanofabrication Facility (NDNF) - provides information on the nanoelectronics fabrication facility.
- Arizona State University - Graduate Research Training Program in Biomolecular Nanotechnology - offers training in biomolecular nanotechnology, molecular electronics, and molecular devices.
- University of Nebraska-Lincoln - Center for Electro-Optics - specializes in optical/electromagnetic measurement and remote sensing technologies research.
- Schubert-Group: Macromolecular Chemistry and Nanoscience - Eindhoven University of Technology - devoted to the combinatorial materials research efforts.
- California Nanosystems Institute (CNSI) - R&D institute working on nanotechnologies and its commercial applications. Jointly established by University of California, Los Angeles, and University of California, Santa Barbara.

- Center for Nanotechnology - NASA Ames Research Center - focuses on experimental research and development in nano and bio technologies as well as computational nanotechnology, nanoelectronics, and optoelectronics.
- Institute for Soldier Nanotechnologies - intent to create a University Affiliated Research Center (UARC), with industry partners, to develop nanometer-scale science and technology solutions for the soldier.
- National Institute of Standards and Technology - Nanometer-Scale Metrology - research projects and measurement services in nanometer scale metrology

World-Wide Nanoelectronics Research

U.S. Government Labs

- Department of Defense (DOD) Laboratories

  DOD Labs conduct some of the most advanced and imaginative research in nanofabrication, nanoelectronics, and nanocomputing. Not all of this research is well represented on the World-Wide Web, however. THE Nanoelectronics & Nanocomputing Home Page has attempted to list some of this research, at least by title, on the Who's Who in Nanoelectronics page, as well as other pages. For further information on DOD Labs and an index of DOD technology departments and projects, see the TechTRANSIT Page. TechTRANSIT connects technology transfer resources and activities to meet the requirements of the Department of Defense Office of Technology Transition.

- Naval Research Laboratory (NRL)

  The Naval Research Laboratory (NRL) is the Navy's corporate research and development laboratory. NRL is the leader among Government Laboratories in nanotechnology and nanocomputing research. This includes very strong efforts in the application of proximal probes, modeling of nanosystems, novel approaches to nanofabrication, as well as other areas. However, only limited information about NRL's nanoelectronics activities is available on their WWW pages. For some information on nanoelectronics work at the NRL, see the NRL Electronics Science & Technology Division.

- Los Alamos National Laboratory (LANL).

  Investigators at LANL are deeply involved in the study of nanoelectronics and nanofabrication as part of the Theory, Modeling, and High Performance Computing core competency at the laboratory. However, only limited information about LANL's nanoelectronics activities is available on their WWW pages. Some of the LANL investigators are represented, however, on the • Who's Who in Nanoelectronics page.

U.S. Universities

- UCLA Nanoelectronics Lab
- The University of Cincinnati Nanoelectronics Lab

  The Nanoelectronics Laboratory is engaged in research at (or around) the nanometer scale in semiconductors and other electronic materials. Specific research areas include focused ion beam processing, integrated optoelectronic devices and waveguides, plasma assisted etching, and deposition.
- Cornell University Nanofabrication Facility (CNF)

  The Cornell Nanofabrication Facility (CNF)has served the US research community for more than 15 years by providing nanofabrication services to over 500 research projects. Subject areas of projects completed are as diverse as Astronomy to Plant Pathology, as well as the more conventional Micro-electronics, Physics and Materials Research, and emerging fields of Integrated Optics and Micro-Electro-Mechanical Systems(MEMS).
- Massachusetts Institute of Technology (MIT) NanoStructures Laboratory

  The NanoStructures Laboratory (NSL) at MIT (formerly the Submicron Structures Laboratory) develops techniques for fabricating surface structures with feature sizes in the range from nanometers to micrometers, and uses these structures in a variety of research projects.
- University of Minnesota Nanostructure Lab

  The Nanostructure Lab's goal is to develop new nanotechnologies for fabricating structures and devices substantially smaller than current technology permits, and to explore innovative nanoscale electronic, optoelectronic, and magnetic storage devices. Research at the lab is headed by Stephen Chou.
- Notre Dame Microelectronics Lab

  The Microelectronics Laboratory is equipped to fabricate integrated circuits and devices in both silicon and compound semiconductors with geometries as small as 0.01 microns (10 nm).
- Penn State Electronic Materials and Processing Research Laboratory (EMPRL)
- Purdue University Nanoscale Physics Laboratory

  This Laboratory is a part of an interdisciplinary Center of Excellence in nanotechnology that has been developed at Purdue 'University as part of a University Research Initiative funded via the Army Research Office. Among the accomplishments of the interdisciplinary, interdepartmental

Purdue team is the fabrication of arrays of functioning nanometer-scale circuit elements and wires using techniques of chemical self-assembly.

- Stanford University Nanofabrication Facility

  The Stanford Nanofabrication Research Facility is located in the Center for Integrated Systems. This state-of-the-art research laboratory consists of over $10M worth of modern processing equipment in a 10,000 sq.ft. clean room, extensive computing and networking capability, established processes for silicon, gallium arsenide and other material, a highly skilled technical staff and some of the most advanced research projects in the world.

U.S. Government Sponsored Research and Development Programs

- Ultra Dense, Ultra Fast Computing Components/Nanoelectronics (ULTRA) Program

  Sponsored by the Defense Advanced Research Projects Agency, the goals of the ULTRA Nanoelectronics program are to explore and develop material, processing technologies, quantum and conventional devices and device architectures for a next generation of information processing systems. The ULTRA program seeks improved speed, density, and functionality beyond that which can be achieved by scaling transistors.

- Nanoelectronics Modeling (NEMO) Program

  The U.S. Government's Project NEMO is developing an advanced software tool for modeling nanoelectronic devices.

U.S. Government Sponsoring Agencies

- Defense Advanced Research Projects Agency (DARPA)

  The Defense Advanced Research Projects Agency (DARPA)is the advanced research and development organization for the U.S. Department of Defense. More information on DARPA's nanoelectronics work is available at the DARPA Electronics Technology Office(ETO)Home Page. The DARPA ULTRA Program (for the development of Ultra-dense and Ultra-fast computers) is the premier nanoelectronics research program in the world.

- Air Force Office of Scientific Research (AFOSR)

  The Air Force Office of Scientific Research (AFOSR) is charged with directing the Air Force's basic research program, which includes extensive efforts in next-generation electronics and related technologies.

- Army Research Office (ARO)

The Army Research Office sponsors a diverse set of research and development projects in nanoelectronics. ARO is led by Dr. Gerry Iafrate, who was a very early supporter and investigator in the fields of quantum electronics and nanoelectronics.

- Office of Naval Research (ONR)

  The Office of Naval Research (ONR) coordinates, executes, and promotes the science and technology programs of the United States Navy and Marine Corps through universities, government laboratories, and nonprofit organizations. The ONR Electronics Division has a nanoscale electronics program. The objective of this program is to study the electronic properties of solid state materials and develop ultra small devices for microelectronic applications.

- The National Science Foundation (NSF)

  NSF's mission is to promote the progress of science; to advance the national health, prosperity, and welfare; and to secure the national defense. NSF funds research at U.S. universities in nanoelectronics, and it also has considerable involvement in the funding of DNA/biomolecular nanocomputation.

Federally Funded Research and Development Centers (FFRDCs)

- The Aerospace Corporation

  Aerospace operates a Federally Funded Research and Development Center serving as an architect-engineer for advanced space systems. Aerospace has been concerned with developing and applying nanoelectronics and nanotechnology to ease the manufacture and, especially, to reduce the size and weight of spacecraft. This could have dramatic impact in reducing launch costs, as well as in improving reliability. Aerospace has worked with NASA to sponsor several meetings on these topics, and they have published technical reports and proceedings in connection with these activities. To learn more about nanoelectronics at The Aerospace Corporation, see the Web pages for their Electronic Systems Division.

- MIT Lincoln Laboratory

  Dr. Gerhard Sollner of the MIT Lincoln Lab is a well-known investigator at the forefront of nanoelectronics research.

- The MITRE Corporation

  Nanoelectronics research at The MITRE Corporation is performed by the MITRE Nanosystems Group. The MITRE Corporation is a Federally Funded

Research and Development Center (FFRDC) for systems engineering and integration serving in the public interest. MITRE applies leading-edge technology to some of the most important systems in the world, including U.S. Military command and control systems and air traffic control systems, as well as similar systems around the world.

- The RAND Corporation

  RAND is a nonprofit institution that helps improve public policy through research and analysis. In 1995, RAND published a report on "The Potential of Nanotechnology for Molecular Manufacturing" by Max Nelson and Calvin Shipbaugh.

Private Foundations

- The Mayo Foundation/Mayo Clinic, Rochester, Minnesota

  Nanoelectronics research at the Mayo Foundation is focused in the Mayo Special Purpose Processor Development Group. This group is led by Dr. Barry Gilbert, who is one of the world's preeminent experts in the field of quantum-effect electronic devices nanoelectronics.

US Industry

- Hughes Electronics

  Hughes is engaged in nanoelectronics research. However, only limited information about Hughes' nanoelectronics activities is available on their corporate WWW pages.

- International Business Machines (IBM)

  IBM is one of the leaders world-wide in nanotechnology and nanoelectronics research. Some of their capabilities are on view in the • I B M Almaden Scanning Tunneling Microscope Image Gallery. This page contains spectacular images of atoms being manipulated with an STM.

Additional IBM nanotechnology research is performed at the IBM-Zurich Lab. One of the most important tools of nanotechnology, the scanning tunneling electron microscope (STM), the first • proximal probe device, was invented at IBM-Zurich. The Lab's Web site has a description of an IBM-Zurich molecular positioning experiment that uses an STM.

- Motorola, Inc.

  Motorola is among the leaders in nanoelectronics research. However, only limited information about Motorola's nanoelectronics activities is available on their corporate WWW pages.

- Texas Instruments, Inc. (TI)

  The Texas Instruments Nanoelectronics Group is one of the world's leaders in nanoelectronics research. However, only limited information about TI's nanoelectronics research activities is available on their corporate WWW pages. Several of the TI nanoelectronics investigators are listed, however, on the Who's Who in Nanoelectronics page.

Japanese Laboratories

- ERATO Single Quantum Dot Project

  Five year, $20 million interdisciplinary project to investigate the basic properties and fabrication of quantum dots. Begun in October 1995. Sponsored by Japan's ERATO (Exploratory Research for Advanced Technology) Initiative, under the auspices of Japan's Science and Technology Agency.
- Fasol Lab at the University of Tokyo

  The Fasol Lab home page has an in-depth nanotechnology index.
- Hitachi Corp. Research and Development, Tokyo, Japan

  Hitachi is engaged in an extensive nanoelectronics research effort, involving a number of preeminent scientists and engineers. However, only limited information about Hitachi's nanoelectronics activities is available on their corporate WWW pages.

European Laboratories

- Cavendish Laboratory, Cambridge University, Great Britain

  As noted by Dr. William Tolles in his 1994 report "Nanoscience and Nanotechnology in Europe," important research on the subject of nanoscience and nanoengineering is ongoing at the Cavendish Laboratory. However, little information about Cavendish's nanoelectronics activities is available on their laboratory WWW pages.
- Project on Nanometre Structures for Future Optoelectronic Applications sponsored by the European Union's Espirit Program for Information Technologies Research
- Project on Organic Materials for Electronics (and Nanostructures) sponsored by the European Union's Espirit Program for Information Technologies Research
- Project on Resonant Tunneling for Novel Electronic Applications sponsored by the European Union's Espirit Program for Information Technologies Research

- University of Glasgow Nanoelectronics Research Center, Glasgow, Scotland
- IBM-Zurich Laboratory, Zurich, Switzerland

  Much of the IBM's nanotechnology research is performed at the IBM-Zurich Lab. The Lab's Web site has a description of an IBM-Zurich molecular positioning experiment.

- National Physical Laboratory, Teddington, Great Britain

  As noted by Dr. William Tolles in his 1994 report "Nanoscience and Nanotechnology in Europe," important research on the subject of nanometrology is ongoing at the National Physical Laboratory. However, little information about the National Physical Laboratory's nanoelectronics activities is available on their laboratory WWW pages.

- University College London, UK

  Professors Terry Fountain and Michael Duff at the University College London are leading experts in the field of architectures for nanoelectronic computers. However, little information about the University College London's nanoelectronics activities is available on their university WWW pages.

Nanotechnology Research - Lab Groups

- Massachusetts Institute of Technoogy (MIT) - Microsystems Technology Laboratories - explores microsystems, microfabrication, and manufacturing technologies.
- Princeton University - NanoStructures Laboratory - explores and develops new nanotechnologies to fabricate structures; innovates nanoscale electronic, optoelectronic, and magnetic devices.
- New York University (NYU) - Seeman Lab - investigating unusual DNA molecules in model systems that use synthetic molecules, with a major effort in DNA nanotechnology.
- University of Cincinnati - Nanoelectronics Laboratory - research at or around the nanometer scale in semiconductors and other electronic materials.
- Academia Sinica - Surface Nanostructure Laboratory - creates, characterizes, and exploits nanostructures (1~50 nm) on solid surfaces.
- Waseda University - Ohdomari Lab - research on the process of nanostructure fabrication, and the physics of silicon surfaces and interfaces.
- University of Pune - Scanning Tunneling Microscopy Group - performs research on nanoparticles, nanolithography, and noise in STM.

NIH Nanotechnology and Nanoscience Information

This page contains information on (1) currently active NIH and BECON research and training opportunities and (2) listings of funded grants for NIH and BECON program announcements related to nanotechnology and nanoscience. Additional information on the NIH bioengineering funding opportunities can be found at http://www.becon.nih.gov/becon_funding.htm.

While much of biology is grounded in nanoscale phenomena, NIH has not reclassified most of its basic research portfolio as nanotechnology. Only those studies that use nanotechnology tools and concepts to study biology; that propose to engineer biological molecules toward functions very different from those they have in nature; or that manipulate biological systems by methods more precise than can be done by using molecular biological, synthetic chemical, or biochemical approaches that have been used for years in the biology research community, are classified as nanotechnology projects. NIH participates in the National Nanotechnology Initiative (www.nano.gov), the U.S. federal multi-agency R&D program for nanoscale science, engineering, and technology.

A list of technical and financial contacts for Nanoscience and Nanotechnology related projects is given at: http://www.becon.nih.gov/nano_contacts.htm.

A list of technical and financial contacts for PA-02-125, Bioengineering Nanotechnology Initiative (SBIR) is given at: http://www.becon.nih.gov/nano_contacts_sbir.htm.

Nanomedicine NIH Roadmap Initiatives:

- Nanomedicine Roadmap Homepage

NIH Reports Related to Nanotechnology:

- NIBIB/DOE Workshop on Biomedical Applications of Nanotechnology, March 17-18, 2005
- Nanoscience and Nanotechnology Symposium - June 25-26, 2000
- NHLBI Nanotechnology Working Group Report
- NCI Cancer Nanotechnology Brochure
- Recent Slideshow on NIH Nanotechnology Funding Opportunities/Examples

Other Trans-NIH Initiatives:

- Nanoscience and Nanotechnology in Biology and Medicine
- Bioengineering Nanotechnology Initiative
- Bioengineering Research Partnerships
- Exploratory/Developmental (R21) Bioengineering Research Grants

National Heart Lung and Blood Institute

- Innovative Technologies for Engineering Small Blood Vessels
- NHLBI Programs of Excellence in Nanotechnology

National Cancer Institute

- NCI Alliance for Nanotechnology in Cancer (including current funding opportunities)
- Innovative Technologies for Molecular Analysis of Cancer
- Application of Emerging Technologies for Cancer Research

National Institute of General Medical Sciences

- Single Molecule Biophysics and Nanoscience - Research on, and development of, new and improved instruments, methods, and technologies for nanoscience, and for the analysis of single protein and nucleic acid molecules and their complexes in vivo and in vitro. Current approaches include optical and fluorescent spectroscopies, scanning probe microscopy, and biomechanical techniques to analyze the behavior and heterogeneity of single molecules and subcellular structures at the nanometer scale. Examples of targets for study include protein or RNA folding, enzyme catalysis, signaling, molecular machines, and the assembly and dynamics of complex cellular structures. A major goal is to develop and enhance existing methods and reagents for the 3-D visualization of cellular processes in living cells in real time at high resolution.

National Toxicology Program (NTP)

Three agencies form the core of the NTP:

- National Institute of Environmental Health Sciences of the National Institutes of Health (NIEHS/NIH)
- National Institute for Occupational Safety and Health of the Centers for Disease Control and Prevention (NIOSH/CDC)
- National Center for Toxicological Research of the Food and Drug Administration (NCTR/FDA)
- The NTP is conducting studies on the toxicology of nanoscale materials.
- Report of the first nanotoxicology workshop http://www.nanotoxicology.ufl.edu/ aimed at developing appropriate experimental strategies for evaluating the safety/toxicity of nanomaterials

Additional NIH Nanoscience and Nanotechnology Research Funding Opportunities:

- DEVELOPMENT OF CELL-SELECTIVE TOOLS FOR STUDIES OF THE BLADDER, PROSTATE, AND GENITOURINARY TRACT - Application Receipt Dates: February 1, 2005
- NOVEL APPROACHES TO CORNEAL TISSUE ENGINEERING - Application Receipt Dates: February 1, June 1, October 1, ending June 1, 2005
- NINDS EXPLORATORY/DEVELOPMENTAL PROJECTS IN TRANSLATIONAL RESEARCH - Application Receipt Dates: February 1, June 1, October 1, ending July 31, 2005
- SMALL BUSINESS BIODEFENSE PROGRAM - Application Receipt Dates: April 1, August 1, December 1, ending Aug. 2, 2005
- CUTTING-EDGE BASIC RESEARCH AWARDS (CEBRA) - Application Receipt Dates: February1, June 1, October 1, ending Oct. 5, 2005
- NEUROTECHNOLOGY RESEARCH, DEVELOPMENT, AND ENHANCEMENT - Application Receipt Dates: February 1, June 1, October 1 ending October 2006

Information on Funded Grants and Projects for NIH/BECON Program Announcements:

- Summary of Funded NIH Bioengineering Nanotechnology Initiative (SBIR)
- Summary of Funded NIH Bioengineering Nanotechnology Initiative (SBIR)
- Summary of Funded NIH Bioengineering Nanotechnology Initiative (SBIR)
- Summary of Funded NIH Bioengineering Nanotechnology Initiative (SBIR)
- Summary of Funded NIH Bioengineering Nanotechnology Initiative (SBIR)
- Summary of Funded Nanoscience and Nanotechnology in Biology and Medicine (PAR-03-045)

NSF - NSES

The potential impact of research on the nanoscale is well recognized, and yet to remain nationally and internationally competitive formidable challenges must be met. Through a unique partnership of academic institutions, industrial partners, and national labs, the NSF Nanoscale Science and Engineering Center (NSEC) for Integrated Nanopatterning and Detection Technologies, headquartered at Northwestern University under the leadership of Chad A. Mirkin, is helping to meet these challenges by providing a cross-disciplinary research program well integrated with innovative educational outreach programs.

The NU-NSEC brings together recognized leaders from Northwestern University, University of Chicago, University of Illinois/Urbana-Champaign, and Argonne National Laboratory, who, building on the nanopatterning tools and synthetic methods unique to this Center, are making advances in the development of advanced nanopatterning techniques and nanoscale sensors. Potential applications in disease and biological/ chemical detection systems arena are expected to be profound. However, this research will also directly impact new technological directions outside of the existing purview of the Center including molecular electronics, catalysis, information storage, and therapeutics.

In a field as dynamic as nanotechnology, collaborations are a vital means to transfer knowledge and an important component of the Center mission. NSEC researchers have numerous meaningful collaborations with industry, national laboratories, and academia. Additionally, the Nanotechnology Corporate Partners (NCP) program provides a forum for companies and business groups with allied interests to develop partnerships with the NSEC and help transition its technology into the private sector. Working with the Kellogg School of Management, the NSEC also offers the Small Business Evaluation and Entrepreneurs (SBEE) program, which gives scientists and engineers with new technologies assistance with development of comprehensive business plans for presentation to potential investors. A $10 million seed fund was established by Lurie Investment Fund to initiate business development based on technology developed at the Center, and three companies have already been successfully launched.

Excellent in-house user facilities are available to NSEC researchers. Additionally, the State of Illinois committed $5,000,000 to support the development of next-generation instrumentation and to establish a multi-user, multi-purpose, open-access Nanoscale Imaging, Fabrication, Testing and Instrumentation (NIFTI) facility to provide soft lithographic writing, imaging, and analysis capabilities.

The NU-NSEC is also committed to educating the scientists, technicians, and teachers of tomorrow. This commitment is realized not only in the research of graduate students, undergraduates, and postdoctoral associates, but also through the Center's innovative *educational programs*. Some of these programs build on current educational strengths and infrastructure, while others launch new initiatives both internally and through exciting new partnerships with the Museum of Science and Industry, Chicago, and Harold Washington College (a primarily minority institution).

Undergraduates and minority undergraduates from around the country have opportunities to engage in research at the Center. High school science teachers, many from schools with large minority populations, participate in an innovative program which combines hands-on research with curriculum development. NSEC researchers are collaborating with pre-college educators on the development of a "Nanoscience

Module" to introduce nanoscience concepts at the high school level through guided hands-on activities and design challenges. The NSEC is also collaborating with the Museum of Science & Industry, Chicago, to produce a design concept for a 13,000 square foot exhibit in nanotechnology, which is expected to open in 2006 and educate millions of museum visitors each year. An interactive web site for children entitled "Mission Spec-tacular" will be launched in spring 2003, and a series of dramatic educational demonstrations entitled "Frontiers of Nanotechnology" has already been premiered with pre-college and lay audiences. NSEC researchers and their students are also actively involved and highly visible in the broader community, through lectures, radio, and television.

Through synergistic research of the highest quality, strong and active collaborations, and innovative outreach programs, the NU-NSEC is advancing scientific knowledge and understanding in the area of integrated nanopatterning and nanoscale detection technologies, while advancing formal education as well as public understanding of nanoscale science and engineering.

Nanoscience and Technology at US - DOE Labs

The Role of DOE in the National Nanotechnology Initiative:

"The DOE has a stunning portfolio of research and scientific user facilities devoted to visualizing, characterizing, and controlling the nanoworld - from atoms and molecules to bulk materials - which makes the Department's research capabilities unique in the world. The DOE is currently making a broad range of contributions in these areas. For example, the enhanced properties of nanocrystals for novel catalysts, tailored light emission and propagation, nanocomposites and supercapacitors are all being explored. Nanocrystals and layered structures offer unique opportunities for tailoring the optical, magnetic, electronic, mechanical and chemical properties of materials, and DOE researchers are have synthesized layered structures for electronics, novel magnets, and surfaces with tailored hardness.

Specific examples of past accomplishments at DOE include:

- Addition of aluminum oxide nanoparticles that converts aluminum metal into a material with wear resistance equal to that of the best bearing steel
- Novel optical properties of semiconducting nanocrystals that are used to label and track molecular processes in living cells
- Nanoscale layered materials that can yield a four-fold increase in the performance of permanent magnets
- Layered quantum well structures to produce highly efficient, low-power light sources and photovoltaic cells

- Novel chemical properties of nanocrystals that show promise as photocatalysts to speed the breakdown of toxic wastes
- Meso-porous inorganic hosts with self-assembled organic monolayers that are used to trap and remove heavy metals from the environment

The DOE also maintains a large array of major national user facilities that are ideally suited to nanoscience discovery and to developing a fundamental understanding of nanoscale processes. Large computational facilities at DOE will also be key contributors in nanoscience discovery, modeling and understanding.

National Nanotechnology Initiative Research at the DOE:

Major new efforts in nanoscale science, engineering, and technology at the Department of Energy will take advantage of opportunities afforded by recent advances. These efforts will be part of the Basic Energy Sciences (BES) program and have the following broad goals:

1. to attain a fundamental scientific understanding of nanoscale phenomena, particularly collective phenomena;
2. to achieve the ability to design and synthesize materials at the atomic level to produce materials with desired properties and functions;
3. to attain a fundamental understanding of the processes by which living organisms create materials and functional complexes to serve as a guide and a benchmark by which to measure our progress in synthetic design and synthesis; and
4. to develop experimental characterization tools and theory/modeling/ simulation tools necessary to drive the nanoscale revolution.

The principal missions of DOE in science, energy, defense, and environment will benefit greatly from developments in these areas. For example, nanoscale synthesis and assembly methods will result in significant improvements in solar energy conversion; more energy-efficient lighting; stronger, lighter materials that will improve efficiency in transportation; greatly improved chemical and biological sensing; use of low-energy chemical pathways to break down toxic substances for environmental remediation and restoration; and better sensors and controls to increase efficiency in manufacturing."

Sandia National Labs

It's no wonder that venturing into the tiny domain of atoms and molecules is catching on. Amazing discoveries are being made, inspiring scientists the world over.

Throughout the scientific community, including Sandia National Laboratories, researchers say building things atom-by-atom or molecule-by-molecule will revolutionize the production of virtually every human-made object. *Exciting* prospects—but they also point out the promise of nanotechnology can *only* be realized if we learn to understand the special rules that control behavior at this small scale and develop the skill needed to integrate these concepts into practical devices.

The excitement stems from the understanding that the behavior of materials at the nanoscale is *nothing* like that at the large scale. Only recently have the necessary tools, such as powerful new microscopes, been developed to let researchers see these surprising behaviors. Sandia National Labs' scientists and engineers are among the leading architects and builders of these tools. To appreciate their unique design features, it helps to get an idea of the size of things at the atomic level. In nanoscience objects are measured in nanometers, 1 billionth of a meter. For comparison, the smallest features on current computer chips, measure about 200 nanometers. And a human hair is 100,000 nanometers thick.

Moving beyond observation, scientists are now poised to make exciting advances in nanotechnology, the creation of materials, devices and systems through the *control* of matter at the atomic level. By understanding and controlling the way molecules organize into nanoscale patterns, scientists are discovering new phenomena and learning to design materials with vastly different sets of properties. As one Sandia scientist put it "Design possibilities are limited only by one's imagination."

Sandia Labs continues to enhance its proficiency in many fields, all in keeping with its Department of Energy mission to unite science and engineering to serve national needs. And along with world class capabilities in materials science, micro fabrication - including 40,000 square feet of clean room space - high performance computing and systems engineering - Sandia is uniquely positioned to be the integrating center for new discoveries in nanoscience.

Sandia is uniquely positioned to be the intergrating center for new discoveries in nanoscience. Nano materials are already helping to attain scientist's vision for new technology such as, the *micro-chemlab*, by increasing a thousandfold the ability to collect the chemicals of interest. Currently in the development stage, this handheld device will carry out all of the functions of a full chemistry laboratory allowing first responders, military, and emergency personnel to rapidly access chemical and biological hazards.

All of these advances are coming from the work of multi-disciplanary teams. Their ability to bring together and integrate micro-scale devices is another key part of the nanotechnology success story Through strong commitment and continued partnerships with other national labs, universities and industry. Sandia scientists believe the promise of nanotechnology applications that touch peoples lives can be realized. Grand prospects based on tiny environments. It will require thinking big in a nano sized world.!

U.S. Naval Research Lab: Nanotechnology Science and Technology

Strategic Research Area

- To achieve dramatic, innovative enhancements in the properties and performance of structures, materials, and devices that have controllable features on the nanometer scale (i.e., tens of angstroms).
- The ability to affordably fabricate structures at the nanometer scale will enable new approaches and processes for manufacturing novel, more reliable, lower cost, higher performance and more flexible electronic, magnetic, optical, and mechanical devices.

Recognized Applications of Nanoscience

- Ultrasmall, highly parallel, computers with multi-teraflop speed
- Image information processors, e.g., extraction and recognition
- Low-power personal and autonomous communication and computation devices
- high-density information storage devices, e.g., terabit/cm2 nonvolatile memory
- Lasers and detectors for weapons and countermeasures
- Optical (infrared, visible, ultraviolet) sensors for improved surveillance an targeting
- Integrated sensor suites for chemical and biological agent detection
- Catalysts for enhancing and controlling energetic reactions
- Synthesis of new compounds (e.g., narrow-bandgap materials)
- Designer materials with combinations of properties that do not currently exist

SRA Representatives

Air Force: Dr. Gernot S. Pomrenke

Army: Dr. Henry Everitt

DARPA: Dr. Christie Marrian

Navy: Dr. James Murday (Chair)

US Office of Naval Research

The Surface Nanoscience and Sensor Technology section of the Naval Research Laboratory specializes in (1) basic and applied research in surface science, the mechanical properties of materials, the molecular basis of adhesion, film nucleation and growth, and chemical/biological sensors; (2) development of tip-based probes for studies in the area of nanostructures and nanomechanics of surfaces and single molecules; and (3) development of new surface/interface analysis techniques for structural and chemical analysis of materials.

Current Areas of Research

Structure of semiconductor surfaces and interfaces

We are studying the atomic-scale structure and reactivity of single crystal silicon, germanium, and III-V compound semiconductor surfaces and interfaces in ultra-high vacuum (UHV) using scanning tunneling microscopy (STM) and other surface analysis techniques. Due to the rapidly shrinking size of electronic devices, such studies are vital to the development of future Navy electronics.

Nanomechanics of surfaces

The nanomechanical properties of materials has become an area of increasing importance as the scale of coatings, multilayer films, particles, components and devices approach atomic dimensions. A goal of our nanomechanics program is to replace the phenomenological understanding of adhesion, tribology and interfacial/fracture mechanics with specific atomic- or molecular-scale mechanisms. Our research objectives include measuring the intermolecular and surface forces of materials with molecular specificity and resolution; elucidating the mechanics of materials or species under compression, tension, or shear; and developing new science-based molecular models for adhesion and tribology. This program also focuses on the relationship between interfacial chemistry and the structural and mechanical properties of interfaces by applying state-of-the-art nanoprobe technology including atomic force microscopy (AFM) and nanoindenation techniques.

Interaction forces of biomolecules

Atomic force microscopy (AFM) offers the unique ability to perform nanomechanical measurements on individual biological molecules. We have developed this ability and are using it to investigate inter- and intra-molecular interaction forces in model systems, between complementary strands of DNA, and between antibodies and their antigens. Molecular modeling studies are carried out to provide insights into the bond rupture mechanisms that underly these experiments.

FABS / BARC / FDA biosensors

Building on our research into the interaction forces between biomolecules, we are developing sensors for biomolecules and other chemicals, viruses, and bacteria. These sensors use microfabricated transducers to detect the minute forces with which antibodies bind to particular molecules. This research will result in the development of portable, highly-sensitive, automated chemical and biological sensors. The Force Differentiation Assay (FDA) and the Force Amplified Biological Sensor (FABS) are expected to deliver orders-of-magnitude sensitivity improvements over methods that are now in common use. The Bead Array Counter (BARC) will operate on similar principles but will trade sensitivity for the ability to simultaneously detect hundreds of types of molecules.

Quantitative surface analysis

The surface analytical techniques of x-ray photoelectron spectroscopy (XPS) and Auger electron spectroscopy (AES) are used to quantitatively determine the surface composition or in-depth distribution of low-level dopants in a variety of advanced materials including semiconductors, polymers, diamond films, and engineering alloys. These approaches are employed also for tribology, surface cleaning, and insulation degradation studies. Experimental and theoretical methods have been developed to overcome problems caused by matrix effects, surface segregation, background, peak overlap, specimen charging, sputtering artifacts, etc. Auger lineshapes and factor analysis methods are used extensively to extract chemical information.

Langmuir-Blodgett and self-assembled films

Research is conducted on physicochemical properties and effects of surface-active organic compounds that 1) form organized monomolecular films and 2) can be transferred to solid surfaces as mono- or multilayer films with precisely controlled properties such as packing density and phase state. Such films are also used in fundamental research studies as reference standards for scanning probe microscopes; as model systems to help understand the physical and chemical properties of biological membranes; and as model systems to increase understanding of the molecular basis of adhesion and lubrication. They have also been used in practical applications that take advantage of the unique properties of organized thin films, for example, as rapidly-equilibrating active coatings for gas sensors. Properties and effects of films that occur naturally on the sea surface have also been studied.

Facilities

Scanning probe microscopes

Scanning tunneling microscopy/spectroscopy (STM/S) enables the surface topography, chemical reactivity, and electronic structure of conductive substrates to be observed with atomic-scale resolution. Atomic force microscopy (AFM) can provide nanometer-scale resolution of surface topography, mechanical properties, and tip-surface interaction forces on both conductive and insulating substrates. The tip-surface interaction forces, including frictional forces, can be measured with nanonewton (single chemical bond) resolution.

- Homebuilt ultra-high-vacuum (UHV) STM/S with facilities for low-energy electron diffraction (LEED) and Auger electron spectroscopy (AES), for characterizing atomic-scale surface properties;

- Homebuilt STM/S with fast, variable-temperature STM and facilities for LEED, AES, temperature programmed desorption, for real-time atomic-scale observations of surface reactions;
- Park Scientific Instruments AutoProbe UHV STM/AFM integrated with the NRL Molecular Beam Epitaxy (MBE) Epicenter for characterizing semiconductor surfaces following MBE and related processing;
- RHK Technology STM 2000 UHV STM/S;
- Homebuilt ambient STM/S with capability of generating high-frequency surface acoustic waves (SAW) for measurement of elastic properties of surfaces;
- Nanoscope IIIa multimode AFM (lateral force, magnetic force, and tapping modes) equipped with breakout box and Hysitron TriboScope nanoindentation instrument for mechanical property measurements;
- Nanoscope IIIa multimode AFM (lateral force, magnetic force, and tapping modes) equipped with breakout box and force-volume mapping system;
- Homebuilt ambient AFM for adhesion and tribology studies;
- Homebuilt liquid-cell AFM for molecular recognition studies;
- Topometrix TMX 2000 AFM, belonging to code 6174, but housed within code 6177.

Surface analytical techniques

These techniques are used to characterize the surfaces of solids with respect to their electronic and geometric structure, mass, molecular weight, or optical properties. The techniques include X-ray photoelectron spectroscopy (XPS) and Auger electron spectroscopy (AES). Both approaches carry information about the electronic structure of the solid and its chemical composition. In addition, a rastered electron beam (AES) or parallel imaging (XPS) can produce images of chemical species.

- Perkin-Elmer 660 50 nm scanning Auger microprobe (SAM) with sputter depth profiling capability and X-ray dispersive analyzer, for analyzing thin films and determining the lateral or in-depth distribution by depth profiling of elements at the <0.5 atomic percent level.
- Surface Science Instruments ESCA301 150 μm monochromatic XPS with depth profiling capability and angle-resolved analysis.
- VG Scientific model 220i-XL XPS with monochromatic and non-monochromatic X-ray sources, depth profiling capability, small area analysis (20 μm) and <5 μm spatial resolution parallel image capability.

- JA Woollan M44 410-750 nm ex situ spectroscopic ellipsometer with ambient and liquid cells for film thickness, surface roughness, optical constant, and composition measurements.

Chemistry laboratories

These facilities are primarily intended for the preparation and characterization of 1) antibody or DNA-derivatized surfaces and 2) Langmuir-Blodgett and self-assembled films.

- Biochemistry and molecular biology facilities with -70° C freezer, UV/vis spectrophotometer, PCR cycler, gel box, microplate reader, incubator, glove box, centrifuge, etc.
- Zeiss Axiovert and Axiotech reflected- and transmitted-light optical microscopes with epifluorescence and differential interference contrast (DIC) attachments; equipped with Narashige three-dimensional hydraulic micromanipulator and video microscopy/digital video capture system.
- Dedicated surface chemistry clean-room laboratory with KSV 5000 Langmuir-Blodgett System.

Magnetics equipment

The following equipment for generating and measuring magnetic fields is part of the FABS/BARC/FDA program.

- Group 3 DTM-141 Digital Teslameter with serial communications and MPT-141 miniature Hall probe.
- Walker Scientific 40 kOe pulsed magnetizing coil with 1 inch bore.

Scanning electron microscopy, Fourier transform IR and Raman spectroscopy, solid state NMR, X-ray fluorescence, Rutherford backscattering, MS/MS, and other supporting techniques are also available within the Chemistry Division.

W.M. Keck Nanostructures Laboratory - University of Massachesetts, Ahmerst

The W.M. Keck Nanostructures Laboratory is a part of the Materials Research Science and Engineering Center (MRSEC) and the Silvio O. Conte National Center for Polymer Research at the University of Massachusetts at Amherst. It is equipped with several Atomic Force Microscopes, Optical Microscopes, Variable Angle Spectroscopic Ellipsometer (VASE), Thin Film Optical Profilometer, Stylus Profilometer, Metal Evaporator etc. for the use of UMass community and as well as from other academic institutions and Industry. The mission of the Keck Nanostructures Laboratory is to provide access to material characterization equipment, technical support, training and consultation, as well as to perform a range of services for users in the area of Atomic

Force Microscopy (AFM), Small Angle X-ray Scattering (SAXS), Variable Angle Spectrocopic Elliposmetry (VASE) and Optical Microscopy. The laboratory capabilities include high-resolution imaging of materials structure, analysis of X-ray Diffraction patterns, Thin Film characterization etc. In addition the Nanostructures Laboratory has access to the Confocal Laser Scanning Microscope from the Optical Microscope facility of the MRSEC

NanoTech User Facility at the UW (NTUF)

Training

Welcome to the NanoTech User Facility (NTUF). NTUF is available to graduate students and industrial researchers as a cost center. NTUF staffs will provide training on instruments upon request. Please register online and we will contact you to schedule training sessions.

Training consists of two sessions. First, we will provide an in-depth introduction on the principles of the instrument and demonstrate capabilities by running standard samples. The second training session will be hands-on and users are encouraged to work with their own samples. We may request additional practice sessions depending on qualification on the use of equipment.

For qualified users who can perform independent experiments and who have agreed with NTUF policy, NTUF is available 24 hours a day and seven days a week.

- Zeiss Confocal Microscope Guided Tour
- Principles of Confocal LSM
- Beam Path Configuration Guide
- Zeiss LSM 510 Operating Procedure
- Volocity Users Guide
- Digital Multimode III Operating Procedure
- Nanoscope III User Manual (© Digital Instruments)
- Nanoscope III schematic (© Digital Instruments)
- Leica Operating Procedure
- Quick Procedure for Raman Spectroscopy
- Raman system schematic
- FEI Sirion Operating Procedure
- E-beam Lithography Training Procedure I
- E-beam Lithography Training Procedure II

- E-beam Lithography Introduction
- E-beam Resist (PMMA) Data Sheet (© MicroChem)
- NTUF E-Beam Resist Information
- NTUF Master Template I
- NTUF Master Template II

Charles B. Musgrave

Quantum Simulations of Molecular Processes in Chemical Engineering

Our research program focuses on using quantum mechanics to simulate molecular processes of important engineering problems. Our approach is fundamental and interdisciplinary and combines quantum mechanics, chemical kinetics, surface chemistry and materials chemistry. Our goals are to develop a fundamental molecular and mechanistic understanding of the processes underlying important new technologies and to establish quantum chemical simulations as a practical engineering tool for computational prototyping of systems with molecular detail. Systems we investigate include atomic layer deposition, dye sensitized nanostructured solar cells, solid oxide fuel cells, hybrid interfaces between semiconductors and organic and biological molecules, molecule-surface electrical junctions and nanotechnology.

Atomic Layer Deposition

ALD is a process that has recently gained enormous attention. It involves exposure of surfaces to alternating pulses of different precursors to form atomic layers of deposited material through self-limiting surface reactions. Currently, the most active area of ALD research is the deposition of high-K gate dielectrics, such as ZrO 2 and HfO 2 because finding a replacement for SiO 2 gate dielectrics is the most critical technical challenge facing the semiconductor industry in the current transition to nanoelectronic devices. Although research on ALD of high-K materials has recently exploded, no detailed mechanisms or first principles studies had been reported until we reported detailed mechanisms for ALD of several important materials including ZrO 2, HfO 2,, HfN, Al 2 O 3, SiO 2 and Si 3 N 4. In addition to being the first group to describe detailed mechanisms of ALD, we are creating the fundamental framework for the understanding of the ALD process which provides a based upon which ALD can be extended and improved. We are also exploring reactions for low-temperature ALD on organics and self-assembled monolayers for area-selective ALD and molecular electronics, ALD for fuel cells, and dye sensitized solar cells.

Dye-Sensitized Solar Cells

The heart of dye-sensitized solar cells is the dye-molecule bound to a metal oxide nanostructured surface who's primary function is to covert photons to electron hole pairs in order to harvest solar energy. The goal of our work on dye-sensitized solar cells (DSC) is to develop new molecular technologies to control the key interfacial properties of DSC with the aim of improving their efficiency, and reducing their cost and the energy required for their manufacture. Our strategy is to use quantum chemical simulations and carrier transport simulations to perform molecular engineering of the DSC interface structures by computationally prototyping both the interfacial physics which governs the operation of the DSC as well as the chemical processes involved in their fabrication. The DSC has great potential for economically harvesting significant amounts of the solar energy incident upon the earth's surface without the production of green house gases. Furthermore, despite being early in their development cycle relative to traditional photovoltaics, much progress has been made in both the fabrication and efficiency of the DSC with efficiencies exceeding 10%.

Our view of DSC is that the surface-tethered dye is an optoelectronic molecular device whose function critically depends on the quantum electronic structure. From our previous work in organic functionalization of semiconductors surfaces and ALD of metal oxides we have become one of the world's leading groups in the electronic structure of organic functionalized surfaces as well as in the electronic structure of organometallic molecules on metal oxide surfaces and we leverage this expertise to enable us to expand the theory and understanding of the DSC with the hope that this will lead to improvements and subsequent faster implementation of DSC technology.

Low Temperature Solid Oxide Fuel Cells

Fuel cells have the potential to play an immensely important role in a potential hydrogen economy. Solid oxide fuel cells (SOFC) use solid electrolyte gates and have many advantages over polymer gate systems. The main disadvantage is that they must be operated at very high temperatures (1000-1200 C). Our goal is to simulate oxygen ion migration through the electrolyte to determine the migration mechanism and to design and computationally prototype new electrolytes with higher ion conductivities with the goal of reducing the operating temperature of the fuel cell by using dislocations, and built-in chemical potential gradients, and ultra thin membranes fabricated using ALD.

Hybrid Bio and Organic Semiconductor Devices

Semiconductors are at the center of many key technologies including computation, communication, data storage and sensing. Inorganic interfaces in semiconductor based devices have been key components in the unique properties that these devices exploit for their operation. We have undertaken an extensive study of how various organic molecules react on semiconductor surfaces and developed the most comprehensive theoretical description of organic reactions on semiconductors. We are extending this leadership position in two ways. First, by moving from studying the reactivity of organics on semiconductors to investigating the unique electronic properties of the interfaces formed between various organic functional groups and semiconductor surfaces.

Second, by extending the types of organic molecules on semiconductors we study to biological molecules, such as amino acids. We have recently found several interesting new types of reactions between biological molecules and semiconductor surfaces which are not only unique in their reactivity, but which also exhibit strong quantum mechanical coupling between the biological molecule and the semiconductor, potentially leading to new types of electrical junctions in molecular electronics. We are also exploring the use of this phenomenon to electrically connect carbon nanotubes to each other.

Molecular Electronics

There has been significant recent interest in using molecules as components in electronic circuits. Critical to this area is the ability to transport electrons between molecules and bulk materials across the interface formed between the molecule and the material surface. This is mostly governed by the electronic structure of the molecule-material interface. Several of our projects involve charge transfer between molecules and surfaces including dye-sensitized solar cells, the oxidation and reduction in fuel cells and our quantum resonance junctions between certain amino acids and Si and Ge surfaces. Our strategy is to use quantum simulations to develop a clear understanding of the charge transfer process so that we may engineer molecules to optimize them as electronic junctions.

Theoretical Studies of a Hydrogen Abstraction Tool for Nanotechnology

Processes that use mechanical positioning of reactive species to control chemical reactions by either providing activation energy or selecting between alternative pathways will allow us to construct a wide range of complex molecular structures. An example of such a process is the abstraction of hydrogen from diamond surfaces by a radical species attached to a mechanical positioning device for synthesis of atomically precise diamond-like structures. In the design of a nanoscale, site-specific hydrogen abstraction tool, we suggest the use of an alkynyl radical tip. Using *ab initio* quantum-chemistry techniques including electron correlation we model the abstraction of hydrogen from dihydrogen, methane, acetylene, benzene and isobutane by the acetylene radical. Of these systems, isobutane serves as a good model of the diamond (111) surface. By conservative estimates, the abstraction barrier is small (less than 7.7 kcal mol^{-1}) in all cases except for acetylene and zero in the case of isobutane. Thermal vibrations at room temperature should be sufficient to supply the small activation energy. Several methods of creating the radical in a controlled vacuum setting should be feasible. Thermal, mechanical, optical and chemical energy sources could all be used either to activate a precursor, which could be used once and thrown away, or alternatively to remove the hydrogen from the tip, thus refreshing the abstraction tool for a second use. We show how nanofabrication processes can be accurately and inexpensively designed in a computational framework.

Tim Harper

Interview with Tim Harper

Founder & CEO of the nanotechnology integrator CMP Cientifica, Tim is a former European Space Agency engineer, who now runs conferences and information services and works closely with the VC industry to help select and guide investments in nanotechnology companies. Tim is co-author of the *Nanotechnology Opportunity Report* and joint editor in chief of *The X Report*. A brief biography of Tim Harper is available on the CMP website.

Steve: What's the story behind CMP-Cientifica, Trends in NanoTechnology (TNT) and your new collaboration agreement with Spark Inversiones?

Tim: What we are trying to achieve with CMP, TNT and Spark are three separate activities, although they are all involved in nanotechnology. In essence they are the technologies, the information and the funding. I should make it clear that CMP Cientifica and Spark Inversiones are two very separate companies that are collaborating in the field of nanotechnology, while the TNT label covers all of CMP Cientifica's information services, whether conferences or publications.

CMP Cientifica has been involved in nanofabrication for a long time, since I was at ESA in fact. Over the last four years we have worked with both clients & suppliers to develop tools and techniques with nanometer precision. Unfortunately, most of our work is covered by NDA's so I can't tell you as much as I would like to.

CMP Cientifica also coordinates the Phantoms network, an EU funded network on Nanoelectronics and NanoSpain, a more local network. Our experience shows us that the best way to advance technology is to bring together a wide range of experts, and get them to talk to each other. That's where the problems get solved, and it helps reduce duplication of effort. The TNT conferences are a great enabler for cooperation, and we have an excellent relationship with the NSF in the US, that allows us to get a truly global perspective on nanotechnology.

TNT Weekly was born out of the realization that most people outside the scientific community had almost no idea what nanotechnology is. In most people's minds, nanotechnology is tangled up with miniature robots, grey goo and terraforming Mars. You may think that the science fiction associations don't matter, but if you are looking for funding for a nanotechnology related business, it means you'll have a tough time convincing anyone to invest. VCs aren't in the business of investing in 30-year R&D projects, they need shorter term returns on their investments, but that's not the way most VCs see nanotechnology.

Essentially what we are doing with TNT Weekly is trying to provide clear, unbiased factual information about nanotechnology. The fact that it is as popular with the investment community as with the scientific community indicates we're doing something right.

Given that we have the technologies, and the information, the only remaining obstacle is getting funded. Most VC companies are staffed by financial professionals, and they can bring in analysts for whatever market sectors interest them. With nanotechnology there is no market to speak of, so how do you get a handle on where to invest? In fact you don't, and lack of information raises the risks, so the investment goes into better-understood markets.

What we are doing with Spark is bringing together our knowledge and experience of nanotechnology, with their expertise in early stage investing. This means that you'll be presenting your business plan to people who understand your market, and that's crucial for any kind of business success. Imagine that, a VC who really understands nanotechnology - for me that's a dream come true, and we hope it will be for European nano entrepreneurs.

Steve: What sort of environment, discussion and progress was there at TNT 2000 and what can be expected from TNT 2001?

Tim: The TNT conferences can be divided into two themes, science & networking. We keep these running together by having reasonably short lectures, followed by long coffee breaks & lunches. Organizing the conferences in old monasteries in medieval walled cities such as Toledo and Segovia certainly helps the atmosphere. What I always like to see is little knots of people frantically discussing fine points of technology, if you can get this going then the conference will be a success. TNT 2000 was our first conference on nanotechnology, and the proceedings have just been published in the IOP journal "Nanotechnology." This year we have a much more global participation, with support from both the EU and the NSF, and we've noticed a lot more industrial participation, which is vital if we are to get nanotechnology out of the lab and onto the market. TNT 2001 will also feature a lot more breakout meetings during the coffee breaks, where people can discuss specific themes.

What is a Nano?

Steve: If at all possible, how can one distinguish between a nanotechnology company and an ordinary company that is only using nano as a marketing scheme? I mean, anyone can say, "Our products are atomically precise."

Tim: Well the 'N' word seems to be in fashion at the moment, and we're not afraid to highlight misuses in TNT Weekly. Some of the marketing guys have already realized that a 'pico' is smaller than a 'nano', and it won't be long before we hear more about femto. Erik Moderegger of AT&S compiled a list of nano trademarks, which we put on our web site, and some of them are mind-boggling. I'd bet that very few of them are registered to anyone connected with nanotechnology.

As nanotechnology will have such an impact on most industries, the only way to distinguish nano from non-nano is by defining nanotechnology so tightly that it excludes everyone except Zyvex. Maybe I should turn the question around and ask whether a cosmetics company manufacturing nanoparticles for use in its products is a nanotechnology company or not?

Steve: In deciding which companies and products to include in my collection of links and how to organize them, I must draw the line somewhere. I consider nanotechnology to require observing, understanding and controlling the structure of matter on the nanoscale in order to get novel effects from the materials and subsequent integration of those materials. For instance, I would consider Nanophase to be a nanotechnology company because their sun-screens incorporate particles of a strategic size such that they absorb UV rays, without absorbing the larger visible frequencies. Thus, they block the harmful rays while remaining invisible to the human eye.

They don't carve each of their nanoparticles with nanomachines, but rather produce them with conventional methods. However, we now have novel analytical tools for observing nanoscale features, allowing us to tune the fabrication process to create the nanostructures that produce desired effects. Together with the theory, it then becomes possible to design devices based upon nanoscale or quantum size effects. Thus, we seem to be able to create nanostructured products based on rational design rather than years of trial and error. A couple more examples of real products resulting from this trend are molecular imprinting, solid lubricants and nearly perfect mirrors, as well as the instruments that allow nanoscale quality control of these objects.

I would not consider Zyvex a nanotechnology company, but rather a "Molecular NanoTechnology (MNT)" company, as they also call themselves. Perhaps MNT is a subset of the more general nanotechnology that you and I seem interested in, but it definitely deserves a class of its own.

Tim: Agreed. This just illustrates the difficulty of separating nano from non-nano. Even the definitions used by the EU and the NSF are slightly different; so, although we have to draw a line somewhere, its position has not yet been agreed on.

Revolutionary Nanotech Products

Steve: I imagine that with nanotechnology it might be possible to create a liquid paint that once it dries could function as a display screen. The paint might either be subsequently plugged in to receive signals and energy, or simply be solar powered and receive wireless signals. Of course, it needn't emit light as our current display screens do. It would be enough just to have pixels that could change color. Might nanotechnology economically produce such a paint?

Tim: Nanotechnology will be bound by the same rules of economics as most other companies, selling various things to other people for a profit. Our experience in the Field Emission Display market has shown the difficulties of taking a great idea to market, ten years of development have yet to yield a product that can be sold at a reasonable price. As an investor, I would be very wary of anyone proposing a paint that functions as a screen, is self powered, communicates with other screens, tap dances and a whole host of other things. For each of those applications, you have a different set of problems, of lifetime, of stability, of interfacing, of driving the thing & so on.

If someone comes along with a plan to produce a display based on nanoparticles, that's fine. When they start adding all kinds of other functions before the core product has been developed, it becomes a problem. This comes down to the age-old fight between engineers and business people. The reason for this is driven by economics rather than science and engineering, and the need to have, at some clearly defined time-scale, a marketable product. Engineers and scientists will always give you something better in a few months time, that's their job. However, companies, and their investors, need to take a decision on how good something needs to be before you can put it on the market. As Guy Kawasaki says, 'Don't Worry, Be Crappy" - meaning get a product on the market in order to generate some sales income to pay for the fixes/improvements. You could argue that Microsoft takes this to extremes by getting an unfinished product onto the market as quickly as possible and then adding the patches at their leisure. If I were to be cynical, I could mention the premium rate lines for technical support in this context.

So the answer is, nanotechnology might produce such a paint, but you would probably have to commercialize the applications one step at a time. There is nothing wrong with looking further ahead in terms of functionality and applications, but at some stage, and preferably an early stage, you have to have some kind of product that you can move from development to sales.

Steve: Perhaps my imagination was running a little too far into the future. The first step towards smart paint would clearly just be a paint that could change color on command. Of course, such materials exist, but there isn't really a market for them. Thus, there is nothing motivating the traditional step-wise development of smart paint - unless, the whole product could be strategically designed in advance, based upon existing data stored in steadily shrinking computer chips... How long do you expect until we begin to see some revolutionary nanotech products?

Tim: Reading the NSF report on the societal implications of nanotechnology, it amazed me that so much of the thinking was linear, and based on our current understanding of technology. While nanotechnology may make some things better it will also create whole classes of new materials, and consequently markets, that we would be crazy to try to predict right now. Because of this wide spectrum of applications, there are some nanotechnology products that have been commercialized for years now, such as nanometer sized gold particles for labeling of oligonucleotides, lipids, peptides, proteins etc., and other applications still bubbling away in University labs that might bear fruit in 20 years. So that's our time-scale, any time from now, until sometime in the future!

Steve: It seems that the demand for smart paint (I want smart paint!) is relatively unimportant compared to the trillion dollar electronics industry or priceless medical industry - ventures that can afford lofty goals with roadmaps. Of course one must beware of snake oil and asbestos, but is it really futile to imagine what new materials and markets might result from the current trends?

Tim: There are so many possibilities it's almost pointless to speculate on the killer applications, it's like medieval theologians debating how many angels can dance on the head of a pin—an interesting debate, but how do you verify your conclusions? The applications will surely come, but at the moment it's difficult to predict where from, simply because nanotechnology is a huge field and at a very early stage. We have identified several areas where we believe nanotechnology will have a significant, and early impact, but you have to keep an open mind, and stay in touch with the science, because things can change very rapidly, just ask any company who bought a 3G mobile telephony license!

Electricity, the Wheel and Nanotechnology

Steve: In which areas do you think nanotechnology will have a significant and early impact?

Tim: The recent Red Herring issue on Nanotechnology identified materials, medicine and optics. I would argue that the optical applications they mention are just exploiting properties of materials anyway, although I can understand their reasons for wanting to include optics, and in particular optical switching, in a magazine aimed at investors.

Nanotechnology is often called the new Internet, but its repercussions will be felt much more widely than that. I would liken it more to the invention of the telephone, or even the discovery of electricity. Now when we talk about future applications of nanotechnology, we should remember that electricity was initially used to replace gaslights & candles, the telephone was not designed to enable you to connect to the Internet. Put the technology in the hands of some bright people, and the results will be unimaginable, given our current perception of technology.

However, I would predict materials, electronics and pharmaceuticals are obvious areas in the short term, as a lot of the groundwork has already been done, and these industries already have the tools and experience of working at sub micron scales. While these may not be the most exciting applications of nanotechnology, they will improve existing materials & processes, and their commercialization will help fund a great deal of research into large scale manufacturing. Nanophase Technologies is a good example of how nanotechnology can have an impact in an unexpected way but this is just the tip of the iceberg. What's better, an army of nanobots mopping up skin cancer in twenty years time, or more efficient sun blocks now?

Steve: Perhaps progressive development and creative dreams are two complementary approaches to nanotechnology - like theory and experiment, top-down and bottom-up or even TNT and NTN. The wheel might be another technology worth remembering - seeing as for so long we thought we had invented it, only to find that biomolecular nanotechnology has had wheels all along. In addition to faster computers, nanomedicine and mucho dinero, what are you looking forward to from nanotechnology?

Tim: Personally, I'm a big fan of any new technology, so of course I'm looking forward to all kinds of devices getting better, faster, lighter, and having less buttons. However, the area where nanotechnology has the greatest scope to change the world is by making things, whether they be mobile phones, computers, or drugs for the treatment of HIV or cancer, just work a whole lot better than they do at present.

5

Nano Education: News, Events and Workshops

NEWS

Frontiers Announces Annual Meeting and First Year Evaluation

September 19 @ 06:19:06: ENSCHEDE, The Netherlands, September 19 /PRNewswire/ — Frontiers, the Network of Excellence in life science related nanotechnology announces its first Annual Meeting (19-21 September 2005 at Karlsruhe University) and evaluation of success a year since its launch in August 2004.

Frontiers is supported by the Sixth Framework Program (FP6) from the European Commission and focuses on the crossover between life sciences and nanotechnology. The network represents twelve renowned nano-technology institutes from across Europe, bringing together the backgrounds and knowledge of around 200 scientists and researchers.

Nano World: Slashing Quantum-dot Costs

September 19 @ 06:16:39: NEW YORK, Sept. 17 (UPI) — Nanotech devices known as quantum dots are growing ever more popular in the electronics and biotech industries, but typically cost more than $2,000 per gram, limiting their large-scale use. Now scientists have developed a new method that could cut quantum-dot costs by 80 percent, experts told UPI's Nano World.

Quantum dots are semiconductor crystals just a few nanometers wide and made up of as few as 10 atoms. They fluoresce brightly when they absorb even tiny amounts of light and are helping scientists image the behavior of cells and organs to a level of detail never seen before. The conventional fluorescent dyes used in the $500 million worldwide market for biological detection agents as tags for molecules normally lose their ability to emit light within seconds. Quantum dots, on the other hand, last far longer, permitting investigators to monitor cells and organs in diseased and healthy conditions on a molecular scale in real time.

(Source: *UPI*)

Korea's Nanotech Feats Outshine High-Profile Biotech

September 16 @ 17:03:55: Korea's largely unsung achievements in nanotechnology are even more impressive in international comparison than those in biotechnology, which are stealing the headlines thanks to Prof. Hwang Woo-suk¡'s breakthrough in cloning human stem cells.

A report released by the Ministry of Science and Technology on Friday shows Korea ranked fifth in the world in the field of nanotechnology last year in terms of the number of papers (1,128) listed in the Science Citation Index (SCI), a global indicator of the level of scientific study in a country. In biotechnology, the country ranked 13th with 3,212 papers published.

(Source: *Chosun*)

Researchers Create DNA-Based Sensors for Nano-Tongues and Nano-Noses

September 16 @ 17:00:49: PHILADELPHIA — Nano-sized carbon tubes coated with strands of DNA can create tiny sensors with abilities to detect odors and tastes, according to researchers at the University of Pennsylvania [profile] and Monell Chemical Sciences Center. Their findings are published in the current issue of the journal Nano Letters, a publication of the American Chemical Society.

According to the researchers, arrays of these nanosensors could detect molecules on the order of one part per million, akin to finding a one-second play amid 278 hours of baseball footage or a single person in Times Square on New Years' Eve. In the report, the researchers tested the nanosensors on five different chemical odorants, including methanol and dinitrotoluene, or DNT, a common chemical that is also frequently a component of military-grade explosives. The nanosensors could sniff molecules out of the air or taste them in a liquid, suggesting applications ranging from domestic security to medical detectors.

'Bowtie Nanoantennas' Could Shed Light on Molecules, Other Nano-sized Objects

September 16 @ 16:16:30: One of the great challenges in the field of nanotechnology is optical imaging—specifically, how to design a microscope that produces high-resolution images of the nano-sized objects that researchers are trying to study. For example, a typical DNA molecule is only about three nanometers wide—so tiny that the contours of its surface are obscured by light waves, which are hundreds of nanometers long.

Now, researchers from Stanford University [profile] have greatly improved the optical mismatch between nanoscale objects and light by creating the "bowtie nanoantenna," a device 400 times smaller than the width of a human hair that can compress ordinary light waves into an intense optical spot only 20 nanometers wide. These miniature spotlights may one day allow researchers to produce the first detailed images of proteins, DNA molecules and synthetic nano-objects, such carbon nanotube bundles.

World Nano-Economic Congress Singapore

The Second Annual World Nano-Economic Congress (WNEC) Singapore is a cooperative effort between Cientifica and the Economic Development Board of Singapore to bring the focus of nanotechnologies and their impact upon industries to the Asian marketplace.

The WNEC Singapore represents the sixth editon of this successful event series with previous event having been held in Washington, DC, London, UK, Munich Germany, Dublin, Ireland, and last year's Singapore event.

The World Nano-Economic Congress is part of the Global Entrepolis @ Singapore Trade event, and represents one of the "Industry Conferences" that make up the event. GES 2005 will see over 10,000 participants, many of whom are top decision makers such as chief executives, venture capitalists, private equity players, thought leaders, entrepreneurs and innovators, attending one or more of the nearly 30 conferences held during the week.

Trends in Nanoscience: Structure and Functions

The International Symposium "Trends in Nanoscience: Structure and Functions" will bring together scientists from both, chemistry and physics, addressing experimental as well as theoretical issues of the topics given below. The meeting aims at promoting our understanding on the organization of complex nanosystems and, by trying to relate structure and function, will advance knowledge leading, ultimately, to new technologies.

- Clusters & Quantum Dots
- Selforganization and Unconventional Lithographies
- Magnetic Nanostructures
- Molecular Electronics
- Surface Structures on the Nanoscale
- Nanobiochemistry
- Transport through Nanostructures
- Nanooptics

Scanning Probe Microscopy in Life Sciences—4th International Workshop

JPK Instruments will hold the fourth international workshop "Scanning Probe Microscopy in Life Sciences" together with the University Clinic CharitÃ© in Berlin in October 2005. The workshop covers the application of scanning probe microscopy (SPM) to areas such as biochemical and medical research. It addresses students as well as advanced researchers. http://www.spm-workshop.jpk.com/

Pascal Couchepin, Swiss Minister for Home Affairs (including Science) Visits ApNano Materials

September 19 @ 19:17:28: Ness Ziona, Israel , September 19, 2005. Pascal Couchepin, Swiss Minister for Home Affairs (including Science), visited ApNano Materials' [profile] subsidiary, NanoMaterials, in Israel, with a delegation of leading figures from Swiss industry and academia.

Mr. Couchepin reviewed ApNano's revolutionary commercial nanotechnology-based solid lubricant – NanoLub, presented by ApNano executives: Mr. Aharon Feuerstein—Chairman and CFO, Dr. Menachem Genut – CEO, and Dr. Niles Fleischer, Vice President of Business Development and Product Development.

Molecular Medicine

September 19 @ 18:49:06: New tiny sensors spot disease before it occurs.

When doctors need to peek at mysterious goings-on inside the human body, they can skip the scalpel and rely on MRIs and CT scans. These state-of-the-art imaging technologies, successors to the X ray, reveal exquisitely detailed pictures of various tissues, organs and other anatomy.

Yet some of the most important details are invisible to MRI (magnetic resonance imaging) and CT (computed tomography). They can't reveal the molecular defects that indicate whether an abnormal lump is cancerous or benign. They can't tell doctors which bad genes are active in a patient's tumor or whether a drug is reaching its intended target. The result, says Stanford University imaging researcher Dr. Sanjiv Gambhir: "Everyone is shooting blindly." (Source: *Forbes*)

Imago's High-performance Nano-scale Metrology System Selected By The Toshiba Corporation

September 19 @ 18:43:04: Madison, WI – Imago® Scientific Instruments [profile] in conjunction with its representatives based in Japan, The Noah Corporation, announced today that they have delivered a LEAP 3000 Metrology System to the Toshiba Corporation solidifying Imago's continued global leadership position in nano-scale metrology.

The LEAP provides critical, 3-dimensional information at the nano-scale allowing positive identification of elements and isotopes within a material system. The true power of the LEAP lies in its ability to tie compositional information to spatial position, this allows manufactures to shorten time-to-market, enhance product yields and maximize capital equipment utilization. This capability has proven valuable in solving a variety of issues facing device manufacturers and researchers in the micro-electronics/data storage field.

Double-edged potential of nanoscience

September 19 @ 06:14:33: Prince Charles once famously warned that nanoparticles could envelop the world in a "grey goo" that could finish us all off.

Two reports, one last year from a German consultancy, Helmut Kaiser, and another from New York-based Innovest, as yet to be published, are more positive about nanotechnology but also outline several potential doomsday scenarios.

Nanotech As Disease Detector

September 16 @ 16:29:16: Startup Nanosphere [profile] may have a technology that can sniff out telltale markers early enough to advance treatment. The challenge: translating potential to real-life results.

(Source: *BusinessWeek*)

Nanocluster catalysts aim to clean up fossil fuels

February 05, 2003: Scientists from the University of Aarhus, Denmark, have developed a technique that could improve the commercial processes used to remove environmentally harmful sulphur from fossil fuels. ... *the Institute of Physics journal Nanotechnology the Danish team explain how they have studied the chemical ... by costly time-consuming trial-and-error methods. Nanotechnology is about to change this, since we can now build ...* http://www.brightsurf.com/news/feb_03/TIP_news_020503.html

Scientists use microscope to view magnetism at atomic level

November 07, 2002: Scientists and engineers build the transistors that run televisions, radios and similar electronic devices based on the moving electric charges of electrons. But the electron also has another key property: a magnetic spin that scientists believe could be exploited to develop faster, smaller and more efficient devices. ... *a very important technique for nanotechnology," he said. Nanomagnetism is a growing area of nanotechnology, Smith said, and scientists in the field expect to begin ... building nanoscale magnetic structures in the next ...* http://www.brightsurf.com/news/nov_02/EDU_news_110702.html

Purdue physicists hone rules for nanotech game

Nanotechnologists could have a firmer handle on the forces at play in their microscopic world thanks to recent physics research at Purdue University. ... *Purdue physicists hone rules for nanotech gameAugust 12, 2003 WEST LAFAYETTE, Ind. ? Nanotechnologists could have a firmer handle on the forces at play ... thorn in the side of nanotechnology manufacturers.* http://www.brightsurf.com/news/aug_03/EDU_news_081203.php

Scientists develop Atomic-Scale Memory

September 4, 2002: Scientists at the University of Wisconsin-Madison have created an atomic-scale memory using atoms of silicon in place of the 1s and 0s that computers use to store data. ... *feat, reported in the journal Nanotechnology, represents a first crude step toward a practical atomic-scale memory ... the senior author of the Nanotechnology paper. Although the memory created by Himpsel and his colleagues ...* http://www.brightsurf.com/EDU_news_090402_b.html

Purdue simulation to help merge molecules with silicon electronics

Engineers at Purdue University have created a nanotech simulation tool that shows how current flows between silicon atoms and individual molecules to help researchers design "molecular electronic" devices for future computers and advanced sensors. ... *LAFAYETTE, Ind. â?? Engineers at Purdue University have created a nanotech simulation tool that shows how current flows between silicon atoms and individual molecules to help researchers design â??molecular electronicâ?? devices for* ... http://www.brightsurf.com/new...view.article.php?ArticleID=20632

Green, leafy spinach may soon power more than Popeye?biceps

For the first time, MIT researchers have incorporated a plant?ability to convert sunlight to energy into a solid-state electronic 'spinach sandwich' device that may one day power laptops and cell phones. ... *the U.S. Naval Research Laboratory, including electrical and biomedical engineers, nanotechnology experts and biologists, collaborated on the worldês first solid-state photosynthetic solar cell. The work was reported in NanoLetters, a publication of ...* http://www.brightsurf.com/news/sept_04/MIT_news_092204.php

Discovery of 'doping' mechanism in semiconductor nanocrystals

Novel electronic devices based upon nanotechnology may soon be realized due to a new understanding of how impurities, or 'dopants,' can be intentionally incorporated into semiconductor nanocrystals. ... *Novel electronic devices based upon nanotechnology may soon be realized due to a new understanding of ... an important advance toward future nanotechnologies. According to Dr. David Norris, an Associate Professor of*

Chemical ... http://www.brightsurf.com/new...view.article.php?ArticleID=20333

Gold bowties may shed light on molecules and other nano-sized objects

One of the great challenges in the field of nanotechnology is optical imaging—specifically, how to design a microscope that produces high-resolution images of the nano-sized objects that researchers are trying to study. ... *2005 One of the great challenges in the field of nanotechnology is optical imagingâ??specifically, how to design a microscope that produces high-resolution images of the nano-sized objects that researchers are trying to* ... http://www.brightsurf.com/new...view.article.php?ArticleID=20777

U. T. Dallas-led research team produces strong, transparent carbon nanotube sheets

University of Texas at Dallas (UTD) nanotechnologists and an Australian colleague have produced transparent carbon nanotube sheets that are stronger than the same-weight steel sheets and have demonstrated applicability for organic light-emitting displays. ... *of Texas at Dallas (UTD) nanotechnologists and an Australian colleague have produced transparent carbon nanotube sheets ... journal Science, scientists from the NanoTech Institute at UTD and a collaborator, Dr. Ken Atkinson from* ... http://www.brightsurf.com/new...view.article.php?ArticleID=20652

Bioengineers Use Computer Model to Predict Evolution of Bacteria

November 21, 2002: Bioengineers used their computer model of E-coli to accurately predict how the bacteria would evolve under specific conditions. The results may have applications for designing tailor-made biological materials for commercial uses or for predicting the evolution of drug-resistant bacteria.... *the controlling cellular function, which could have important applications in nanotechnology and gene and cell therapy. Hasty's long-term goal is to build synthetic genetic networks which could be inserted into a patient's* ... http://www.brightsurf.com/news/nov_02/EDU_news_112102.html

Moon-based systems could supply solar power to earth

October 07, 2002: The key to a prosperous world is clean, safe, low-cost electrical energy, according to University of Houston physicist David Criswell. And his idea for how to get it is literally out of this world. ... *Texas Center for Superconductivity and Advanced Materials (TcSAM) are developing nanotechnology techniques that could transform the lunar soil into solar cells. "The raw materials needed to make solar cells are present in* ... http://www.brightsurf.com/news/oct_02/EDU_news_100702_c.html

Chemist's technique enables creation of novel carbon nanoparticles

Using a technique pioneered by Washington University in St. Louis chemist Karen Wooley, Ph.D., scientists have developed a novel way to make discrete carbon nanoparticles for electrical components used in industry and research. ... *really illustrates a particularly attractive strategy in the evolution of nanotechnology," said Kowalewski, principal investigator on this research and a postdoctoral researcher at Washington University in the late 1990s, and long-time collaborator* ... http://www.brightsurf.com/news/may_04/EDU_news_051104_d.php

Experiment could reveal 'extra dimensions,' exotic forces

October 30, 2002: Physicists have devised a new experiment that will be used in the quest for exotic forces in nature and additional spatial dimensions. ... *happen. "In order to go from fundamental physics to applied nanotechnology, you really will have to understand the laws that govern what happens at a very small scale. This research helps to* ... http://www.brightsurf.com/news/oct_02/EDU_news_103002.html

Chemists learn to build curved structures with nanoscale building blocks

The natural world is full of curves and three dimensions, but the ability to deliberately and rationally construct such complex structures using nanoscale building blocks has eluded nanotechnologists who are eager to add curved structures to their toolbox. ... *nanoscale building blocks has eluded nanotechnologists who are eager to add curved structures to their toolbox. ... director of Northwestern's Institute for Nanotechnology, who led the research team. "Much like proteins which must* ... http://www.brightsurf.com/news/jan_04/EDU_news_011604_d.php

Magnetic forces may turn some nanotubes into metals

A new study, published in today's issue of the journal Science, finds that the basic electrical properties of semiconducting carbon nanotubes change when they are placed inside a magnetic field. ... *applications," said paper co-author Richard Smalley, director of Rice's Carbon Nanotechnology Laboratory. "These findings remind us that there are still unique and wonderful properties that we have yet to uncover about nanotubes."* ... http://www.brightsurf.com/news/may_04/EDU_news_052104.php

Air bubbles in breakfast syrup illustrate potential pathway to new technology

The behavior of air bubbles in ordinary breakfast syrup demonstrates how scientists might be able to make vanishingly thin tubes and fibers for biomedical and other applications. ... *impact in fields from biomedicine and biotechnology to the hot nanotech industry." This line of research began for Zhang in the laboratory of Nagel, the Stein-Freiler Distinguished Service Professor in Physics. Nagel* ... http://www.brightsurf.com/news/nov_04/EDU_news_110304_c.php

NYU physicists find way to create three-dimensional quasicrystals

New York University physicists have applied a ground-breaking nanotechnology method to create three-dimensional quasicrystals, highly ordered structures that, unlike conventional crystals, never repeat themselves. ... *12, 2005 New York University physicists have applied a ground-breaking nanotechnology method to create three-dimensional quasicrystals, highly ordered structures that, unlike conventional crystals, never repeat themselves. Metallic quasicrystals created from exotic alloys* ... http://www.brightsurf.com/new...view.article.php?ArticleID=20353

Biomedical Scientist Testing Nanoparticles as Early Cancer Detection Agent

October 23, 2002: Biomedical scientist Shuming Nie is testing the use of nanoparticles called quantum dots to dramatically improve clinical diagnostic tests for the early detection of cancer. ... *Æ and director of cancer nanotechnology at Emory"s Winship Cancer Institute. His research focuses on the ... focuses on the field of nanotechnolgy, in which scientists build devices and materials one atom or* ... http://www.brightsurf.com/news/oct_02/EDU_news_102302_b.html

Diatom genome reveals key role in biosphere's carbon cycle

The first genetic instruction manual of a diatom, from a family of microscopic ocean algae that are among the Earth's most prolific carbon dioxide assimilators, has yielded important insights on how the creature uses nitrogen, fats, and silica to thrive. ... *ornate silica frustule. "Diatoms can manipulate silica in ways that nanotechnologists can only dream about. If we understood how they can design and build their patterned frustule as part of their biology,* ... http://www.brightsurf.com/news/oct_04/JGI_news_100404.php

UCI scientists use nanotechnology to create world's fastest method for transmitting information in cell phones and computers

UC Irvine scientists in The Henry Samueli School of Engineering have demonstrated for the first time that carbon nanotubes can route electrical signals on a chip faster than traditional copper or aluminum wires, at speeds of up to 10 GHz. *UCI scientists use nanotechnology to create worldâ??s fastest method for transmitting information in cell ... to the growing enthusiasm about nanotechnologyâ??s revolutionary potential. â??Our prior research showed that nanotube transistors can ... operate* ... http://www.brightsurf.com/new...view.article.php?ArticleID=20152

Tiny computers go where no computer has gone before

A major breakthrough in the use of molecules as information processors is to be announced at this year's BA Festival of Science in Dublin. ... *announced at this yearâ??s BA Festival of Science in Dublin. Nanotechnology experts are exploring the capabilities of molecules that act like conventional computers but can operate in tiny places where no silicon-based* ... http://www.brightsurf.com/new...view.article.php?ArticleID=20859

Researchers develop 'fingerprinting' for biological agents

September 1, 2002: Scientists at Northwestern University have developed a powerful new method for detecting infectious diseases, including those associated with many bioterrorism and warfare threats such as anthrax, tularemia, smallpox and HIV ... *led by Chad A. Mirkin, director of Northwestern's Institute for Nanotechnology, has invented a technique for creating thousands of DNA detection probes made of gold nanoparticles with individual molecules attached. Much like* ... http://www.brightsurf.com/EDU_news_090102.html

Scientists devise tiny liquid crystal devices for telecommunications

December 24, 2002: Scientists have built tiny liquid crystal devices on the tips of optical fibers—the plastic or glass cables used to carry high-speed signals from television, computer, telephone and radar—to correct signal distortions in high-speed optical communications. ... *achieve the necessary speeds," explains John Rogers, director of the Nanotechnology Research Department at Bell Laboratories in Murray Hill, New Jersey, and a coauthor of the study. "Our work shows not only* ...

http://www.brightsurf.com/news/dec_02/BNL_news_122402.html

Purdue scientists treat cancer with RNA nanotechnology

Using strands of genetic material, Purdue University scientists have constructed tiny delivery vehicles that can carry anticancer therapeutic agents directly to infected cells, offering a potential wealth of new treatments for chronic diseases. ... *scientists treat cancer with RNA nanotechnologySeptember 14, 2005 WEST LAFAYETTE, Ind. â?? Using strands of genetic ... Engineering. â??Physicians have hoped that nanotechnology might provide a solution to the problem, and itâ??s possible* ... http://www.brightsurf.com/new...view.article.php?ArticleID=20941

Hygienic surfaces, biocidal and self-cleaning coatings

Microbian evolution on a wide variety of surfaces can produce phenomena such as corrosion, dirt, smells and even serious hygiene and health problems. ... *of coatings and treatments on specific surfaces, and in which nanotechnology plays a key role. Most of these coatings acquire their biocidal / self-cleaning capacity by incorporating specific nanoparticles: basically silver (Ag) and titanium* ...

http://www.brightsurf.com/new...view.article.php?ArticleID=20932

New Machining Technology Drills Nano-diameter Holes

Santa Cruz , CA—NanoMatrix, Inc., the developer of a new approach for performing small-scale fabrication, offers its customers Micro and Nano machining services and equipment. The company's proprietary technology includes the control of high-energy liquid jets to cut and shape features with sizes below 50 nanometers in a wide variety of materials. "At the scale we are working, the jet's stream of individual molecules seems to produce a machining effect similar to sand blasting" says John Thomas, Director of Business Development for NanoMatrix. The processes developed by NanoMatrix are environmentally friendly. Effects are created mechanically and no harsh chemicals are required to support the process. The process is also very tolerant of contaminants so it can be used in environments with less stringent cleanliness controls. Currently, NanoMatrix is focusing its efforts on joint development projects with customers and business partners. "We have found that there are a number of companies that are interested in exploring alternatives to building next generation products since their current tool kits are very limited."

NanoMatrix is working with a variety of MEMS/NEMS, photonic devices, and optics companies to develop solutions to problems associated with fabricating products with feature sizes that are often smaller than the wavelengths of visible light. In optics applications, for example, our process can be used to selectively change the surface profile of a lens while leaving the overall optical performance unchanged. NanoMatrix, Inc. is also developing a general-purpose workstation for customers who wish to work independently on development and prototyping projects. *For more information please visit www.nanomatrix.com.*

Toumaz Technology Secures Future Funding for Product Development and Commercialisation Through Acquisition By Nanoscience Inc.

Toumaz Technology Limited has announced that its acquisition by the nanotechnology investment vehicle Nanoscience Inc., has formally completed, with the associated placing of Nanoscience shares raising £7.1 million (approximately £6.4 million net of expenses). Toumaz is now a wholly-owned subsidiary of the AIM listed Nanoscience. The consideration payable in respect of the acquisition has been satisfied through the issue of 96,337,210 Nanoscience shares, valuing Toumaz at approximately £17.7 million. Toumaz will retain its independent commercial identity while being able to take advantage of the benefits of being part of a stronger enlarged publicly listed entity.

The strategic focus for Toumaz remains unchanged but the transaction will enable Toumaz to accelerate the ongoing development of its ultra-low power Sensium family of devices based on the company's proprietary AMx™ advanced mixed signal technology. Toumaz's close links with Imperial College, London – with which it has a technology pipeline agreement – have recently been strengthened by the co-founding of the new

Institute of Biomedical Engineering by the company's Chairman and CTO, Professor Chris Toumazou, giving Toumaz the opportunity to access the major global pharmaceutical and medical device manufacturers.

Professor Chris Toumazou, newly appointed Non-Executive Director of Nanoscience, commented on the strengthened relationship with Imperial College, "With the New IP Option Agreement with Imperial Innovations, Toumaz has been granted the exclusive right to explore the commercial potential of intellectual property generated within the new Institute of Biomedical Engineering in the field of wireless communications and low power circuit design."

Toumaz plans to invest part of the funds raised in a number of key areas, including ongoing new product development, support for further investigations into advanced semiconductor processes and the strengthening of the company's engineering and commercial teams.

Commenting on today's announcement, Richard Rose, Non-Executive Chairman of Nanoscience, said, "It is my belief that this acquisition marks a significant and substantial progressive step for both parties. Toumaz will bring to Nanoscience two key elements: the first being a world-leading technology with a number of established and identified target markets ranging from healthcare to consumer electronics; and the second being a significant base of industry and academic contacts that will be key to the procurement of further technologies and the establishment of collaborative product development projects."

Keith Errey, CEO of Toumaz Technology, added, "This acquisition combines a fantastic technical team in Toumaz with a very accomplished commercial team in Nanoscience. By enabling us to leverage Nanoscience's wide commercial expertise and investor experience, this route will allow Toumaz to remain focused on our technology development and the execution of that technology – our core business. We believe that the combined strengths of the companies will create an entity of significant value."

Source: http://www.toumaz.com/

Carbon Nanotube Membranes Allow Super-Fast Fluid Flow

Membranes composed of manmade carbon nanotubes permit a fluid flow nearly 10,000 to 100,000 times faster than conventional fluid flow theory would predict because of the nanotubes' nearly friction-free surface, researchers at the University of Kentucky report in the Nov. 3 issue of Nature.

In their study, Mainak Majumder, Nitin Chopra and Bruce J. Hinds of UK's Chemical and Materials Engineering Department, and Rodney Andrews of UK's Center for Applied Energy Research found the flow dynamics of carbon nanotube (CNT) membranes with pores measuring 7 nanometers in diameter permit a fluid flow exceeded the flows predicted by conventional hydrodynamic predictions.

In their study "Enhanced Flow in Carbon Nanotubes," the researchers note an "aligned CNT membrane has fast transit approaching the extraordinary speed of biological channels. The membrane fabrication is scalable to large areas, allowing for industrially useful chemical separations.

"(E)ach side of the membrane can be independently functionalized. These advantages make the aligned CNT membrane a promising large-area platform to mimic protein channels for sophisticated chemical separations, trans-dermal drug delivery and selective chemical sensing," the researchers say.

Source: http://www.uky.edu/

3D Barcodes to Identify Stolen Valuables

DIAMONDS and valuable works of art could be protected against theft using a microscopic barcode that stores encrypted information about the provenance of the items, making ownership easy to prove if they are stolen. The barcode, which takes the form of a cube 30 micrometres across, is being developed by a team at the National Physical Laboratory (NPL) at Teddington, near London. The cube is made of silicon coated with a 100-nanometre-thick layer of polymethyl methacrylate, a transparent plastic. It can be attached to hard surfaces using adhesive, or woven into the canvas of paintings. To create the barcode an electron-beam lithograph drills 90,000 small squares into the plastic coat of each face at five different depths. The position and depth of each square is unique, so data can be encrypted using a key-based code and stored digitally. The cube is scanned line by line using an electron force microscope, which can detect differences in the depth of the squares. This scanning process takes around a minute, says Alexandre Cuenat, one of the team of researchers that developed the cube. The technology has huge potential for storing information securely, says Cuenat. "You could get two copies of the bible, the King James version, on the sharp end of a pin." However, in practice storing information on the provenance of an object would not require anything like this capacity. The nanobarcode has three levels of security to protect objects against theft and counterfeit. First, thieves and fences are unlikely to realise that the item is protected. "You cannot see or feel the cube, even if you roll it between your finger and thumb," says Cuenat. Second, most fraudsters are unlikely to have the specialist equipment to read or write the nanobarcode, and so will be unable to duplicate it. And finally the encryption will be practically unbreakable. "Theoretically all codes can be broken," says David Mendels, who led the NPL team. You just have to test all the different permutations. In this case, it would be possible to make the encryption extremely complex because of the cube's high storage capacity. "You'd need a computer doing a million calculations a second and it would still take more than 187 billion years," says Mendels. For day-to-day use, the encryption could be far less complex and still be virtually unbreakable, says Cuenat.

Billions of Tiny Mechanical Levers Could be Used to Store Songs on Future MP3 Players and Pictures on Digital Cameras

As bizarre as the idea might sound, researchers at a Dutch company have already demonstrated that miniscule mechanical switches can be used to store data using less power than existing technologies and with greater reliability. Nanomech memory, developed by Cavendish Kinetics in the Netherlands, stores data using thousands of electro-mechanical switches that are toggled up or down to represent either a one or zero as a binary bit. Each switch is a few microns long and less than a micron wide—roughly one-hundredth the width of a human hair. Existing computer memory typically stores data as an electrical or magnetic charge. Cavendish Kinetics claims Nanomech memory can read and write data using 100 times less power than such systems, and works up to 1000 times faster. It is also much more resilient to both temperature and radiation, the company claims.

Nanomech memory incorporates hundreds of thousands of conductive metal levers, each just a few microns long. These are created by lithographically etching a design onto a template and then chemically dissolving away unwanted layers. Introducing a miniscule voltage to an electrode below a lever causes it to bend forwards until it makes contact. Thanks to intermolecular forces on this scale, once flipped, a lever will also maintain its position, even when the voltage is switched off. The trick can be reversed by applying voltage to an electrode on the other side of the lever. And the state of each switch flips can be sensed easily by the electrodes themselves. Repeating the feat across thousands of switches makes it possible to store something as complex as a computer program in Nanomech memory. Cavendish Kinetic has so far developed a unit capable of storing 256 kilobytes of data, or around two million individual bits. This is suitable for simple microcontrollers—the tiny computers used to operate commonplace electric motors found, for example, in cars and consumer electronics products.

But eventually the company hopes to stack many more levers together, boosting memory capacity to several gigabytes. "That's the long term aim," says Charles Smith at Cambridge University, UK, who is a chief technology officer at Cavenish Kinetics. "We want to put millions and millions on an individual chip." This should make Nanomech memory a viable solution for portable music players and digital cameras. And it could improve battery life dramatically. "The big growth area for large scale non-volatile memory devices is portable consumer applications, such as iPods and cameras," says Jim Miles, another nanotechnology researcher at Manchester University, UK. "And for these devices, power consumption is the big issue." Miles, however, points out that it is vital to be able to repeatedly read and write to this type of memory. "It's important to know rewrite speeds as well as how many rewrite cycles they can survive," he says. Smith says the first increased capacity Nanomech chips should be available by the end of 2006. He adds that the memory technology's resilience to radiation should also make it suitable for use aboard satellites and other spacecraft.

This is because cosmic radiation can cause space-borne computers to suddenly malfunction by flipping bits stored in memory.

World's Smallest Toothbrush Scrubs Capillaries Clean

Nano-sized toothbrushes that can clean very small surfaces have been developed by researchers. Fabricated out of millions of carbon nanotubes, the minuscule brushes could even paint the inside of capillaries thinner than a human hair. But besides being small, the nanotubes have other advantages over traditional bristles, says one of their creators Pulickel Ajayan at Rensselaer Polytechnic Institute in Troy, New York. The materials typically used for making brush bristles include animal hairs, synthetic polymer fibres and metal wires. But each has its limitations. Metals corrode and weaken, hair is not very strong and synthetic fibres melt. Carbon nanotubes, on the other hand, are 30 times stronger than steel, yet five times less dense. They are highly elastic, resistant to heat, have large surface areas and even conduct electricity. The latter property makes them highly suitable for the contact brushes used in electric motors, says Ajayan. He and Anyuan Cao at Rensselaer, in collaboration with colleagues at the University of Hawaii in Honolulu, US, have designed brushes consisting of a silicon carbide fibre base on which the carbon nanotubes are grown in a single row of bristles like a toothbrush, or in groups of bristles more like a toilet brush. One end of the base is coated with gold, which acts like a "handle" and also inhibits nanotube growth at that end. The dimensions can be varied dramatically, says Ajayan, but typically the bundle of nanotubes is no smaller than a few micrometres in diameter. The brushes could have a variety of applications. They have already been used to sweep particles measuring just 50 nanometres in diameter off a plain surface, and off another surface pitted with microscopic grooves. The researchers have also attached the gold handle of a three-pronged brush to the shaft of a small electric motor. This allowed them to rotate the bristles inside a 300-micrometre-wide capillary tube, cleaning contamination off its walls. Then, by coating the brush with a red dye, they were able to paint the inside of the cavity using the same approach. The brushes' ability to clean small particles appears to be extremely good, says Ajayan, although he is still not sure why. "Perhaps the cleaning occurs by electrostatic interactions, and since the surface area of the nanostructures is large, the cleaning is possibly more effective." The aim now is to apply the brushes to more specific microelectronic and biomedical applications. Journal reference: *Nature Materials* (DOI: 10.1038/nmat1415)

Harmful Bacteria Shown up by Nanoparticles

A new nanoparticle test for dangerous bacteria such as *Escherichia coli* O157:H7 is so sensitive it can detect a single bacterial cell within minutes. The food industry, medicine and the fight against bioterrorism could all eventually benefit from it, researchers say. Even a few cells of the *E. coli* strain in food can be dangerous so it is important to be able to detect them in low numbers. Current tests, however, need a higher number of bacteria to be present before they can detect it, which can lead to long delays. The new test, developed by Weihong Tan and colleagues from the University of Florida, US, could offer a sharper and faster way of detecting contamination. "If you can give us one bacterial cell in a sample then we can detect it," says Lisa Hilliard, one of the team. "Most people have to grow it and then detect it." The whole test can be carried out in just 20 minutes, compared with up to 48 hours for conventional tests. "Other tests could be as sensitive but you would need to perform an enrichment first," says Mike Peck from the Institute of Food Research in the UK. "You would have to pop the food into a growth media for 24 hours," he explains.

Waiting a day or two for the results can be costly and inconvenient, says Andrew Brabban, who works on *E. coli* at the Evergreen State College, Washington. "One of the basic problems at the moment is what is called 'Test and Hold'. Samples are held at US plants until they are shown to be free of O157:H7. This is obviously expensive for the industry, having large quantities of beef as shelved stock," he told New Scientist. The new test consists of silica nanoparticles, each housing thousands of fluorescent dye molecules, and each attached to an antibody for a given bacterium. The nanoparticles are added to a solution of the test sample, such as ground beef. If the bacterium sought is present the nanoparticles will quickly attach to it. The sample is then separated by weight in a centrifuge. The target bacteria, being heavier than the nanoparticles, will separate away from them. But those dye molecules already attached will fluoresce in this heavier sample, identifying the bacteria.

The new test differs from other dye tests in that thousands of dye molecules will fluoresce if only one bacterium is present because they are all attached to the same antibody. Other tests contain only a few dye molecules for each test antibody used so that one bacterial cell will not cause enough fluorescence to be observed. Using samples so diluted that only one in four of them contained any bacteria at all, the researchers showed their test could detect a single bacterial cell. Although the test was developed using *E. coli* O157:H7, it could be adapted to many different bacteria. The group are already looking at ways to identify more than one type of bacteria at a time, using different coloured dyes for each. If this can be achieved quickly and accurately it could have great potential. "The need to detect single cells is real," says Brabban. "Any method that is fast, accurate and has [such a sensitive] detection level would certainly be very useful," he says. Journal reference: *Proceedings of the National Academy of Sciences* (DOI: 10.1073/pnas.0404806101)

Buckyballs Made Safer for Fumans

Carbon molecules called "buckyballs"—which hold great promise for nanotechnology—but have been shown to harm fish have been made safer by scientists. The soccer-ball-shaped carbon nanoparticles were shown to cause brain damage in fish and kill water fleas in a study in March 2004. But now a team at Rice University in Houston, Texas, US, has come close to understanding how buckyballs – more formally known as fullerenes—kill cells and how their toxicity can be lowered in human cells. Although the toxic nature of the carbon-60 nanoparticles may be useful in medicine, for example in fighting cancer, there are concerns that their potentially widespread use in fuel cells, drug delivery and cosmetics could mean they find their way into the environment, and so into animals and humans. "There are a couple of different manufacturers that will, and are, mass producing fullerenes," says Christie Sayes, one of the team. "They could make it into consumer based products: fuel cells and batteries or make-up," she says.

The team, led by Vicki Colvin, looked at the effects of buckyballs on human cells. They found that even at quite low concentrations in water the buckyballs killed human skin cells. However, when they attached chemical groups such as hydroxyl groups to the buckyballs, their toxicity was greatly reduced. The higher the number of groups that were attached, the less toxic the fullerenes became, so that a buckyball with 24 hydroxyl groups attached showed a toxicity seven orders of magnitude lower than the original buckyball. The researchers believe the buckyball is toxic because in water it leads to the formation of an oxygen free radical which reacts with lipid molecules forming the cell membrane surrounding a cell. This causes the lipid molecules in turn to become free radicals – and when these try to interact with the water outside the cell, the membrane ruptures leading to cell death.

The team were able to show the hole in the membrane using dyes that only fluoresce when in the cell. Dyes that are usually too big to enter a cell fluoresced after the buckyballs had been added, showing the permeability of the cell had increased to such an extent that even large molecules could move freely through the membrane. More research is needed to determine whether the same result would be seen in the body, says Sayes. "Cells in culture act differently to those in the body," she says. The research could eventually be use to make buckyballs safer, so that they do not cause an environmental or health threat as they become more commonly used. Or even to make them more toxic for certain medical treatments. Eva Oberdõrster at Southern Methodist University in Dallas, Texas, who led the buckyball study in fish, called the new work "an excellent first step" in reducing the toxicity of buckyballs. Journal reference: *Nano Letters*.

U-turn on Goo

GREY goo is no more. Eric Drexler, the futurist who dreamed up the vision of self-replicating nanomachines spreading across the planet has publicly repudiated his idea. In his 1986 book Engines of Creation, Drexler suggests that microscopic machines might one day be able to manipulate individual molecules to build any desired structure. If such a machine contained its own blueprint and could scavenge raw materials, it could take over the planet in a chain reaction of self-replication, he warned. "The self-replicating machine idea has had a profound effect on nanotechnology, or rather its perception—mostly for the bad," says Mark Welland, editor-in-chief of the journal Nanotechnology. Now, in an article in the June issue of Nanotechnology, Drexler says that nanofactories will be desktop-sized machines that can manipulate large numbers of molecules. Although such a factory could be directed to build a copy of itself, says Drexler, it would not be able to do so on its own. And neither would it be able to spread around the planet, Hollywood style, eating everything its path.

Carbon Nanotubes Used in Computer and TV Screens

The next generation of computer and television screens could be built using carbon nanotubes. Next week a prototype high-definition 10-centimetre flat screen made using this technology will be launched in Boston at the Society for Information Display conference. The new screen, called a nano-emissive display or NED, is made from two sheets of glass, one covered by a layer of nanotubes standing on end, the other by a layer of blue, red or green phosphors similar to those used in conventional cathode ray tube screens. When charged, the nanotubes direct electrons at the phosphors, making them light up. Because the electrons have only a short distance to travel, even a 105-centimetre NED would use relatively little power, says maker Motorola. A screen that size will also have a wide viewing angle and could sell for less than $400, the company claims.

Building nanomachines out of living bacteria

LIVE bacteria could one day act as reconfigurable components for nanoscale electronic circuits, or even a scaffold for building nanomachines. "Nature has developed these fantastic building blocks," says Robert Hamers of the University of Wisconsin-Madison. "Our approach is to simply grab onto them very gently." His team have been using electrodes to manipulate individual bacterial cells, they will report in a future issue of Nano Letters. Their chosen bacterium was Bacillus mycoide, 5 micrometres long and 0.8 wide—large enough to be visible under an optical microscope. The trick is to apply an electric field to the microbes, so they become polarised and stick to the electrodes. But the idea should also work with smaller bacteria, as a change in the electrical current of the electrode would show they were attached. The researchers even persuaded the bacteria to form a conducting bridge between two electrodes, a knack that could perhaps lead to reconfigurable nano-circuitry. "The idea of using cells as part of a more complex electronic circuit is very important," says Hamers. In his experiments the bacteria attached themselves to a long electrode before being shunted along its length by the motion of the solution in which they are suspended. When the bacteria reach a junction with another electrode, the polarities of the two electrodes securely fix the bugs in place. At the moment, nanostructures have to be put together manually. But it might be possible to automate the process using bacteria, as components tagged with particular biological molecules will stick to complementary surface proteins on the bacteria. Another use for Hamers-type electrodes would be in biosensors that could detect biological agents such as anthrax from changes in an electrode's current as spores become attached.

Mechanical Chip Promises Huge Data Storage

A super-dense memory chip that stores data in the form of nanoscale holes in a plastic film has made its public debut at the CeBIT electronics exhibition in Hanover, Germany. Storing data in the form of holes is not new—CDs use pits in a polycarbonate disc, for instance, and 19th-century looms stored patterns on punched cards. But the "Millipede" technology from IBM's Zurich lab promises very high capacity thanks to its use of holes just 10 nanometres wide. This means that a square chip measuring 2.4 centimetres on a side should be able to store 125 gigabytes, says the company, equivalent to 25 DVDs. The Millipede chip achieves this by having an array of tens of thousands of silicon cantilevers, explains IBM researcher Evangelos Eleftheriou. Each has a pointed tip that writes data by poking holes—representing a digital 0 or 1—in the soft polymer below. The cantilever also reads the data when needed, by measuring a change in its electrical resistance when it drops into a hole. Electromagnetic actuators within the chip package move the polymer film beneath the cantilevers so that each tip can read and write within a 100-micrometre-square area. Data is erased using a heater in each cantilever which melts the polymer locally, smoothing the pits over for re-use.

That is the theory, but to make Millipede a commercial reality IBM admits it needs an industrial partner with expertise in the manufacture of microelectromechanical systems (MEMS). "We have no MEMS product line at IBM. Companies who make accelerometers or microscale actuators will have the kind of production capabilities Millipede needs," says Eleftheriou. The technology's appearance at CeBIT was geared towards finding that partner. At the show, IBM used a video microscope to demonstrate the micro-cantilevers going about their work on a 10 gigabyte version of a Millipede chip. If IBM can commercialise the memory chip and get it reading and writing data at speeds of 20 to 30 megabits per second—like today's "flash" memory chips—it hopes the technology could form the heart of future digital cameras, cellphones and USB memory sticks. However, some analysts are sceptical of Millipede's prospects, questioning the need for a mechanical memory device when consumer electronics firms already have electronic, optical and magnetic technologies in use.

Nanoscopic 'Ruler' Could Provide Microchip Benchmark

A nanoscopic measuring device that uses atomic lattices to gauge tiny distances could enable microelectronics engineers to build better components, say US scientists. The "nanoruler" was developed by researchers at the US government's National Institute of Standards and Technology (NIST) in Maryland, and industry collaborators. They say it will enable researchers and engineers working on nanoscale objects to calibrate their instruments more precisely, resulting in greater accuracy in their work. To construct the instrument, NIST researcher Michael Cresswell says the team exploited the way in which silicon atoms align themselves into uniform lattice structures. They had to etch the raw silicon in the correct orientation to ensure the lattices were perpendicular to the top and bottom surfaces. If not, the distance between each lattice would not be standard. The nanoscopic slab can provide precise measurements by counting the number of silicon lattice lines crossed by an object. An advantage silicon is that the width of each atomic lattice is known to within about one-tenth of a nanometre, says Cresswell, at 0.3 nanometres. Nanorulers varying in size from 40 to 275 nanometres can be constructed. These could, for example, be used by nano-engineers to improve the calibration of instruments such as atomic force microscopes, scanning tunnelling microscopes and electron microscopes.

Researchers currently use a variety of materials to calibrate their instruments, the best of which is still seven times less accurate than the silicon ruler, Cresswell told *New Scientist*. He adds that the nanoruler could prove especially useful in microelectronics fabrication, where engineers check the dimensions of components after each step of their work. Peter Hatto, chairman of the European Committee for Standardization's working group on nanotechnology, says a precise standard for nano-measuring would be a very useful research tool. "You really need to use a fundamental measurement standard but you can't use the metre without an artefact to calibrate against," he says. The NIST nanoruler has the advantage that the properties of silicon are already well understood, he adds.But Hatto also notes that it will be crucial to provide researchers with easy access to the material to ensure it is used. The ruler was developed by NIST in collaboration with three US companies; Sematech, in Texas, US, and VLSI Standards and Accurel, both based in California, US.

Mechanical Chip Promises Huge Data Storage

A super-dense memory chip that stores data in the form of nanoscale holes in a plastic film has made its public debut at the CeBIT electronics exhibition in Hanover, Germany. Storing data in the form of holes is not new—CDs use pits in a polycarbonate disc, for instance, and 19th-century looms stored patterns on punched cards. But the "Millipede" technology from IBM's Zurich lab promises very high capacity thanks to its use of holes just 10 nanometres wide. This means that a square chip measuring 2.4 centimetres on a side should be able to store 125 gigabytes, says the company, equivalent to 25 DVDs. The Millipede chip achieves this by having an array of tens of thousands of silicon cantilevers, explains IBM researcher Evangelos Eleftheriou. Each has a pointed tip that writes data by poking holes—representing a digital 0 or 1—in the soft polymer below. The cantilever also reads the data when needed, by measuring a change in its electrical resistance when it drops into a hole. Electromagnetic actuators within the chip package move the polymer film beneath the cantilevers so that each tip can read and write within a 100-micrometre-square area. Data is erased using a heater in each cantilever which melts the polymer locally, smoothing the pits over for re-use.

That is the theory, but to make Millipede a commercial reality IBM admits it needs an industrial partner with expertise in the manufacture of microelectromechanical systems (MEMS). "We have no MEMS product line at IBM. Companies who make accelerometers or microscale actuators will have the kind of production capabilities Millipede needs," says Eleftheriou. The technology's appearance at CeBIT was geared towards finding that partner. At the show, IBM used a video microscope to demonstrate the micro-cantilevers going about their work on a 10 gigabyte version of a Millipede chip. If IBM can commercialise the memory chip and get it reading and writing data at speeds of 20 to 30 megabits per second—like today's "flash" memory chips—it hopes the technology could form the heart of future digital cameras, cellphones and USB memory sticks. However, some analysts are sceptical of Millipede's prospects, questioning the need for a mechanical memory device when consumer electronics firms already have electronic, optical and magnetic technologies in use. The CeBIT exhibition runs from 10 to 16 March.

Charge a Battery in Just Six Minutes

A rechargeable battery that can be fully charged in just 6 minutes, lasts 10 times as long as today's rechargeables and can provide bursts of electricity up to three times more powerful is showing promise in a Nevada lab. New types of battery are badly needed. Nokia's chief technologist Yrjö Neuvo warned last year that batteries are failing to keep up with the demands of the increasingly energy-draining features being crammed into mobile devices (New Scientist print edition, 28 February 2004). The highest energy-per-weight ratio in today's batteries is provided by lithium-ion (Li-ion) batteries. They are also cheaper in terms of energy delivered per unit of weight than alternative types of battery such as nickel-metal-hydride (NiMH) and nickel-cadmium (Ni-Cd) types. But Li-ion cells have their drawbacks too. They eventually wear out, and they cannot discharge energy quickly enough for applications requiring power surges, such as camera flashguns and power tools. They may soon be able to. Altair Technologies of Reno has created a new type of Li-ion cell in which the anode has an exceptionally high surface area. This allows electrons to enter and leave it quickly—making fast recharging possible and providing high currents when needed.

Li-ion batteries work by forcing lithium ions from a lithium cobalt oxide cathode to migrate to a carbon anode via an electrolyte solution. Altair's patented modification is to make the anode surface out of lithium titanate nanocrystals, using chemical tricks to give it a surface area of about 100 square metres per gram, compared with 3 square metres per gram for carbon. The firm is keeping the chemistry that allows it to do this pretty close to its chest for commercial reasons. But the patent (US 6689716) reveals that the increased surface area is achieved using a carefully controlled sequence of evaporative steps when making the lithium titanate crystals. The high current that this modified electrode is able to carry means power-hungry devices can be installed in mobile phones, which until now have been denied them. For instance, camera phones might now have enough power to run a flashgun.

Altair says the battery will have other advantages, too. The crystalline surface of a carbon anode is susceptible to damage by the repeated temperature changes that occur as the battery is used and recharged. This limits its life to around 400 charging cycles. The more rugged lithium titanate anode should make it possible to recharge the battery as many as 20,000 times says Roy Graham, development director at Altair. A longer lifespan should also be better for the environment, he says. "The continual use of polluting cobalt oxides is questionable." Altair plans to develop its batteries for power tools, which have till now required more expensive Ni-Cd or NiMH batteries to provide the large currents these devices need. The company hopes to license its technology to major battery-makers, who could have the device on the market in two years' time. Altair says it eventually wants to produce batteries for a broad range of devices, from phones to hybrid electric vehicles.

Nanotechnology may Create New Organs

Scientists have built a minute, functioning vascular system—the branching network of blood vessels which supply nutrients and oxygen to tissues—in a significant step towards building whole organs. Conventional tissue engineering methods have successfully grown structural tissues such as skin and cartilage in the lab. But not being able to create the supporting vascular system has proved a major stumbling block preventing scientists from creating large functioning organs such as liver or kidneys. Now, researchers from Massachusetts Institute of Technology and Harvard Medical School have used computers to design branching networks of venous and arterial capillaries, which start at three millimetres wide and reach a fineness of just 10 microns. "We used living vessels as a guide to model factors such as the angle and size ratio between branching vessels. But we optimised our design to improve it," said lead researcher Mohammad Kaazempur-Mofrad, from MIT¹s department of mechanical engineering and division of biological engineering.

The networks were etched on to 15 centimetre-wide silicon wafers and the paths were then used as a mould to set a layer of biodegradable polymer. Two of these were then sealed together with a microporous membrane sandwiched between them, producing a mini artificial vascular system. Endothelial cells—which are flat cells lining the walls of blood vessels in a single layer—were injected into the network on one side of the membrane and either liver or kidney cells were injected on the other side. The endothelial cells coated the inside of the polymer nanotubes. These nanotubes biodegraded to leave a living shell of vessels similar to a natural vascular network. This method would provide an efficient means of supplying the liver or kidney cells with enough oxygen and nutrients to survive. The one-layer systems of kidney and liver cells were successfully implanted into rats for two weeks—95 per cent of the cells survived. "The next step is to work on bigger animals, such as a pig or rabbit, using more layers," Kaazempur-Mofrad told New Scientist. "Eventually, we want to be able to replace whole organs with several layers of these constructs. The critical mass for liver is one-third, probably 30 to 50 stacked layers." "So in the next 10-15 years, we will hopefully have reached a point where we can do this procedure clinically in human patients," Kaazempur-Mofrad added. The research was presented at the American Society for Microbiology Conference on Bio-, Micro-, and Nanosystems, in New York City on Tuesday.

Nanoparticles Clearly Finger the Culprit

Oil-seeking nanoparticles could give police the clearest fingerprints yet, suggests new research. Law enforcement officers currently search for prints by dusting a crime scene with fluorescent powder. This sticks to the oily residue left by the fingertip, showing up the whorls and ridges. But sometimes the prints are not clear enough to finger a suspect. The new dust made of sticky nanoparticles could help. The powders used today work because oily prints have a natural tackiness. But the nanoparticle dust being developed at the University of Sunderland in the UK will actively seek out any oil. The nanoparticles are tiny glass spheres between 200 and 600 nanometres in diameter. As well as being speckled with a fluorescent dye, they are coated with hydrophobic molecules, which are repelled by water and attracted to oil. So they fix tightly to the fingerprint. Fred Rowell, who is leading the research, says the nanoparticles should pick out even the faintest of fingerprints because they stick to tiny traces of oil. And the prints should be much sharper, providing fine detail that can be crucial to identifying a print, such as how ridges in the print branch and finish.

Nano-transistor Self-assembles Using Biology

A functional electronic nano-device has been manufactured using biological self-assembly for the first time. Israeli scientists harnessed the construction capabilities of DNA and the electronic properties of carbon nanotubes to create the self-assembling nano-transistor. The work has been greeted as "outstanding" and "spectacular" by nanotechnology experts. The push to shrink electronic circuits to ever smaller dimensions is relentless. Carbon nanotubes, which have remarkable electronic properties and only about one nanometre in diameter, have been touted as a highly promising material to help drive miniaturisation. But manufacturing nano-scale transistors has proved both time-consuming and labour-intensive. The team, at the Technion-Israel Institute of Technology, overcame these problems with a two step process. First they used proteins to allow carbon nanotubes to bind to specific sites on strands of DNA. They then turned the remainder of the DNA molecule into a conducting wire.

"DNA is very good at building things in molecular biology, but unfortunately, it does not conduct electricity. We had to get a metal conductor on the DNA," explains physicist Erez Braun, who led the research. "This is spectacular work," says Cees Dekker, a nanoscience expert at Delft University in the Netherlands. "It demonstrates that it's possible to use biology to build an inorganic device that works." "But while it is a first step towards molecular computing based on this type of DNA configuration, we are still many years way from large scale self-assembly electronic devices, such as computers," Dekker cautions.

Braun's team began their manufacturing process by coating a central part of a long DNA molecule with proteins from an *E. coli* bacterium. Next, graphite nanotubes coated with antibodies were added, which bound onto the protein. After this, a solution of silver ions was added. The ions chemically attach to the phosphate backbone of the DNA, but only where no protein has attached. Aldehyde then reduces the ions to silver metal, forming the foundation of a conducting wire. To complete the device, gold was added. This nucleates on the silver and creates a fully conducting wire. The end result is a carbon nanotube device connected a both ends by a gold and silver wire. The device operates as a transistor when a voltage applied across the substrate is varied. This causes the nanotubes to either bridge the gap between the wires—completing the circuit—or not. Out of 45 nanoscale devices created in three batches, almost a third emerged as self-assembled transistors. They work at room temperature and the only restriction for future devices is that the components must be compatible with the biological reactions and the metal-plating process. The team have already connected two of the devices together, using the biological technique. "The same process could allow us to create elaborate self-assembling DNA sculptures and circuitry," says Braun. Journal reference: *Science* (vol 302, p 1380)

'Bio-barcoding' Promises Early Alzheimer's Diagnosis

Combining magnetic and gold nanoparticles with strands of DNA could allow the early detection of Alzheimer's disease. If successful, future treatments could then be used to prevent symptoms from ever appearing. Identifying and tracking Alzheimer's disease currently relies on brain imaging and psychological testing. A firm diagnosis can only be made by autopsy. But recent studies have revealed several biochemical markers that may provide the basis for a living diagnosis. These include tiny proteins called amyloid-beta-derived diffusible ligands (ADDLs), which exist at elevated levels in the brains of Alzheimer's patients. Dimitra Georganopoulou and her colleagues at Northwestern University in Chicago, Illinois, US, suspected ADDLs might also be present in the cerebrospinal fluid, which can be collected using a spinal tap. But the concentration of the proteins would be incredibly tiny, if they were there at all. So the team turned to a recently developed technique called bio-barcoding, shown to be thousands of times more sensitive for protein detection than conventional tests. "It's the only tool in town with the necessary sensitivity," says team member Chad Mirkin, whose lab developed the technique.

The bio-barcoding technique involves the ADDLs becoming sandwiched between two different antibodies. One antibody is attached to a 1000-nanometre-diameter magnetic particle and the other is fixed to a 30-nanometer-diameter gold particle. The gold nanoparticle also has thousands of short segments of DNA attached to it—the so-called barcoding DNA. Once the ADDLs are firmly attached to both types of the nanoparticle, they are drawn off using the magnetic particle. The separated mixture will contain all of the magnetic nanoparticles initially added but only those gold particles that became attached to the other side of an ADDL protein. The DNA snippets attached to the gold nanoparticles are then chemically released from the gold particles and counted, giving a hugely amplified signal of the original protein. The Northwestern University team used the barcoding method on cerebrospinal fluid from 15 people with Alzheimer's disease—as confirmed after

The team will now test the technique on blood samples, which are far less difficult to collect, although the concentration of ADDLs is expected to be even lower in blood than in cerebrospinal fluid. But the researchers are confident that the barcoding approach should still work. "We're nowhere near the limits of the test's sensitivity," says team member William Klein. ADDLs have been shown to attack the connections between neurons, called synapses, far earlier than currently observable symptoms appear in patients. So early detection could allow future treatments to prevent symptoms appearing in the first place, says Klein. Experiments using antibodies against ADDLs have even reversed memory loss in transgenic mice, says Klein, a scientific advisor to Chicago-based Acumen Pharmaceuticals. The company is working in collaboration with US company Merck & Co., Inc. to develop Alzheimer's disease vaccines based on ADDL research. Journal reference: *Proceedings of the National Academies of Sciences* (DOI: 10.1073/pnas.0409336102)

'Smart Bombs' to Deliver Fatal Blast to Tumours

Nanoscale polymer capsules could one day be used to deliver chemotherapy direct to tumours, leaving adjacent tissue unscathed. The capsules would be designed to rupture when heated by a low-energy laser pulse, delivering their payload right where it is needed. Anti-cancer drugs would be more effective, and the side effects less severe, if they could home in on a tumour and be delivered in a single burst. This would allow the drug to reach the concentrations needed to kill cancer cells, while minimising damage to surrounding tissue. So Frank Caruso and his team at the University of Melbourne, Australia, are developing an ingenious way of doing this. Their trick is to enclose the drug in polymer capsules that are peppered with gold nanoparticles and attached to tumour-seeking antibodies. When injected into the bloodstream, the capsules will concentrate inside tumours. When enough capsules have gathered there, a pulse from a near-infrared laser will melt the gold, which strongly absorbs near-infrared wavelengths. This will rupture the plastic capsules and release their contents.

To make the capsules, the researchers repeatedly add the polymer to a suspension of drug particles around 1 micrometre wide until the polymer forms multi-layered spheres containing the drug payload. They then add gold particles around 6 nanometres in diameter to the mix, which become embedded in the polymer. Finally they add a lipid, which forms an outer layer, and the antibodies that will target tumour cells. In lab tests, the capsules were ruptured by a 10-nanosecond pulse from a near-infrared laser. While bulk gold has a melting point of 1064°C, gold nanoparticles melt at far lower temperatures—between 600°C and 800°C. The brief pulse was enough to melt the nanoparticles, which could be seen under an electron microscope to swell to up to 50 nanometres in diameter as they coalesced. The pulse was too short to damage the contents of the capsule. Caruso showed that a lysozyme enzyme did not lose its activity after being released from the capsules in this way.

In clinical use, the laser would be able to penetrate a few millimetres of tissue. It could be shone through the skin, or be beamed inside the body via an endoscope. The 100 millijoules per square centimetre of infrared energy that would be needed to rupture the capsules is well within safety limits. "It's way below that used to remove tattoos," Caruso says. Clinical use is still some way off, however, and even animal tests are several years away. The next refinement will be to make the capsules a lot smaller. Caruso plans to shrink them from around 1 micrometre in diameter to a couple of hundred nanometres by starting with smaller drug particles. Caruso thinks his team's key innovation has been in making the capsules react to a laser that is harmless to the body. Gold usually absorbs light from the visible to ultraviolet part of the electromagnetic spectrum, which can burn tissue. But electromagnetic interactions between the gold nanoparticles in the capsules change the properties of the metal, making it absorb light from the near infrared instead. This light is the most transparent to tissue. Observers are impressed. "This work is tremendously cool," says Matt Trau of the Nanotechnology and Biomaterials Centre at the University of Queensland in Brisbane. Photo-activating the capsules without damaging surrounding tissue is the particularly innovative part, he says. Journal reference: *Advanced Materials* (vol 16, p 2184)

DNA 'Velcro' binds nanoparticles

Strands of DNA have been used to fasten—and then separate—nanoparticles in experiments that could lead to the development of fundamentally "self-constructing" materials. Christof Niemeyer and colleagues at the University of Dortmund, Germany, used sections of artificially synthesised DNA to attach gold nanoparticles together before separating them again. DNA consists of two complementary strands that bind together depending on the chemical "bases" on either strand. Attaching nanoparticles by appending them to complementary strands of DNA is a tried and tested technique, already used in some protein sensing systems. But, until now, no one has shown that it is possible to separate the strands again afterwards. Selectively binding nanoparticles could provide a way to construct complex nanostructures piece by piece, using different DNA strands to add different nanoparticles. These materials could have novel electrical and optical properties that cannot currently be obtained using conventional chemistry.

Third strand

Being able to separate these materials again would offer even greater flexibility. Niemeyer says it could conceivably be used to modify nanostructures after construction. “It could be used to build [self-constructing] materials,” he told New Scientist. “Although that is very far off at the moment.” The team uses artificial DNA, fabricated to have particular bases. Each gold particle—measuring around 15 nanometres (15 billionths of a metre) across—is attached using sulphur to the centre of a DNA strand. The strand extends out on either side. Two gold particles are then joined together by adding a third strand of DNA that complements half of each gold-bound DNA strand and attaches to both, forming a bridge. This just left the problem of how to separate the gold particles once the bridge was firmly in place.

So Niemeyer’s team developed the bridging strand of DNA to have one end longer than the other with the tip of the long end refusing to attach to the gold-bound strands, leaving it loose. The trick, finally, was to use an additional fourth piece of DNA to bind to this loose end of the bridging strand before “peeling” it away completely—like pulling off a piece of Velcro—leaving the gold particles separated once again. Chad Mirkin, who pioneered the use of DNA as a nanoscale construction material says the approach is interesting, although the potential applications remain unclear. “It’s a step towards creating a structure where you could have triggerable changes,” says Mirkin, now based at Northwestern University in Chicago, US. “But it’s not yet clear where these would be useful.” Journal reference: *Angewandte Chemie* (vol 43, p 6469)

The Global Market for High-Brightness LEDs

This new report, brought to you jointly by NanoMarkets and CIR, analyzes the world wide market for high brightness LEDs (HB-LEDs). Although there have been recent rumors of industry maturity in this sector recently, our new study finds that HB-LEDs continue to grow rapidly, both in terms of revenues and in terms of new applications. This report is geared to the needs of both managers and investors who must understand the future prospects for the HB-LED markets and where the emerging business opportunities are to be found. Among the areas on which this report provides in-depth guidance are the following: Key design and manufacturing improvements that are enabling HB-LED manufacturers to distinguish themselves in the market and improve yields; The impact of new materials platforms such as quantum dots and silicon photonics for the HB-LED market; Emerging applications in signage, intelligent transport systems, rural lighting and dentistry and how HB-LEDs are enabling entirely new kinds of products to be designed and manufactured; The changing geography of HB-LED markets. In which part of the world are HB-LEDs being bought? In which parts of the world are they being manufactured?. Strategies and structure. In this report, we critically appraise the strategies of the major HB-LED manufacturers including their manufacturing and outsourcing strategies and ask the questions — is there room for start-ups in HB-LED space anymore? And, will there be consolidation in the HB-LED busines?; CIR has been tracking the optoelectronics business for more than two decades while NanoMarkets has produced over a dozen reports that offer a unique perspective on the impact of new materials on electronics and semiconductor markets. This report was based on interviews conducted by an international team carrying out interviews on three continents and will be invaluable reading for manufacturers of electronics and lighting products of all kinds as well as the HB-LED manufacturers themselves — and, of course, investors. The report includes detailed forecasts broken out by type of product and application.

HP Quantum Science Researchers Win Two More Awards

Researchers in molecular electronics at HP Labs, already the recipients of more than a half-dozen awards, have been honored by The Wall Street Journal and by Small Times magazine. The Wall Street Journal gave a runner-up Global Technology Innovation Award in its semiconductor category to the team's molecular crossbar latch, a technology that may be able to replace transistors – the fundamental building blocks of computers for the past 40 years – leading to a new way to build computers in the future. In addition, Phil Kuekes, senior computer architect, was named Best of Small Tech researcher of the year award by the nanotechnology publication Small Times for his contributions to the crossbar latch and other advances. Kuekes accepted the award at the publication's NanoCommerce conference in Chicago Nov. 2. The Wall Street Journal solicited nominations from all over the world through advertisements in the paper and through the Innovation Awards Web site. The newspaper received some 750 applications from approximately two dozen countries. Applicants completed a rigorous form that requested extensive details relating to their entry and why it constituted a true innovation. Wall Street Journal editors then screened the applications and selected 104 semifinalists. Journal editors looked for innovativeness, clarity of explanation and whether the innovation was covered by patents or had achieved some kind of track record. "Innovation is the key for any company to have continued success, and through these awards The Wall Street Journal recognizes these technological breakthroughs throughout the world," said Karen Elliott House, Journal publisher and senior vice president, Dow Jones & Company. "This year's winners showed the creative thinking that has allowed them to rise to the top of their respective industries."

Wake Forest Improve Efficiency of Organic or Flexible Solar Cells

At a time when oil prices are reaching record highs and people are bracing for winter heating bills, researchers at *Wake Forest University's Center for Nanotechnology and Molecular Materials* have made significant strides in improving the efficiency of organic or flexible solar cells. Traditional silicon solar panels are heavy and bulky and convert about 20 percent of the light that hits them to useful electrical power. For years, researchers have worked to create flexible, or "conformal," organic solar cells that can be wrapped around surfaces, rolled up or even painted onto structures, but the best scientists have been able to do is about 3 percent efficiency, until now.

Researchers at Wake Forest, with the help of researchers at New Mexico State University, have achieved an efficiency rate for organic solar cells of almost 6 percent. In order to be considered a viable technology, the solar cells must be able to convert about 10 percent of the energy in sunlight to electricity. Wake Forest researchers hope to reach 10 percent by October 2006, said David Carroll, director of the nanotechnology center at Wake Forest. "The consumer market would be really open to having these conformal systems if you could, for instance, roll them up and put them away," said Carroll, who is also an associate professor in Wake Forest's physics department. "Imagine a group of hikers with a tent that when you unrolled the tent and put it up, it could generate its own power. Imagine if the paint on your car that is getting hot in the sun was instead converting part of that heat to recharge your battery." Carroll said flexible, organic solar cells also offer several possibilities for military use. "The military would obviously want something like that because you could only put maybe tens of those big solar panels on a transport, but you could put hundreds of ultra-thin flexible ones on a transport and supply half the army," he said. Most experts have estimated that flexible, solar cell technology for consumers was about a decade away, but Carroll said the new breakthrough at Wake Forest and NMSU means that consumers could be using this technology in the next five years. Using a set of polymer coatings, researchers at Wake Forest constructed a nanophase within the polymer called a "mesostructure." The "mesostructure" changes the properties of the plastic and makes it better for collecting light. The researchers also removed the current from the polymer coating, Carroll said. A test system at Wake Forest's nanotechnology center was used to simulate the sun, Carroll said, and the simulated spectrum was precisely measured and shot onto the organic solar cell, which appeared as a thin coat of paint. Devices at the center have registered almost 6 percent efficiency. This breakthrough was announced in October at the Santa Fe Workshop on Nanoengineered Materials and Macro-Molecular Technologies, which was sponsored by Wake Forest's nanotechnology center.

Nanotechnology Kills Cancer Cells

Nanotechnology has been harnessed to kill cancer cells without harming healthy tissue. The technique works by inserting microscopic synthetic rods called carbon nanotubules into cancer cells. When the rods are exposed to near-infra red light from a laser they heat up, killing the cell, while cells without rods are left unscathed. Details of the Stanford University work are published by Proceedings of the National Academy of Sciences. Researcher Dr Hongjie Dai said: "One of the longstanding problems in medicine is how to cure cancer without harming normal body tissue. "Standard chemotherapy destroys cancer cells and normal cells alike. "That's why patients often lose their hair and suffer numerous other side effects. "For us, the Holy Grail would be finding a way to selectively kill cancer cells and not damage healthy ones."

The carbon nanotubules used by the Stanford team are only half the width of a DNA molecule, and thousands can easily fit inside a typical cell. Under normal circumstances near-infra red light passes through the body harmlessly. But the Stanford team found that if they placed a solution of carbon nanotubules under a near-infra red laser beam, the solution heated up to about 70C in two minutes. They then placed the tubules inside cells, and found they were quickly destroyed by the heat generated by the laser beam. Dr Dai said: "It's actually quite simple and amazing. We're using an intrinsic property of nanotubes to develop a weapon that kills cancer." The next step was to find a way to introduce the nantubules into cancer cells, but not healthy cells. The researchers did this by taking advantage of the fact that, unlike normal cells, the surface of cancer cells is covered with receptors for a vitamin known as folate. They coated the nanotubules with folate molecules, making it easy for them to pass into cancer cells, but unable to bind with their healthy cousins. Exposure to the laser duly killed off the diseased cells, but left the healthy ones untouched.

The researchers believe it should be possible to refine the technique still further, for instance by attaching an antibody to a nanotubule to target a particular kind of cancer cell. They have already started work on tailoring the technique to target lymphoma in mice. Dr Emma Knight, of the charity Cancer Research UK, said: "Nanotechnology has a lot to offer biomedical science, and the results of this paper suggest yet another way in which it may help in the fight against cancer. "However, this work is still at a very early stage. The researchers have shown that near-infra red light can cause nanotubes to produce heat that can kill cancer cells. "But their work so far has focused on cells that have been grown in culture in the laboratory. "Further research will be crucial to see whether these effects can be reproduced in the more complex environment of a tumour and, ultimately, the human body."

Particles Home in On Diseased Cells

Nanoparticles that reveal pancreatic cancer cells in mice have been developed in the laboratory of Ralph Weissleder at Harvard Medical School. Tools for early detection of pancreatic cancer in humans are lacking, and the nanoparticles may be a step toward better diagnosis of this deadly disease.

Weissleder and coworkers thought that decorating certain fluorescent magnetic nanoparticles—already being used clinically for magnetic resonance imaging—with small molecules would enable the nanoparticles to be taken up selectively by specific cells. The researchers have now shown that this idea works. For example, nanoparticles studded with isatoic or 5-chloroisatoic anhydride are selectively taken up by pancreatic cancer cells in mice. Their presence inside those cells can be revealed by whole-body fluorescence or magnetic resonance imaging (*Nature Biotechnol.,* published Oct. 23, dx.doi.org/10.1038/nbt1159).

Weissleder's approach is novel, says Robert S. Langer, professor of chemical engineering at Massachusetts Institute of Technology. "It could lead to a whole new way of targeting cells," he adds. The work is a step toward giving nanomaterials more biological function—in this case, the ability to access specific cells or biological processes, Weissleder says. "One long-term goal is to use materials like these to detect diseases much earlier than currently possible," he adds. The researchers prepared a library of 146 nanoparticles, each conjugated to an average of 60 copies of a small molecule. Screening revealed 14 with high affinity for pancreatic cancer cells. Among those, the particles decorated with isatoic acid or 5-chloroisatoic acid are poorly taken up by macrophages, which scavenge unmodified nanoparticles, and so were prepared on a large scale for the in vivo studies with mice.

Nanowires May Provide Innards for Quantum Computers

SEMICONDUCTING nanowires could one day store and process data in quantum computers. A team of researchers led by Silvano De Franceschi of the Delft University of Technology in the Netherlands made nanowires out of indium arsenide, each about 100 nanometres in diameter and 100 times as long. By attaching aluminium electrodes to the ends of the wires, they could probe the wires' behaviour at temperatures close to absolute zero. At such temperatures aluminium becomes superconducting, able to conduct electrical current with zero resistance. The team found that the superconducting electrodes induced similar behaviour in the nanowires because the superconducting properties leaked into the wire—the so called "proximity effect". By applying a voltage to the substrate on which the nanowires rested they could alter the strength of the proximity effect and so turn the superconductivity on and off at will (*Science*, vol 309, p 272). Because superconductivity is an inherently quantum effect, superconducting nanowires could form part of the quantum circuitry necessary for a future quantum computer. They could store information, or help read out the result of a calculation.

Revolutionary New Nanotechnology-Based Lithium-Ion Battery

A123Systems, developer of a new generation of Lithium-ion batteries, today unveiled its technology and announced that it is delivering batteries with unprecedented power, safety, and life as compared to conventional Lithium technology. A123Systems' first battery is now in production and being delivered to the Black & Decker Corporation. It will be first utilized by the corporation's DeWALT brand, a leading manufacturer of power tools.

A123Systems' battery technology delivers up to 10X longer life, 5X power gains and dramatically faster charge time over conventional high power battery technology, as validated by independent testing at Motorola and government research labs. A123Systems' batteries use proprietary nanoscale electrode technology built on research at Massachusetts Institute of Technology (MIT) and exclusively licensed from MIT.

"A123's revolutionary technology will enable manufacturers to improve the performance and form factor of existing high-power portable devices and to transform products currently dependent on power cords and sockets into a new class of portable devices," said David Vieau, CEO and president of A123Systems. "We expect that our technology will have the same impact on high-power products as the introduction of first generation Lithium-ion technology had on the development and commercialization of consumer electronics in the 1990s."

A123Systems' initial family of batteries is targeted at applications that require high power, high levels of safety, and longer life. These include power tools, advanced medical devices, hybrid electric vehicles, mobility products such as electric scooters, robotics, and consumer electronics.

DeWALT is using A123systems' battery in a new, innovative, heavy-duty 36-volt power tool platform. The 36-volt power tools will provide users with increased levels of power and runtime, at a similar or less weight than their corded counterparts. Available in 2006, the new power tool line was designed from the ground up to ensure durability and reliability and to make optimum use of the increased power available from A123's innovative battery technology. The line will include a hammerdrill, reciprocating saw, circular saw, impact wrench, rotary hammer, jigsaw, flashlight and combo kits.

A123Systems has seen considerable interest in its advanced battery technology across many applications. The company is working with the U.S. Department of Energy as part of a major undertaking to develop battery materials for future use in hybrid electric vehicles. As part of its work with the DOE, A123Systems has already completed two development phases of advanced battery technology. Preliminary performance results show a technology with great promise for the future of electric and hybrid electric vehicles.

Power, Safety and Life – What Makes A123Systems' Batteries Unique:

High Power. A123Systems' first product packs up to five times the power density of current rechargeable, high power batteries. In addition, the battery has the ability to recharge to 90% of its capacity in five minutes.

Intrinsic Safety. Unlike conventional Lithium-ion batteries, A123Systems' batteries employ new thermally stable, non-combustible active materials, enabling a safer cell and allowing cost reductions such as the elimination of unnecessary battery pack components. In addition, A123Systems uses an environmentally friendly chemistry.

Long Life. With up to 10X improvement in life over existing rechargeable batteries, A123Systems' batteries can deliver thousands of cycles at high rates. Cycles refer to the number of times a battery can be charged and discharged before it no longer has any power remaining.

In addition to its high power products, A123Systems is working on a family of high-energy products that significantly improves upon the performance of existing Lithium-ion batteries used in notebook computers and cellular phones. This line of products will be introduced in 2006.

With more than 300 person-years of battery industry experience, the staff at A123Systems is made up of technology industry veterans and pre-eminent scientists from some of the world's leading battery companies and research institutions. A123Systems is led by David Vieau, and was founded by Dr. Yet-Ming Chiang, Dr. Bart Riley, VP of research and development, and Ric Fulop, VP of business development and marketing.

Having raised more than $32 million in funding, A123's investors include Desh Deshpande (chairman of the A123 board), Qualcomm, Sequoia Capital, Motorola, North Bridge Venture Partners, MIT, YankeeTek and OnPoint Technologies, a strategic private equity firm funded by the United States Army.

Source:L http://www.a123systems.com

AML Cracks Chinese, French, Irish & German MEMS, Microelectronics & Opto Markets

After breaking into the German market with a sale to Fraunhofer IZM in Berlin earlier this year, AML has now delivered one of the latest generation of Aligner – Wafer bonding machines to China with a sale to Xiamen University. It has been a good few weeks for breaking into new markets, as well as Germany and China, the first machine was delivered last week to France namely CNRS-LAAS and the first machine will be delivered to Ireland's Tyndall National Institute (formerly NMRC) in a few weeks time.

Rob Santilli director of AML declared " we now have machines in almost every major industrialised country including, UK, France, Germany, Holland, Denmark, Ireland, USA, Japan, Korea, India, Taiwan and China. "

All this follows the recent announcement that AML is now the key wafer bonding centre node in the new multi million pound UK MNT infrastructure with investment in facilities to aid the development of new bonding and associated processes, an applications database, training and technology transfer. This will support the wafer bonding service provided by AML on a global basis. AML's equipment arm also ensures that machines able to run the new bonding processes developed can be made available to industry.

Rob Santilli added, " This further demonstrates how different we are to our major competitor who is basically just an engineering company. Recently at COMMS 2005 they stated that they "do not carry out internal R&D but just implement what their customers ask of them" AML feels internal R&D and the development of processes is an integral part of being a wafer bonding company – it gives us a quite a big edge to the extent that customers now prefer to pay for our machines than receive an almost free one from our competitors!"

Tony Rogers – Technical director said " The increased sales show that the MNT industry now recognises that AML's unique technology of in-situ align and bond is the industry's preferred method for aligned wafer bonding"

Nick Aitken, AML's senior applications engineer also added, "AML is in a unique position to offer consultancy, process development and low volume production runs to emerging and established optoelectronic and MEMS companies as well as research institutions. Given the ever increasing diversity in optoelectronic materials and MEMS device requirements, AML's ability to take the risk out of the packaging development and deliver a working process along with proven bonding equipment, will be of benefit to the industry as a whole."

Source: http://www.aml.co.uk

Wake Forest Improve Efficiency of Organic or Flexible Solar Cells

At a time when oil prices are reaching record highs and people are bracing for winter heating bills, researchers at Wake Forest University's Center for Nanotechnology and Molecular Materials have made significant strides in improving the efficiency of organic or flexible solar cells.

Traditional silicon solar panels are heavy and bulky and convert about 20 percent of the light that hits them to useful electrical power. For years, researchers have worked to create flexible, or "conformal," organic solar cells that can be wrapped around surfaces, rolled up or even painted onto structures, but the best scientists have been able to do is about 3 percent efficiency, until now.

Researchers at Wake Forest, with the help of researchers at New Mexico State University, have achieved an efficiency rate for organic solar cells of almost 6 percent. In order to be considered a viable technology, the solar cells must be able to convert about 10 percent of the energy in sunlight to electricity. Wake Forest researchers hope to reach 10 percent by October 2006, said David Carroll, director of the nanotechnology center at Wake Forest.

"The consumer market would be really open to having these conformal systems if you could, for instance, roll them up and put them away," said Carroll, who is also an associate professor in Wake Forest's physics department. "Imagine a group of hikers with a tent that when you unrolled the tent and put it up, it could generate its own power. Imagine if the paint on your car that is getting hot in the sun was instead converting part of that heat to recharge your battery."

Carroll said flexible, organic solar cells also offer several possibilities for military use.

"The military would obviously want something like that because you could only put maybe tens of those big solar panels on a transport, but you could put hundreds of ultra-thin flexible ones on a transport and supply half the army," he said.

Most experts have estimated that flexible, solar cell technology for consumers was about a decade away, but Carroll said the new breakthrough at Wake Forest and NMSU means that consumers could be using this technology in the next five years. Using a set of polymer coatings, researchers at Wake Forest constructed a nanophase within the polymer called a "mesostructure." The "mesostructure" changes the properties of the plastic and makes it better for collecting light. The researchers also removed the current from the polymer coating, Carroll said. A test system at Wake Forest's nanotechnology center was used to simulate the sun, Carroll said, and the simulated spectrum was precisely measured and shot onto the organic solar cell, which appeared as a thin coat of paint. Devices at the center have registered almost 6 percent efficiency.

This breakthrough was announced in October at the Santa Fe Workshop on Nanoengineered Materials and Macro-Molecular Technologies, which was sponsored by Wake Forest's nanotechnology center.

Source: http://www.wfu.edu/

Nanion Receives Bavarian Innovation Award 2005

The German nanobiotech company Nanion Technologies GmbH was recently awarded the Bavarian Innovation Award 2005 and is Bavarias company of the year (small/ medium enterprise (SME)).

Nanion received the award for its chip-based patch clamp technique, which allows to perform sophisticated electrophysiological measurements from cells in an automated and parallel manner. Nanions entry level device, the Port-a-Patch©, supersedes the common pipette based patch clamp technique in ease of use and throughput and doesn't require a microscope, vibration isolation or a micromanipulator. The Port-a-Patch© uses an all electrical read-out for automatically positioning and electrically contacting cells on the chip. With the Port-a-Patch©, Nanion offers the world's smallest patch-clamp workstation. This innovative drug discovery technology makes patch clamp available also to non-electrophysiologists.

This innovation and its successful market introduction has been award the prestigious prize, which clearly is an important step and great commendation for Nanion.

Source: http://www.nanion.de

The Quantum Dot History Project from Evident Technologies

Evident Technologies has announced that it launched a living online history of early work in the field of quantum dot technology, "The Quantum Dot History Project," which is a representative overview of key research and researchers from 1960-1996.

Evident welcomes the submission of additional entries and will award a $500 Amazon gift certificate for the earliest published research entry accepted and an Apple iPod nano for each of the next two earliest published entries accepted. In appreciation for the efforts of individuals submitting entries, the company will provide them with a $10 Amazon gift certificate for each novel history addition accepted. Deadline for submission is January 31, 2006, and is open to all.

"Quantum dots are a versatile form of semiconductors that mean scientists and engineers are no longer shackled by the fixed properties of nature. With quantum dots, new materials can be engineered to fit a need, solve a problem and create new products or even markets," said Clinton Ballinger, Ph.D., Chief Executive Offi cer of Evident Technologies.

"We are publishing this overview as our commitment to pioneers who did the early work in this fi eld and to future scientists and developers who want to use quantum dots to create revolutionary change," said Ballinger. "The early semiconductor work of Bell Labs in 1947 led to a whole revolution in electronics, forever changing our world. An open and free flowing exchange of ideas and knowledge in our field, which our overview celebrates, could help improve so many areas of our life."

"I am very pleased to see this history of nanocrystals on-line as a ready resource for current and future researchers," said Dr. Alexander Efros, currently at the United States Naval Research Laboratory and one of the earliest discoverers of quantum dots. "When we discovered quantum confinement in nanocrystals in the early 80's, we knew this would be an exciting area of scientific exploration and discovery. I am personally excited to see today's pioneers at Evident translating quantum dot technology into product applications ranging from life science research to solar cells to anticounterfeiting."

Dr. Margaret Hines, Head of Nanomaterials R&D at Evident Technologies, commented: "Explorations into the science of quantum dots began decades ago. The technology is finally gaining momentum. It is inspiring to see this presentation of the rich history of quantum dots. The web site is easily accessible and will interest scientists at all levels and spur further innovation."

To view the history or enter the contest visit www.evidenttech.com/history.

Source: http://www.evidenttech.com

Fuel Efficiency Increases For Vehicles Using Fuel Borne Envirox Cerium Oxide Catalyst Additive From Oxonica

Envirox has been extensively tested both in the laboratory and under commercial operating conditions in long term trials. Envirox has demonstrated a significant statistically tested fuel economy benefit as high as 11.4%. The highly active catalyst contained in Envirox, cerium oxide has long been recognised for its catalytic properties and is used in many other industrial applications, including use in automotive gasoline catalytic systems. Cerulean has used nanotechnology expertise developed by its parent company Oxonica to allow the material to be dispersed directly into diesel fuel.

Envirox delivers fuel economy by two separate mechanisms. Cerium oxide is a redox catalyst that helps to promote the optimum amount of oxygen within the combustion chamber, leading to a more complete and cleaner burn and increased power. In addition the material significantly lowers the temperature at which carbon deposits are burnt off allowing the potential to restore and retain the engine as close as possible to its original design state.

Following initial work in the laboratory Cerulean saw that it was necessary to prove that the benefits seen there would also be apparent under actual commercial operating conditions. To test this an extensive range of field evaluations have been initiated.

Trials have been designed to allow fuel economy comparison between groups of vehicles in commercial operation. Since fuel efficiency is dependent on many variable factors: weather, load, age of vehicle, route etc, care has been taken to minimise these variables as far as possible.

Trials have been run under commercial operating conditions, wherever possible with additised and control group of a similar age with the same engine and transmission types, running similar routes and using the same base fuel. Mileage readings and the amount of fuel used is recorded regularly for each vehicle. Lubricant samples are taken periodically and details of routings, individual maintenance carried out on the vehicles etc are recorded for the duration of the evaluation.

This trial comprised 80 vehicles, 40 with Volvo engines and 40 with Cummins engines. Of these 20 were allocated to the additised group and 20 to the control group. All of the vehicles were approximately seven years old at the start of the trial, all routes were city operation. The trial lasted for six months and data were collected for a further 2 months after the dosing ended.

Analysis of pre-trial data demonstrated that the fuel economy of the additised and control groups of buses was similar prior to the start of dosing. When use of the additive ceased the fuel economy of the additised group quickly returned to that of the control group. The data from the Hong Kong trial were analysed by an independent statistician. The main trial fuel economy benefit after the first 6 weeks was decided to be the best representation of the ongoing effect of the additive on fuel economy. The additive has a cumulative effect on fuel economy and the full effects are seen after 6-8 weeks. The difference in mean fuel economy after the first 6 weeks between the Cummins additised fuel group and the Cummins control group is 0.15 km/L (95% Confidence Interval). This corresponds to a fuel economy benefit of 11.4% for the additised Cummins group, compared to the control Cummins group, after the first 6 weeks of use.

The difference in mean fuel economy after the first six weeks between the Volvo additised fuel group and the Volvo control group is 0.11 km/L (95% Confidence Interval). This corresponds to a fuel economy benefit of 9.9% for the additised Volvo group compared to the control Volvo group after the first 6 weeks of use. The conclusion of this analysis was that "This trial has demonstrated a clear and highly significant fuel economy benefit for buses which used the diesel additive over the control buses which used unadditised fuel...The benefit of the use of additised fuel for both Cummins and Volvo engines was found to be robust to the method of analysis used and to methods of handling odd data points."

A range of commercial evaluations are currently underway with a number of fleet customers operating under different conditions. Evaluations include further work in buses, trucks and service fleets across a number of countries including the Philippines, Singapore, Australia, Taiwan and the United Kingdom. Approximately twelve hundred vehicles are currently being run with Envirox. Evaluations continue to show an increase in fuel economy, in some cases a greater fuel economy benefit than Cerulean demonstrated in the Hong Kong trial. Further 'real time' evidence for cerium oxide's known function of lowering the temperature at which carbon combusts has been demonstrated by a reduction in black smoke and a lowering of DPF regeneration temperature by 30-40°C. Evidence for a more complete and cleaner burn has also been demonstrated by the results of several test programmes in independent engine test houses. These programmes were designed to look at the emissions benefits when using Envirox and as such were short duration tests which did not allow for the 'bedding-in' period required to see the full fuel economy benefit of Envirox. However, even in these tests an immediate small but statistically relevant fuel economy benefit was observed.

Envirox is currently available to commercial fleet operators in the United Kingdom, Southern Europe and the Asia-Pacific region. The technology is scheduled to be rolled out in other regions and in other sectors including marine and power applications.

Malvern to display new particle characterization systems at Powtech 2005

By: mcleodc (mcleodc) 2005.07.15: Malvern Instruments will introduce a new particle size and shape analyzer based on image analysis at Powtech 2005 (Nürnberg,

11-13 October, 2005). The system will be displayed alongside other new products and flagship instruments from Malvern's comprehensive portfolio of rheometers, viscometers, and on-line, at-line and laboratory-based particle characterization systems.

The breadth of Malvern's product range enables the company to support materials characterization applications across all areas of industry and academia, from laboratory analysis through to in-process measurements. Being able to provide instrumentation, technical and applications support for both rheology measurement and particle characterization is especially important as the links between particle characteristics and a material's bulk properties are increasingly examined and understood.

As well as new product introductions, highlights on the Malvern stand will include: the recently launched Insitec APS at-line particle size analyzer for use in process applications; the industry-leading Mastersizer 2000 laser diffraction system; and the award-winning Zetasizer Nano system for the characterization of nano-sized particles and molecules.

2nd International Conference and Exhibition on Very Small (Nano) Technologies

Your company and industry WILL be affected by the new Nano technologies and by attending this industry conference you will find out:

- Exactly how this will happen.
- How Nanotechnology is already helping overseas companies produce goods that are taking away your markets.
- What can be done, and what is being done to help companies like yours counteract this.

Attend this conference and also have DIRECT input into the Federal Government's decision making process into the future of Nanotechnology aimed to assist Australia's manufacturers become more competitive.

Who Should Attend?

Every decision-maker in EVERY manufacturing company in Australia which produces/ manufacturers goods and products simply cannot afford not to attend this very difference kind of conference.

Why You Should Attend?

Because every one of you reading this page will have heard the words "Nanotechnology", but very few of you will fully understand what this means, and how, if at all, this has anything to do with you or your company.

The Conference

- Will explain what Nanotechnology is: what it can and is capable of doing to any industry; why you ignore finding out about this new technology at your peril; and, once you know what Nanotechnology can do, how, where and from whom, can you get access, information, advice and assistance about it?
- Will also provide you with a direct opportunity to have a say about your ideas and views regarding issues, problems, solutions and opportunities facing Australia's takeup of Nanotechnologies. The Federal Government will have representatives in attendance who will be interested to hear your views, because they are developing policies for Nanotechnology.
- Will give you direct access to experts from all forms of governments, research and industry who will explain and provide you and your company with practical solutions and real answers to what can be done to help Australian companies be capable of holding their own in an increasingly competitive environment.

If you are usually too busy or simply do not attend conferences, think again for THIS one.

This conference WILL change the way you think about your business; about your future competitiveness and the way you see nanotechnology and your business.

Trends in NanoTechnology (TNT) 2001

Electronics, Photonics, Nanotubes, Mechanical Devices, Theoretical Models, Self-Assembly, Networks and Funding

September 3-7, 2001: I was lucky enough to attend TNT 2001 in Segovia Spain, where 250+ nano scientists and engineers met to discuss the latest Trends in NanoTechnology. While most scientific meetings are held in modern hotels or lecture halls, TNT 2001 took place in the Isabelino-gothic monastery 'Santa Cruz'. The old, yet aesthetically stable surroundings reminded me that mankind has been fabricating remarkable structures on the macroscale for millennia. Thus, the setting seemed an inspiring place to learn how to do that on the nanoscale.

On the more modern side of the conference were the trends, with some of the cooler subjects being Electronics, Photonics, Nanotubes, Mechanical Devices, Theoretical Models, Self-Assembly, Networks and Funding. Here I will briefly describe why each of these trends appears to be in fashion in the world of nanotechnology.

Electronics

There doesn't seem to be any question that our microelectronics industry will soon have to deal with the finite nature of matter found at the nanoscale. This imminent problem (known by the semiconductor industry as "the red brick wall") provides a major motivation for a better understanding of electronic behavior at the nanoscale, for reliable nanofabrication processes, and the development of novel computational architectures. While purely top-down silicon processing can likely continue improving at a steady rate for 5-10 years, nanoelectronics is expected to bring about a new paradigm in computing, and is a very hot application for nanotechnology.

Photonics

Photonic Band Gap (PBG) Materials are similar to semiconductors, where the electrons are replaced by photons (also known as light). By creating periodic structures out of materials with contrast in their dielectric constants, it becomes possible to guide the flow of light through the PBG material in a way similar to how electrons are directed through doped regions of semiconductors. The period of the structure is related to the wavelength of light for which a PBG will exist, for instance a few hundred nanometers for visible light. These materials are a relatively new discovery, with potential applications ranging from efficient light emitting diodes to quantum computers.

Nanotubes

One of the most stylish materials resulting from nanotechnology is the stable phase of carbon known as carbon nanotubes. Discovered ten years ago by Sumio Iijima, one of the participants of TNT 2001, the number of different types of nanotubes (and related materials) is as diverse as their potential applications. A few of those discussed at the conference were for scanning probe microscopy (SPM) tips, field emission displays, nanocircuitry, and even catalysts. With so much potential for utility, nanotubes may just be the wheel of nanotechnology. However, because of their nanoscale diameters there are still plenty of inventions to be made in nanotube research and development.

Mechanical Devices

Macroscale mechanical engineers have already demonstrated the utility of mechanical devices such as cars, assembly lines, plumbing, clocks and other machines. Micromechanics and nanomechanics on the other hand is only now being developed. A typical example of a useful micromechanical system is a cantilever (a miniature diving board). As a key component in many types of SPMs, cantilevers bend in response to forces so small that they enable one to create images of surfaces with nanoscale resolution. While SPM is one success story for micromachines, several others are likely on their way with two realistic applications being very sensitive sensors and intelligent drug delivery. While research involving synthetic (meaning non-biological) nanomechanical devices seems motivated more by curiosity than immediate applications, it's worth remembering that the first microscope was created by a toy-maker.

Theoretical Models

Mathematical and computational modeling of nanoscale systems requires a unique combination of quantum mechanics with classical models. Ab initio (meaning computed from first principles without any empirical input) quantum mechanical models require an immense amount of computational power, which severely limits the size of a system that can be modeled by accurate quantum mechanical models to about 50 atoms with today's best super computers (of course nanotechnology is helping to improve our computers). Classical models, on the other hand can neglect important quantum effects that give nanoscale devices their unique properties. The development of efficient and accurate mathematical models of nano-systems is an essential and rewarding effort.

Self-Assembly

As features decrease in size from micro to nano, they become more expensive to create using conventional fabrication technologies such as photolithography or electron beam lithography. Thus, bottom-up methods that take advantage of the natural tendency for atomically precise devices to self-organize are a highly sought after approach to nanotechnology. Furthermore, conventional lithography is inherently two dimensional, while self-assembly can make full use of three dimensions. An example of a success story for self-assembly is the photonic band gap material known as "inverse opal". This material can be created by filling in the spaces between nanoscale spheres which have been self-assembled into a 3-D hexagonally packed array. As an added bonus, nature provides a plethora of self-assembling molecules to serve not only as inspiration but also as a tool-box for molecular self-assembly.

Networks and Funding

Now that nanotechnology has been accepted by the scientific community, funding interdisciplinary nanotech research and collaborative networks is becoming increasingly trendy. Governments, universities and capitalists are gaining interest in nanotechnology, which means better labs, well defined goals, and more money for nanoscientists and nanoengineers. Nanotechnology in the European Union appears to be a high priority. For instance, the PHANTOMS Network is a European initiative which provides funding to European seniors and young students involved in nanoelectronics. The UK based Institute of Nanotechnology helps to spread the word via the internet. Nanotechnology in the Ibero-American community is facilitated by the CYTED Sub-Program IX: Microelectronics. In the USA, NASA and the National Science Foundation (NSF) help to fund and organize nanotech research. For those considering going commercial, the Joint Venture CMP Cientifica / Spark Inversiones provided tips on how to obtain venture capital.

The utility of interdisciplinary nanotech conferences such as TNT for sparking new ideas was apparent. For instance, on the first day Juan José Sáenz gave a theoretical lecture about how a strong force should be exerted on a neutral particle of nanoscale dimensions inside of a waveguide. Questions about how to measure the effect, and even one about how it might be useful followed. On the next day Christian Joachim gave a talk entitled, "Mono-Molecular Mechanical Machines at the Nanoscale" in which molecules were being pushed around on a surface with a scanning tunneling microscope tip. Changes in the tunneling current could then be correlated to mechanical movements of sub-molecular components. So, the moral of this story is that it might be worth trying to roll nanomachines down a nano-hill by placing neutral nanomachines in a waveguide. Just one idea of many...

Acknowledgements

Thank you Antonio Correia (+Tim, Adriana & Nacho) of CMP Cientifica, Pedro Serena of CSIC, Juan José Sáenz of the Universidad Autónoma de Madrid, and everyone who helped to organize the excellent conference. I'm also grateful to all of the participants and am looking forward to meeting again at TNT2002, or sooner.

NSTI Nanotechnology Conference and Trade Show - Nanotech 2005

Nanotech Program Highlights

- 600+ Nano & Bio Presentations
- 50+ Nanotech Venture Companies
- 200+ Investment/IP/Funding Talks
- 16 Industrial Focused Tutorials

Nanotech 2005 Conference Program — May 8 - 12

- Physical Sciences, Electronics, Materials, Microsystems, Food, Energy
- Life Sciences, Medicine, Personal Care, Biotech, Health Care
- Business, Investment, Government Programs and Nanotech Ventures

Nano Industrial Impact Workshop — Sunday May 8

One-Day Intensive Program to educate participants in the current state of the art in a range of nanotechnologies and impacted industries. All courses are presented by leading experts in their respective fields.

Nanotech and Bio Nano Expo — May 9 - May 11

Full Nanotech Trade Show with over 150 exhibiting companies May 9-10! One-Day BioNano/Nanotech exhibit and poster session on May 11th. View participating companies and floor plans.

Nanotech Ventures Emerging Company Review — May 9 - 11

Join the Nanotech Ventures community in participating in the world's largest nanotechnology venture forum. 50 Early Stage Companies Presenting in most every high-tech industrial sector.

National Cancer Institute Symposium — Tuesday May 10

NSTI is proud to collaborate with the National Cancer Institute (NCI) and Northeastern University in presenting a Special Symposium on Nanotechnology for Cancer Prevention, Diagnosis and Treatment.

Conference Program by Industry

Health Sciences

- Cancer Treatment
- Drug Delivery
- Clinical Diagnostics
- Medical Imaging
- Medical Applications

Materials

- Nano Particles
- CNTs, Carbon Structures
- CNT Devices
- Self Assembly

Transportation

- Fuel Cells
- Sensing
- Advanced Materials

Energy

- Optoelectronics
- Fuel Cells & Batteries
- Oil Extraction
- Catalysis

Environmental

- Protection & Remediation

- Sensor Networks
- Health Systems
- Bio and Chem Sensors

Telecommunications

- MEMS, NEMS
- Nanoscale Devices
- Active Devices

Displays & Optics

- Organics
- MOEMS & MEMS
- Optoelectronics
- Nano Photonics

Polymers

- Surfaces & Interfaces
- Additives & Particulates
- Soft Materials
- Composites

Nano Biotechnology

- Bio Molecular Analysis
- Microfluidics
- Bio Sensors, Arrays
- Cell Handling

Surface Science

- Scanning Probe Microscopy
- Functionalized Surfaces
- Coatings

Personal Care

- Nanostructured Fluids
- Condensed Matter
- Soft Nanotech

Food

- Food Industry Nanotech
- Nanostructured Colloids
- Surfaces and Flims
- Sensors

Electronics

- Devices, Processes
- Nano Photonics
- Spins,
- Nanowires

Computational

- Atomic Simulation
- Mesoscale Modeling
- Comp Biology

Design & Modeling

- Microsystems (MSM)
- Compact Modeling (WCM)
- Microfluidics

Coatings

- Thin Films, Electronics
- Paints & Smart Coatings
- Self Cleaning & Healing

Confirmed Speakers

National Nanotechnology Coordination Office (NNCO)

Clayton Teague, *Director, NNCO*

Nanotechnology and Transcending Moore's Law

Steve Jurvetson, *Draper Fisher Jurvetson, US*

Nanotechnology at Motorola Labs

Vida Ilderem, *Motorola, US*

National, DOD, Navy - Nanotechnology Program Perspectives
James S. Murday, *NSET Executive Secretary, NNI, Office of Naval Research, US*

Mauro Ferrari
Davis Heart & Lung Research Institute, *Ohio State University*

Bio-Nano-Informatics Fusion
Chih-Ming Ho, *University of California, Los Angeles*

Addressing Challenges in Cancer Diagnosis and Therapy in the 12st Century
Thomas J. Kipps, *University of California, San Diego*

New Nanocrystal Assemblies for Use in Cancer Research
Paul Alivisatos, *University of California Berkeley*

Carbon Nanotube based Nanotechnology
Meyya Meyyappan, *NASA Ames Research Center, US*

Semiconductor Nanocrystal Probes for Human Chromosomes Peter E. Barker, NIST, Rebecca and John Moores UCSD Cancer Center

Medical Devices: What does Nanotechnology enable?
Michael Helmus, *Boston Scientific Corporation*

Cubosome Formation Processes - Kinetic Effects and Consumer Product Implications Patrick T. Spicer, Procter & Gamble, US

Emerging Nanoelectronic Devices and their Benchmarking Methodology
Robert Chau, *Intel Corporation, US*

Introduction to Polymer Nanocomposites
Gregory S. Blackman, *DuPont, US*

Nano and Molecular Electronics
Mark Reed, *Yale University, US*

Nanotechnology at ChevronTexaco
Waqar Qureshi, *ChevronTexaco Technology Ventures, US*

Systems of Microsystems: wireless sensor networks
Deborah Estrin, *University of California at Los Angeles, US*

Functional Nanoparticulate Synthesis and Applications

Lutz Mädler, *University of California, Los Angeles*

Nanostructured Fluids, Soft Materials, and Self-Assembly

Tim Lodge, *University of Minnesota, US*

Semiconductor Nanowires for Future Nanoelectronics

Lars Samuelson, *Lund University, SE*

Nanoscale Simulation Tools

Wolfgang Windl, *Ohio State University, US*

Nanotechnology in Soft Materials

Fiona Case, *Case Scientific, US*

Nanoscale Photonics: Optics beyond the diffraction limit

Wolfgang Bacsa, *UMR-CNRS, FR*

Fundamentals and Applications of Micro, Bio and Nano Fluidics

Steven T. Wereley, *Purdue University, US*

From Micro- to Nanosystems: mechanical sensors go nano

Christopher Hierold, *ETH Zürich, CH*

Self-Assembly and Directed Assembly of Amphiphilic Block Copolymers,Paschalis Alexandridis, State University of New York at Buffalo, US

Simulation of Nanostructured Fluids, Soft Materials, and Self-Assembly

Toshihiro Kawakatsu, *Tohoku University, JP*

WINS: protocols design and implementation

Ian Marshall, *University of Kent, UK*

Higher-order Sigma-Delta Modulator Interfaces for MEMS

Michael Kraft, *Southampton University, UK*

Node design for pervasive sensor networks: a top down view

Robert M. Newman , *Coventry University, UK*

National Nanotechnology Policy

Richard M. Russell, *Associate Director, OSTP, Executive Office of the President, US*

Emerging Technologies and Corporate Investment

Keith R. Larson, *Intel Capital, US*

The National Nanotechnology Initiative from a Policy Perspective: Where are we and where are we going?

Celia Merzbacher, *NRL Representative, OSTP, Executive Office of the President, US*

Cell Adhesion & Extracellular Matrix Biology

Erkki Ruoslahti, *The Burnham Institute, US*

Electronics and Optoelectronics with Single Carbon Nanotubes

Phaedon Avouris, *IBM, US*

Nanoparticles in Cell Imaging, John V. Frangioni, Beth Israel Deaconess Medical Center, Harvard University

Application of Self-assembly Principles in Foods

Martin Michel, *Nestlé Research Center, Lausanne, CH*

Gregory J. Downing

Director, Office of Technology and Industrial Relations, *National Cancer Institute*

Nanoscale Electronic Processing

Douglas Resnick, *Molecular Imprints, US*

Nanotechnology for Medical Imaging and Therapy

Mansoor Amiji, Department of Pharmaceutical Sciences, *Northeastern University, US*

Spin manipulation in semiconductors and spintronic ultrafast nanodevices

A.M. Bratkovsky, *Hewlett-Packard Laboratories, US*

Applications of Ultra-Dense Nanowire Circuitry

James R. Heath, *California Institute of Technology, US*

Smart Dust: Concept ? Macro ? Micro ? Nano

Kris Pister, *Berkeley Sensor & Actuator Center, Dust Networks, US*

Timothy D. Sands

Basil S. Turner Professor of Engineering, *Purdue University, US*

Application of Polymer Self Assembly to Semiconductor Microelectronics

Charles T. Black, *IBM, York Town, US*

A History of MOS Transistor Compact Modeling

Chih-Tang Sah, *University of Florida, Gainesville, Florida, US*

Vladimir Torchilin

Bouvé College of Health Sciences, *Northeastern University, US*

Nanoscale Bio Analysis

Srinivas Iyer, *Los Alamos National Laboratory, US*

Nanotechnology in Drug Delivery, Hayat Onyuksel, O. Koo, A. Krishnadas, V. Sethi and I. Rubinstein, University of Illinois at Chicago, US

Bio-Nano Materials

Thomas J. Webster, *Purdue University, US*

"Intelligent sensors: systems or components?"

Neil White, *Southampton University, UK*

Soft Glasses and Gels - From Colloid Physics to Food Technology

Peter Schurtenberger, *Dept. of Physics, Univ. of Fribourg, CH*

Food Goes Nano -Novel Nano Sized Liquid Vehicles and Technology for Solubilization of Nutraceuticals in Foods

Nissim Garti, *Nutralease and the Hebrew University, IL*

Recent Applications of Mesoscale Modeling to Nanotechnology and Drug Delivery, Amitesh Maiti, Accelrys, San Diego, US

Data centric wireless sensor networks

Paul Havinga, *University of Twente, NL*

Integrated Resonant Microstructures for Sensor and Timing Applications

Ashwin A. Seshia, *Cambridge University, UK*

Y-nano X-micro technologies: nanometric optical control

Davies William De Lima Monteiro, *Federal University of Minas Gerais, BR*

Consumer and Automotive Embodiments of Smart Sensors

Benedetto Vigna, *ST Microelectronics, IT*

Conference Quick Links

- Conference Program
- Industrial Impact Workshop
- Nanotech Business & Investment
- Exhibit & Sponsorship Opportunities
- Registration
- Submit IP/Tech Transfer

Sponsors

Nanotech, BioNano, ICCN, and MSM are sponsored and or endorsed by:

Gold Sponsors

- Motorola
- National Cancer Institute
- Nanotech Japan
- Australian Government - Invest Australia
- Swiss Nanotech

Silver Sponsors

- Nanophase Technologies
- Veeco Instruments
- Keithley Instruments
- Leica Microsystems
- Advance Nanotech, Inc.
- Wilmer Cutler Pickering Hale and Dorr LLP
- Dorsey & Whitney L.L.P.
- CBEN Center for Biological and Environmental Nanotechnology
- IEE Inspec

Gold Key Room Sponsors

- Taiwan Business Alliance Forum
- Finetex Technology
- Yield Engineering System, Inc.

Nanotech Ventures Sponsors

- Dorsey & Whitney L.L.P.
- Wilmer Cutler Pickering Hale and Dorr LLP
- Trilogy Capital Partners
- ChevronTexaco
- Brinks Hofer Gilson & Lione

Sponsors

- General Electric Company
- ChevronTexaco
- Olympus Industrial America
- Ciphergen Biosystems, Inc.
- National Science Foundation
- Northern Ireland Nanotechnology
- Nanotec NI
- Italian Trade Commission
- Freescale Semiconductors
- Texas Instruments
- Agilent Technologies
- SEMI
- Zygo Corporation
- Burns, Doane, Swecker & Mathis, L.L.P.
- Accelrys, Inc.
- Hielscher
- NanoInk, Inc.
- Raith
- Cantion
- piezosystem jena
- ALCATEL - Adixen Vacuum Technology
- Evans Analytical Group
- Edax, Inc.
- Carl Zeiss SMT
- Jenoptik Mikrotechnik GmbH

- LEISTER Technologies, LLC
- Intelligent Enclosures
- ChemImage
- PolyInsight, LLC
- First Nano, Inc.
- Kinetic Systems, Inc.
- HERZAN, LLC
- BioForce Nanosciences, Inc.
- Malvern Instruments, Inc.
- Atomistix
- NIST Advanced Technology Program
- nanoTITAN
- FEI Company
- Georgia Technode - NNIN
- Nanometrix, Inc.
- LioniX BV
- Mad City Labs
- SUSS MicroTec
- Sigma-Aldrich
- Bühler AG
- NanoWorld AG
- Applied Nanotech
- Asylum Research
- COMSOL
- Nanosurf AG
- NANOSENSORS
- NanoMarkets LLC
- Infinitesima, Ltd.
- Center for Nanoscale Materials, Argonne National Laboratory
- NanoGram Corporation
- NanoDynamics, Inc.
- mPHASE Technologies, Inc.

- Phoeni*X*
- Nanostellar, Inc.
- Wyatt Technology Corporation
- Particle Sizing Systems, Inc.
- HORIBA Jobin Yvon
- Seiko Instruments Inc.
- Nanotechnology Researchers Network Center of Japan
- Foley & Lardner L.L.P.
- NaturalNano, Inc.
- PI (Physik Instrumente) L.P.
- U.S. Civilian Research & Development Foundation
- ANSYS, Inc.

Global Partner

- nano tech 2005, Tokyo, Japan

Supporting Organizations

- National Institutes of Health (NIH)
- The Whitaker Foundation
- Earth and Sky Radio Series
- TIMA-CMP Laboratory, France
- Ecos Corporation
- Boston University
- Greenberg Traurig
- The Global Emerging Technology Institute, Ltd. (GETI)
- The Electrochemical Society, Inc.
- SAGE Crossroads

Media Sponsors

- Nature Publishing Group
- Science / AAAS
- Springer
- Institute of Physics
- Photonics Spectra

- nanotechweb.org
- IEEE Computer Society
- Nanotechnology Law & Business
- EurekAlert! / AAAS
- SciTechDaily Review
- PhysOrg.com
- A2C2 Magazine
- NanoBiotech News
- Computing in Science & Engineering
- D&MD Publications
- ivcon.net
- Controlled Environments
- Excite Media Group
- Nanotechnology Now
- NanoInvestorNews
- Foresight Institute
- tinytechjobs
- AZoNano.com
- Asia Pacific Nanotechnology Forum (APNF)
- Fuel Cell Magazine
- Materials Engineering News
- NanoSPRINT
- Battery Power Products & Technology
- Equipment Protection Magazine
- Antenna Systems & Technology
- The Nanotech Company
- R&D Magazine
- Industrial Hygiene News
- nanoparticles.org
- Laser Focus World
- Military & Aerospace Electronics
- nanovip.com

- Small Times Media
- The Scientist
- Southern California Physician
- BioTechniques
- Journal of Experimental Medicine
- Materials Today
- BioProcess International
- BioExecutive International
- The Journal of Cell Biology
- mst | news
- Berkeley Nanotechnology Club
- Springer Microfluidics and Nanofluidics
- Springer Microsystem Technologies
- International Business Forum Conferences

Publishing Sponsors

- Nanotech, NanoBio, ICCN, MSM Conference Technical Proceedings, Nano Science and Technology Institute
- Molecular Simulation, Taylor and Francis Group

Sponsorship Opportunities

To become a sponsor, please follow.

Exhibiting Companies

Confirmed Exhibitors

- Accelrys, Inc.
- Accurion
- Advance Nanotech, Inc.
- Advance Nanotech, Inc.
- Agilent Technologies
- Alcatel Micro Machining Systems
- Ambios Technology, Inc.
- Andeen-Hagerling, Inc.
- Applied MicroStructures, Inc.

- Applied MicroStructures, Inc.
- Applied Nanotech
- Applied Surface Technologies
- ARC Seibersdorf Research
- Artech House Publishers
- Asylum Research
- Asylum Research
- Atomistix
- Australian Government — Invest Australia
- Austrian Research Promotion Agency (FFG)
- Austrian Trade Commission
- AZoNetwork Group
- BioForce Nanosciences, Inc.
- BioForce Nanosciences, Inc.
- Bruker AXS
- Burns, Doane, Swecker & Mathis, L.L.P.
- Bühler AG
- Cantion
- Carl Zeiss SMT
- Center for Nanoscale Materials, Argonne National Laboratory
- Centice
- Charles Evans & Associates
- Cheap Tubes
- Cheap Tubes
- ChemImage
- COMSOL
- Contact East
- CRC Press
- Dorsey & Whitney L.L.P.
- Dorsey & Whitney L.L.P.
- Edax, Inc.
- Edmonton Economic Development Corporation

- EV Group
- Evans Analytical Group
- EXAKT Technologies, Inc.
- Excite Media Group
- FEI Company
- Finetex Technology
- First Nano, Inc.
- Flow Science Inc.
- Foley & Lardner L.L.P.
- Halcyonics
- Harvard University
- Heidelberg Instruments
- Herzan
- Hielscher
- Hielscher
- HORIBA Jobin Yvon, Inc.
- HTA Photomask
- IBF – International Business Forum
- IMS Nanofabrication
- Infinitesima, Ltd.
- Inspec
- Institute of Medical Technologies and Health Management - Joanneum Research
- Integrated Engineering Software
- Intelligent Enclosures
- Intelligent Enclosures
- Interferomet Ltd
- IOP Publishing
- Italian Trade Commission
- Italian Trade Commission
- Japan Pavilion
- Jenoptik Mikrotechnik GmbH

- Jensen Tools
- JEOL
- Keithley Instruments
- Kinetic Systems, Inc.
- Leica Microsystems
- LEISTER Technologies, LLC
- LEISTER Technologies, LLC
- Mad City Labs
- Malvern Instruments, Inc.
- Mantis Deposition
- Micromeritics
- Microtrac, Inc.
- MiNTeC
- Minus K Technology, Inc.
- Molecular Imaging
- Molecular Imprints
- mPHASE Technologies, Inc.
- nano tech Japan
- nano tech Japan
- Nano-S Biotechnoloy
- NanoDynamics, Inc.
- NanoGram Corporation
- NanoInk, Inc.
- NanoInk, Inc.
- NanoMarkets LLC
- Nanometrix, Inc.
- Nanomotion, Inc.
- Nanophase Technologies
- Nanoscience Instruments
- NANOSENSORS
- NanoSPRINT
- Nanostellar, Inc.

- Nanosurf AG
- Nanotec NI
- Nanotechnology Researchers Network Center of Japan, National Institute for Materials Science
- nanoTITAN
- NanoVantage, Inc.
- NanoWorld AG
- National Cancer Institute
- National Cancer Institute
- National Institute for Nanotechnology
- National Nanotechnology Infrastructure Network — Georgia Technode
- National Science Foundation
- Naval Research Laboratory
- Nion Technology Corporation
- NIST Advanced Technology Program
- Norsam Technologies, Inc.
- Northern Ireland Nanotechnology
- Ocean Optics
- Olympus Corporation
- Olympus Corporation
- Olympus Industrial America
- Pacific Nanotechnology
- Particle Sizing Systems, Inc.
- PhoeniX / Intellisense
- Photonics Spectra
- PI (Physik Instrumente) L.P.
- piezosystem jena
- piezosystem jena
- PolyInsight, LLC
- Precision Process Equipment, Inc.
- Quesant Instrument Corporation
- Raith USA, Inc.

- RGC Jenkins
- SAMCO International, Inc.
- Scientific Systems
- Seki Technotron
- semi
- Sigma-Aldrich
- SII NanoTechnology Inc.
- Silex Microsystems AB
- Smart Imaging Technologies Co.
- SoftMEMS
- Sokang Nano Technology
- SPECS Scientific Instruments, Inc.
- Springer
- Strem Chemicals, Inc.
- Surface Imaging Systems
- SUSS MicroTec
- Swiss Business Hub USA
- Synkera Technologies
- Taiwan Business Alliance Forum
- Telic Company
- Tethis
- The Nanotech Company
- Transfer Devices, Inc.
- TSL
- U.S. Army Corrosion Office
- U.S. Civilian Research & Development Foundation
- UCI Office of Technology Alliances & Integrated Nanosystems Research Facility
- Veeco Instruments
- Wilmer Cutler Pickering Hale and Dorr LLP
- Wyatt Technology Corporation
- Yield Engineering Systems, Inc.

- Yield Engineering Systems, Inc.
- Zygo Corporation

Confirmed Participating Companies Include:

- 3DM, Inc.
- 3i
- 3M
- Accelrys, Inc.
- Accurion
- Ace Korea, Inc.
- Activator Corporation
- Advanced Diamond Technologies
- Advanced Micro Devices
- ALD NanoSolutions, Inc.
- Ambri Ltd
- AMRI-UNO
- Apax Partners
- Aphios Corporation
- Applied Materials, Inc.
- Applied Nanotech, Inc.
- BioForce Nanosciences, Inc.
- Biophan Technologies, Inc.
- Biozentrum
- Blue Heron Biotechnology, Inc.
- Buhler AG
- Burns Doane
- Cabot Corporation
- Cadence Design Systems
- Cambrios Technologies Corporation
- CapiliX BV
- Cargil Investments
- Center for Textile Technologies

- CFD Research Corporation
- Charles Evans & Asscciates
- ChevronTexaco
- ChevronTexaco Technology Ventures
- Ciphergen Biosystems, Inc.
- CMEA Ventures
- Colossal Storage Corporation
- ConverTec Corporation
- Cypress Semiconductor
- CytImmune Sciences, Inc
- Degussa AG
- Doresy and Whitney
- Draper Fisher Jurvetson
- Duke Scientific Corporation
- DuPont
- DuPont Ventures
- Eastman Kodak Company
- Echelon Ventures
- EV Group
- FEI Company
- Fluent, Inc.
- Freescale Semiconductor
- Fujitsu Laboratories
- Gatan, Inc.
- GE
- GeneFluidics, Inc.
- General Electric
- GMA Industries, Inc.
- Greenberg Traurig
- Harris and Harris Group
- HECUS X-Ray Systems GmbH
- Hewlett Packard Laboratories

- Hielscher
- HighGround
- Honeywell
- HRL Labs
- IBM
- IDC
- Imation Corporation
- Infinitesima
- Inframat Corporation
- Intel
- Intel Capital
- Jenoptik
- Keithley
- Kinik Company
- Konarka
- Kopin
- Leica
- LTX Corporation
- Lumera Corporation
- Malvern Instruments, Inc.
- MD Anderson Cancer Center
- Mesophotonics Ltd.
- Mohr, Davidow Ventures
- Molecular Imaging, Inc.
- Molecular Imprints
- MolecularDiamond Technologies
- Motorola
- Motorola Ventures
- mPhase Technologies
- Nano Interface Technology, Inc.
- NanoBioMagnetics, Inc.
- NanoDynamics, Inc.

- NanoInk, Inc.
- Nanometrix, Inc.
- Nanomix Inc
- Nanophase Technologies Corporation
- Nanoplex Technologies
- NanoSciences, Inc.
- NanoSPRINT
- Nantero
- Nestlé Research Center
- NTT
- NTT DoCoMo
- Nutralease
- ON Semiconductor
- Philips Research Laboratories
- Platypus Technologies
- PolyInsight, LLC
- Procter & Gamble
- Quantum Sphere
- Quid Technologies, Inc.
- Samsung
- SENSIA
- Siliconlinx
- Silvaco Japan
- Solaris Nanosciences Corporation
- SolidWorks Corporation
- SRI International
- ST Microelectronics
- Sumitomo Electric
- SurModics, Inc.
- Suss
- Synkera Technologies, Inc.
- Synopsys, Inc.

- The Aerospace Corporation
- The Boeing Company
- UBS
- Veeco
- VTI Technologies
- VTT Information Technology
- Wilmer Cutler Pickering Hale and Dorr
- Zyvex Corporation

Nanotech Ventures — Early Stage Presenting Companies

Biotech

- Biophan Technologies, Inc.
- Cambridge Biomagnetics, UK
- CytImmune Sciences, Inc.
- Midatech Limited, UK
- Nano S Biotechnologie, AT
- NanoBioMagnetics, Inc.
- NanoTechLogic Corporation
- Nanotrope, Inc.
- Raustech Pty.Ltd., AU
- Russian Biotechnologies, RU
- ViaLogy Corporation

Electronics

- arccure technologies GmbH, DE
- Nanometrix, Inc., CA
- NanoOpto Corporation
- Nanopower Technologies, Inc.
- Optimer Photonics, Inc.
- XCom Wireless, Inc.

Energy

- Akermin
- Blu-Si

- Collossus Nanotech
- QuantumSphere

Health Care

- 3DM Inc.
- Alpha-O Peptides, CH
- AptaSense, AU
- Inovatia Lab.
- Nano Interface Technology

Instrumentation

- Digital Matrix Corp
- Focal Point Microsystems
- Iatia Ltd, AU
- ionscope limited, UK
- Molecular Imprints
- Prodyne Corporation
- Quantum Precision Instruments, AU

Materials

- Advanced Diamond Technologies, Inc.
- Applied MicroStructures
- Atomistix A/S, DK
- Ceimig Ltd, UK
- ENSbio
- Nanothinx, GR
- Solaris Nanosciences

Sensors

- Applied Standard
- BioNano Systems
- Cantion A/S, DK
- CENTICE Corporation
- MET Tech
- SENSIA, ES

Glossary

Active shield: A defensive system with built-in constraints to limit or prevent its offensive use.

Active site: Region of an enzyme surface to which a substrate molecule binds in order to undergo a catalyzed reaction.

Binding: The process by which a molecule (or ligand) becomes bound, that is, confined in position (and often orientation) with respect to a receptor. Confinement occurs because structural features of the receptor create a potential well for the ligand; van der Waals and electrostatic interactions commonly contribute.

Bio-assemblies or Biomolecular Assemblies: Containing several protein units, DNA loops, lipids, various ligands, etc.

Biochauvinism: The prejudice that biological systems have an intrinsic superiority that will always give them a monopoly on self-reproduction and intelligence.

Biocompatible coated materials: Biocompatible materials usually used in dental and bone implants that enhance biologic fixation, thereby increasing the bond strength between the coated material and bone, and minimize possible biological effects that may result from the implant itself. MeSH, 1999

Biocompatible materials: Synthetic or natural materials, other than drugs, that are used to replace or repair any body tissue or bodily function. MeSH, 1973

Body-centered cubic (BCC): A common crystal structure found in some elemental metals. Within the cubic unit cell, atoms are located at corner and cell center positions.

Bond: Two atoms are said to be bonded when the energy required to separate them is substantially larger than the van der Waals attraction energy. Ionic bonds result from the electrostatic attraction between ions; covalent and metallic bonds result from the sharing of electrons among atoms; hydrogen bonds are weaker and result from dipole interactions and limited electron sharing. When used without modification, "bond" usually refers to a covalent bond.

Bone substitutes: Synthetic or natural materials for the replacement of bones or bone tissue. They include hard tissue replacement polymers, natural coral, hydroxyapatite, beta-tricalcium phosphate, and various other biomaterials. The bone substitutes as inert materials can be incorporated into surrounding tissue or gradually replaced by original tissue.

Breakdown: Failure of a material resulting from an electrical overload. The resulting damage may be in the form of thermal damage (melting or burning) or electrical damage (loss of polarization in piezoelectric materials). The failing of a physical principle or unit of measurement as the scale is changed.

Brownian Assembly: Brownian motion in a fluid brings molecules together in various position and orientations. If molecules have suitable complementary surfaces, they can bind, assembling to form a specific structure. Brownian assembly is a less paradoxical name for self-assembly (how can a structure assemble itself, or do anything, when it does not yet exist?).

Brownian Motion: Motion of a particle in a fluid owing to thermal agitation, observed in 1827 by Robert Brown. (Originally thought to be caused by vital force, Brownian motion in fact plays a vital role in the assembly and activity of the molecular structures of life).

Corrosion: Deteriorative loss of a metal as a result of dissolution environmental reactions.

Coulomb [C]: Measure of electrical charge: 1 C is an amount of charge equal to that of about 6.24×10^{18} electrons.

Coupling: A connection between more than one object or energy pathway so that together they function as a single unit.

Covalent bond: A bond formed by sharing a pair of electrons between two atoms. A primary interatomic bond that is formed by the sharing of electrons between neighboring atoms. An attachment between two atoms resulting from the sharing of a pair of electrons

Covalent radius: Given a set of N elements that can form covalent single bonds in molecules, with $N(N - 1)$ possible elemental pairings, it has proved possible to define a covalent radius for each element such that the actual bond length between any two elements that form a covalent single bond is roughly equal to the sums of their covalent radii.

CPU: The central processing unit of a computer, responsible for executing instructions to process information.

Creep: The time-dependent permanent deformation that occurs under stress; for most materials it is important only at elevated temperatures.

Crosslinked polymer: A polymer in which adjacent linear molecular chains are joined at various positions by covalent bonds.

Cross-linking: A process forming chemical bonds between two separate molecular chains.

Cross-sensitivity: The influence of one measurand on the sensitivity of a sensor, another measurand.

Crosstalk: Electromagnetic noise transmitted between leads or circuits in close proximity to each other.

Crybiology: The science of biology at low temperatures; research in cryobiology has made possible the freezing and storing of sperm and blood for later use.

Crystal Lattice: The regular three-dimensional pattern of atoms in a crystal.

Crystal structure: For crystalline materials, the manner in which atoms or ions are arrayed in space. It is defined in terms of the unit cell geometry and the atom positions within the unit cell.

Crystallescence: In medical nanorobotics, the crystallization of solid solute that is offloaded by nanorobot sorting rotors at a concentration that exceeds the solvation capacity of the surrounding solvent.

Curie temperature (also Curie point) (T_c): The temperature above which a ferromagnetic or ferrimagnetic material becomes paramagnetic. For iron the Curie point is 760oC and for nickel 356°C.

Current [A]: Measure of rate of flow of electric charge: a one-ampere current is a flow of 1 C of charge per second.

Cutoff: Condition in a diode or bipolar junction transistor in which the potential across a p-n junction prevents current flow.

Cyclic: A structure is termed cyclic if its covalent bonds form one or more rings.

Cycloaddition: A reaction in which two unsaturated molecules (or moieties within a molecule) join, forming a ring.

Cyclotron: A type of particle accelerator in which an ion introduced at the center is accelerated in an expanding spiral path by use of alternating electrical fields in the presence of a magnetic field.

Cytocarriage: In medical nanorobotics, the commandeering of a natural motile cell, by a medical nanorobot, for the purposes of in vivo transport (of the nanorobot), or to perform a herding function (of the affected cell), or for other purposes.

Deoxyribonucleic acid (DNA): A huge nucleotide polymer having a double helical structure with complementary bases on two strands. Its major functions are protein synthesis and the storage and transport of genetic information.

Deoxyribonucleic acid: Deoxyribonucleic acid are long chains consisting of four kinds of nucleotides; the order of these nucleotides encodes the information needed to construct protein molecules. These in turn make up much of the molecular machinery of the cell. DNA is the genetic material of cells.

Dermal Zippers: In medical nanorobotics, a theorized medical nanorobot that can rapidly perform incision-wound repairs to the dermis and epidermis; dermal zippers.

Design Ahead: The use of known principles of science and engineering to design systems that can only be built with tools not yet available; this permits faster exploitation of the abilities of new tools.

Design Diversity: A form of redundancy in which components of different design serve the same purpose; this can enable systems to function properly despite design flaws.

Design: To plan and delineate with an end in mind and subject to constraints.

Desmosome: Specialized cell-cell junction, usually formed between two epithelial cells, characterized by dense plaques of protein into which intermediate filaments in the two adjoining cells insert.

Detergent: Type of small amphipathic molecule that tends to coalesce in water, with its hydrophobic tails buried and its hydrophilic heads exposed; widely used to solubilize membrane proteins.

Determination: Commitment by an embryonic cell to a particular specialized path of development; it reflects a change in the internal character of the cell.

Development: Succession of changes that take place in an organism as a fertilized egg gives rise to an adult plant or animal.

Devitrification: The process in which a glass (noncrystalline or vitreous solid) transforms to a crystalline solid.

Dialysis: A phenomenon in which a semipermeable membrane allows transfer of both solvent molecules and small solute molecules and ions.

Diode: two-terminal device that conducts current well in one direction and poorly in the other.

Dipole (electric): A pair of equal yet opposite electrical charges that are separated by a small distance.

Directed-Assembler: A specific type of assembler that makes use of directed-assembly, such that the assembly process requires external energy or information input.

Disassembler: A system of nanomachines able to take an object apart a few atoms at a time, while recording its structure at the molecular level. (2) An instrument able to take apart structures a few atoms at a time, recording structural information at each step. (3) In molecular nanotechnology, a nanomachine or system of nanomachines able to take an object apart while at each step recording the structure and composition of that object at the molecular level.

Disasterbation: Idly fantasizing about possible catastrophes (ecological collapse, full-blown totalitarianism) without considering their likelihood or considering their possible solutions and preventions. [David Krieger, 1993]

Disequilibration: In medical nanorobotics, maintenance or inducement of a state of perpetual ionic, chemical, or energetic disequilibrium in a living cell by a medical nanorobot, usually for the purpose of inducing cytocide.

Dislocation: A linear crystalline defect around which there is atomic misalignment. Plastic deformation corresponds to the motion of dislocations in response to an applied shear stress. Edge, screw, and mixed dislocations are possible.

Disruptive Technology: Technology that is significantly cheaper than current, is much higher performing, has greater functionality, and is frequently more convenient to use. Will revolutionize markets by superseding existing technology. "Paradigm shifting" is a well-worn connotation. Although the term may sound negative to some, it is in fact neutral. It is only negative when businesses who are unprepared for change fail to adapt, only to fall behind and fail. The results are not *evolutionary*, they are *revolutionary*.

Disruptive techonlogies: In his book, *The Innovator's Dilemma*, Harvard professor Clayton Christensen used the term "disruptive technologies" to describe developments capable of displacing established market leaders.

Dissolution: Deterioration in an organism such that its original structure cannot be determined from its current state.

Distributed Intelligence: An intelligent entity which is distributed over a large volume (or inside another system, like a computer network) with no distinct center. This is the opposite to the strategy of Concentrated intelligences. Distributed intelligences have much longer communications lags, but are more flexible in their structure and can survive damage to their parts.

DNA (Deoxyriobonucleic Acid): DNA molecules are long chains consisting of four kinds of nucleotides; the order of these nucleotides encodes the information needed to construct protein molecules. These in turn make up much of the molecular machinery of the cell. DNA is the genetic material of cells.

DSP: Digital Signal Processing; a process by which a sampled and digitized data stream (real-time data such as sound or images) is modified in order to extract relevant information. Also, a digital signal processor.

DumbSizing: Apealing to the least common denominator by explaining difficult concepts in such a manner so they loose meaning. Also, talking down to someone less informed or learned. [uhf]

Dynamic characteristics: A description of an instrument's behavior between the time a measured quantity changes value and the time the instrument obtains a steady response.

Electromotive force (emf) series: A series of chemical elements arranged in order of their electromotive force. The electromotive force is the greatest potential difference that can be generated by a particular source of electric current. In practice this potential may be observable only when the source is not supplying current, because of its internal resistance.

Enabling science and technologies: Areas of research relevant to a particular goal, such as nanotechnology. Also, technology that "enables" other technology to advance, such as the transistor enabled the computer chip revolution, as did photolithography.

Endocrine cell: Specialized animal cell that secretes a hormone into the blood; usually part of a gland, such as the thyroid or pituitary gland.

Endocytosis: Uptake of material into a cell by an invagination of the plasma membrane and its internalization in a membrane-bounded vesicle.

Endoplasmic reticulum (ER): Labyrinthine, membrane-bounded compartment in the cytoplasm of eucaryotic cells, where lipids are synthesized and membrane-bound proteins are made.

Endosome: Membrane-bounded organelle in animal cells that carries materials newly ingested by endocytosis and passes many of them on to lysosomes for degradation.

Endothelium: Single sheet of highly flattened cells (endothelial cells) that forms the lining of all blood vessels. Regulates exchanges between the bloodstream and surrounding tissues and is usually surrounded by a basal lamina.

Energy [J]: Capacity for performing work or to cause heat flow. Like work itself, it is measured in Joules.

Energy Conversion: Changing energy from one form to another.

Engineering: The use of scientific knowledge and trial-and-error to design systems.

Engulf formation: In medical nanorobotics, a configuration that may be adopted by a metamorphic nanorobot, in which the nanorobot reshapes itself to create an interior cavity capable of trapping a living cell, virion, or other biological particle.

Exponential Growth: Growth that proceeds in a manner characterized by periodic doublings. Inaccurately referred to as "self-replication," exponential growth refers to the process of growth or replication involving doubling within a given period.

Extrinsic: Characterizes doped, rather than pure, semiconductor.

Fabricator: A small nano-robotic device that can use supplied chemicals to manufacture nanoscale products under external control. Fabricators could work together to build macroscale products by convergent assembly. Similar to assemblers, but less complex, easier to build, and probably more efficient.

Face-centered cubic: A crystal structure found in common elemental metals also FCC. Within the cubic unit cell, atoms are located at all corner and face-centered positions.

Fail-stop: Describes a component or subsystem that, in the event of a failure, produces no output (e.g., of material or data) rather than producing a damaged or incorrect output.

Faraday: A constant representing the charge on one mole of electrons; 96, 485 coulombs.

Far-Field Diffraction Limit: A physical limit to how focused a beam of light can be.

Fatty acid: Compound such as palmitic acid that has a carboxylic acid attached to a long hydrocarbon chain. Used as a major source of energy during metabolism and as a starting point for the synthesis of phospholipids.

Fault-tolerant: Describes a system that can suffer failure in a component or subsystem, yet continue to function correctly.

Femtoengineering: Will involve engineering using mechanisms within a quark.

Femtometer: [abbr: fm] A unit suitable to express the size of atomic nuclei. One quadrillionth (10 to minus 15) of a meter.

Femtosecond: Femtosecond is one quadrillionth of a second, and is to a second what a second is to 32,700,000 years. At 186,000 miles per second, in one femtosecond light travels only far enough to traverse about 1,000 silicon atoms. When used to time a laser pulse, it allows for ultra-precise micromachining, with virtually no damage to surrounding material.

Femtotechnology: Femtotechnology the art of manipulating materials on the scale of elementary particles (leptons, hadrons, and quarks). [CA-B] The next step smaller after picotechnology, which is the next step smaller after nanotechnology.

Ferromagnetism: Permanent and large magnetizations found in some metals (e.g. Fe, Ni and Co), resulting from the parallel alignment of neighboring magnetic moments.

Fertilization: Fusion of a male and a female gamete (both haploid) to form a diploid zygote, which develops into a new individual.

FET: Field-Effect Transistor — semiconductor device whose insulated gate electrode controls current flow.

Fiber-optic: Relates to transmission of information as modulated light in tiny transparent fibers instead of copper wires.

Fibrin sealant: A biological tissue glue derived from human plasma. Sometimes referred to as fibrin glue, it contains the components necessary for the last step of coagulation: i) fibrinogen and other coagulation factors and ii) thrombin. It is used in surgical procedures to arrest bleeding and as an adjunct to wound healing.

Fibrin: A protein derived from FIBRINOGEN in the presence of THROMBIN, which forms part of the blood clot. [MeSH]

Foglet: A mesoscale machine. A discreet component of utility fog. [J. S. Hall 1994]

Follicle cell: One of the cell types that surround a developing oocyte or egg.

Follower: A component in a cam system that is driven through a pattern of displacements as it rests against a moving contoured surface.

Free Radical: A molecule containing an unpaired electron, typically highly unstable and reactive. Free radicals can damage the molecular machinery of biological systems, leading to cross-linking and mutation.

Free-energy change (DG): Change in the free energy during a reaction: the free energy of the product molecules minus the free energy of the starting molecules. A large negative value of DG indicates that the reaction has a strong tendency to occur.

Frequency response: Two relations between sets of inputs and outputs. One relates frequencies to the output-input amplitude ratio; the other relates frequencies to the phase difference between the output and input.

Frequency: Number of times per second that a quantity representing a signal, such as a voltage, changes state. Also, the number of waves (cycles) per second that pass a given point in space.

FUD: Fear, Uncertainty, Doubt.

Fullerene: A class of cage-like carbon compounds composed of fused, pentagonal and/or hexagonal sp^2 carbon rings. A new allotrope of carbon characterized by a closed cage structure consisting of an even number of three coordinate carbon atoms devoid of hydrogen atoms. This class was originally limited to closed-cage structures with twelve isolated five- membered rings, the rest being six- membered rings. [IUPAC Provisional Recommendations for the Nomenclature for the C_{60}-I_h and C_{70}-$D_{5h(6)}$ Fullerenes, 2001]

Fullerenes: Fullerenes are a molecular form of pure carbon discovered in 1985. They are cage-like structures of carbon atoms, the most abundant form produced is buckminsterfullerene (C60), with 60 carbon atoms arranged in a spherical structure. There are larger fullerenes containing from 70 to 500 carbon atoms. [Wid] What are fullerenes? A very good explanation, even though some of the translation is a bit rough.

Functional Group: Part of a molecule (group) with the potential to take part in a chemical reaction

Functional navigation: In medical nanorobotics, a form of nanorobotic navigation in which nanodevices seek to detect subtle variations in their environment, comparing sensor readings with target tissue/cell profiles and then congregating wherever a precisely defined set of preconditions exists.

G protein: One of a large family of heterotrimeric GTP-binding proteins that are important intermediaries in cell-signaling pathways. Usually activated by the binding of a hormone or other signaling ligand to a seven-pass transmembrane receptor protein.

G0 phase: (G-"zero" phase) State of withdrawal from the eucaroytic cell-division cycle by entry into a quiescent G1 phase; often seen in differentiated cells.

G1 phase: Gap 1 phase of the eucaryotic cell-division cycle, between the end of cytokinesis and the start of DNA synthesis.

G2 phase: Gap 2 phase of the eucaryotic cell-division cycle, between the end of DNA synthesis and the beginning of mitosis.

GAG (glycosaminoglycan): Long, linear, highly charged polysaccharide composed of a repeating pair of sugars, one of which is always an amino sugar. Mainly found covalently linked to a protein core in extracellular matrix proteoglycans. Examples include chondroitin sulfate, hyaluronic acid, and heparin.

Gain: The ratio of the amplitude of an output to input signal.

Galvanic corrosion: The preferential corrosion of the more chemically active of two metals electrically coupled and exposed to an electrolyte.

Gamete: Specialized haploid cell, either a sperm or an egg, serving for sexual reproduction.

Ganglion (plural ganglia): Cluster of nerve cells and associated glial cells located outside the central nervous system.

Ganglioside: Any glycolipid having one or more sialic acid residues in its structure. Found in the plasma membrane of eucaryotic cells and especially abundant in nerve cells.

Gap junction: Communicating cell-cell junction that allows ions and small molecules to pass from the cytoplasm of one cell to the cytoplasm of the next.

Gastrula: Animal embryo at an early stage of development where cells are invaginating to form the rudiment of a gut cavity. (From Greek *gaster*, belly.)

Gate: Circuit whose logical output variables are determined by its inputs. In digital logic, a component that can switch the state of an output dependent on the states of one or more inputs.

Gauss: The cgs unit used in measuring magnetic induction.

Gene: A given segment of the DNA molecule that contains the code for a specific protein.

Genegeneering: Genetic engineering.

General-purpose molecular manufacturing: A manufacturing technology that will find many applications across many segments of society. Its extreme flexibility, precision, high capacity, and low cost will cause rapid adoption almost everywhere, and therefore will have disruptive effects in many industries.

Genetic Algorithm: Any algorithm which seeks to solve a problem by considering numerous possibilities at once, ranking them according to some standard of fitness, and then combining ("breeding") the fittest in some way. In other words, any algorithm which imitates natural selection. [AS]

GENIE: An AI combined with an assembler or other universal constructor, programmed to build anything the owner wishes. Sometimes called a Santa Machine. This assumes a very high level of AI and nanotechnology. [AS]

Genomics: "The study of the full complement of genes that make up an organism."

Giant Magnetoresistance (GMR): It results from subtle electron-spin effects in ultra-thin 'multilayers' of magnetic materials, which cause huge changes in their electrical resistance when a magnetic field is applied. GMR is 200 times stronger than ordinary magnetoresistance.

Gibbs free energy: The Gibbs free energy is the Helmholtz free energy plus the product of the system volume and the external pressure. Changes in the Gibbs free energy at a constant pressure thus include work done against external pressure as a system undergoes volumetric changes. This proves convenient for describing equilibria in gases and liquids at a constant pressure (e.g., at one atmosphere), but is of little use in describing machine-phase chemical processes. Changes in the Gibbs free energy caused by a change in the applied pressure (at constant volume) have no direct physical significance.

Glass transition temperature (T_g): The temperature at which, upon cooling, a noncrystalline ceramic or polymer transforms from a supercooled liquid to a rigid glass.

Glass: An amorphous solid obtained when silica is mixed with other compounds, heated above its melting point, and then cooled rapidly.

GNR technologies: (Genetic Engineering, Nanotechnology, and Robotics)

Golden Goo: Another member of the grey goo family of nanotechnology disaster scenarios. The idea is to use nanomachines to filter gold from seawater. If this process got out of control we would get piles of golden goo (the "Wizard's Apprentice Problem"). This scenario demonstrates the need of keeping populations of self-replicating machines under control; it is much more likely than grey goo, but also more manageable. [AS - Originated on sci.nanotech 1996]

Grain boundary: The interface separating two adjoining grains having different crystallographic orientations.

Grain growth: The increase in average grain size of a polycrystalline material: for most materials, an elevated temperature heat treatment is necessary.

Grain size: The average grain diameter as determined from a random cross section.

Gray Goo or Grey Goo - destructive nanobots [AKA: "gray dust"]: Opposite of Blue Goo. See Star Trek scenario. Vast legions of *destructive* nanites. Typically, created by accident. Left unchecked, they will basically convert everthing they contact into more of themselves, or consume and digest it for energy. Either way, its pretty much bad news. Also - Self-replicating (von Neumann) nanomachines spreading uncontrolably, building copies of themselves using all available material. This is a commonly mentioned nanotechnology disaster scenario, although it is rather unlikely due to energy constraints and elemental abundances. More probable disaster scenarios are the green goo, golden goo, red goo, khaki goo scenarios. As a protection blue goo has been proposed.

Gray goo problem: Nanotech Now's Goo and Paste Glossary. The following quote was taken from Eric Drexler's book Engines of creation. The early transistorized computers soon beat the most advanced vacuum-tube computers because they were based on superior devices. For the same reason, early assembler-based replicators could beat the most advanced modern organisms. "Plants" with "leaves" no more efficient than today's solar cells could out-compete real plants, crowding the biosphere with an inedible foliage. Tough, omnivorous "bacteria" could out-compete real bacteria: they could spread like blowing pollen, replicate swiftly, and reduce the biosphere to dust in a matter of days. Dangerous replicators could easily be too tough, small, and rapidly spreading to stop - at least if we made no preparation. We have trouble enough controlling viruses and fruit flies.

Gray goo: The name given to free-range self-replicating miniature machines that could, in theory, run out of control and cause severe damage to the biosphere. The actual threat is generally overrated, as we explain here.

Green ceramic body: A ceramic piece, formed as a particulate aggregate, that has been dried but not fired.

Green Goo: Nanomachines or bio-engineered organisms used for population control of humans, either by governments or eco-terrorist groups. Would most probably work by sterilizing people through otherwise harmless infections. See Nick Szabo's essay Green Goo — Life in the Era of Humane Genocide. [AS] "A scary concept dreamed up by Erik K Drexler whereby tiny *assemblers*, or molecular machines, that are capable of making copies of themselves, are let loose and proceed to replicate uncontrollably, consuming everything in their path and turning it into a grey goo."

Ground state: The lowest-energy state of a system. The electronic ground state of a system cannot reduce its energy by an electronic transition, but may contain vibrational energy (kinetic and potential energy associated with the motions and positions of its atoms); extended systems at ordinary temperatures are always vibrationally excited, and so "ground state" is often taken to mean "electronic ground state."

Ground: To make electrical connection to the earth or to the chassis of a device (verb); the connection point so used (noun).

Group velocity: In wave propagation, the speed of the waveform (e.g., of a peak) can be different from the speed of a group of waves (e.g., of a set of ripples in water). The latter is the *group velocity*, and is the speed of propagation of information and wave energy. The waveform speed is the *phase velocity*.

Group: A set of linked atoms in a molecule; a defined substructure. Typically, a set that is usefully regarded as a unit in chemical reactions of interest.

GUI: Graphical User Interface — hardware, software, and firmware that produces the display on modern personal computers.

Guy Fawkes Scenario: If nanotechnology becomes widely available, it might become trivial for anyone to committ acts of terrorism (such as making nanomachines build a large amount of explosives under government buildings a la Guy Fawkes). This would either force strict control over nanotechnology (hard) or a decentralized mode of organization.

Hair cell: Specialized sensory epithelial cell in the ear with bundles of giant microvilli (stereocilia) protruding from its apical surface. Sound vibrations tilt the stereocilia, evoking an electrical change in the hair cell, which thus acts as a sound detector.

Hall effect: The phenomenon whereby a force is applied to a moving electron or hole by a magnetic field that is applied perpendicular to the direction of motion. The force direction is perpendicular to both the magnetic field and the particle motion directions.

Haploid: Having only one set of chromosomes, as in a sperm cell or a bacterium, as distinct from diploid (having two sets of chromosomes).

Hardness: The measure of a material's resistance to deformation by surface indentation or by abrasion. There are various scales in use to express hardness. The Mohs scale is qualitative and somewhat arbitrary and ranges from 1 on the soft end for talc to 10 for diamond. Quantitative scales are the Rockwell (HR), Brinell (indicated by HB), Knoop (HK) and Vickers (HV). Knoop and Vickers are referred to as microhardness testing methods on the basis of load and indenter size.

Harmonic oscillator: A system in which a mass is subject to a linear restoring force, like an ideal spring. A harmonic oscillator vibrates at a fixed frequency, independent of amplitude.

Heat capacity (C_v at constant volume and C_p at constant pressure): The quantity of heat required to produce a unit temperature rise per mole of material. The ratio of the heat input to the temperature increase in a system. Note that this definition does not imply that a system *contains* heat, despite the name *heat capacity*.

Heat shock protein (stress-response protein): Protein synthesized in response to an elevated temperature or other stressful treatment; usually helps the cell to survive the stress.

Heat: As defined in thermodynamics, heat is the energy that flows between two systems as a result of temperature differences (a system *contains* neither heat nor work, but can *produce* heat or *do* work). Heat thus differs from thermal energy.

Heisenberg Uncertainty Principle: A quantum-mechanical principle with the consequence that the position and momentum of an object cannot be precisely determined. The Heisenberg principle helps determine the size of electron clouds, and hence the size of atoms.

HeLa cell: Line of human epithelial cells that grows vigorously in culture. Derived from a human cervical carcinoma.

Heme: An iron complex. Cyclic organic molecule containing an iron atom that carries oxygen in hemoglobin and carries an electron in cytochromes.

Hemidesmosome: Specialized cell junction between an epithelial cell and the underlying basal lamina.

Hemopoiesis (hematopoiesis): Generation of blood cells, mainly in the bone marrow.

HEMT: High electron mobility transistor.

Henry (H): Unit of inductance (see"inductance"). One henry (H) is the inductance of a closed circuit in which an electromotive force of 1 volt is produced when the electric current in the circuit varies uniformly at the rate of 1 ampere per second.

Henry's law: The amount of gas dissolved in a solution is directly proportional to the pressure of the gas above the solution.

Heterocaryon: Cell with two or more nuclei produced by the fusion of two or more different cells.

Heterochromatin: Region of a chromosome that remains unusually condensed and transcriptionally inactive during interphase.

Heterodimer: Protein complex composed of two different polypeptide chains.

Heteronuclear: Consisting of different elements. [TNTWeekly]

Heterozygote: Diploid cell or individual having two different alleles of a specified gene.

Heuristics: Rules of thumb used to guide one in the direction of probable solutions to a problem.

Highest Occupied Molecular Orbital (HOMO): The highest energy molecular orbital of an atom or molecule that contains an electron.

Histone: A protein present in the eukaryotic nucleus that is bound to the DNA at regularly spaced intervals.

Homogeneous and Heterogeneous Assays: A homogeneous assay does not require a separation step to remove free antigen from bound antigen and relies upon the fact that the function of the label is modified upon binding, leading to a change in signal intensity. Because of high background signal a heterogeneous approach incorporating a separation step of bound and unbound makes the detection limit lower, approaching the values obtained by RIA. The homogeneous assay is less technically demanding.

Human Genome Project: A research initiative that has mapped the entire human genome.

Human tissue engineering No universally agreed definition exists. But a Commission working hypothesis suggests a human tissue engineered product means any autologous (emanating from the patient himself) or allogeneic (coming from another human being) product which: contains, consists of, or results in engineered human cells or tissues; and has properties for, or is presented as having properties for, the regeneration, repair or replacement of tissue, where the new tissue or cells, in whole or in part, are structurally and functionally analogous to the original tissue that is being regenerated, repaired or replaced.

Hydrogen Bond: A bond formed between hydrogen and two other atoms. A hydrogen atom covalently bound to an electronegative atom (e.g., nitrogen, oxygen) has a significant positive charge and can form a weak bond to another electronegative atom; this is termed a hydrogen bond.

Hydrophobic force: Water molecules are linked by a network of hydrogen bonds. A nonpolar, nonwetting, surface (e.g., wax) cannot form hydrogen bonds. To form their full complement of hydrogen bonds, the nearby water molecules must form a more orderly (hence lower entropy) network. This both increases free energy and causes forces that tend to draw hydrophobic surfaces together across distances of several nanometers.

Hydrophobic: Literally - water fearing, from the Greek *hydro* - "water" and *phobo* - "fear".

Hypertext: A computer-based system for linking text and other information with cross-references, making access fast and criticisms easy to publish and find.

Hypsithermal limit: The maximum amount of energy that may be released at Earth's surface, as a result of human technological activities, without significantly altering the natural global energy balance; estimated as 10^13-10^15 watts.

Hysteresis: The difference in the output when a specific input value is approached first with an increasing and then with a decreasing input. This phenomenon occurs in ferroelectric materials and results in irreversible loss of energy through heat dissipation.

Imaging Contrast Agent: A molecule or molecular complex that increases the intensity of the signal detected by an imaging technique, including MRI and ultrasound. An MRI contrast agent, for example, might contain gadolinium attached to a targeting antibody. The antibody would bind to a specific target—a metastatic melanoma cell, for example – while the gadolinium would increase the magnetic signal detected by the MRI scanner.

Imaging Contrast Agent: A molecule or molecular complex that increases the intensity of the signal detected by an imaging technique, including MRI and ultrasound. An MRI contrast agent, for example, might contain gadolinium attached to a targeting antibody. The antibody would bind to a specific target—a metastatic melanoma cell, for example – while the gadolinium would increase the magnetic signal detected by the MRI scanner.

Immortalization: Production of a cell line capable of an unlimited number of cell divisions. Can be the result of a chemical or viral transformation or of fusion with cells of a tumor line.

Immune Machines: Medical nanomachines designed for internal use, especially in the bloodstream and digestive tract, able to identify and disable intruders such as bacteria and viruses.

Immune response: Response made by the immune system of a vertebrate when a foreign substance or microorganism enters its body.

Immune system: Population of lymphocytes and other white blood cells in the vertebrate body that defends it against infection.

Immunoglobulin (Ig): An antibody molecule. Higher vertebrates have five classes of immunoglobulin - IgA, IgD, IgE, IgG, and IgM - each with a different role in the immune response.

Immunoglobulin like (Ig-like) domain: Characteristic protein domain of about 100 amino acids that is found in antibody molecules and in many other proteins that form the Ig superfamily.

Inductance [in Henry, H]: That property of an electric circuit which tends to oppose change in current in the circuit. One henry (H) is the inductance of a closed circuit in which an electromotive force of 1 volt is produced when the electric current in the circuit varies uniformly at the rate of 1 ampere per second.

Induction (embryonic): Change in the developmental fate of one tissue caused by an interaction with another tissue.

Inductor: Energy storage circuit component consisting of a coil of wire and possibly a magnetic material.

Inflammatory response: Local response of a tissue to injury or infection. Caused by invasion of white blood cells, which release various local mediators such as histamine.

Infrared: Invisible electromagnetic radiation having a longer wavelength, and lower frequency, than visible red light.

Inhibitor: A chemical substance that, when added in relatively low concentrations, slows down a chemical reaction.

Inline Universities: (As opposed to online universities), nanocomputer implants serving to increase intelligence and education of their owners, essentially turning them into walking universities.

Inmessaging: In medical nanorobotics, conveyance of information from a source external to the human body, or external to working nanodevices, to a receiver located inside the human body.

Insertion point (in lithography context): Adaptation of a new lithography technique is referred to as the insertion point of that technique.

Insulator: A substance, object or material that does not conduct electricity. Material that conducts electricity very poorly.

Integrated Circuit (IC): An electronic circuit consisting of many interconnected devices on one piece of semiconductor, typically into 10 millimeters on a side. ICs are the major building blocks of today's computers. Semiconductor circuit, typically on a very small silicon chip, containing microfabricated transistors, diodes, resistors, capacitors, etc.

Interagency Working Group on Nanoscience, Engineering and Technology IWGN, National Science and Technology Council NSTC, US

Interface: In physical terms, an interface is the boundry between two phases, for instance between a solid and liquid or between a liquid and gas.

Intermolecular: Between more than one molecule. Describes an interaction (e.g., a chemical reaction) between different molecules.

Internal energy: The sum of the kinetic and potential energies (including electromagnetic field energies) of the particles that make up a system.

Internet: Worldwide digital communication network in which packets of information travel between senders and recipients.

Interstitial diffusion: a diffusion mechanism that causes atomic motion from interstitial site to interstitial site.

Intramolecular: Describes an interaction (e.g., a chemical reaction) within a single molecule. Intramolecular interactions between widely separated parts of a molecule resemble intermolecular interactions in most respects. (2) Within a single molecule.

Intrinsic: Characterizes pure undoped semiconductor; electrical conductivity depends only on temperature and the band gap energy.

Ion: An atom or molecule with a net charge. **(2)** An atom with more or fewer electrons than those needed to cancel the electronic charge of the nucleus. An ion is an atom with a net electric charge.

Ionic bond: A chemical bond resulting chiefly from the electrostatic attraction between positive and negative ions. A coulombic interatomic bond existing between two adjacent and oppositely charged ions.

Ionophore: A macro-organic molecule capable of specifically solubilizing an inorganic ion of suitable size in organic mediums.

Junna: Minor Negentropy Alliance world. During the period from 3267 to 3421 was center for the Institute of Baseline Psychotypology. Suffered during the Version War and became an MPA garrison. Since the age of separative empires the locals have drifted increasingly towards Zoeticism

Jupiter-Brain: A posthuman being of extremely high computational power and size. This is the archetypal concentrated intelligence. The term originated due to an idea by Keith Henson that nanomachines could be used to turn the mass of Jupiter into computers running an upgraded version of himself.

Karyotype: Full set of chromosomes of a cell arranged with respect to size, shape, and number.

Keratin (cytokeratin): Member of the family of proteins that form keratin intermediate filaments, mainly in epithelial cells. Some specialized keratins are found in hair, nails, and feathers.

Ketone: Organic molecule containing a carbonyl group linked to two alkyl groups.

Kevlar (TM): A synthetic fiber made by E. I. du Pont de Nemours & Co., Inc. Stronger than most steels, Kevlar is among the strongest commercially available materials and is used in aerospace construction, bulletproof vests, and other applications requiring a high strength-to-weight ratio.

Kilobyte (kB): 2^{10} (= 1024, or about one thousand) bytes of information.

Kilocalorie (kcal): Unit of heat energy equal to 1000 calories. Often used to express the energy content of food or molecules: bond strengths, for example, are measured in kcal/mole. An alternative unit in wide use is the kilojoule, equal to 0.24 kcal.

Kilohertz (kHz): One thousand cycles per second.

Kilojoule: Standard unit of energy equal to 1000 joules, or 0.24 kilocalories.

Kinesin: One type of motor protein that uses the energy of ATP hydrolysis to move along a microtubule.

Kinetic energy: Energy resulting from the motion of masses.

Kinetic molecular theory: A model that assumes that an ideal gas is composed of tiny particles (molecules) in constant motion.

Kinetochore: Complex structure formed from proteins on a mitotic chromosome to which microtubules attach and which plays an active part in the movement of chromosomes to the pole. The kinetochore forms on the part of the chromosome known as the centromere.

Knowbots: Knowledge robots, first developed Vinton G. Cref and Robert E. Kahn for National Research Initiatives. Knowbots are programmed by users to scan networks for various kinds of related information, regardless of the language or form in which it expressed. "Knowbots support parallel computations at different sites. They communicate with one another, and with various servers in the network and with users."

Label: Chemical group or radioactive atom added to a molecule in order to follow its progress through a biochemical reaction or to locate it spatially. Also, as a verb, to add such a group or atom to a cell or molecule.

Lagging strand: One of the two newly made strands of DNA found at a replication fork. The lagging strand is made in discontinuous lengths that are later joined covalently.

Lambda bacteriophage (l bacteriophage): Virus that infects *E. coli;* widely used as a DNA cloning vector.

Lamin: Extracellular matrix protein found in basal laminae.

Lamins: Intermediate filament proteins that form the fibrous matrix (nuclear lamina) on the inner surface of the nuclear envelope.

Langmuir-Blodgett (LB) Film: A film of organic material (often surfactant molecules) assembled at the liquid-gas interface and subsequently transferred onto a solid substrate.

Langmuir-Blodgett: The name of a nanofabrication technique used to create ultrathin films (monolayers and isolated molecular layers), the end result of which is called a "Langmuir-Blodgett film".

Laser trimming: A method for adjusting the value of thin- or thick-film resistors by using a computer-controlled laser system.

Laser tweezers: Laser tweezers and related approaches are important for manipulation and isolation of subcellular *organelles* and structural components. [National Center for Research Resources "Integrated Genomics Technologies Workshop Report" Jan 1999]

Laser: Light Amplification by the Stimulated Emission of Radiation — quantum device that produces coherent light.

LCD (Liquid Crystal Display): LCD is the predominant technology used in flat panel displays. The principle that makes the display work is this: A crystalís alignment can be altered with an electric current. If the crystal is lined up one way ñ it will allow the light waves to pass through a polarized filter, but if the electric current alters the crystalís alignment, it will guide light so that the polarized filter blocks the light. By densely packing red, blue and green light emitting crystals next to each other on a sheet (ìcalled a substrateî), one can create a full color display. The great thing about LCD is that the crystals can be packed together closely, allowing for a higher-resolution, finer-detail display. The con is that LCDs are somewhat fragile, require a lot of power and are relatively less bright. Liquid Crystal Display — display device employing light source and electrically alterable optically active thin film.

Leading strand: One of the two newly made strands of DNA found at a replication fork. The leading strand is made by continuous synthesis in the 5´-to-3´ direction.

Leakage: The loss of all or parts of a useful agent, as of the electric current that flows through an insulator or the magnetic flux that passes outside useful flux circuits.

Lectin: Protein that binds tightly to a specific sugar. Abundant lectins derived from plant seeds are often used as affinity reagents to purify glycoproteins or to detect them on the surface of cells.

LED: Light-Emitting diode — semiconducting diode that produces visible or infrared radiation.

LEDs (Light Emitting Diodes): LEDs work on a completely different concept. Traditionally LEDs are created from two semiconductors. By running current in one direction across the semiconductor the LED emits light of a particular frequency (hence a particular color) depending on the physical characteristics of the semiconductor used. The semiconductor is covered with a piece of plastic that focuses the light and increases the brightness. These semiconductors are very durable, there is no filament, they donít require much power, theyíre brighter and they last a long time. By densely packing red, blue and green LEDs next to each other on a substrate one can create a display. The disadvantage of LEDs is that they are much larger ñ therefore the resolution is not nearly as good as LCD displays. Thatís why most LED displays are large, outdoor displays, not smaller devices, like monitors.

Leucine zipper: Structural motif seen in many DNA-binding proteins in which two a helices from separate proteins are joined together in a coiled-coil, forming a protein dimer.

Life (lifetime): The length of time the sensor can be used before its performance changes.

Life Science Applications of Deep Red T2-MP EviTag Quantum Dots: The new T2-MP EviTags are available in the deep red wavelengths used in life science research – Macoun Red (660nm) and Jonamac Red (680nm) – with carboxyl, amine, biotin and non-functionalized surfaces. Having a biologically compatible quantum dot label like the T2-MP EviTags allows further exploration and development of in vivo assays, following cells inside living organism to get a new understanding of life. In particular, it offers the possibility to create new classes of assays for use where current testing materials could produce cytotoxicity or cell death. Potential applications include live cell and even whole animal imaging, blood cancer assays, and numerous other applications.

Ligand: In protein chemistry, a small molecule that is (or can be) bound by a larger molecule is termed a ligand. In organometallic chemistry, a moiety bonded to a central metal atom is also termed a ligand; the latter definition is more common in general chemistry.

Light Emitting Diodes (LEDs): Light Emitting Diodes work on a completely different concept. Traditionally LEDs are created from two semiconductors. By running current in one direction across the semiconductor the LED emits light of a particular frequency (hence a particular color) depending on the physical characteristics of the semiconductor used. The semiconductor is covered with a piece of plastic that focuses the light and increases the brightness. These semiconductors are very durable, there is no filament, they donít require much power, theyíre brighter and they last a long time. By densely packing red, blue and green LEDs next to each other on a substrate one can create a display.

Lightsail: A spacecraft propulsion system that gains thrust from the pressure of light striking a thin metal film.

Limit of detection: The smallest measurable input. This differs from resolution, which defines the smallest measurable change in input. For a temperature measurement, this would provide an indication of the lowest temperature a sensor could generate an output in response to.

Limited Assembler: An assembler with built-in limits that constrain its use (for example, to make hazardous uses difficult or impossible, or to build just one thing). Assembler capable of making only certain products; faster, more efficient, and less liable to abuse than a general-purpose assembler.

Linde Scenario: A scenario for indefinite survival of intelligent life. It assumes it is possible to either create basement universes connected to the original universe with a wormhole or the existence of other cosmological domains. Intelligent life continually migrates to the new domains as the old grow too entropic to sustain life. [AS/Mitch Porter, 1997. The name refers to Linde's chaotic inflation cosmology, where new universes are continually spawned.]

Linear: Aside from its geometric meaning, *linear* describes systems in which an output is directly proportional to an input. In particular, a linear elastic system is one in which the internal displacements are (at equilibrium) directly proportional to applied forces.

Linearity: The degree to which the calibration curve of a device conforms to a straight line.

Lipids: Water-insoluble substances than can be extracted from cells by nonpolar organic solvents.

Liposome: A type of nanoparticle made of lipids, or fat molecules, surrounding a water core. Liposomes, several of which are widely used to treat infectious diseases and cancer, were the first type of nanoparticle to be used to create therapeutic agents with novel characteristics.

Liposome: A type of nanoparticle made of lipids, or fat molecules, surrounding a water core. Liposomes, several of which are widely used to treat infectious diseases and cancer, were the first type of nanoparticle to be used to create therapeutic agents with novel characteristics.

Lone pair: Two valence electrons of an atom that share an orbital but do not participate in a bond.

Low-dimension Structures: Quantum wells, quantum wire and quantum dots.

Lowest Unocupied Molecular Orbital (LUMO): The lowest energy molecular orbital of an atom or molecule that does not contain an electron.

Machine-phase chemistry: The chemistry of systems in which all potentially reactive moieties follow controlled trajectories (e.g., guided by molecular machines working in vacuum).

Macroscale: Larger than nanoscale; often implies a design that humans can directly interact with; too large to be built by a single assembler (one cubic micron of diamond contains 176 billion atoms).

Macrosensing: In medical nanorobotics, the detection of global somatic states (inside the human body) and extrasomatic states (sensory data originating outside of the human body) by in vivo nanorobots.

Magnetic field strength (designated by H) [A/m]: Magnetic field produced by a current, independent of the presence of magnetic material. The units of H are ampere-turns per meter, or just amperes per meter.

Magnetic flux density or magnetic induction (designated by B): The magnetic field produced in a substance by an external magnetic field. The units of B are tesla (T). One tesla is the magnetic flux density given by a magnetic flux of 1 weber per square meter. One weber is a magnetic flux that, linking a circuit of 1 turn, would produce in it an electromotive force of 1 volt if it were reduced to zero at a uniform rate in 1 second. Both B and H are field vectors. One henry (H) is the inductance of a closed circuit in which an electromotive force of 1 volt is produced when the electric current in the circuit varies uniformly at the rate of 1 ampere per second. The magnetic field strength and flux density are related according to: $B = \mu H$, where μ is the permeability (see under permeability).

Magnetic Force Microscopy (MFM): A method for observing local magnetic fields near a surface by scanning the surface with a magnetic probe.

Magnetic susceptibility ($\div_m$): The proportionality constant between the magnetization M (see under"magnetization") and the magnetic field strength H. The magnetic susceptibility is unitless.

Magnetization (M): The total magnetic moment per unit volume of material. Also, a measure of the contribution to the magnetic flux by some material within an H field. The magnitude of M is proportional to the applied field as: $M = \div_m \times H$, with $\div_m$ the magnetic susceptibility.

Magnetostrictive material: A material that changes dimension in the presence of a magnetic field or generates a magnetic field when mechanically deformed.

Manufacturing: The ability to make products, in this case ranging from clothing, to electronics, to medical devices, to books, to building materials, and much more.

Martensite: A metastable iron phase supersaturated in carbon that is the product of a diffusionless (athermal) transformation from austenite.

Mask: Pattern on glass, like a photographic negative, for producing integrated-circuit elements on semiconductor wafer.

Mass to Charge Ratio: A number defining how a particle will respond to an electric or magnetic field that can be calculated by dividing the mass of a particle by its charge.

Massometer: In medical nanorobotics, a nanosensor device for measuring the mass of individual molecules or small physical objects to single-proton resolution.

Materials science: Science of ceramics, glass, metals, plastics, semiconductors.

MatML Materials Markup Language: Materials property data distributed on the World Wide Web in documents using hypertext markup language.

Matter as Software: "Autonomous, motile microdevices clearly are on the horizon. They may be regarded as the first step in the evolution of a technology for "programming" the structure and properties of material objects at the microscopic and the submicroscopic levels. As this evolution progresses, *the physical and economic properties of such programmable matter are likely to become much like those of present day software."* [MITRE Corporation]

MCM: MultiChip Module; the interconnection of two or more semiconductor chips in a semiconductor-type package.

Measurand: A physical quantity, condition, or property that is to be measured.

Meat Machine: AKA Cabinet Beast. A box containing assemblers and raw material, within which is formed meat [or whatever else it was programmed to make].

Mechanical: Pertaining to the positions and motions of atoms, as defined by the positions of their nuclei; see electronic. A purely mechanical device can be described in terms of atomic positions and motions without reference to electronic properties, save through their effect on the potential energy function.

Mechanochemistry: Chemistry accomplished by mechanical systems directly controlling the reactant molecules; the formation or breaking of chemical bonds under direct mechanical control. In this volume, the chemistry of processes in which mechanical systems operating with atomic-scale precision either guide, drive, or are driven by chemical transformations. In general usage, the chemistry of processes in which energy is converted from mechanical to chemical form, or vice versa.

Mechanosynthesis: Chemical synthesis controlled by mechanical systems operating with atomic-scale precision, enabling direct positional selection of reaction sites; synthetic applications of mechanochemistry. Suitable mechanical systems include AFM mechanisms, molecular manipulators, and molecular mill systems. Processes that fall outside the intended scope of this definition include reactions guided by the incorporation of reactive moieties into a shared covalent framework (i.e., conventional intramolecular reactions), or by the binding of reagents to enzymes or enzymelike catalysts. Molecular tools with chemically specific tip structures can be used, sequentially, to modify a work piece and build a wide range of molecular structures.

Mechatronics: The study of the melding of AI and electromechanical machines to make machines that are greater than the sum of their parts. The synergistic combination of precision mechanical engineering with electronic control.

Megabyte (MB): 2^{20} (= 1,048,576, or about one million) bytes of information.

Megahertz (Mhz): One million cycles per second

Meme: An idea that replicates through a society as it is propagated through person-to-person interaction, both direct and indirect. Memetics is a field of study that focuses on memes' role in the evolution of a culture. Self-reproducing idea or other information pattern which is propagated in ways similar to that of a gene. [Richard Dawkins, 1976]

MEMS MicroElectro Mechanical Systems: MEMS is less an application area in itself than a manufacturing or fabrication technique that enables other application areas. Many authors use MEMS as shorthand to imply a number of particular application areas. As it is used here, MEMS is a "top- down" fabrication technology that is especially useful for integrating mechanical and electrical systems together on the same chip. It is grouped in the category of integrated microsystems because these same MEMS techniques can be extended in the future to also help integrate biological and chemical components on the same chip, as discussed below. Thus far, MEMS techniques have been used to make some functional commercial devices such as sensors and single- chip measurement devices. Many researchers have used MEMS technologies as analytical tools in other areas of nanotechnology [Central Intelligence Agency, US The Global Technology Revolution, Chapter Two Technology Trends, Genomics, 2001] Stood originally for Micro-ElectroMechanical System — microscopic mechanical elements, fabricated on silicon chips by techniques similar to those used in integrated circuit manufacture, for use as sensors, actuators, and other devices. Today almost any miniaturized device (based on Si technology or traditional precision engineering, chemical or mechanical) is referred to as a MEMS device generic term to describe micron scale electrical/mechanical devices.

MEMS: Stood originally for Micro-ElectroMechanical System — microscopic mechanical elements, fabricated on silicon chips by techniques similar to those used in integrated circuit manufacture, for use as sensors, actuators, and other devices. Today almost any miniaturized device (based on Si technology or traditional precision engineering, chemical or mechanical) is referred to as a MEMS device.

Mesoscale: A device or structure larger than the nanoscale (10^-9 m) and smaller than the megascale; the exact size depends heavily on the context and usually ranges between very large nanodevices (10^-7 m) and the human scale (1 m). [AS]

Messenger molecule: A chemically recognizable molecule which can convey information after it is received and decoded by an appropriate chemical sensor.

Metal nanoshells: A new type of *nanoparticle* composed of a *semiconductor* or dielectric core coated with an ultrathin conductive layer.. By adjusting the relative core and shell thicknesses, metal nanoshells can be fabricated that will absorb or scatter light at any wavelength across the entire visible and infrared range of the electromagnetic spectrum.

Metamorphic: In medical nanorobotics, capable of adopting multiple physical configurations via smooth changes from one configuration to another.

Micellar nanocontainers: Block copolymer micelles are water- soluble biocompatible nanocontainers with great potential for delivering hydrophobic drugs. An understanding of their cellular distribution is essential to achieving selective delivery of drugs at the subcellular level. R. Savic, L. Luo, A. Eisenberg, D. Maysinger, Micellar nanocontainers distribute to defined cytoplasmic organelles Science 300 (5619): 615- 618, Apr. 25, 2003

Micelles: Micelles are small, spherical structures composed of molecules that attract one another to reduce surface tension. The head of the molecule is hydrophilic, meaning it likes water, while the interior portion is hydrophobic, meaning it avoids water.

Microbiotagraphics: Mapping the microbiotic populations present in the human body.

Microbubbles: Very small encapsulated gas bubbles (diameters of micrometers) that can be used in diagnostic and therapeutic applications. Upon exposure to sufficiently intense ultrasound, microbubbles will cavitate, rupture, disappear, release gas content, etc. Such characteristics of the microbubbles can be used to enhance diagnostic tests, dissolve blood clots, and deliver drugs or genes for therapy. MeSH 2004

Microchemistry: The development and use of techniques and equipment to study or perform chemical reactions, with small quantities of materials, frequently less than a milligram or a milliliter. MeSH 2003

Microdevices: Narrower terms: integrated microdevices, nanodevices, SED Single Electron Devices. Related terms microelectronics, microfluidics, micro- TAS.

Micro-Electromechanical Systems (MEMS): "The fabrication or micro-machining of materials to make stationary and moving structures, devices and systems of a nominal size scale from a few centimeters to a few micrometers."

Microelectronics: Narrower terms: MEMS, nanoelectronics, optoelectronics, SED Single Electron Devices. Related terms: molecular electronics, semiconductors

Microencapsulation: Individually encapsulated small particles.

Microfabrication: This technology includes techniques used to manufacture integrated circuits (ICs), discrete microelectronic devices, MEMS devices such as sensors and actuators, and various electro-optic devices. [University of Louisville Microfabrication Lab] is currently serving as a center for research activity in the areas of micromachined sensors and actuators, electro- optic devices, special- purpose microelectronic devices, planar waveguides, chemical transducers, microstrip and microgap radiation detectors, micromachined nozzles, and micromachined ink- jet printheads. Microfabrication holds the key to developing monolithically integrated devices capable of extremely high throughput. ... Microfabrication of surface characteristics provides a means of controlling the cell-environment interaction. Microfabricated structures also present new opportunities for cell, protein and nucleic acid separations and analysis- including high throughput single protein molecule sequencing. New approaches to analysis utilizing conductance techniques or tunneling probe microscopy merely emphasize the breadth of the potential of nanofabrication for future technology development relevant to biological issues. [National Center for Research Resources "Integrated Genomics Technologies Workshop Report" Jan 1999]

Microfluidics: A multidisciplinary field comprising physics, chemistry, engineering and biotechnology that studies the behavior of fluids at volumes thousands of times smaller than a common droplet. Microfluidic components form the basis of so-called "lab-on-a-chip" devices that can process microliter and nanoliter volumes and conduct highly sensitive analytical measurements. The fabrications techniques used to construct microfluidic devices are relatively inexpensive and are amenable both to highly elaborate, multiplexed devices and also to mass production. In a manner similar to that for microelectronics, microfluidic technolcgies enable the fabrication of highly integrated devices for performing several different functions on the same substrate chip. Microfluidics is a critical component in gene chip and protein chip development efforts. The study of fluid channels and chambers of tiny dimensions of tens to hundreds of micrometers and volumes of nanoliters or picoliters. This is of interest in biological MICROCIRCULATION and used in

MICROCHEMISTRY and INVESTIGATIVE TECHNIQUES [MeSH 2004] Enables the fabrication of networks of channels, chambers, and valves for the flow of liquids as minute as one picoliter. These systems have no moving parts and require little assembly. [Coventor "About Coventor" 2002] Within the microelectromechanical systems (MEMS) and biological and chemical detection communities, *microfluidics* refers to the research and development of microscale devices that handle very small volumes of fluids (nano- and picoliter volumes). These devices may perform tasks such as DNA analysis or the separation of human blood cells. [Josh Molho "What is microfluidics?" Stanford Univ., US, 2000]

Microinjection: The insertion of a substance into a cell through a microelectrode. Typical applications include the injection of drugs, histochemical markers (such as horseradish peroxidase or lucifer yellow) and RNA or DNA in molecular biological studies. To extrude the substances through the very fine electrode tips, either hydrostatic pressure (pressure injection) or electric currents (ionophoresis) is employed. [OMD] A technique for introducing a solution of DNA, protein, or other soluble material into a cell using a fine microcapillary pipet. [Life Sciences Dictionary]

Micromachines: Mechanical objects that are fabricated in the same general manner as integrated circuits. They are generally considered to be between 100 nanometres to 100 micrometres in size, though that is debatable. The applications of micromachines include accelerometers that detect when a car has hit an object and trigger an airbag. Complex systems of gears and levers are another application. The fabrications of these devices is usually done by one or both of two techniques: surface micromachining and bulk micromachining.

Micromachining: Techniques used to fabricate **MEMS**. **Broader term: microengineering**

Micron: 10^{-6} Symbol is υ. One millionth of a meter, or about 1/25,000 of an inch.

Microparticles: Applications include calibration of *flow cytometers*, particle and hematology analyzers, *confocal laser scanning microscopes* and zetapotential measuring instruments; flow measurements in gases and liquids like Laser Doppler Anemometry (LDA); Particle Dynamics Analysis (PDA), and Particle Image Velocimetry (PIV); medical diagnostics; separation phases for *chromatography*; support for immobilized enzymes; spacer in liquid crystal displays (LCD's); peptide synthesis; cell separation; *tracers* in environmental science; model systems in medicine, biochemistry, colloid chemistry, and aerosol research. [Microparticles GmbH, Berlin Germany]

Microphone: Device that produces voltage or current in response to a sound wave.

Microprocessor: Chip containing the logical elements for performing calculations and carrying out stored instructions.

Microspectrophotometry: Analytical technique for studying substances present at enzyme concentrations in single cells, in situ, by measuring light absorption. Light from a tungsten strip lamp or xenon arc dispersed by a grating monochromator illuminates the optical system of a microscope. The absorbance of light is measured (in nanometers) by comparing the difference between the image of the sample and a reference image. MeSH 1990

Microspheres: Microspheres are generally defined as small spheres made of any material and sized from about 0.5 μm to 100 μm. Similar, but smaller spheres sized 10 to 500 nm are called *nanospheres.* Ideally, microspheres are completely spherical and homogeneous in size, although less perfect particles are often termed microspheres as well. Depending on the preparation method and material used, microspheres show a typical size distribution which often deviates from the mono-sized ideal.

Microstructure: In material engineering the structural features of a material such as grain boundaries, grain size and structure, subject to observation under a microscope, selective etching etc. In MEMS microstructure unfortunately is also used to designate a micromachined feature. The last decade has seen rapid developments in the fabrication, characterization and conceptual understanding of synthetic microstructures in many different material systems including silicon, III-V and II-VI semiconductors, metals, ceramics and organics. The objective of this journal [Superlattices and Microstructures] is to provide a common interdisciplinary platform for the publication of the latest research results on all such "nanostructures" with dimensions in the range of 1 - 100 nm; the unifying theme here being the dimensions of these artificial structures rather than the material system in which they are fabricated.

Microsystem: A microscale machine that can sense information from the environment and act accordingly. Outside the U.S., it can also refer to microelectromechanical systems (MEMS). [smalltimes glossary, 2002]

MicroTas, micro Total Analysis Systems, uTAS: Although initial research dates back to the early 1970's, the field of micro- TAS formally started in 1990, when Manz et al described the possibility of creating microsystems that would take care of many or all the traditional analytical steps involved in a biochemical analysis (sample introduction, handling, extraction, purification, concentration, filtration, analysis, detection) .... Micro-TAS offer many advantages over traditional analysis systems. Low power consumption and small reaction volumes, faster analysis, ultrasensitive detection, and minimal human intervention are key parameters in the development of micro- TAS. Most biochemical reactions take place in liquid environments. Hence, the development of MicroTAS is intrinsically linked to the design of liquid handling micro- devices. [Biomedical Applications Group (GAB) Centro Nacional de Microelectronica (CNM- IMB) Bellaterra, Spain, 2000]

Microtubule: A hollow cylindrical tube used for the transport of materials inside cells.

Miller indices: A set of 3 integers (4 for hexagonal) that designate crystallographic planes, as determined from reciprocals of fractional axial intercepts.

Millimeter: One thousandth of a meter, or about 1/26 of an inch.

MIMIC: [Micromoulding in capillaries] one-step rapid prototyping technique.

Miniaturization: Desirable for many technologies for overall cost reduction (including reduction in the amount of reagents and analytes). Important to remember that building space is often the least available and most expensive component of an overall laboratory budget.

MIPS: Millions of Instructions per Second — a measure of computing power.

Misreaction: A chemical reaction that fails by yielding an unwanted product.

MM2/CSC: A molecular mechanics program developed by Cambridge Scientific Computing that closely follows the MM2 model, adding a graphical user interface and other features.

MM2: A molecular mechanics program developed by Norman Allinger and coworkers; the "MM2 model" is the molecular potential energy function described by the equations, rules, and parameters embodied in that program.

MNT: An abbreviation for molecular nanotechnology; refers to the concept of building complicated machines out of precisely designed molecules.

Mobility (electron, and hole): The proportionality constant between the carrier drift velocity and applied electric field.

Modulus of elasticity (E) : The ratio of stress to strain when deformation is totally elastic. Also the Young's modulus.

MOEMS MicroOpticalElectroMechanical systems: In addition to mechanical and electrical components, integrate waveguides or other optical features into the body of the silicon chip. Intellisense Corp. MEMS The Next Small Thing Google = about 10 Aug. 8, 2002; about 12 June 23, 2004; about 10 May 2, 2005

Moiety: A portion of a molecular structure having some property of interest.

Molality: The molality or molal concentration (symbol m) is the amount of substance per unit mass of solvent or mol kg^{-1}. Concentration in a liquid solution (symbol c), in terms of the number of moles of a solute dissolved in 10^6 mm^3 (10^3 cm^3) of solution in mol l^{-1}.

Molding (plastics): Shaping a plastic material by forcing it, under pressure at a high temperature, into a mold cavity.

Mole: A number of instances of something (typically a molecular species) equaling ~6.022 1023. *Mole* ordinarily means gram-mole; a kilogram-mole is ~6.022 1026.

Molecular - Nanotechnology: The technology of precisely- constructed molecular-scale machines; from nanometer: a billionth of a meter. For a good introduction, see What is Nanotechnology? or Ralph C. Merkle's nanotechnology page.

Molecular assembler: A general-purpose device for molecular manufacturing, able to guide chemical reactions by positioning individual molecules to atomic accuracy (e.g. mechanosynthesis) and to construct a wide range of useful and stable molecular structures according to precise specifications. Also known as an assembler, a molecular assembler is a molecular machine that can build a molecular structure from its component building blocks.

Molecular Beam Epitaxy: Process used to make compound (multi-layer) semiconductors. Consists of depositing alternating layers of materials, layer by layer, one type after another (such as the semiconductors gallium arsenide and aluminum gallium arsenide).

Molecular Integrated Microsystems (MIMS): microsystems in which functions found in biological and nanoscale systems are combined with manufacturable materials. See Molecular Integrated Microsystems

Molecular machine: A mechanical device that performs a useful function using components of nanometer scale and defined molecular structure; includes both artificial nanomachines and naturally occurring devices found in biological systems. Any machine with atomically precise parts of nanometer dimensions; can be used to describe molecular devices found in nature.

Molecular Recognition: A chemical term referring to processes in which molecules adhere in a highly specific way, forming a larger structure; an enabling technology for nanotechnology.

MUX: Device for combining several signals or data streams into a single flow.

Myosin: Myosin is one of the two major proteins responsible for contraction of muscle. In muscle cells, myosin is arranged in long filaments that lie parallel to the microfilaments of actin. In muscle contraction, filaments of actin and myosin alternately unlink and chemically link in a sliding action. The energy for this reaction is supplied by adenosine triphosphate.

NAD+ (nicotine adenine dinucleotide): Coenzyme that participates in an oxidation reaction by accepting a hydride ion (H-) from a donor molecule. The NADH formed is an important carrier of electrons for oxidative phosphorylation. Coenzyme closely related to NAD+ that is used extensively in biosynthetic, rather than catabolic, pathways.

Nanapheresis: In medical nanorobotics, the removal of bloodborne medical nanorobots from the body using aphersis-like processes.

Nanarchist: Someone who circumvents government control to use nanotechnology, or someone who advocates this. [Eli Brandt, October 1991]

Nanite: Machines with atomic-scale components. (Popularized by the Star Trek episode "Evolution") As to their weight, a popular question: "Do you 'feel' heavier after you drink a mouthful of water? A mouthful of water, roughly 5 cm^3, would have the same mass as a ~2 terabot (2 trillion nanites) dose of 1 micron^3 nanorobots. You'll never feel it." Robert A. Freitas Jr. "Nanobot" and "Nanorobot" usually mean the same thing. Slang term for a nanomachine (see nanotechnology), esp. a machine able to replicate itself. [Popularized by the Star Trek episode "Evolution"]

Nano Cubic Technology: An ultra-thin layer coating that results in higher resolution for recording digital data, ultra-low noise and high signal-to-noise ratios that are ideal for magneto-resistive (MR) heads. It is capable of catapulting data cartridge and digital videotape to one-terabyte native (uncompressed) capacities and floppy disk capacities to three gigabytes. To help visualize the potential, 1TB can store up to 200 two-hour movies.

Nanoarray: an ultra-sensitve, ultra-miniaturized array for biomolecular analysis. BioForce Nanosciences' Nanoarrays utilize approximately 1/10,000th of the surface area occupied by a conventional microarray, and over 1,500 nanoarray spots can be placed in the area occupied by a single microarray domain.

Nanoassembler: the Holy Grail of nanotechnology; once a perfected nanoassembler is availble, building anything becomes possible, with physics and the imagination the only limitation (of course each item would have to be designed first, which is another small hurdle). See Bootstrapping a Nanofactory "So how hard is it to build a nanofactory? You need to start with a working fabricator, a nanoscale device that can combine individual molecules into useful shapes. But once you have that, the rest is pretty straightforward." —CRN

Nanobarcode: SurroMed's Nanobarcode™ technology uses cylindrically-shaped colloidal metal nanoparticles, in which the metal composition can be alternated along the length and the size of each metal segment can be controlled. Intrinsic differences in reflectivity between the metal segments allow individual particles to be identified by conventional optical microscopy.

Nanobeads: Polymer beads with diameters of between 0.1 to 10 micrometers. Also called nanodots, nanocrystals and quantum beads. *Impregnating fluorescent crystal chips into these beads allows simultaneous measurement of thousands of biological interactions, a stepping stone for breakthroughs in the diagnosis and treatment of disease. ... with the potential to accelerate drug discovery and clinical diagnostics."*

Nanobot: Machines with atomic-scale components. (Popularized by the Star Trek episode "Evolution") As to their weight, a popular question: "Do you 'feel' heavier after you drink a mouthful of water? A mouthful of water, roughly 5 cm^3, would have the same mass as a ~2 terabot (2 trillion nanites) dose of 1 micron^3 nanorobots. You'll never feel it." Robert A. Freitas Jr. "Nanobot" and "Nanorobot" usually mean the same thing. Slang term for a nanomachine.

Nanobubbles: tiny air bubbles on colloid surfaces. Thought to reduce drag, such as would be of benefit to swimmers wearing a suit coverd in them.

Nanocantilever: The simplest micro-electro-mechanical system (MEMS) that can be easily machined and mass-produced via the same techniques used to make computer chips. The ability to detect extremely small displacements make nanocantilever beams an ideal device for detecting extremely small forces, stresses and masses. Nanocantilevers coated with antibodies, for example, will bend from the mass added when substrate binds to its antibody, providing a detector capable of sensing the presence of single molecules of clinical importance.

Nanocell Surgery: Modifying cellular structures by the use of medical nanomachines.

Nanocentrifuge: In medical nanorobotics, a proposed nanodevice that can spin materials at very high speed, imparting rotational accelerations of up to one trillion gravities (g's), thus permitting rapid sortation.

Nanocentrifuge: In medical nanorobotics, a proposed nanodevice that can spin materials at very high speed, imparting rotational accelerations of up to one trillion gravities (g's), thus permitting rapid sortation.

Nanochemistry: The goal of the Nanochemistry group is to develop chemistry beyond the realm of individual molecules by exploring how systems of several molecules or other electronically confined entities self-organize into distinct structures on the nanometer scale. The new insight is used to study and manipulate biological nanosystems and to create novel nanoarchitectures with unique structural, electronic, optical and magnetic properties. Specific research areas include bionanotechnology and self- assembling molecular electronics.

Nanochip: In Friday's issue [Aug. 1, 2001] of the journal *Science,* physicists from IBM's Thomas J. Watson Research Center [Philip G. Collins, Michael S. Arnold and Phaedon Avouris] announce their fabrication of the world's first array of transistors made from carbon nanotube. We are approaching the limits of standard microchip technology; thus, the "nanochip" — a next-smaller microchip. [ed] They are also a next-gen device for mass storage, of significantly higher density, with greater speed, and much lower cost.

Nanochondria: Nanomachines existing inside living cells, participating in their biochemistry (like mitochondria) and/or assembling various structures.

Nanochronometer: In medical nanorobotics, a proposed clock or timing mechanism constructed of nanoscale components.

Nanocircles: Stanford scientists have synthesized a molecule of DNA that is capable of shutting off specific genes in living bacteria. Dubbed the "nanocircle," the new nanometer-size molecule might one day give researchers the ability to target harmful genes that cause cancer and other diseases in humans. A nanocluster or nanocrystal is a fragment of solid comprising somewhere between a few atoms to a few tens of thousands of atoms. Nanoclusters are therefore a novel state of matter with properties that are neither those of a bulk crystal nor those of individual atoms and molecules. Over the past 10 years huge advances have been made both in the synthesis of size- tunable, monodisperse (i.e. size-selected) nanoclusters of various chemical compositions and in the development of techniques for their assembly into nanostructured solids (facilitating the synthesis of what have been termed "designer materials"). Nanoparticles, Nanoclusters, Nanoscience Group, Nottingham Univ., UK, 2002

Nanocomputer: A computer made from components (mechanical, electronic, or otherwise) built at the nanometer scale. These computers could be many orders-of-magnititude faster than today's, which enables software to take proportional leaps. (2) A computer whose fundamental components measure only a few nanometers in size. State of the art current computer components are no smaller than about 350 nm. [Mnemosyne Mnews 21 (3) January 2001 " The Nanotechnology Initiative and Future Electronics" Presentation by Gail J. Brown, Air Force Research Laboratory, Wright- Patterson Air Force Base, November 16, 2000] (3) A computer with parts built on a molecular scale.

Nanocones: Nonplanar graphitic structures. Carbon-based structures with five-fold symmetry that form due to disclination defects in two-dimensional graphene sheets. They have been observed as nanotube caps and as freestanding structures. [North Carolina State University].

Nanocontainers: "Micellar nanocontainers" or "Micelles," these are nanoscale polymeric containers that could be used to selectively deliver hydrophobic drugs to specific sites within individual cells.

Nanocrit (Nct): In medical nanorobotics, volume-fraction or bloodstream concentration of medical nanorobots, expressed as a percentage.

Nanomachining: like traditional machining, where portions of the structure are removed or modified, nanomachining involves changing the structure of nano-scale materials or molecules.

Nanomanipulation: The process of manipulating items at an atomic or molecular scale in order to produce precise structures.

Nanosieving: In medical nanorobotics, a nanodevice that can sort molecules or other nanoscale objects by physical sieving.

Nanosieving: In medical nanorobotics, a nanodevice that can sort molecules or other nanoscale objects by physical sieving.

Nanosome: Nanodevices existing symbiotically inside biological cells, doing mechanosynthesis and disassembly for it and replicating with the cell. Similar to nanochondria. [AS January 1998]

Nanosome: Nanodevices existing symbiotically inside biological cells, doing mechanosynthesis and disassembly for it and replicating with the cell. Similar to nanochondria. [Anders Sandberg January 1998]

Nanoswarm: UFog and Goo

Nanosystem: A eutactic set of nanoscale components working together to serve a set of purposes; complex nanosystems can be of macroscopic size.

Nanotech: Slang for nanotechnology.

Nanotechism: the religion of nanotech, as opposed to the science of nanotech

Nanoterrorism: using MNT derived nanites to do damage to people or places.

Nernst equation: An equation relating the potential of an electrochemical cell to the concentrations of the cell components: $E = Eo + RT/zF \ln C_1/C_2$ with z the charge exchanged at the electrode and C_1 and C_2 concentrations of two electro-active compounds.

Neroprosthesis: Implanted cybernetic brain augmentation.

Neural Simulation: Imitating the functions of a neural system—such as the brain—by simulating the function of each cell.

Neurocomputation: The study of how natural and artificial neural networks process information.

Neuron Star: A neutron star used as a basis for a densely packed mind (similar to a jupiter brain or a omegon) exploiting neutronium or quark matter for computation [Damien Broderick, 1996]

Neuron: A nerve cell, such as those found in the brain.

Neuronaut: A person who explores their own neural functioning and internal mentational processes by various means, including deep introspection and meditation, psychoactive drugs, mind machines, and neuroscientific understanding.

Neurosuspension: Cryonically suspending only the head or brain of a person.

NMOS: An acronym for n-*channel metal-oxide-semiconductor*, as in NMOS transistor and NMOS logic.

Nono: A prefix meaning ten to the minus ninth power, or one billionth.

Nootropic: A cognition-enhancing drug that has no significant side-effects.

NOR: NOT-OR — logic gate whose output is the negation of that of the OR gate.

NOT: Logic gate whose output is binary 1 when its input is 0, and whose output is a 0 when its input is a 1.

Now shock: Being shocked or confused by the rapid changes that has already taken place, a kind of future shock before the present. [Michael Rothschild 1995, "Cornucopia Or Black Hole?", *Upside*]

NRAM™ - Nanotube-based/Nonvolatile RAM, developed by Nantero, using proprietary concepts and methods derived from leading-edge research in nanotechnology.

n-type: Characterizes a semiconductor containing predominantly mobile electrons (also see"p-type").

Nuclear Magnetic Resonance (NMR): An analytical method that can detect subatomic and structural information of molecules by measuring the adsorption of radio-frequency electromagnetic radiation by nuclei under the influence of a magnetic field.

Nucleation: The initial stage in a phase transformation, evidenced by the formation of small particles (nuclei) of the new phase, which are capable of growing.

Okazaki fragments: Short lengths of DNA produced on the lagging strand during DNA replication, discovered by R. Okazaki. They are rapidly joined by DNA ligase to form a continuous DNA strand.

Peptide nanotubes: NANOTUBES formed from cyclic peptides (PEPTIDES, CYCLIC). Alternating D and L linkages create planar rings that self assemble by stacking into nanotubes. They can form pores through CELL MEMBRANE causing damage. MeSH 2004

Permeability [µ]: From the relation between magnetic induction and magnetic field ($B = \mu H$); for free space, $\mu_0 = 1.26 \times 10^{-6}$ H/m.

Permitivity [å]: From the relation between polarization charge and electric field; for free space, $å_0 = 8.85 \times 10^{-12}$ F/m.

Peroxisome: Small membrane-bounded organelle that uses molecular oxygen to oxidize organic molecules. Contains some enzymes that produce and others that degrade hydrogen peroxide (H2O2)

Phagocyte: General term for a professional phagocytic cell - that is, a cell such as a macrophage or neutrophil that is specialized to take up particles and microorganisms by phagocytosis.

Phagocytosis: Process by which particulate material is endocytosed ("eaten") by a cell. Prominent in carnivorous cells, such as *Amoeba proteus*, and in vertebrate macrophages and neutrophils. (From Greek *phagein*, to eat.)

Pharmacyte: In medical nanorobotics, a theorized (nanorobotic) device capable of delivering precise doses of biologically active chemicals to individually-addressed human body tissue cells (e.g. cell-by-cell drug delivery).

Phase shift: A time difference between the input and output signals.

Phase space: A classical system of N particles can be described by its $3N$ position and $3N$ momentum coordinates. The phase space associated with the system is the $6N$ dimensional space defined by these coordinates.

Phase transformation: A change in the number and/or character of the phases that make up the microstructure of an alloy.

Photolithography: Writing or creating patterns by means of light.

Photon: A particle of light. The energy E of a photon is given by the Einstein relation, $E = hf$. Here f is the frequency of the light and h is Planck's constant. The momentum p of a photon is given by the de Broglie relation, $p = h/$ë. Here ë is the wavelength of the light.

Photonic Band Gap (PBG): An energy range (and corresponding wavelength range) for which a material neither absorbs light nor allows light propagation.

Phototransistor: Transistor that, when powered, produces amplified voltage (current) in response to illumination.

Pico Technology: (trillionth of a meter) — The next step smaller, after Nano-technology. The art of manipulating materials on a quantum scale.

Point defect: A crystalline defect associated with one or several atomic sites.

Poisson's ratio (í): For elastic deformation, the negative ratio of lateral and axial strains that result from an applied axial stress.

Poisson's ratio: A bar of an isotropic, elastic material ordinarily shrinks laterally when it is stretched longitudinally. The lateral contracting strain divided by the applied tensile strain is Poisson's ratio, which varies from material to material.

Probability density function: Consider an uncertain physical property and a corresponding space describing the range of values that the property can have (e.g., the configuration of a thermally excited N particle system and the corresponding $3N$ dimensional configuration space). The probability density function associated with a property is defined over the corresponding space; its value at a particular point is the probability per unit volume that the property has a value in an infinitesimal region around that point.

Protein Design, Protein Engineering: The design and construction of new proteins; an enabling technology for nanotechnology.

Protein Folding: "The process by which proteins acquire their functional, preordained, three-dimensional structure after they emerge, as linear polymers of amino acids, from the ribosome." [The Scientist]

Pyroelectricity: The property of certain crystals, such as tourmaline, of acquiring opposite electrical charges on opposite faces when heated.

Q factor: A rating, applied to coils, capacitors, and resonant circuits, equal to the reactance divided by the resistance. The ratio of energy stored to energy dissipated per cycle in an electrical or mechanical system.

Quantization: The concept that energy can occur only in discrete units called quanta.

Qubit: The quantum computing analog to a bit. Qubits exhibit superposition. Thus, unlike normal bits, qubits can be both 1 and 0 at the same time.

Radiation damage: Chemical changes (bond breakage, ionization) caused by high-energy radiation (e.g., x-rays, gamma rays, high-speed electrons, protons, etc.).

Representative point: The point in a configuration space that represents the geometry of a system.

Reproducibility: A property of an experiment or process where one tends to receive consistent results from following a specific procedure.

Residual stress: A persisting stress in a material free of external forces or temperature gradients.

Resist: A material or coating that can protect a surface from chemical reactions.

Resistance [Ω-ohm]: Characteristic of a resistor: in a 1-ohm resistance a current of 1 A produces a voltage drop of 1 V.

Resistivity (ñ): The reciprocal of electrical conductivity, and a measure of a material's resistance to passing electric current.

Resistor: Energy dissipative element consisting of a poor conductor in series with connecting wires.

Resolution: The minimum distance between two objects that can be distinguished in microscopy or the minimum spacing between two features that can be fabricated with lithography. The smallest measurable change in input that will produce a small but noticeable change in the output. In the context of chemical separations, defines the completeness of separation.

Resonant frequency: The frequency at which a moving member or a circuit has a maximum output for a given input.

Respirocrit: In medical nanorobotics, the volume-fraction or bloodstream concentration of respirocyte nanorobots, expressed as a percentage.

Respirocyte: In medical nanorobotics, a theorized bloodborne spherical 1-micron (nanorobotic) device having a 1000-atm pressure vessel with active pumping powered by endogenous serum glucose, that serves as a mechanical artificial red blood cell.

Response time: The time it takes for the sensor's output to reach its final value. A measure of how quickly the sensor will respond to changes in the environment. In general, this parameter is a measure of the speed of the sensor and must be compared with the speed of the process.

Restriction Enzyme: An enzyme that cuts DNA at a specific site, allowing biologists to insert or delete genetic material.

Reverse bias: The insulating bias for a p-n junction rectifier; electrons flow into the p side of the junction.

RF: Radio Frequency — refers to alternating voltages and currents having frequencies between 9 kHz and 3 MHz.

Ribonuclease: An enzyme that cuts RNA molecules into smaller pieces.

SAMFET: (self assembled monolayer field effect transistor). Where a few molecules act as FETs, exhibiting both very strong gain, and extraordinarily rapid response. [Mark Ratner & MT 5(2) p. 20]

Sanguinatation: In medical nanorobotics, locomotion (especially swimming by a nanorobot) through the bloodstream.

Sarcomere: Repeating unit of a myofibril in a muscle cell, composed of an array of overlapping thick (myosin) and thin (actin) filaments between two adjacent Z discs.

Sarcoplasmic reticulum: Network of internal membranes in the cytoplasm of a muscle cell that contains high concentrations of sequestered Ca2+ : that is released into the cytosol during muscle excitation.

Satellite DNA: Regions of highly repetitive DNA from a eucaryotic chromosome, usually identifiable by its unusual nucleotide composition. Satellite DNA is not transcribed and has no known function.

Star Trek scenario: Someone builds potentially dangerous self-replicating devices that spread disastrously.

Steady-state diffusion: The diffusion condition for which there is no net accumulation or depletion of diffusing species. The diffusion flux is independent of time.

Step response: The response of a system to an instantaneous jump in the input signal.

Steric hindrance: Slowing of the rate of a chemical reaction owing to the presence of structures on the reagents that mechanically interfere with the motions associated with the reaction, typically by obstructing the reaction site.

Superconductor: An object or substance that conducts electricity with zero resistance.

Superintelligence: An intellect that is much smarter than the best human brains in practically every field, including scientific creativity, general wisdom and social skills. This definition leaves open how the superintelligence is implemented: it could be a digital computer, an ensemble of networked computers, cultured cortical tissue or what have you. It also leaves open whether the superintelligence is conscious and has subjective experiences. [Bostrom]

Superlattice Nanowire Pattern (transfer): [SNAP] a technique for producing "Ultra High Density Nanowire Lattices and Circuits".

Superlattice: Nanowire interwoven bundles of nanowires using substances with different compositions and properties.

Surfactant: Short for 'SURFace ACTive AgeNT' - a surfactant is a molecule that lowers suface tension.

Synapse: A structure that transmits signals from a neuron to an adjacent neuron (or other cell).

Synthesis: The production of a specific molecular structure by a series of chemical reactions.

Synthespian: An artificial actor, for example a 3D model animated by motion capture from a real actor or a computer program. [AS]

Temperature sensitive (ts) mutant: Organism or cell carrying a genetically altered protein (or RNA molecule) that performs normally at one temperature but is abnormal at another (usually higher) temperature.

Temperature: A system in which internal vibrational modes have equilibrated with one another can be said to have a particular temperature. Two systems A and B are said to be at different temperatures if, when brought into contact, heat flows from (say) A to B, increasing the thermal energy of B at the expense of the thermal energy of A.

Template: An external source of information or structure often used in manufacturing and mass production. As a nanoscale example, biological systems are known to use nucleic acids as a templates for protein synthesis.

Templating: The directed assembly of materials through interaction with a template.

Tertiary structure: Complex three-dimensional form of a macromolecule, especially a protein.

Tesla [T]: Unit of magnetic induction: 1T = 1 weber/m^2 (also, 1T = 10^4 gauss).

Tetrahymena: Genus of ciliated protozoa used in studies of ciliary axonemes, self-splicing RNA, and telomere reproduction.

TFT: Thin film transistor.

The Atom: Atoms are the building blocks of matter. All things are made of arrangements of atoms. There are 118 (with some very short lived ones) different elements in the periodic table, which means that we know of 118 different kinds of atoms.

Virtual Nanomedicine: Using VR to perform surgery and other functions inside the body.

Virtual reality system: A combination of computer and interface devices (goggles, gloves, etc.) that presents a user with the illusion of being in a three dimensional world of computer-generated objects.

Index

Climate Change, Bio-Diversity and Green Economy

About the Editors

Prof. H. S. Sharma (b. 1945) holds post-graduate and doctoral degrees in Geography from the University of Saugar, Madhya Pradesh. Prof. Sharma is a former Professor and Head, Department of Geography, University of Rajasthan, Jaipur. He is also a former Director of the Indira Gandhi Centre for Human Ecology, Environment and Population Studies (HEEPS), Academic Staff College and Dean, Faculty of Science, University of Rajasthan.

Prof. Sharma has over 100 research papers in National and International journals like *Nature* and 16 books to his credit. He has supervised more than three dozen Ph.D.s Prof. Sharma has travelled widely across U.K., U.S.A., Germany, Holland, Canada, Norway, Seychelles and Malaysia to attend and chair technical sessions in International Conferences and Exchange Programmes.

Prof. Sharma has successfully conducted several research projects sponsored by the U.G.C., Ministry of Environment and Forests, DRDO, New Delhi, NORAD, Norway and ISRO (SAC) Ahmedabad.

Prof. Sharma has been President of IIG, IGI and NAGI. He is also a President of the Commonwealth Human Ecology Council (India Chapter). Recently, he has been awarded Senior Fellowship by Indian Council of Social Science Research, New Delhi to work on Climate Change and its Impact on Land and Water Resources of Rajasthan.

Prof. S. Padmaja is a geographer with long established career spanning over 34 years. She has specialized in the field of geomorphology. She has the distinction of serving University of Rajasthan, Jaipur, Jawahar Lal Nehru University, New Delhi and Osmania University, Hyderabad. She has to her credit more than 60 research publications in National and International journals of repute. She has also completed major projects sponsored by UGC, ICIR, ICSSR. She has been honoured with citation by CLUMA for her outstanding contributions to teaching and research in 2003 and 2005 successively. She has served Osmania University as a Head (1996-98), Chairperson (BOS, 1998-2000) and the SAP Coordinator (2000-07). She has now superannuated from Osmania University, Hyderabad.

Dr. Ganesh Sharma, is serving as lecturer at the Department of Geography, Government P.G. College, Ajmer. He joined Government College in the year 2002 and has been engaged in teaching and research both at Beawer and Ajmer. Earlier he served HCM Institute of Public Administration as a Research Officer in Disaster Management Cell.

Dr. Ganesh has published seven research papers in national journals of repute. He has been awarded Research Award by U.G.C. to work on Drought Management in Indian Desert. Dr. Sharma has a keen interest in the field of geomorphology and disaster management.

Climate Change, Bio-Diversity and Green Economy

Edited by
H.S. Sharma
S. Padmaja
Ganesh Sharma

CONCEPT PUBLISHING COMPANY PVT. LTD.
NEW DELHI-110059

ISBN-13: 978-81-8069-995-5

Edition 2013

Published and Printed by

Concept Since 1974

Concept Publishing Company Pvt. Ltd.
A/15-16, Commercial Block, Mohan Garden
New Delhi-110059 (India)
T : 25351460, 25351794, *F* : 091-11-25357109
E: publishing@conceptpub.com, W: www.conceptpub.com
Editorial Office: H-13, Bali Nagar, New Delhi-110 015, India.

Cataloging in Publication Data--*Courtesy:* D.K. Agencies (P) Ltd. <docinfo@dkagencies.com>

Climate change, bio-diversity and green economy / edited by H.S. Sharma, S. Padmaja, Ganesh Sharma.
p. cm.
Includes bibliographical references.
ISBN 978-81-8069-995-5

1. Climatic changes—India—Congresses. 2. Biodiversity—India—Congresses. 3. Environmental policy India—Congresses. 4. Economic development Environmental aspects—India—Congresses. I. Sharma, H. S.,1945- II. Padmaja, S. III. Sharma, Ganesh.

DDC 363.738740954 23

Preface

The climate system is a complex interactive system consisting of the atmosphere, hydrosphere, lithosphere, cryosphere and outer space. A recent report (2007) of the Intergovernmental Panel on Climate Change (IPCC) has clearly indicated that the earth's climate is changing. In order to show the climate change, several studies have chosen temperature and rainfall as primary parameter. The IPCC has reported that global temperature has increased by 0.6°±0.2° C over the late twentieth century. While annual rainfall over land has continued to increase in the middle and high latitudes except over Eastern Asia. Where as over the Sub Tropic Zone (10°N-30°N) the land surface rainfall decreases on an average (0.3% per decade), although this has shown a sign of recovery in recent times. The CO_2 has increased from 280 ppm in 1750 to 367 ppm in 1999.

The scientific studies on climate change present contradictory view points. The natural scientists specially geoscientists believe that both paleo and present climatic changes are due to several natural forces like changes in the obliquity of earth's axis, its rotational speed, solar variability or sunspot, volcanic activities. On the contrary, the scientists who believe in global warming say that increase in the emission of green house gases after industrial revolution has been mainly responsible for global warming and climate changes. It is beyond doubt that since 1850 AD the temperature is rising at such a rate that it was never experienced during last 800,00 years during geological past. Some scientists are of the opinion that global warming phase is likely to end in the end of 2015 and the cooling may hamper to life system. Accordingly they state that symptoms of cooling are visible in northern hemisphere specially in Europe and North America which have been experiencing heavy snowfall during 2012 and 2013 winters. Peru in March 2013 is experiencing heavy flooding; and temperature in March at Bombay is touching 40° C due to anti-cyclonic system. All these unusual weather phenomena are warning signals of climate uncertainty in future. It may, therefore, be safely said that climate change is happening in which natural forces have been predominant but during last 150 years the process of climate change has been accelerated due to anthropogenic activities.

Climate change concerns in India include the rising temperature, the projected pattern of precipitation, its intensity and frequency and the enhanced incidences and intensity of extreme events such as cyclones, flash heavy rains, droughts and frequent floods.

Acknowledgements

Many scientists and academicians have contributed advice and help in the course of preparing this volume. Particularly, we have been greatly benefited from the interaction with Professor T.K.N. Unnithan, founder President of Chec-India, Professor R.P. Mishra, Professor K.R. Dikshit and Prof. Ian Douglas. We owe our gratitude to all of them for their suggestions and their generosity of innovative ideas and sacrifice of valuable time. We owe our sincere thanks to Prof. R. D. Gurjar, Prof. R. N. Mishra, Prof. R. D. Doi, Prof. Hemlata Joshi and Dr. P. R. Binda who offered their help and cooperation at all stages. Most importantly, the compilation of this volume has been possible because of the support we received from the authors. We would like to offer our thanks to all authors.

We express our deepest gratitude to Shri Ashok Kumar Mittal of Concept Publishing Company Pvt. Ltd., New Delhi for his support in the activities of Chec-India and bringing out this volume in an impressive text design.

We are confident that the present volume will generate and stimulate a new innovative thinking on climate change and biodiversity in India and would be popular among the students and scholars who have interest on the theme of climate change, biodiversity and related issues.

Contents

Contributors

Ashok, Anshul, Trident Softech, Jaipur.

Ashok, Smriti, MNIT, Jaipur.

Chattopadhyay, Guru Prasad, Professor & Head, Department of Geography, Visva-Bharati, Santiniketan-731235.

Chopra, Anand, Kalptaru Power Transmission, 101, Sun Tower, SB-158A Gandhi Nagar Mod, Jaipur.

De, Sunil Kumar, Department of Geography & Disaster Management, Tripura University, Suryamaninagar-799022.

Dhayal, Yogpal, Atmospheric Chemistry Laboratory, University of Rajasthan, Jaipur-302004.

Dikshit, K.R., Founder President, Institute of Indian Geographers, Pune.

Douglas, Ian, University of Manchester, UK, and Commonwealth Human Ecology Council, Manchester (U.K.).

Gupta, K.S., Atmospheric Chemistry Laboratory, University of Rajasthan, Jaipur-302004.

Khan, T.I., Indira Gandhi Centre for H.E.E.P.S., University of Rajasthan Jaipur-302004.

Kumawat, Rakesh, Biodiversity Research Laboratory, Zoology Department, Maharshi Dayanand Saraswati University, Ajmer-305009.

Misha, R.P., Former Vice Chancellor, University of Allahabad, Allahabad (UP)

Nathawat, M.S., School of Sciences, IGNOU, Maidan Garhi, New Delhi-110068.

Padmaja, S., Department of Geography, Osmania University, Hyderabad.

Pavanaguru, R., Department of Biochemistry, Osmania University, Hyderabad-500007.

Raka Kamal, Formerly Professor and Head, Department of Botany, University of Rajasthan, Jaipur-302004

Ravi, Research Scholar, Department of Geology, Osmania University, Hyderabad.

Sarina Kalia, Department of Geography, University of Rajasthan, Jaipur-302004

Sharma Deepika, Research Scholar, Department of Geography University of Rajasthan, Jaipur-302004.

Sashidhar Rao, B., Department of Biochemistry, University College of Science, Osmania University, Hyderabad-500007, Andhra Pradesh.

Sawant, Nandkumar, Associate Professor, Parvatibai Chowgule College, Margoa, Goa.

Sharma, Dhiraj, Lecturer L.B.S. PG College and Research Scholar, Department of Accountancy & Business Statistics, University of Rajasthan.

Sharma, H.S., Formerly Prof. & Head, Department of Geography, University of Rajasthan, Jaipur-302004.

Sharma, Krishan Kumar, Biodiversity Research Laboratory, Zoology Department, Maharshi Dayanand Saraswati University, Ajmer-305009.

Sharma, L.K., Department of Remote Sensing, BIT, Mesra, Ranchi-835215.

Sharma, M.L., Professor, Department of Accountancy and Business Statistics, University of Rajasthan.

Sharma, Neha, Biodiversity Research Laboratory, Zoology Department, Maharshi Dayanand Saraswati University, Ajmer-305009.

Sharma, Vivek, Biodiversity Research Laboratory, Zoology Department, Maharshi Dayanand Saraswati University, Ajmer-305009.

Shashikala, A.V., Department of Geography, Osmania University, Hyderabad.

Sinha, Suman, Department of Remote Sensing, BIT, Mesra, Ranchi-835215.

Trivedi, P.C., Vice Chancellor, DDU University, Gorakhpur (U.P.).

Vali Pasha, K., Department of Biochemistry, Yogi Vemana University, Kadapa-516003.

Vinod, V.T.P., Department of Biochemistry, University College of Science, Osmania University, Hyderabad-500007, Andhra Pradesh.

Introduction

Earth, life and climate have always changed and climate change itself may have been crucial for the evolution of humans. The long term perspective is needed in debating and communicating climate change and for converging the importance of sustainability. The constant drum beats of Gloom and Doom have often led to wait and see attitudes and decision-paralysis in the developed countries and a sense of hopelessness for the youth in the developing countries.

There is very accurate and reliable data for atmospheric CO_2 and temperatures and instrumental records show unambiguous increase in CO_2 and global mean temperature. But is the cause and effect that obvious? More and more evidences are causing the warming. The phase relationship between temperature and CO_2 depends on the driver. Orbital forcing leads to warming first and then CO_2 responds. It should also be noted that Nature makes the rules and biology finds the loop holes. Microbes have survived through all climate changes and meteorite crashes! The planet will survive global warming and life will survive. The only question is will We??.

The studies of the International Geosphere-Biosphere Programme (IGBP), the US National Oceanic and Atmospheric Administration, IPCC and others confirm that carbon-dioxide, the chief green house gas, is accumulating in the atmosphere faster than expected.

Beginning with the Industrial Revolution in the 1850s and accelerating ever since, the human consumption of fossil fuels has elevated CO_2 levels from a concentration of ~280 ppm to more than 380 ppm today. These increases are projected to reach more than 560 ppm before the end of the 21st century. It is known that carbon dioxide levels are substantially higher now than at any time in the last 800,000 years. Along with rising methane levels, these changes are anticipated to cause an increase of 1.4 to 5.6 °C between 1990 and 2100.

The evidences for climate changes are now overwhelming. Emissions of carbon dioxide and other pollutants into the atmosphere are causing changes to the global climate in ways that we do not fully understand, and the consequences of which we comprehend even less.

Melting of Ice-caps and glaciers, causing sea-level rise, ocean temperatures are rising, causing stronger hurricanes and killing coral reefs. Rainfall is much heavier in some regions, much less frequent in others, causing drought, flood and shifts in agricultural productivity that can spell disaster for subsistence farmers.

The effects of climate change are global, and climate change will affect every part of the planet earth. No-one can stand by and afford to do nothing. Efforts to reduce carbon emission are essential but are happening too slowly to overcome the problem. The roads from Rio (1992) to Kyoto (1997), Bali (2007), Copenhagen (2009) and Montreal (2011) clearly demonstrate the political will and commitments of developed and developing countries. Bali roadmap set stage for an agreement beyond the Kyoto protocol and for commitments by developing countries to the measurable and verifiable national mitigation actions.

The issue of climate change is often portrayed as a battle between the developed and the developing world. Though the origin of climate change lies principally in the developed economies. But its impacts will be equally great in the developing world and in India probably the greater; the retreat of glaciers; rising sea levels; new difficulties being faced by those farming in arid regions like Rajasthan; a less predictable monsoon-on which the lives of Indians depend.

The problem of climate change can not be resolved by the developed world alone. Even if these countries reach zero emissions by 2030, the world's overall level of CO_2 would still be increasing due to green house gases emission growth in developing country including India.

So, there is a need to have a partnership in particularly four areas like **adaptation, technology, energy efficiency** and **reduction commitments.** There are critical areas for collaboration between the developed and developing countries. Involving the developing countries does not mean allowing developed countries to escape from their responsibility to undertake the main burden of adjustment. However, the main financial burden has to be born by the developed country like U.S.A. Canada, Russia, U.K., Japan etc., as they are responsible for more than 60% emission of CO_2 in the atmosphere. In these circumstances India has a critical leadership role to play in taking this vital agenda forward. The 11th meeting of South Asia Cooperative Environment programme (SACEP) held in Jaipur on 21st and 22nd May, 2008 is a welcome step for a common strategy and working in more effective manner to find out solutions for environment related problems in South Asian countries.

The climate change is emerging as a major area of interest for civic organizations, universities, research institutes and public bodies in India. Many foreign NGO's and others are setting up base in India convinced that the Indian sub-continent could play a major role in addressing global climate change.

The UNDP report released in India in Nov. 2007 reveals that the impact of climate change on the poor is **"significantly underestimated"**. The report reveals that INDIA'S poor would be among the world's 40 million poor people to be adversely hit if the 10 years window for initiating major programme on climate change mitigation and adaptation are missed. In view of this the Govt. of India has taken some steps to mitigate climate change and reduce carbon emissions. Some of the steps taken include interalia-setting up the Bureau of Energy Efficiency and encouraging energy conservation, promoting use of renewable energy, use of cleaner and lesser carbon intensive fuels for transport, fuel switching to cleaner energy, afforestation and conservation of forests, promotion of clean coal technology and environmental quality management for all sectors.

In India there is little doubt that climate will become more unstable in the years ahead and will witness more unpredictable and extreme weather patterns. In India the poorest of the poor possess the most marginal land and depend heavily on the climate sensitive factors. They are vulnerable to flooding, soil erosion, drought and crop failure.

The Prime Minister of India, Dr. Mohan Mohan Singh in his Inaugural Address at the 95th Indian Science Congress on 3rd January 2008 shared his anxiety that **"Climate change poses a great and new challenge to our development prospects and to the livelihood of our people. We need a global response, a national response and a local response."** Today, climate related and environmental aspects are an integral part of 11th Five Year Plan. With regard to climate change the Govt. of India is setting up a working group on climate change adaptation to prepare for providing technology required to address the problem of adjustment.

Biodiversity

The climate of earth is changing. It is warming up, and there is now over whelming scientific consensus that it is happening a result of human action. With global warming on the increase and species and their habitat on the decrease, chance for ecosystems to adapt naturally are diminishing. Many are agreed that climate change may be one of the greatest threats facing the planet. Recent years show increasing temperature in various regions including India desert/ and/ or increasing extremities in weather patterns.

The planet Earth is literally teeming with life. An amazing variety of habitats, people, plants and animals are all connected in a frazile web of life we call "Bio diversity" and every member is essential to keeping this web in balance.

Today about 1.4 million species have been identified and researches estimate that million more have yet to be counted. Thousands of species may be vanishing each year as a result of pollution, over harvesting, habitat degradation, and other human actions.

Extinction is not just a loss for science, it is loss for us. We depend of bio diversity in our everyday lives to supply us with a healthy environment and any natural materials that are sources for food, medicine, and other economically important products.

Indian sub-continent is uniquely rich in all aspect of bio diversity-ecosystems, species and wildlife. It has perhaps the largest array of environmental conditions by virtue of its tropical location, varied physical feature and climate types. It is estimated that over 75,000 species of fauna and 45,000 of flora are found in India. Over the centuries, people of India have had a fascination and respect of this natural heritage, which is evident from sentiments expressed in the **"Isopanishad"**, which dates back to over 2000 years. World's first recorded conservation for wild life were enacted in India during the third century B.C. Emperor Ashok's benevolence extended to all living beings. In achieving this, common people were also involved. Bishnois of Rajasthan and Chipko movement in 1973 by Shri Sunder Lal Bahuguna are good example of people's movements.

Indian desert has a long tradition of conservation of bio diversity. It has a variety of bio diversity with extensive grassland like "Sewan". The Great Indian Bustard is highly endangered bird found in Indian Desert.

Thus, there is a need to conserve the bio diversity of the Planet Earth, the year 2010 is a year of 'Bio diversity Conservation'. Keeping this in view, educational events, exhibition and workshops are being organized for examining some of the key environmental problems and possible solutions facing our planet at a global/ national and local level.

Green Economy

As a consequence of climate change the new noval concept of green economy has come up in recent years. The green economy as we know is one that results in "improved human well-being and social equity, while significantly reducing environmental risks and ecological scarcities". In its simplest expression, a green economy is low carbon, resource efficient, and socially inclusive. In a green economy, growth in income and employment should be driven by public and private investments that reduce carbon emissions and pollution, enhance energy and resource efficiency, and prevent the loss of biodiversity and ecosystem services. The green economy approach is an effort to focus sustainable development and poverty reduction efforts on transforming economic activities and economies. The term Green Economy also applies "decoupling economic growth from the use of fossils fuels and non-renewable energy sources".

The green economy can take different forms, depending on the context of each country, its level of development and its geographic location in particular. Some will begin by greening their transport infrastructure, others will prefer to promote ecotourism or organic agriculture.

India is increasingly being looked as a hub for research and development for green solutions. This is to its innovation in technology, low priced manufacturing and services, and traditional knowledge and processes. The past few years have seen a rise in green innovation and increasing amounts of venture capital is flowing in, with India being rated as the third most attractive country for renewable energy investment. The transition has begun; Indian industry and MNCs have already started looking at corporate environmental impact, due to profitable opportunities, to avoid business risk, because of international investment pressure and their own corporate social responsibility.

According to UNEP India has shown leadership by, for example instituting a tree planting system to combat desertification. India has successfully introduced projects that track the health of the nations plants, animals, water and other natural resources including Sunderbans- the largest deltaic mangrove forest in the world, and home of Indian most iconic wild species: the tiger.

India is increasingly at the forefront of some of the 'green shoots' of a Green Economy that are emerging across the globe. From its manufacturing of solar and wind turbines to its Rural Employment Guarantee Act. which underwrites paid work for millions of households via investments in area ranging from water conservation to sustainable land management, foundation are laid down in India towards a fundamental and for reaching new development path.

Keeping above conceptual framework in view, the volume has been designed in three Sections. Section 1 deals with **Climate Change Issues** containing 8 papers. Section 2 is on **Biodiversity** and includes 6 papers and Section 3 on **Green Economy** contains 8 research papers. In all 22 papers have been included in the volume.

In the chapter **Facing Environmental Challenges**, R. P. Misra has covered a broad canvas highlighting major environmental threats like, climate change, gaia and environment principles of nature, instability of modern civilization increasing population, industrialization, urbanization, dehumanized technology. The paper addresses the contrasting views on climate change and discussed various steps at global level ad national level to combat environmental pollution. The paper advocates that the Gandhian way is the only path through which sustainability of environment may be achieved.

Ian Douglas in **Climate Change and Green Economy with reference to India** has discussed that with a rising of global population and perhaps more significantly a rising relatively affluent middle class with high rates per capita consumption, the emission of green house gases continues to rise. The paper examines the issue of climate change with reference to India and suggests strategies that can be adopted for sustainable water energy and food supply and to help urban settlements to adjust to climate change. The paper further explores the notions of inter dependence and collective responsibility to achieve the goals of human ecology.

H.S. Sharma in his paper on **Climate Change, Desertification and water Resource Management in India** has discussed the issue of climate change its natural and anthropogenic factors and no analogue state of Earth, The paper brought out palae-climatic and present climatic changes, desertification and impact of climate change on water resources. A strategy for water management has also been attempted.

Changes in the **Glacial Environmental around the Southeast facing slopes of Kanchenjunga in the Sikkim Himalaya** have been attempted by Guru Prasad Chattopadhyay. He assessed the impact of the global warming on the glaciers of Sikkim Himalaya. Detiled investigation in the recent years by Chattopadhyay through Satellite image processing along with field investigation reveal that the Rathong and Onglaklang glaciers on the south east facing slopes of Kanchenjunga summit complex in the Sikkim Himalaya have been retreating at a very alarming rate. This rapid rate of glacier retreats in this part of the Sikkim Himalaya can be attributed to the increase of temperature due to environmental degradation and global warming.

T.I. Khan in his paper on **Preparedness of Climate Change** has given a theoretical framework of the various aspects of climate change. He also attempted to explain the green house effect, causes of climate change, impacts of climate change, Kyoto Protocol, IPCC, thermohaline circulation and cooling and warming of the earth system.

A.V. Shashikala and S. Padmaja in the chapter on **Rainfall Variability and Agricultural Productivity in Andhra Pradesh** have indicated that the rainfall in Andhra Pradesh is highly variable and the agricultural systems, are affected by variability of rainfall because agriculture in the state is dependent on rainfall.

K.S. Gupta and Yogpal Dhayal in the chapter on **Chemistry of Stratospheric Ozone Depletion and its possible link with Climate Change** have discussed the chemistry of stratospheric ozone and its depletion. The paper also examines Arctic Ozone Hole and also suggests control strategies.

S. Padmaja, R. Pavanaguru and Ch. Ravi in the chapter on the **Impact of Coal Mining on Climate Change** have examined that economic growth, energy supply, thermal electricity generation and national CO_2 emission have profound linkages on climatic changes in India. The increased rate of exploitation and production of coal to meet the energy demand in India, is causing mechanization and automation which is associated with complex issues concerning environment in the country. In view of the enlarged mining activities and for sustainable development, the paper suggests that it is essential to develop environmental information system which may facilitate proper planning and development of the area.

The climate change has a significant impact on biodiversity. R. P. Misra in his paper on **Problem of Conservation and Management of Biodiversity in Indian Desert** has discussed the history and culture of the Thar desert and examined the biodiversity, its conservation and strategy for management. He stated that the pressure of population on the land and biosphere is getting unbearable in Indian desert. The present policy on biodiversity is not working well and hence there is a need to review over policy in the light of pressure of population.

P.C. Trivedi in his chapter on **Biodiversity Conservation in Traditions and Rituals of Rajasthan** has described various traditions, rituals and practices, recognition of trees by government, wildlife conservation and efficient use of water through ingenious methods and techniques with reference to Rajasthan.

The chapter **Tiger Habitat Suitability and Potential Relationship with Forest Biomass in Sariska Wild Life Reserve, India** by M.S. Nathawat, L.K. Sharma and Suman Sinha examines application of Land Sat ETM+ in locating the tiger habitat suitability. The paper recommends Sariska to have great potential for quality tiger habitat provided proper effective conservation and management strategies are implemented.

Sharma K. K., Vivek, Kumawat Rakesh and Sharma Neha in the chapter on **Some Major Threats to Herpeto fauna of Central Aravalli Range and Thar Desert of**

Rajasthan, India have analysed major threats to biodiversity of herpetofauna. They indicated that major threats are climate change, urbanization, mining, habitat destruction poaching and road accidents. These threats have resulted in species loss or shift to any unknown sites which may or may not provide suitable feeding and breeding grounds to new inhabitants, thus local extinction of valuable herpetofaunal species takes place. The paper also suggests conservation strategies.

Raka Kamal in her chapter on **Medicinal Plants-Biodiversity, Conservation and Bioprospecting** has studied medicinal plants resource base and quality standards for medicinal plants. The paper stated that further research is deserve to isolate the compounds responsible for the observed biological activity.

K. Vali Pasha, R. Pavanguru and B. Sashidhar Rao in their chapter on **Environmental Uranium and Human Health** have examined and reviewed the problem from an cpidemiological and biochemical perspective. The environmental effects of uranium mining on human health have been analysed.

K.R. Dikshit in the chapter on **Limits to Green Economy, Limited Potential and Many Conflict Solutions** has raised certain questions on destruction of forests, and biodiversity, organic forming option and green energy. The paper states that the key to Green Economy lies in preserving the greener, we have, multiplying it, but above all in conserving energy and finding alternative sources of non polluting energy resources.

The problem of **Bioremidiation and Toxic Metal Irons Polling the Environment-Green-Approach Based on Utility of a Natural Biopolymer (Gum Kondagogu)** has been discussed by B. Sashidhar Rao and VTP Vinod. The paper states that Indian forests are major source of large number of non wood products. One of such product is an exudates tree gum called a Kondagogu. The potential application of gum kondagogu as a noval biosorbent for removal of the industrial by relevant toxic heavy metals forms effluents.

The issue of **Can Ecotourism Practices Support Green Economy? A Conceptual Framework** has been discussed by Smiti Ashok and Anshul Ashok. The chapter brings out the concept of green economy and the place of ecotourism in green economy sector. The paper is based on the field work conducted by the authors in Sikkim.

Sarina Kalia and Deepika Sharma have studied **Sustainable Renewable Energy-Transition towards Green Economy.** The chapter has discussed Bio-energy with reference to Rajasthan. The paper states that the residue generated from the crops like mustard and bajara husk, cotton stalks, groundnet shells and sugarcane trash are generally not used as a fodder for animals. These large quantity of surplus biomass can be utilized for power generation. The development of bio-energy will help in resolving the power crisis in Rajasthan and open the path for sustainable development and boost the green economy.

Beach Tourism and Green Economy-An Environmental Performance of Hotel Industry in Coastal South Goa, India has been studied by Nand Kishore Sawant. The chapter envisages the impact of increasing accommodation due to tourism and the consequences on the environment.

Sunil Kumar De in the paper on **Impact of Rubber Plantation on Environment taking a case study of some selected areas of Tripura** has analysed various aspects of the issue. The chapter brings out that although rubber plantation looks like a profitable one in the short run but the overall environmental deterioration in terms of climate change, destruction of indigenous plant and animal species will put not only the state but also the whole region under sever threat in the long run.

M.L. Sharma and Dhiraj Sharma in the chapter on **Carbon Credit Accounting** have examined that the impact of global warming on accounting policy. Carbon credits, as the paper states, are a key components of national and international emission trading schemes that have been implemented to mitigate global warming. Carbon can be used to finance carbon reduction schemes between trading partners and around the world.

Last chapter by Anad Chopra deals with **Climate Change and Green Power Generation in Rajasthan** with an example of Kalptaru Power Transmission Limited (KPTL) in Tonk district. The purpose of this project is to use biomass products for generation of electricity and provide employment to rural people of the area.

1
Facing Environmental Challenges
The Gandhian Way

R.P. Misra

"Viewed from the distance of the moon, the astonishing thing about the earth, catching the breath, is that it is alive. The photographs show the dry pounded surface of the moon in the foreground, dead as an old bone. Aloft, floating free beneath the moist, gleaming membrane of bright blue sky is the rising earth, the only exuberant thing in this part of the cosmos. If you could look long enough, you would see the swirling of the great drift of white cloud, covering and uncovering the half-hidden masses of land. If you had been looking a very long, geologic time, you could have seen the continents themselves in motion, drifting apart on the crustal plates, held aloft by the fire beneath. It has the organized, self-contained look of a live creature, full of information, marvellously skilled in handling the sun."

—*Lewis Thomas, 'The Lives of a Cell'*[1]

Prithivi, the daughter of Prithu is Mother Earth, so say the *Puranas*. She offered the necessary living conditions for the *audi-manava* (Homo sapiens) to survive, grow in number, and spread over the earth. The story of life on planet earth began with its separation from the sun some 5 billion years ago; no other celestial body has so far been found to be bubbling with life as earth is. The idea of Prithvi being the mother of all that exists in nature – organic or inorganic – has been an integral part of the Indian knowledge system ever since the Vedic times (3-4 millenniums BC). It proclaims that the evolution of the species did not stop with the emergence of *Homo sapiens*. Humans continue to evolve mentally and spiritually to attain higher levels of consciousness (wisdom). For Europeans, steeped as they were in the Biblical notion of Creation, the theory of evolution of species was a myth created by the scientist. Darwin challenged this myth when he wrote the *Origin of Species* in 1844. He could not, however, muster the courage to make it public. A short paper giving the summary of the book was first published in the *Journal of the Linnaean Society of London, Zoology* under the joint authorship of Darwin and Wallace. Wallace was a friend of Darwin and had similar views on the evolution of species. Darwin was not present when the paper was communicated to the Society on the evening of July 1, 1858. He was busy attending to his sick children. Since the meeting was called to offer condolences on the death of Robert Brown, it aroused little interest. The book came out in 1859 under the title *Origin of Species.*[2]

It is over a century and half since Darwin postulated his theory of evolution in *Origin of Species*, but as revealed by a recent survey reported in the *National Geographic Magazine*, even today a majority of people in the world subscribe to the view that God created all life forms including human beings. Some scientists too have tried to prove that Darwin was wrong. The recent developments in genetics and related fields have, however, confirmed once again that Darwin was right even if not in all details. James Lovelock's Gaia theory proposes that various inorganic and organic components of the earth, the lithosphere, hydrosphere, cryosphere, atmosphere, and biosphere are inextricably linked with each other and they constitute a complex interlocking and interacting system to maintain the climatic and geochemical conditions of the earth in homeostasis. Earth must be seen not in bits but in its totality, as a single organism. He has renamed the earth as Gaia after the Greek supreme goddess of Earth.[3]

"The idea that the earth is alive is probably as old as humankind." Atharva Rishi in the "Prithvi Sukta" of *Atharva Veda* says:

> *Mata bhumih putroham prithvyah*
> (The Earth is My Mother and I am Her Son)

Apart from the *Atharva Veda*, *Bhagawat* and *Vishnu Puranas*, and *Mahabharat* too declare earth to be live with body, mind and soul. The people who made the earth suitable for human habitation were the descendants of the early man represented by King Prithu. He was the son of King Vena, a wicked man. Prithu was a kind man; he changed the face of the earth, saw to it that no harm was done to the *panchbhootas* (land, water, air, energy, and ether) of the earth. The earth, due to his care and protection was made habitable by his descendants. That is why it was named as *Prithvi*, the daughter of Prithu. Prithu represented genus *Homo sapiens* who 400,000 to 250,000 years ago arose in Africa and migrated from there to other continents and replaced the *Homo erectus* in Asia and *Homo neanderthalensis* in Europe, 50,000 to 100,000 years ago.

The descendants of Prithu settled down in a hospitable regions, flattened the rocky surface, ploughed land, cultivated food and other crops, domesticated cattle, exchanged goods and services with neighbours and built villages and towns. Prior to Prithu's reign, people moved from place to place; there was no agriculture, and no commerce. Human civilization took form in his time.

According to Mahabharata, Prithu pursued *dharma* (righteousness), *shri* (good and beautiful), and *artha* (material pursuits). D.R. Patil suggests that the *Rig* Vedic Prithu was a vegetarian deity, associated with Greek god Dionysus and Vedic god Soma.

After governing his kingdom for a long time, Prithu, in his last days, left with his wife Archi, and settled down in a forest to perform penance.[4] O'Flaherty interprets the myth of Prithu—his transformation from a hunter to herdsman-farmer, from eating beef to rearing cows for milk, and cultivating vegetables and food crops.[5] David Shulman has compared Prithu with Shiva; he was an ideal king, but with a violent temper.[6] Chinese scholar Hiuen tsang (*c.* 640 AD) records the existence of a town called Pehowa, named after Prithu, "who is said to be the first person who obtained the title of Raja (king)". Another place associated with Prithu is *Prithudaka* (Prithu's Pool), a town on the bank of river Saraswati. It is referred to as the boundary between Northern and Central India. Patanjali has equated it with modern Pihowa.

It was a Scottish scientist, James Hutton who gave expression to this ancient wisdom in the year 1785, in a meeting of the Royal Society of Edinburgh. He declared that the earth as a super-organism and hence the scientific field that studies it should be akin to physiology. Hutton compared the movement of soil nutrients, and of water from oceans to land with the circulation of blood in human body. He is rightly remembered as father of geology, but his idea of the earth being a live organism was discarded by his fellow scientists. Reductionism, the hallmark of modern scientific inquiry was so deeply entrenched that only philosophers like Korolinko followed up the idea of a living earth and kept it alive.

Following his lead, Lovelock in his 1979 book *Gaia: A New Look at Life on Earth*[7] suggests two new disciplines: "geophysiology" to study the processes involved in the efficient functioning of the earth as a living system and "planetary medicine" to diagnose and cure the diseases of the Gaia. Lovelock says, "I am happy with the thought that the Universe has the properties that make the emergence of life and Gaia inevitable. But I react to the assertion that it was created with this purpose. It might have been, but how the Universe and life began are ineffable questions." He asserts that "unless life takes charge of its planet, and occupies it extensively, the conditions of its tenancy are not met. Planetary life must be able to regulate its climate and chemical state. Part-time or incomplete occupancy or mere occasional visits will not be enough to overcome the ineluctable forces that drive the chemical and physical evolution of a planet."[8]

Gandhi, following India's ancient wisdom, considered earth as a living organism. He expressed his idea in terms of two fundamental laws: Cosmic Law, and Law of Species. Gandhi's Cosmic Law bound the whole universe into a single entity so deeply and intricately that nothing could malfunction outside the threshold limits inbuilt in the grand system that included both living and non-living phenomena. As for the evolution of the species, Gandhi believed, that various species of animals ultimately evolving into human beings would have not been possible without cooperation and sacrifices, unconscious or conscious, of both inorganic and inorganic parts of the earth. He believed that man should not violate the Cosmic Law, for it is this Law that keeps the universe and its elements like our Earth performing their assigned duties. Gandhi derived his concept of Cosmic Law from ancient Indian literatures and traditions.

Understanding Gaia/Mother Earth

Applying the theory of 'thermodynamics of the unsteady state', Erich Jantsch has concluded that life was not a chance event; it was an inevitable consequence.[9] Evolution of earth from an inert planet to a live planet and of life from microorganisms to Homo sapiens was sequential: it was designed to be so. Gandhi called it Truth; sages call it God; scientists call it beyond the laws of science. Thus science and religion have come closer on the presence of God, which is invisible but all pervasive.

Hinduism considers God as something that has no beginning or end, no shape or form; He cannot be seen, heard, smelt, or even known. So do the other religions. For Gandhi: God is Truth; and Truth is God. The Truth is the unknowable larger mind of which each little mind that humans and other carry is a part; and that is the cause as well as the effect. The concepts of Incarnation (*avatar*), Son of God, or Messenger of God is only a vehicle to present the laws of nature and human behaviour in concrete form and to carry home the idea of something that is beyond science and technology, and which can be understood only by following a moral code of conduct. Thus the earth can be both scientific and spiritual at the same time. This is what the Indian civilization proclaimed over 5000 years ago.

"In fact, the belief that life had its origins in a single basic substance is so widespread among the various peoples of the world, primitive or civilized, that it can be considered one of the few universal themes in the history of ideas."[10] It was primeval watery chaos (churning of the ocean in Hindu mythology) from which evolved various gods and goddesses who produced everything else. This universal belief of antiquity is not just a primitive myth. The "deities are. actually personifications of Nature and their activities, predictable and unpredictable, determine what life will be like on Earth."[11] Dr. Stanley Jaki has confirmed the universality of ancient pagan theory of evolution. "All of these ancients were pagan. The essence of paganism, old and new, is that the universe is eternal, that its motions are without beginning and without end. Belief in creation out of nothing is the very opposite of paganism.[12]

The earth was conceived as the Mother of all living entities. The Greek goddess Gaia and Hindu goddess Prithvi became 'Mother Earth' or 'Mother Nature' in various religious traditions. Lovelock's Gaia hypothesis has revived that old belief and converted it into a scientific theory. The evolution of the species and the evolution of their environment are rightly considered coupled as a single and inseparable process. Hurting the environment would hurt all the species including human beings; it would hurt all those who hurt it.

'What is Life?' No one has so far answered this question satisfactorily. Scientists have raised their hands up, for they are trained to analyze only material things. They do not have the tools to know it. And the religionists have propounded so diverse and contradictory hypotheses that the confusion gets more confounded. What is clear, however, is that life is evolutionary and social. Everything is linked to everything else. To begin with the cells might have had independent and separate existence; they now exist in communities and form grosser living entities. This is called homeostasis. Life has homeostasis properties. To know life, one has to study the universe as a whole; and to understand the earth, one has to study it as a single entity linked to the universe as a whole. "Life and its environment are so closely coupled that evolution concerns Gaia, not the organisms or the environment taken separately." says Lovelock.[13] Further explaining the Gaia theory, Lovelock says: "Gaia theory is about the evolution of a tightly coupled system whose constituents are the biota and their material environment, which comprises the Atmosphere, the oceans, and the surface rocks. Self-regulation of important properties, such as climate and chemical composition, is seen as a consequence of this evolutionary process. Like living organisms and many closed loop self-regulating systems, it would be expected to show emergent properties; that is, the whole will be more than the sum of its parts."[14]

All life forms are open as well as closed systems at the same time. During entropy, the internal environment of animals and humans produces carbon dioxide while that of the plants produces oxygen. The two outputs are balanced because plants consume the carbon dioxide and animals and humans consume oxygen. Thus the living earth has no entropy. The environment in which life exists on earth was created by life itself in conjunction with the inanimate objects like rocks, air, water, etc. This is now well recognized by ecologists and oceanographers, but biologists and geologists have yet to accept the truth that "Life has not adapted to an inert world determined by the dead hand of chemistry and physics."[15] We live in a world built by our predecessors, ancient and modern and it is maintained in balance jointly by all the life forms of the earth. Organisms are adapting to a world whose material state is determined by the activities of their neighbours.

Gaia and Environment

The constituents of the earth's atmosphere during the Achaean Eon (3.8-2.5 billion years ago) were different from today's constituents (Table 1.1). It is life that made it livable.

Table 1.1: Constituents of the Atmosphere: Before and After Life

Gases	*Before Life*	*After Life*
Carbon dioxide	Dominant	0.3 per cent
Nitrogen	Unknown	99.00 per cent
Oxygen	0	1 ppm
Methane	0	100 ppm
Hydrogen	Some	1 ppm

The earth would have become a dead planet in about 1.5 billion years since its birth, had the trees and plants not pumped water from the soil to their leaves and then on to the atmosphere through evapotranspiration. Achaean was the period of continental building and emergence of highly mobile and motile bacteria. Between the Achaean and the Proterozoic (2.5 billion to 543 million years ago), emerged oxygen as a major atmospheric gas. Surprisingly, it happened because of CO_2 (carbon dioxide), which the plants consume to produce oxygen. It also envelops the earth as if to keep it warm even when it was cooler than what it is today. It keeps the earth warm enough to support life.

Recent increases in temperature have resulted partly from changes in the luminosity of the sun but largely due to anthropogenic activities. Scientists predict that the present rate of concentration of CO_2 and other green house gases is comparable to their natural increase that closed the last ice age some 18,000 years ago. During the Ice Ages much of the North and South Pole regions were covered with ice. There were icebergs (mountains of ice) floating on the ocean surface. Increase in the green house gases in the atmosphere raised the temperatures of the earth's surface. It happens because the earth becomes a massive green house, which allows sun light to enter but checks the heat emitted by the earth from escaping into the atmosphere. This is what we call 'Green house Effect'. The polar ice melts and the ocean level rises.

For the last 10,000 years or so, sea level has fluctuated so slowly and slightly that it has been treated as constant. This being the case, human and other species have inhabited coral atolls, river deltas, barrier islands, and ocean beaches where the balance between life and sea level is delicate. The global warming resulting from anthropogenic activities now appears to disrupt that balance. It has already raised the sea level by a metre during the twentieth century and is expected to raise it further in the next two centuries if corrective steps are not put in place soon. Many low-lying areas along the coast would be submerged, and island countries like Maldives would disappear.

It is not only a question of coastal areas being swallowed by the sea; no life form would survive anywhere on the earth if the temperature increases high enough to kill vegetation and micro organisms. The steep fluctuations in the distribution of heat and rainfall during the last few decades point to dangers ahead. Unlike the past, the anthropogenic activities like industrialization and urbanization have played a major role in polluting the environment and consequent increase in atmospheric temperature. Knowing that nature cannot accommodate our ever-growing demands on it, we have to change or scale down our demands to avoid catastrophe. If we fail to halt this drift, we would see the doomsday in the lifetime of our own children. Many of our great coastal cities would be swallowed by rising sea level; and ultraviolet light passing through the depleting ozone layer in the stratosphere would cause diseases like skin cancer, and cataracts, and damage plant and plankton population.

The Earth gives enough scope for progress provided that its culture and nurture go together and it is used with deference and within the limits of what its carrying capacities. When the people were few and their technology was human-size, the anthropogenic changes in the environment were within the threshold limits of the earth system. Any abnormalities were set right by the negative feedback mechanism. Had there been only one billion people on the earth, the earth would have remained healthy even in the present technological regime. With around 7 billion people owning over 700 million motor vehicles increasingly galloping non-renewable fossil fuels and releasing CO_2 in the atmosphere, it is no longer possible for the earth to keep itself healthy enough to support life. Had we appreciated that the earth is a unified living whole and adjusted our activities to the requirements of its homeostasis, the crises caused by global warming and climate change, would have been avoided.

When a physician graduates, he takes the Hippocratic Oath, swearing to practice medicine ethically. A similar oath is needed for planetary doctors so that they do not apply cures that harm the earth. Time has come when a person graduating from a college is administered an oath to desist from doing anything that injures the health of Mother Earth. In fact each individual from his early childhood has to be educated and trained to follow such an ethical principle. The whole system of governance from global to local levels has to be brought in tune with it, and those who violate it should be punished for committing crime against humanity.

Organizational Principles of Nature

Over the last 6,000 years, human beings have projected the principles on which the biological world functions together with its structures, functions and technologies output into the physical world. These principles were translated into various kinds of technologies even if imperfectly to create artificial environments. The pumps, filters, batteries, aerodynamics and even writing and printing were copied from nature in general and human body in particular. So long as these inventions and innovations were small in size and in conformity with the biology of the earth, they did not interfere with the earth system beyond its carrying capacity. During the last three centuries population growth, industrialization and urbanization have made the earth system so dense that the feedback mechanisms which kept it stable have become too weak to keep it functioning efficiently.

Man has not yet learnt from the earth the way it organizes itself. He does not know the implications of the technologies he uses for biology of the earth system. Unmindful of these implications, he goes on producing and accumulating goods and services to meet his ever growing needs and ego satisfying comforts and luxuries. He neither looks at 'before' and 'after' of the things he produces, nor at the 'beginning and end' of its interactions and feedbacks with the environment. He looks straight and aims at individual targets ignoring the interconnections, interdependencies and circulatory processes.

The belief that causal-logical thinking is 'scientific' and, therefore, capable of solving all problems all the times is so deeply ingrained in our minds that we tend to over depend on it and ignore the feedback control mechanism that keeps a system functioning smoothly. We forget that in all living and health systems the causal-logical thinking and the feedback control mechanism exist interdependently. The causal logical thinking can easily explain how an inanimate matter functions. In living systems like the earth, it can explain only the tiny steps in the long-term cyclical changes. The systems theory integrates the physics (logical thinking) and biology (feedback mechanism) and helps solve the problems that living systems face.

Earth must be seen as a living system. Its internal organization consists of many closely interlinked and cross-linked sub-systems. A heap of sand is not a system because each of its particles exists independently of the other particles. As against this, a tree is a system, and so is a factory, a city and human body. A human cell is a system; it becomes the subsystem when a large number of cells join to create a higher level system called tissue. Once the tissues join hands to give rise to an organ of the body, the organ becomes a system and the tissues its subsystems. And the process continues till the organs form the human body. Thus we see that the cell is the building block of the human body. How well the body system would work depends on how efficiently the sub-systems that make it work. The *Upanishads* are right when they say: *Yatha pinde tatha Brahmande* (as is the cell so is the Universe). Let us now see the hierarchy of subsystems that constitute our living earth.

Earth consists of subsystems like lithosphere, hydrosphere, atmosphere and stratosphere. The biosphere consists of individual eco-systems. The building blocks of these ecosystems are biotopes and the biotope consists of populations, which in turn consist of individuals living beings. Each of individual life form consists of innumerable organs; the organs consist of cellular tissues, which in turn consist of individual cells. The individual cells consist of many minute organelles and bio-machines. All subsystems of a larger system have one common characteristic: they follow the principle of interlocking self-regulation to keep themselves in equilibrium.

In manmade engineering control systems the guide values and nominal values are given from outside. Since the control system is external, all man-made systems are rigid and centralized. In bio-cybernetics the guide values and nominal values are built in the system. The systems that follow the bio-cybernetic laws are, therefore, stable. They are marked with following characteristics:

1. They are governed by negative feedback mechanism.
2. They are marked with efficient functioning, not by continuous growth.
3. They follow the principle of multiple-use.
4. They try to control and divert already-existent forces and energies in desired directions instead of combating the force applied by others.
5. They are not subservient to the products, which are formed to meet the needs of their permanent functioning.
6. They do not have wastes; everything is recycled.
7. They follow the principle of symbiosis.
8. They are compatible with the biology of man and nature.

Larger a system, more complex it becomes and, therefore, more difficult to manage. It is easier to manage villages, for example, than cities. Small scale industries are easier to manage than large industries. So is the case with irrigation projects, economies, and institutions. In order to function efficiently, larger systems must be decentralized. In the real world situation, the sub-systems that breach the fundamental rules and disturb the higher level systems disqualify themselves after a brief period of explosive growth. They either enter a state of 'deep freeze' or disassociate themselves from the subsystem in disharmony and move towards new stability. When the subsystems of the biosphere get disturbed by anthropogenic or natural processes, the biosphere as a system tries to free itself from the non-conformist sub-systems and returns to a new level of stability. That is how the biological world has survived for so many millenniums.

MODERN CIVILIZATION IS UNSUSTAINABLE

The so-called modern civilization constitutes a very dense network of sub-units unable to form a well-functioning system. It has become huge and highly complex. Unless it conforms to the laws of nature and evolves feedback mechanism to get over systemic problems as and when they arise, it is sure to crash, sooner or later. Let us first look at its main characteristics:

Increasing Population

World population has been steadily growing since ancient times. In 10,000 BC, there were just 4 m people; their number increased to 100 m by 500 BC, 200 m by AD 500, 425 m by AD 1500, 900 m by AD 1800, and 2500 m by AD 1950.[16] It took 123 years (1804-1927) for the world population to double to 2 billion, and 33 years (1927-60) to treble to three billion; and in the next 40 years (1960-1999) it increased to over 6 billion and is projected to be 7 billion by 2011, and 8 billion by 2025. By 2045-2050, the world population is expected to be around 9 billion, plus/minus 1.5 billion. Coming to the recent times, the growth rate of population peaked at 2.2 per cent in 1963 and ebbed to 1.10 per cent in 2009. At the same time, annual births have come down from 163 million in 1990s to 134 million in 2009. They are likely to stand still for the time being. Annual deaths stand at 57 million per annum and are likely to increase to 90 million per annum by 2050. If the present rate of growth continues for even next 30 years, we may confidently predict that the earth system would lose its balance and unexpected developments like global warming and climate change would jeopardize life on planet earth.

Of the 6.83 billion people of the world, 3.8 billion (60 per cent) live in Asia; Africa accounts for 840 million (12 per cent); Europe, 710 million (11 per cent); North America, 514 million (8 per cent); South America, 371 million (5.3 per cent); and Australia, 21 million (0.3 per cent). Each year India adds 18 million people, roughly another Australia to its heavy load of 1,176 million people (January 2010 estimates). China and India together account for 2.53 billion people of the world (40 per cent). China with its current population of 1.34 billion adds 7.5 m people each year, while

India adds around 17 million people to its current population of 1.12 billion. If India continues on the projected demographic path, it will overtake China by 2040, and become the world's most populous country. If India continues on the projected demographic path, it will overtake China by 2040, and become the world's most populous country. By 2050, China and India would respectively have 1,475 m and 1,620 m population. By that time the world population is likely to be 9 billion.

Increase in world population has serious implications for food security and environment. The problems get more serious when we look at the population growth not only in terms of number of people but also in terms of growing trend of conspicuous consumption. Poverty, pollution and violence are the three major problems the world faces today. One-fifth of the world population (1.2 billion people) lives on less than $1 per day. Global warming and resulting climate change have already become a major concern of the world. Shrinking arable land, disappearing forests, and pastures, falling watertables, and declining quality of land, water and air make the future of the world uncertain. And globalization of violence in the form of terrorism is making the world unsafe for the rich and poor alike.

Let us look at the problems India faces in some details. India has tripled its foodgrain output over the last five decades. Rice land productivity has doubled while that of wheat has more than tripled. Early maturing, high-yielding wheat and rice, combined with the tripling of irrigated area, have enabled farmers to double crop: winter wheat and summer rice in north and two crops of rice in south. But this unprecedented increase in productivity has barely kept pace with population growth. Further, by 1990s, the increase in land productivity came down and land availability per head decreased. In 1960, each Indian had an average of 0.21 ha of cropland; by 1999, it dropped to 0.10 ha per head, or less than half as much. And by 2050, it is projected to decrease further to 0.07 ha per head. It means that a family of five members will have to produce its food on 0.35 ha of land or the size of a building plot in a middle class Delhi suburb.

Falling watertable also threatens India's food security. According to International Water Management Institute (IWMI), withdrawal rate of underground water is double the aquifer recharge rate. As a result, watertable is going down almost everywhere. In a country where irrigated land accounts for 55 per cent of foodgrain harvest and where the major share of irrigation water comes from underground, falling watertables add to decreasing food security. The IWMI estimates that aquifer depletion could reduce India's grain harvest by one-fourth. Falling watertables may lead to rising grain prices on a scale that could destabilize not only grain market, but possibly the government too.[17] with 53 per cent of its children being underweight, any drop in food supply can quickly become life threatening.

With a staggering 338 million children under 15 years of age India is also facing a major challenge on the educational front. Despite efforts to educate its people, nearly one-third (34 per cent in 2009) of adults in the world's largest democracy cannot read and write. As in 2009, the dropout rate in the primary schools was almost 41 per cent and in some States like Bihar and UP, as high as 60 per cent. Failure to provide even primary education to children is undermining the efforts of the country to meet any of the eight Millennium Development Goals by 2015.

Providing work opportunities to the 10 million additional hands each year is a daunting task. Nowhere is this more evident than in agriculture, where number of farms increased from 48 million in 1960 to 105 million in 199o while their size shrank by 42 per cent (from 2.7 ha to less than 1.6 ha. By 2020, when the land passes to next generation the size of holdings would further decrease. Naturally it will threaten the ability of those living on the land to earn their livelihood, and trigger migration that could inundate Indian cities.

After several decades of development planning, the government and people of India are overwhelmed by population growth but unable to follow the Chinese or Gandhian way to halt the drift. Efforts are on to educate children, find work in other sectors of the economy, increase land productivity, and industrialize the country. But population continues to grow and environment destruction continues unabated. There is growing tendency on the part of the governments to offer loans, subsidies and grants but much of these fall in the hands of those who need and deserve least. The official machinery is so corrupt, that nothing meaningful can be expected from it.

The specter of rising death rates as a result of aquifer depletion, shrinking farm land and decreasing farm productivity does not haunt India alone. In most African countries the situation is alarming. When a few farmers commit suicide in India, print and electronic media goes into action and forces the central and State governments to do something to check the menace. It is not only food shortages, unemployment, natural calamities, and ethnic violence that are pushing death rates up; HIV epidemic has started taking a heavy toll of life. In South Africa, Botswana and Zimbabwe adult infection rates of HIV now exceed 20 per cent. In Zimbabwe, life expectancy fell from 60 years in 1990 to 44 years in 2000, and is expected to drop to 39 years by the end of 2010.

While the human cost of development is increasing all over the world, governments are unable to spend more on education, health, and community development. The spectre of violence in general and terrorism in particular is forcing them to transfer more resources to physical security, leaving social security in the hands of God.

Industrialization

In mid-nineteenth century, two-thirds of the people of the USA were engaged in agriculture and related activities; the share decreased to 40 per cent in 1900, 15 per cent in 1950, 5.2 per cent in 1970, and 2 per cent in 2002. Decrease in the workforce engaged in agriculture is universal (Tables 1.1 and 1.2). While the developed countries have almost reached a saturation point, the developing countries are trying to follow their seniors slowly but steadily. In 1900, 40 per cent of American lived in cities; by 2000 the percentage increased to 79.

Table 1.2: Major Economies; Workforce engaged in Agriculture

(in per cent)

Country	1960	1970	1990	2002
USA	10.0	5.2	2.0	2.0
UK	–	5.2	–	1.5
France	22.0	14.3	5.0	1.5
Japan	33.0	20.2	7.0	5.0
Brazil	52.0	–	23.0	10.0
Mexico	55.0	–	28.0	18.0
Republic of Korea	66.0	49.14	17.0	10.5
China	71.0*	78.33	73.52	49.8
India	75.0	70.63	64.10	57.0

*1978, +1996, @2007-08, Government of India.

Source: World Bank, World Development Indicators, 1997 and 2005.

Also see, T.J. Byres (2003), *Structural Change: The Agrarian Question and Possible Impact of Globalization* in Jayanti Ghosh and C.P. Chandrasekhar (ed.) 'Work and Wellbeing in the Age of Finance.' Tulika;

Table 1.3. Major Economies: Employment by Sectors: (2002)

Country	Agriculture	Industry	Service
UK	1.5	23.5	75.0
USA	2.0	22.0	76.0
France	1.5	23.5	75.0
Japan	5.0	29.0	65.0
Germany	2.5	31.0	66.0
Italy	5.5	29.5	65.0
Australia	4.5	20.0	75.5
China	49.8	23.5	26.7
Indonesia	55.5	14.0	31.0
Thailand	49.0	18.5	32.5
Philippines	35.0	15.0	50.0
Malaysia	17.5	31.5	51.0
Republic of Korea	10.5	26.5	55.5
Pakistan	58.5	14.5	27.0
India	57.0	21.0	22.0

Source: World Development Indicators (WDI), 2005, pp 56-58; data for Indonesia pertain to 1990-92; for China to 1998. (See: *Eco-reform and labour market regulation in China* by Ying Zhu.)

If the trends set during the last two decades continue, by 2021 only half of the total workforce of India and China would be engaged in agriculture. This means that these major economies together having more than one-third of the population of the world would pollute the environment in years to come far more than they do it today. The only way to reduce the pollution load is to use green technology. But the use of green technology demands more cooperation among the developed and developing countries. Further, the green technology too has its limitations. It must be backed by measures to conserve and reuse resources, and to promote need based life style.

Urbanization

Industrialization and urbanization go together. In 2005, 3.18 billion people constituting 49 per cent of the world population (6.46 billion) lived in cities. This year (2010), for the first time in the history of our species, urban population has exceeded rural population. Shrinking opportunities for work in rural areas and increasing investment in manufacturing and services in urban areas, rural-urban migration has become a major trend in developing countries. Europe and Americas are already urban. The day is not far off when a great majority of the people of the world will live in cities. We would need rural reserves so that the generations may come to know how their forefathers lived.

While industrialization and urbanization have improved the standards of living in the west, the standards of life have gone down all over the world. Violence against nature and fellow human beings has increased. At least 40 per cent of the urban population of developing countries lives in slums and squatter settlements. The European countries saved themselves from this negative development by mass export of people to the New World. The process of industrialization and urbanization in the developing countries today is the same, but the conditions are different. Hence, the developing countries must look for a new model of development, different from the one Europe and Anglo-America followed.

There is another urban trend that deserves attention. Large cities are becoming larger while the smaller ones are decreasing in number as well as growth rate. As a result, the urban-rural divide has widened; and quality of life in large cities has deteriorated. Overcrowding, poverty, unemployment, environmental pollution, and violence have become the main traits of the metropolitan cities of the developing world. The rural migrants to these cities are largely semi-skilled labourers looking for work in the informal sector of urban economy. They live in slums within the inner city or in squatter settlements in the periphery. The global cities like Mumbai and Karachi are known as much for world-class skyscrapers as for world-famous slums of Dharawi and Orangi.

Dehumanized Technology

Modern technology is self-indulgent; it cares not for what happens to human beings or nature; it aims at saving time, labour, and cost. It benefits a few often at the cost of the many. The fault, however, does not lie with technology *per se*; it lies with the way it is used or abused. Its adverse effects on nature and human beings have now become so apparent and visible that they cannot be ignored any further. Among these effects are: mass unemployment, environmental pollution, escalating violence, and unhealthy lifestyles.

Modern technology is not labour intensive. This in itself is not a serious flaw if it is accompanied by social, political and economic policies and strategies that can take care of those left out of the job market. This could happen only if the economy is decentralized and self-employment becomes the main stay of the people. But we know that it would not be a feasible proposition in a globalized economy. The developed countries could prosper because they had three continents (North and South America, and Australia) to rehabilitate their fast growing population. The densely populated developing countries of Asia and Africa do not have that advantage. They have no other place to go. As such each new advance in technology makes more people jobless. Unemployment breeds poverty, frustration, and violence. Earlier, people used hands and sticks to settle scores; now they use guns and missiles, the products of modern technology.

Waste of Resources

Developing countries, despite the jerks their economies get from time to time, are globalizing and 'modernizing' fast. Development means material prosperity and conspicuous consumption even at the cost of human development. High standard of living means more goods and services unmindful of their implications for environment, health and hygiene. To turn away from the basic human values such as truth, non-violence, love, compassion, human brotherhood, and respect for nature on which most great civilizations were built, is another trait of modernity. Nature becomes the resource for exploitation as do the fellow human beings.

By 2025, China and India would be leading economies of the world in absolute terms. But would that also mean leading nations in terms of standard of life measured by gross national happiness (GNH)? Want based conspicuous consumption does not produce happiness; and if it does, it is not for all. The need based consumption on the other hands leads to Sarvodaya (development of all). Bhutan, a traditional country has the highest GNH in South Asia

We have seen how nature is being robbed of its inherent capacity to regenerate. If we continue to tread on the path set by modern civilization, we are sure to destroy ourselves. Would there be enough food in the world to avoid food riots? Would there be enough water for domestic and irrigation needs to ward off water wars? Would the general tendency to resort to violence to settle disputes would abate to enable people to live in peace? Would temperatures of the earth continue to rise and roast everything biotic or a deep freeze would herald the end of our civilization. We can either opt for higher GNH and usher in a new civilization or continue to pursue the mirage of higher GDP and be the dying generation of the so-called modern civilization.

The Issue of Global Warming and Climate Change

A fierce debate is on for the last two decades on implication of increasing use of fossil fuels for the health of the earth system. On the basis of the analysis of the data on temperatures, rainfall, droughts, and floods, scientists foresee an era of global warming very much on hand. The temperatures of the earth are increasing, they say and this would result in melting of the ice caps and mountain glaciers which in turn will raise the level of the sea. Further, the rising temperatures would threaten food security and in due course of time make planet earth inhabitable. Although there are many dissenting voices, a majority warns that global warming and climate change is not a myth; it is something that is already here and unless some urgent and concrete steps are taken by the global community to stop it, the very survival of life on planet earth would be jeopardized.

Pro-Warming Lobby

The land, water and air on the earth surface can support life because they are kept warm by the sun. When sunlight reaches the earth surface, one-third of it is reflected back into space by the clouds, atmospheric dust, and earth surface. The rest heats the land, water and atmosphere of the earth, and makes life possible. Solar energy so received by the earth does not stay there forever. The earth surface (land, air, and sea) emits thermal radiation, or infrared heat out into the space. Some of this outgoing radiation, however, is reabsorbed by water vapour, carbon dioxide, and other gases in the atmosphere (called green house gases (GHGs) because of their heat-trapping capacity) and is then radiated back to the earth. Had there been no green house gases and clouds in the atmosphere, the earth's average surface temperature would have been 18°C instead of 15°C that it is today.

Thus the earth's warming occurs not only in response to external forcing such as variations in its orbit around the sun (orbital forcing); changes in solar luminosity, and volcanic eruptions, but also due to release of GHGs in the atmosphere by anthropogenic activities especially the burning of fossil fuels. The increased concentration of the GHGs in the atmosphere obstructs the heat of the earth surface radiating to the outer atmosphere. And thus the earth is warms up further. As the number of fossil fuel using engines and cars grows, CO_2 concentration in the atmosphere increases. The present rate of increase of GHGs in the atmosphere amounts to around 8 billion tonnes of CO_2 per year. A part of it is absorbed by the 'sinks' like forests and oceans, but the rest stays in the atmosphere. When we allow our trash to decompose in landfills and breed methane-belching cattle, we unconsciously produce millions of pounds of GHGs. The nitrogen-based fertilizers and other so-called modern soil enriching practices release nitrous oxide into the atmosphere. And then adding to all these are HFCs, PFCs, and SF6.

Once the GHGs enter the atmosphere, they stay there for decades. According to the Intergovernmental Panel on Climate Change (IPCC), since the beginning of the industrial revolution in mid-eighteenth century, CO_2 level in the atmosphere has increased by 35 per cent and methane level by 148 per cent. Paleoclimate readings taken from ice cores and fossil records show that the concentration of CO_2 and Nitrous Oxide in the atmosphere is at its highest level today since the last 650,000 years, i.e., the later phase of the Pleistocene epoch (1,808,000 to 11,550 years before present). Since the GHGs in the atmosphere obstruct the heat of the earth surface radiating to the outer atmosphere, temperature of the earth are going up.

Even if GHG concentration is stabilized at the level it is today, earth will continue to warm by about 0.6°C over this century because it takes years for it to put its feedback mechanism in action and to fully react to increases in GHGs. As temperature of the earth surface increases, the upper layers of the ocean also get heated. The ocean currents then carry some of this excess heat from surface layers to lower layers.

Global warming leads to increase in both average and extreme temperature and precipitation and consequent melting of ice caps in the Arctic and Antarctic regions, retreat of glaciers, rise in sea level, increasing coastal erosion, lengthening growing seasons, and increasing range of many infectious diseases. There are more hot days and less cool days. Rainfall is more in the high latitudes and less in the tropical regions. The duration of the rainfall gets shorter but the amount increases. Further, rain is accompanied by massive storms and dry spells. The frequency of droughts increases.

According to the IPCC, global warming has already started affecting the weather and climate although it is not possible at this early stage to pin any particular weather event on it. Heat waves have increased in frequency and so have intense rain events since 1950s. Rainfall measurements using artificial satellites indicate that tropical regions got more rain from bigger storms between 1979 and 2003. The drought affected areas, and the number of intense tropical cyclones have also increased since 1970s. Around 10 per cent of the world's population lives in coastal areas less than 10 metres above the mean sea level. If the sea levels, as estimated by IPCC, rise between 0.18 and 0.59 metres by 2099, many low-lying areas along the coasts will be drowned. The Himalayan glaciers will, it is estimated, disappear in less than a century if the present speed of melting continues and thus the very source of River Ganga and its tributaries will be lost.

Higher temperatures have already affected seasonal rhythms in many parts of the globe. Spring sets in earlier forcing the migrating animals and birds to move earlier to follow food sources. With the lengthening of the growing season, plants need more water to keep growing. Shorter and milder winters multiply insects that destroy crops and infect human body. Growing seasons would increase maximum daily temperatures, which will in turn kill many plants and force wild animals to migrate towards the poles. Those species that cannot migrate will face extinction. According to the estimates of the IPCC, 20-30 per cent of plant and animal species will be at risk of extinction if temperatures increase by 1.5° C or more. Incidentally, it is the people of poorer countries who will be hit hardest.

Anti-Warming Lobby

While most scientists subscribe to the view that global warming and climate change poses serious threat to human survival and development, and that it is being caused by human activities, there are those who consider it a temporary and a part of the normal natural cycle. They cite examples of weather changes in the past and challenge the data put forward by IPCC and fellow scientists on melting of ice caps, and glaciers.

On April 6, 2006, a group of 60 scientists wrote to the Canadian Prime Minister questioning the so-called 'consensus' that the Earth faces a 'climate emergency.' "Observational evidence does not support today's computer climate models, so there is little reason to trust model predictions of the future...Significant (scientific) advances have been made since the (Kyoto) protocol was created, many of which are taking us away from a concern about increasing green house gases. If, back in the mid-1990s, we knew what we know today about climate, Kyoto would almost certainly not exist, because we would have concluded it was not necessary.

"It was only 30 years ago that many of today's global-warming alarmists were telling us that the world was in the midst of a global-cooling catastrophe. But the science continued to evolve, and still does, even though so many choose to ignore it when it does not fit with their predetermined political agenda."[18] The arguments these scientists advance are varied but they all converge on one point: global warming is not imminent.

'Climate Change is Nothing New': Giving the details of a new study on earth's climate history, the Washington Post on October 16, 2006 published an article by its reporter Christopher Lee under the caption "Climate Change is Nothing New" which referred to the findings of Simon Brassell, an Indiana University geologist asserting that climate change occurred during the Mesozoic Era when the earth inhabited by dinosaurs and there were no human beings on the earth. "If there are big, inherent fluctuations in the system, as paleoclimate studies are showing, it could make determining the Earth's climatic future even harder than it is."[19]

It is Global Cooling, not Global Warming: There are some who think that it is the prospect of global cooling rather than global warming, which should concern us. In August 2006 Khabibullo Abdusamatov, the head of the space research sector of the Russian Academy of Sciences' Astronomical Observatory, predicted that "Global cooling could develop on Earth in 50 years and have serious consequences before it is replaced by a period of warming in the early twentysecond century.... On the basis of our (solar emission) research, we developed a scenario of a global cooling of the Earth's climate by the middle of this century and the beginning of a regular 200-year-long cycle of the climate's global warming at the start of the twentysecond century."[20]

According to Khabibullo Abdusamatov a period of global cooling similar to one seen in the late seventeenth century – when canals froze in the Netherlands and people had to leave their dwellings in Greenland – could start in 2012-2015 and reach its peak in 2055-2060. "The Kyoto initiatives to save the planet from the green house effect should be put off until better times,.... "The global temperature maximum has been reached on Earth, and Earth's global temperature will decline to a climatic minimum even without the Kyoto protocol", said Abdusamatov.[21]

According to the data released by the National Oceanic and Atmospheric

Administration (NOAA) of the USA on July 14, 2006 Alaska's state-wide January to June average temperature was "0.55°F (0.30°C) cooler than the 1971-2000 average."[22] A September article in the Geophysical Research Letters reports that over 20 per cent of the heat gained in the oceans since the mid-1950s was lost in just two years. The sudden cooling of the oceans, says Roger Pielke, Senior, a former climatologist for the state of Colorado, USA "certainly indicates that the multi-decadal global climate models have serious issues with their ability to accurately simulate the response of the climate system to human- and natural-climate forcings."[23] Low temperatures in the present (2009-10) winter season in the northern hemisphere make these claims credible. India had the coldest January in 2010.

Global Warming is Not Global: Recent studies point to the Antarctic getting colder and the ice cover increasing. Another study of 2006 also published in Geophysical Research Letters has concluded that the sun was responsible for up to 50 per cent of warming in twentieth-century.[24] A satellite tropospheric temperature data put forward by NASA reveals that the Southern Hemisphere has not warmed in the past 25 years contrary to "global warming theory" and modelling. The present concern should more appropriately be referred to as 'Northern Hemisphere warming.'[25]

Global Warming Stopped in 1998: Bob Carter, a noted paleoclimate researcher and geologist at James Cook University in Australia wrote in April 2006 that "There is a problem with global warming. it stopped in 1998."[26] Another Paleoclimate scientist of the same university has corroborated Carter's findings.[27] Official temperature records of the Climate Research Unit at the University of East Anglia in the UK show that the global average temperature did not increase during 1998-2005. "...this eight-year period of temperature stasis did coincide with society's continued power station and SUV-inspired pumping of yet more carbon dioxide into the atmosphere,"[28] There are studies to prove that predictions made by global warming proponents did not come true. In a paper published in Geophysical Research Letters by Konstantinos Andreadis and Dennis Lettenmaier finds droughts in the U.S. becoming "shorter, less frequent and cover a small portion of the country over the last century."[29] Ola Johannessen and his colleagues showed in 2005 study that the interior of Greenland is gaining ice mass.[30] Also, according to the International Arctic Research Institute the Arctic Ocean was warmer in the 1930's than today.[31]

RESPONSES TO GLOBAL WARMING AND CLIMATE CHANGE

Without going into the merits of the claims of the pro- and anti-warming scientist, there is a bit of exaggeration on both sides. First, there is clash of national interests. Second, data presented by the scientists are selective, not comprehensive and comparative. And third, scientists too get swayed by their national interests. Neither developed nor developing countries to do anything that will slow down their economic growth, the latter because economic growth is a must to eradicate hunger and poverty, and the former, because they do not want to continue to rule over the destiny of the world. That is why, some developing countries like Bolivia, Cuba, Sudan and Venezuela termed the Copenhagen consensus as a document of powerful countries aimed at killing the Kyoto Protocol, which had put the onus and responsibility for cleaning the mess on developed countries.[32]

The question before us is: what should we do as individuals, communities, and societies in the face of contradictory scientific evidences, and dilly-dally by the developed countries to take any concrete and time bound corrective action before it is too late? We face a danger if global warming is on; we also face a danger if global cooling is on. But even if neither of these extremes occurs, we are not out of danger because the pollution of the soil, water, land, air is increasing so fast that no life will survive on the earth in not too distant a future. We would not have clean water to drink, clean air to breathe, and clean food to eat. Noise pollution is already a serious problem in our cities. Much of the foods we consume are polluted and contaminated. Many species of birds and animals are on the verge of extinction. If the health of the earth continues to deteriorate at this rate, it will reach a point of no return within this century. The health of the human beings is deteriorating so fast that that medicine has become a major industry of the world.

Irrespective of the occurrence of global warming or global cooling:

1. We would not have enough of petroleum to run our power plants, cars, buses, trains, ships and airplanes beyond this century.
2. Our land, water and air would be so polluted that no one would lead a normal healthy life by 2050.
3. We have created a civilization which treats earth as a resource for exploitation not as the home of all life forms humans included.
4. We have lost the culture of love and compassion that Rama, Buddha, Christ and Gandhi taught and practiced.
5. We have become so materialistic that we are not ashamed of living in an island of plenty in the midst of the sea of filth, poverty and death.
6. We have become so cruel that we do not mind killing ourselves in order to kill many others.

So, while scientists and UNFCCC debate whether global warming and climate change is in the offing or not, our very survival and development is at stake. If global warming does occur, we are doomed as individuals, communities, nations and civilizations. In case it does not, we would not be safe either. Let the debate on global warming and climate change continue for we do want incontrovertible scientific evidence on it, we must mount a frontal attack on environmental pollution engulfing us all. While do so, we must see environment in much broader perspectives. It should certainly cover land, water, air and noise pollution but above all it should include cultural pollution that leads to the pollution of thc mind, body and heart of us all.

Culture has to be seen in two perspectives: inner and outer. Inner connotes harmony within while outer connotes harmony among fellow beings and between human beings and nature. A harmonious inner environment leads man to higher consciousness and thus self-realization. The idea of falsehood, hatred and violence does not penetrate the minds of those who are cultured from within. Nature and human society constitutes the outer environment. It can be clean only if we live within the carrying capacity of nature, and we have compassion for all life forms all the way from plants to human beings. If we must cross the limits of nature, we should give back to it what we have taken from it. If we hurt other living beings, we must undergo penance for it and devise ways and means to check its recurrence. If we cut a tree, we must plant more trees.

This means that we should not stop policies and programmes initiated so far to combat global warming, for the steps proposed to be taken in this regard are also the necessary steps to check environmental pollution, and to ensure food and energy security. Our environment must be clean so that life can exist and prosper on planet earth.

COMBATING ENVIRONMENTAL POLLUTION

Global Initiatives

The efforts to arrive at an international consensus on environment and development took a concrete form when the representatives of 165 countries converged in Rio in 1992 for the Earth Summit and signed the Rio Declaration. It said that 'right to development must be fulfilled so as to equitably meet developmental and environmental needs of the present and future generations.' The Summit dwelt on following four issues:

1. Systematic scrutiny of patterns of production—particularly the production of toxic components, such as lead in gasoline, or poisonous waste including radioactive chemicals;
2. Alternative sources of energy to replace the use of fossil fuels which produce GHGs;
3. Greater reliance on public transportation systems in order to reduce vehicular emissions, congestion in cities, and health problems caused by polluted air and smog; and
4. Adequate supply of potable water.

The Summit produced the following documents:

1. Rio Declaration on environment and development;
2. Agenda 21
3. Convention on biological diversity;
4. Forest principles; and

5. UN Framework Convention on Climate Change (UNFCCC).

Among the measures suggested to ward off global warming and climate change, was the imposition of quota on fossil fuel production, as there was a direct link between fossil fuel production and CO_2 emissions. The concept of green house gas emission trading which became a part of Kyoto Protocol in 1997 under the auspices of UNFCCC was one of the byproducts of the Rio Conference. The Protocol was signed by 160 countries, and covered more than 55 per cent of global green house gases emissions. The USA, the world's largest emitter of green house gases, however, did not ratify the treaty as it considers the Protocol unfair and an ineffective means of solving the problem. The solution lies in improved energy technology. Acceptance of the Kyoto Protocol would mean 'serious harm' to the US economy. USA also objects to the exemption of '80 per cent of the world; including major population centres' like China and India from the treaty obligations. The Kyoto Protocol is to expire in 2012.

To work out future commitments for developed countries, the Conference of the Parties to the Kyoto Protocol established an Ad Hoc Working Group (AWG-KP) in December 2005. The UNFCCC met in Bali on December 2-15, 2007 and decided to jointly step up international efforts to combat global warming and climate change. They met again in Bangkok in March 2008. Delegates from 162 countries and over 1,000 others representing NGOs participated in this Conference. It was decided to cap the GHG emissions at the level of 2010 by 2020-25, and dramatically reduce overall emissions by 2050. The next round of negotiations took place in Accra, Ghana, on August 21-27, 2008.

In December 2009, the member countries to the Kyoto Protocol met at Copenhagen, Denmark to cap the emission of GHGs at an agreed level. It was supposed to prepare the ground for a new international climate change treaty to be signed in Mexico in December 2010—one that would include the pledges of the world's biggest GHG emitters, in a legally-binding agreement. It did try to get commitment from the participating countries to keep the rise in global temperature under 2° C, but failed to specify the amount of emissions reduction needed to achieve this goal. The agreement also promises to mobilize $100 billion a year by 2020 to help developing countries change over to green energy. The developing nations agreed to open their emissions reduction activities to oversight by international bodies. The developing countries were, however, disappointed with the outcome of the conference. The chief negotiator for the G77 group of developing nations, Lemma Did-Aping, said the Copenhagen Accord represented a "gross violation of the tradition of the United Nations" and "locks developing nations into poverty forever."

INDIAN RESPONSE

India signed the UNFCCC on 10 June, 1992 and ratified it on 1 November, 1993. It has constituted a Working Group on the UNFCCC and Kyoto Protocol. The Kyoto

Protocol requires developed countries and the economies in transition (Annexure B of the Protocol) to reduce their GHG emissions by an average of 5.2 per cent below 1990 levels. India is thus not bound to lower its green house emissions under the UNFCCC.

India has sponsored a number of research projects to assess methane emissions from paddy cultivation, and CO_2 emissions from coal, etc., and prepared an inventory of GHG and the steps taken to reduce their emissions with 1994 as the base year covering energy, manufacturing, agriculture, forestry, and waste disposal. The Department of Science and Technology (DST) has constituted Technology Information, Forecasting and Assessment Council (TIFAC) to promote the use of environmentally sound technology. Efforts are also on to conserve biodiversity, to minimize the use of coal and to upgrade technology of coal washing, combustion and recovery of coal-bed methane. Similar steps are being taken to conserve oil and gas. Gas has already replaced coal in domestic sector. The share of gas in the power sector has increased from 2 to 8 per cent. At present, hydropower accounts for only a quarter of the total installed capacity. In the area of renewable energy, the solar energy based photovoltaic system is being used in rural electrification, railway signalling, microwave repeaters, power to border outposts, TV transmission and reception, home and street-lighting systems, stand-alone power plants, and pumping systems, among others.

India is among the five leading nations in wind power generation. It has installed capacity to produce 1,507 MW of wind energy. It manufactures 250-600 KW capacity wind power generators. Among the other achievements of India in clean energy sector are: 358 MW of installed biogas power generation capacity; and gasification systems of a total capacity of 42.8 MW for decentralized energy uses. In rural areas, over 3.2 million biogas plants and 33 million improved stoves have been installed. Total installed capacity of small hydropower projects is 1,423 MW. Further, projects with an aggregate capacity of about 15 MW have been completed to use energy recovered from urban, municipal and industrial wastes. For energy efficiency and conservation, Bureau of Energy Efficiency (BEE) has been set up; it is charged with the responsibility of putting into operation, such conservation measures as energy standards, labelling of equipment/appliances, building energy codes, and energy audits.

An expert committee on climate change was constituted by the Ministry of Environment and Forest (MoEF) in 2008 under the chairmanship of R. Chidambaram, Principal Scientific Adviser to the Government of India. The job of the committee is to study the impact of human activities on climate and to suggest measures to minimize the impact. The Union Budget 2008 notes that India is neither a significant contributor to GHG emissions "nor will it be so in the foreseeable future." It nevertheless recognized the country as one of the most vulnerable to climate change and recommended the setting up of an expert committee to study the impact of climate change on India.

A National Action Plan on Climate Change was launched in June 2008. It includes strategies to mitigate global warming and reduce green house gas emissions and thereby to lead India towards sustainable development. The strategies include national missions on solar energy and energy efficiency, water use, forestation, agriculture, sustainability in urban environments, the Himalayan ecosystem and scientific research. But none of these strategies have so far produced assuring results. What is lacking is people's participation in the efforts being made by the government channels.

Let us not forget that environmental pollution is coupled with cultural and mental pollution caused by so-called modern civilization. Modernity is not something that is borrowed from elsewhere; it is the product history, geography and culture of a people. No forward looking culture is static; it evolved itself through time to meet the challenges posed new exigencies over which it has no control. But its roots do not change and, hence, it remains distinct from other cultures. Modernization is thus a built in process in all cultures. Hence, each culture develops its own model of modernity. Modernization is thus not borrowed; it evolves from within.

Mahatma Gandhi was deeply rooted in Indian culture but he was highly influenced by other cultures too. He kept his windows open to get fresh breeze from outside but kept his door closed so that he was not swept away by the storms. He experimented with a model of modernization that cared for the human beings as well as nature. He called it Sarvodaya. "There is enough on this earth to meet everyone's need but not everyone's greed." said he. He stood for economic development but only to strengthen moral development. The time has come to return to the Gandhian way for that is the only way to save the earth from becoming lifeless and thus to ensure a prosperous future for humanity.

THE GANDHIAN WAY

Almost a century ago in 1909, Gandhi wrote in *Hind Swaraj* that insatiable and unending pursuit of material prosperity in the west was a threat to human progress. He defined *swaraj* as self-control; self-realization; and freedom from the evils that afflict human beings. Once a person has attained *swaraj*, no one can enslave him. It applies to nations and civilizations too. When India lost *swaraj*, the British came and occupied it. In fact, the British did not enslave India; we the people of India invited the British to enslave us, asserted Gandhi. He asked his countrymen to return to the fundamental values of Indian civilization. National independence will follow once people attain *swaraj*.

Gandhi stood for harmony, harmony among people of all countries, races, religions, social levels. All deserve their self-respect, freedom and opportunities for moral and material advancement. "There is orderliness in the universe, there is an unalterable law governing everything and every being that exists or lives. It is no blind law, for no blind law can govern the conduct of human beings."[34] This law must be followed by all, for the violation of it leads to disorder, which in turn leads to unhappiness. He extended the principle of harmony beyond the sphere of human beings to cover relations between man and nature. Man-nature relationship too should be harmonious; nature is not a resource for exploitation but the very source of life on planet earth. It must be treated kindly and should be so used that its capacity to meet the present as well as the future needs of all life forms, humans included. Coming to ecology, he said, "The earth, the air, the land and the water are not an inheritance from our forefathers but on loan from our children. So we have to hand over to them at least as it was handed over to us." And he warned "what we are doing to the forests of the world is but a mirror reflection of what we are doing to ourselves and to one another."[35]

While asking people to lead a simple life, so that all can meet their needs, he said "the earth provides enough to satisfy every man's needs, but not every man's greed." He was totally against conspicuous consumption. "A certain degree of physical harmony and comfort is necessary, but above a certain level it becomes a hindrance instead of help."[36] Branding modern industrial civilization highly destructive of ecology and environment, he said, "God forbid that India should ever take to industrialism after the manner of the West. The economic imperialism of a single tiny island kingdom is today keeping the world in chains. If an entire nation of 300 million took to similar economic exploitation, it would strip the world bare like locusts."[37] On the question of movement to protect the environment, he said, "we cannot have ecological movement designed to prevent violence against nature, unless the principle of non-violence becomes central to the ethics of human culture."

Gandhi was a systems man. In his scheme of things the earth too was a system with lithosphere, hydrosphere, atmosphere, and biosphere as its first level subsystems. Each of these sub-systems was formed by still lower level sub-systems. And finally going down to the sub-atomic substances whose number is beyond human comprehension. These various levels of systems and sub-systems constitute the living earth. They cannot exist alone; isolated from the system above and sub-systems below, they would wither away the same way as our limbs separated from the body become lifeless. How these systems evolved, we just do not know but we know that they are there and are governed by a Cosmic Law whose violation destroys system or makes it dysfunctional.

Similarly, he considered humanity as a system bound by the Law of Species which calls for cooperation and coordination. Human being must care for their fellow beings; they must care for lower level life forms too for they all are interdependent. Man depends on nature for his survival and development. Gandhi throughout his life demonstrated how human beings could build a harmonious relationship with other human beings and nature. He followed the time honoured principle of *vasudhaiva kutumbakam*. He never treated the British as his enemies; and he was an ardent lover of nature. It is unfortunate that we no longer adhere to Gandhian principles. His principle of 'simple living and high thinking' has now been converted into principle of 'ostentatious living and narrow thinking'. We believe in exploiting nature as well as our fellow human beings for our selfish ends.

No statesman or political leader of Gandhi's stature ever devoted so much time and energy to inculcate in the people around him conservation of natural resources, cleanliness, and healthy living. His *ashrams* were ecologically live and environmentally energizing. This is in contrast to what our modern day leaders do. The question that haunts us today is: How to determine the trade off between material advancement and human well-being and between inner development and outer development. If we continue to advance materially, we are sure to reach a dead end within the next few decades. If we moderate our material advancement and keep it within the carrying capacity of the earth system and focus on making human being better human beings, we will not only ensure our future but also the future of the generations to come. Let us learn a lesson from what Chief Seattle told the then President of America, Mr. Franklin Pierce in 1854 when the latter offered a Reservation elsewhere if he and his people would sell their land in Washington territory:

> "Teach your children what we have taught our children; that the Earth does not belong to humanity; humanity belongs to the Earth. This we know. All things are connected like the blood that unites one family. Whatever befalls the Earth befalls the sons and daughters of the Earth. Men and women did not weave the web of life; they are merely a strand in it. Whatever they do to the web, they do to themselves." And he added "How can you buy or sell the sky, the warmth of the land? The idea is strange to us. If we do not own the freshness of the air and the sparkle of the water, how can you buy them?"
>
> "You must teach your children that the ground beneath their feet is the ashes of our grandfathers. So they will respect the land, tell your children that the earth is rich with the lives of our kin. Teach your children what we have taught our children that the earth is our mother. Whatever befalls the earth befalls the sons of the earth. Man did not weave the web of life; he is merely a strand in it. Whatever he does to the web, he does to himself. Even the white man, whose God walks and talks with him as friend to friend, cannot be exempt from the common destiny. We may be brothers after all. We shall see. One thing we know, which the white man may one day discover—our God is the same God. You may think now that you own Him as you wish to own our land; but you cannot. He is the God of man and his compassion is equal for the red man and the white. This earth is precious to Him, and to harm the earth is to heap contempt upon its Creator. The whites too shall pass; perhaps sooner than all other tribes. Contaminate your bed, and you will one night suffocate in your own waste." [38]

The President must have been amused to hear these words from a native Chief, steeped in primitivism and ignorant of the world beyond his place of birth. The Chief was defeated and with him his words of wisdom were also defeated even if only temporarily. Today Americans are the greatest polluters; they are not prepared to reflect on the word of wisdom the Chief Seattle uttered over a century and half back. And taking the cue from Americans the whole world is set upon a destructive path of modernization and development. Nature is increasingly alienated to satisfy our ever-growing hunger for body comforts and luxuries. We gleefully watch the biodiversity of nature disappearing and global warming and climate change knocking at our doors.

We have taken what has been given to us in trust, and have used and abused it for our selfish and futureless ends.

There is no religious tradition that does not enjoin on its followers to tread the path of peace and harmony. Indian tradition treats Earth as Mother, which it really is, deserving veneration. Western tradition too is pro-nature. St. Francis of Assisi was well-known for his love of nature and environment. His love for brotherhood included all that was created by God right from small flowers, birds, and animals to human beings. For him, his love for birds was equivalent to his love for the Pope. He endeared himself to birds and animals so deeply, says one of the many fables about him that a flock of sparrows he preached, flew only when he asked them to do so. According to another fable, he calmed a terrifying and ferocious wolf, which was known to devour men and animals alike. Yet, another fable recounts that the saint thanked his donkey for carrying him and helping him on his deathbed. He was so prophetic when he said: "If you have men who will exclude any of God's creatures from the shelter of compassion and pity, you will have men who will deal likewise with their fellow men."[39]

In the very first chapter of the *Bible* it is said that in the beginning, when things were as God meant them to be, animals were not created to be food for humans. The animals were to eat grass, foliage and cereals; and human beings to live on fruits and nuts. Only later, when sin was rampant in the world, were animals granted to humankind for food, with the sinister words: 'The fear of you and the dread of you shall fall upon all wild animals and birds and fish; they are given into your hands.' William Ralph Inge echoed the same feelings when he said: "We have enslaved the rest of animal creation and have treated our distant cousins in fur and feathers so badly that beyond doubt, if they were to formulate a religion, they would depict the Devil in human form."[40]

Pythagoras, the Greek philosopher who is known as a geometrician, has said: "For as long as men massacre animals, they will kill each other. Indeed, he who sows the seed of murder and pain cannot reap joy and love."[41] Thomas Edison has said: "Non-violence leads to the highest ethics, which is the goal of all evolution. Until we stop harming all other living beings, we are still savages."[42] And Arthur Schopenhauer opined "compassion for animals is intimately connected with goodness of character and it may be confidently asserted that he who is cruel to animals cannot be a good man."[43]

Ahimsa (non-violence) and compassion are the most basic tenets of Hinduism, Buddhism and Jainism, the three religions of Indian origin. Mahatma Gandhi was highly influenced and inspired by these and other religions like Christianity and Islam. He did not say much about environmental problems as they had not emerged as serious problems in his time but if we take his life as his message, he was an environmentalist par excellence. He prescribed the following eleven vows or *vratas*. If we adhere to them, all the environmental problems we face today would disappear within no time.

1. *Ahimsa* (Non-violence),
2. *Satya* (Truth),

3. *Asteya* (Non-stealing),
4. *Brahmacharya (*sex for procreation alone*)*,
5. *Aparigraha* (Non-avarice)
6. *Shareer- shram* (Bread Labour /Physical labour),
7. *Aswad (*Control of palette)
8. *Sarva Dharma Sambhava* (Religious harmony)
9. *Abhaya (*Fearlessness)
10. *Swadeshi (*self-reliance*)*, and
11. *Asprishyata nivaran* (Abolition of untouchability)

In his autobiography he says "to my mind the life of a lamb is no less precious than that of a human being. I should be unwilling to take a life of a lamb for the sake of the human body. I hold that, the more helpless a creature, the more entitled it is to protection by man from the cruelty of man." Elsewhere he has said, "man has no power to create life and, therefore, he has no right to destroy life....Man has been endowed with higher faculties so that he can be compassionate to lower beings."[44] His Holiness the XIV Dalai Lama of Tibet says: "It is only some carnivorous animals that have to subsist on flesh. Killing animals for sport, for pleasure, for adventures, and for hides and furs is a phenomenon which is at once disgusting and distressing."[45] *Yin-chih-wen*, a Confucian-Taoist treatise, says: "Buy captive creatures and set them free. Hold fast to vegetarianism and abstain from taking life. Whenever taking a step, always watch for ants and insects. Prohibit the building of fires outside (lest insects be killed) and do not set mountain woods or forests ablaze. Help people in distress as you would help a fish in a dried up rut. Free people from danger as you would free a sparrow from a fine net."[46] And look at what Quakers have to say in this regard: "We shall respect that of God in all creations. We shall live in loving harmony with the earth. Humankind shall be a joyful gardener of the world given to us by God, and shall use its fruits wisely and moderately."[47] And the Talmudic sages (Judaism) say: "Just as He clothes the naked, so shall you clothe the naked. Just as He is merciful, so shall you be merciful."[48] Compassion for all creatures is not only God's business; it is a virtue that all human beings must emulate. Compassion is central to our entire approach to life.

And Islam, in whose name the *Jehadis* kill people, is no less a religion of non-violence and peace. God does not love *fasad* (violence). God abhors violence which pervades all walks of life today, says *Qur'an*. This is supported by other statements in this sacred book such as peace is one of God's names (59:23); those who seek to please God should follow "the path of peace" (verse 5 of 16th *Surah*); and Paradise is "the home of peace" (89:30).

Gandhi called himself a *Sanatani* Hindu; he was deeply rooted in Indian tradition derived from a mixture of Hinduism, Buddhism and Jainism and influenced by Zoroastrianism, Islam and Christianity. His education in England helped him to compare life-styles and achievements of India and the West and came to the conclusion that western civilization paraded being welcomed in India as modern civilization was

'Bhasmasuri' (devilish). It is materialistic and devoid of spirituality. It cannot be called modern.

All those who wanted to be inmates of his *ashrams* (hermitage) had to clean public latrines by turn, and clean their own clothes and utensils. While taking bath in the freely flowing Sabarmati river on which his Ashram is located, he used minimum quantity of water. When asked about using abundantly available river water so sparingly, he replied that the water flowing in the river did not belong to him. It was a telling example of *aparigraha* (non-possession) and exemplary way of conserving natural resources. Following him, many of his associates used scraps of papers for writing notes and reusing old envelopes while replying to letters Gandhi received in plenty.

Gandhi did not invent new principles and theories. He was a man of practical wisdom. For him, the principles on which Indian civilization was based or for that matter even western civilization were good enough to keep this world peaceful and progressing morally and economically without encountering such dangers like global warming and climate change. Nature should be approached with a sense of gratitude; it should not be hurt for luxuries or to amass wealth. Man, the most powerful living being on the earth should desist from hurting other sentient beings including plants and animals except for satisfying their biological and cultural needs. This would ensure ecological balance. Man has been endowed with a brain that can take him to far greater heights of creativity than he has achieved so far and a heart that has the potentials of leading him to moral heights that would take him closer to Rama, Krishna, Christ, Mohammad and other great men and women of wisdom.

Gandhi was an environmentalist even before environmental problems we face today. If the world follows him, there would be no problem of global warming or cooling. Gandhian way does not ask us to go back to primitive times; it rather asks us to be modern in the real sense, and to be rational and scientific with human touch.

CONCLUDING REMARKS

Environmental pollution is no longer a problem of the distant future; it is at here and now. We need vision, strategies, and action to combat it. We have to bring down GHGs in the atmosphere to a level the earth can handle without jeopardizing the biota including human life. There is consensus on this objective. But there is disagreement on who should bell the cat first. Both developed and developing countries are wedded to the style of development that has been promoted by the west during the last two centuries with great advantage to itself. The developed countries agree that this style of development is not the one that is sustainable but want to defer action lest they lose their economic superiority. The developing countries on the other hand do not want to check pollution at par with developed countries as their contribution to present day pollution level is relatively small.

The present style of development is anti-nature and anti-human values. It has built-in inequity in its end results. It benefits those who have access to advanced technology and control over earth's resources at the cost of many. In the process, it pollutes the environment which affects all equitably. Today we have reached a state of development in which there are rich continents and poor continents; rich nations and poor nations, rich regions and poor regions, rich cities and poor cities, and rich households and poor households. The haves are reluctant to change the existing pattern, while the poor want to catch up with the rich using the same strategies that the rich used to bring the world to the present state of social, economic and environmental chaos. If the benefits of modern development processes are in favour of the rich, we cannot expect the rich and poor to make equal sacrifices to ward off global warming and climate change. A wide economic gulf divides the developed and developing countries; it would be unreasonable to expect developing countries like India to compromise their higher rates of economic growth to arrest environmental pollution caused largely by the West. The historical polluters of the industrialized West must come forward to make more sacrifices. But that should not be the ground for the developing countries to do nothing. They too must contribute their mite for the deluge would be non-discriminatory; and poisonous gases would not spare anyone rich or poor.

If developed countries which are the major emitters of GHGs do not reduce their emissions, developing countries like India will be forced by circumstances to further increase their GHGs emission in future even though it may amount to joint hara-kiri for both. To save the earth from imminent disaster, both developed and developing countries have to change over to green technology. The investment that the developed countries would make in green technology would not only reduce their own GHGs emission but also help developing countries move in that direction. At present, an average American accounts for more than 15 times the carbon emissions of an average Indian, and 7 times of average Briton. In absolute terms, India's GHGs emissions are expected to surpass those of the US by 2040. India must ensure that its per capita emission does not exceed that of the developed countries.

Environmental problems began when the West treaded on industrialization path over 200 years ago. It forced the rest of the world to adopt its way of economic development. The source of the problem is thus the West. It is its moral responsibility to bear much of the cost of solving the problem. The developing countries should join hands with them but they cannot take on themselves yet another colonial burden on their shoulders. The danger is imminent. We can succumb to our love for comforts and luxuries and allow our civilization to crash, or we can choose a better way of life and living wherein love, affection and compassion are as important if not more as material well-being. The question is: should we opt for Adam Smith who advocated multiplication of wants, or for Mahatma Gandhi who advocated limitation of wants? The choice is ours !

If we are serious about our future, we should start thinking and working together. We do not own the earth and, therefore, we cannot use it to satisfy our greed. The time has come to stop squabbling over who should pay the bill for saving the earth from becoming inert. We all would be dead—friends and adversaries, rich and poor, developed or developing – if the situation deteriorates further. All those who desecrate the earth by adding green house gases in the atmosphere, wasting and polluting water, and destroying the natural fertility of the soil must be treated as enemies of humanity. To remain neutral would amount to colluding with 'these merchants of greed in their insane ambitions that will eventually end in global suicide !'

We have to seek out the goodness that is inherent in the human soul. When someone asked Abraham Lincoln, the newly elected President of USA, what he was going to do about his enemies, he replied, "I am going to destroy them. I am going to make them my friends." Mahatma Gandhi in our own times transformed his enemies into friends. Let us all individually and collectively act and make this earth a better place to live. All of us have a very significant role, no matter where we live and what we do. This is our moment of Truth !

NOTES

1. Thomas, Lewis (1974), *Lives of a Cell: Notes of a Biology Watcher*, New York: The Viking Press, pp. 5, 82.
2. Darwin, Charles (1872), *The Origin of Species by Means of Natural Selection, or The Preservation of Favoured Races in the Struggle for Life* (6th ed.), London: John Murray.
3. Lovelock, James (1995), *the Ages of Gaia: A Biography of Our Living Earth,* New York: Norton; see also Lovelock, James (2000), *Gaia: A New Look at Life on Earth*. Oxford: Oxford University Press.
4. Singh, Nagendra Kumar (1997). *Encyclopaedia of Hinduism,* Anmol Publications.
5. O'Flaherty, W.D. (1995), *Other People's Myth: The Cave of Echoes,* Chicago: University of Chicago Press. See also Singh, Nagendra Kumar (1997), *Encyclopaedia of Hinduism*, New Delhi: Anmol Publications; and *Srimad Bhagawatam*, ISKON Publication.
6. Schulman, David and Don Handelman (1997), *God Inside Out: Siva's Game of Dice*, New York: Oxford University Press.
7. Lovelock, J. (1979), *Gaia: A New Look at Life on Earth* (Third Ed.) Oxford: Oxford University Press.
8. *Ibid.*
9. Jantsch, E. (1980), *The Self Organising Universe: Scientific and Human Implications of the Engineering Paradigm of Evolution*, New York: Pergamon Press.
10. Abel, E.L. (1973), *Ancient Views on the Origin of Life,* Fairleigh: Dickinson University Press, p.15.
11. *Ibid*
12. Jaki, S.L. (1990) "Science: Western or What?" *Intercollegiate Review,* Vol. 26, Fall 1990.
13. Lovelock, James (1995) *op. cit.* p.19. Also see Stephen H. Schneider, James R. Miller, Eileen Crist and Penelope J. Boston (eds.) *Scientists Debate Gaia: The Next Century,* Cambridge, MIT Press
14. Lovelock, James (1991), Geophysiology: The Science of Gaia, in Stephen Schneider and. Penelope Boston, eds., *Scientists on Gaia*, Cambridge: MIT Press.
15. Lovelock, James (2004), *Earth in the Balance: Ecology and the Human Spirit*, Boston: Houghton: see also Al Gore (2007), *An Inconvenient Truth: The Crisis of Global Warming*, New York: Penguin (paperback)

16. Sources: McEvedy, Colin and Richard Jones (1978), *Atlas of World Population History*, New York: Facts on File," ISBN 071391033; also see, Stanley H. Ambrose (1998), "Late Pleistocene Human Population Bottlenecks, Volcanic Winter, and Differentiation of Modern Humans". *Journal of Human Evolution,* 34 (6), pp. 623-651.
17. *National Post*, April 11, 2006.
18. en.rian.ru/russia/20060825/53143686.html
19. http://www.publicaffairs.noaa.gov/releases2006/jul06/noaa06-065.html
20. www.firstenercastfinancial.com/./7-global-warming-thread.html–United States
21. www.nrdc.org/globalwarming/2006.asp
22. www.tornadochaser.net/globalwarming.html
23. epw.senate.gov/pressitem.cfm?party=rep&id=264777
24. wattsupwiththat.com/...breaking-news-scientist-admits-ipcc-used-fake-datas-to-pressure-policy-makers
25. disposablewisdom.com/global-warming/...page-3.htmi-United States
26. *World Climate Report* (The Web's Longest-Running Climate Change Blog) October 13, 2006, *Where are the Droughts?* Also see *Hydrocycle Looking Better than Ever*, Filed under: Droughts, Floods, Precipitation, January 20, 2010.
27. www.tornadochaser.net/globalwarming.html
28. Arctic Council (2004), *Arctic Climate Impact Assessment Report*
29. epw.senate.gov/public/index.cfm?fuseAction...id...
30. www.news.mongabay.com/2006/1116_inhofe.html
31. *earthsky.org/earth/ice-free-arctic-summers-within-a-decade*
32. belfercenter.ksg.harvard.edu >Analysis
33. Houser, Trevor, Peterson Institute for International Economics Policy Brief No. PB 10-5, March 2010
34. www.gandhiserveorg/.angven_spiritual_message.html
35. *www.lifepositive.com/spirit/masters/mahatma./gandhi.asp*
36. Moolakkattu, John S (2010), 'Gandhi as Human Ecologist' in *Journal of Human Ecology* 29 (3), pp. 151-5.
37. *Harijan*, January 28, 1939.
38. *www.great_spiritmother.org/-/chiefseatle_a_plea_to_cherish_mother_earth.html*
39. *www.natureofanimals.com/stfrancis.html*
40. Inge, William Ralph (1922), Outspoken Essays.
41. *www.myspace.com/spiritual_hype*
42. en.wikipedia.org/wiki/Thomas_Edison
43. en.wikipedia.org/wiki/Arthur_Schopenhour
44. Gandhi, MK, *The Story of My Experiments with Truth*, Ahmedabad; Navajiwan, Pat III: A Month with Gokhale.
45. *www.happycow.net/famous/dalai_lama/-*
46. *www.allaboutanimals.org.uk/ST_quotes.asp*
47. *www.allaboutanimals.org.uk/STstage4_RE.asp*
48. *www.jewishveg.com/DSrespet.html*

REFERENCES

Cazenave, A. (2006), 'How Fast Are the Ice Sheets Melting?' *Science,* 314, pp. 1251-52.

Dyurgerov, M. and Meier, M. (2005), Glaciers and the Changing Earth System: A 2004 Snapshot, *Occasional Paper 58.* Boulder, CO: Institute of Arctic and Alpine Research, University of Colorado.

Emanuel, K. (2005), Increasing Destructiveness of Tropical Cyclones over the Past 30 Years, *Nature*, 436, pp. 686-88.

Foucal, P.; Frölich, C.; Spruit, H.; and Wigley, T. (2006). 'Variations in solar luminosity and their effect on the Earth's climate'. *Nature*, 443, pp. 161-66.

Hansen, J.; Nazarenko, L.; Ruedy, R.; Sato, M.; Willis, J.; Del Genio, A.; Koch, D.; Lacis, A.; Lo, K.; Menon, S.; Novakov, T.; Perlwitz, J.; Russell, G.; Schmidt, G. A.; and Tausnev, N. (2005), Earth's energy imbalance: Confirmation and implications. *Science*, 308, pp.1431-35.

Intergovernmental Panel on Climate Change. (2007), *Climate Change 2007: The Physical Science Basis Summary for Policymakers, A Report of Working Group I to the Fourth Assessment Report of the Intergovernmental Panel on Climate Change.* See also their Climate Change 2007: *Climate Change Impacts, Adaptation and Vulnerability Summary for Policymakers, A Report of Working Group II to the Fourth Assessment Report of the Intergovernmental Panel on Climate Change.*

Intergovernmental Panel on Climate Change (2007), *Climate Change,* 2007, Contribution of Working Group I to the Fourth Assessment Report of the IIPCC (Solomon, S.D. Qin, M. Manning (Eds.*)*

Joint Science Academies (2005), *Joint Science Academies' Statement: Global Response to Climate Change,* June 2005.

Kiehl, J. T.; and Trenberth. K.E. (1997), Earth's Annual Global Mean Energy Budget, *Bulletin of the American Meteorological Society*, 78, 197-208.

Lau, K. M.; and Wu, H.T. (2007). Detecting trends in tropical rainfall characteristics, 1979-2003, *International Journal of Climatology,* 27.

Luthcke, S. B.; Zwally, H. J.; Abdalati, W.; Rowlands, D. D.; Ray, R. D.; Nerem, R. S.; Lemoine, F. G.; McCarthy, J. J.; and Chinn, D.S. (2006). Recent Greenland ice mass loss by drainage system from satellite gravity Observations, *Science*, 314, pp.1286-89.

McGranahan, G.; Balk, D.; and Anderson, B. (2007). The rising tide: Assessing the risks of climate change and human settlements in low elevation coastal zones, *Environment and Urbanization*, 19(1). International Institute for Environment and Development (IIED).

Sabine, C. L.; Feely, R. A.; Gruber, N.; Key, R. M.; Lee, K.; Bullister, J. L.; Wanninkhof, R.; Wong, C. S.; Wallace, D.W.R.; Tilbrook, B.; Millero, F. J.; Peng, T. H.; Kozyr, A.; Ono, T.; Rios A.F. (2004), The Oceanic Sink for Anthropogenic CO_2 *Science*, 305, pp. 367-71.

Shepherd, A., and Wingham, D. (2007). Recent sea-level contributions of the Antarctic and Greenland Ice Sheets, *Science*, 315, pp.1529-32.

U.S. Climate Change Science Program, *Temperature Trends in the Lower Atmosphere*, 2006.

U.S. Environmental Protection Agency. (2007), *Climate Change,* Reuters, July 18, 2008 09:21 AM.

Velicogna, I., and Wahr, J. (2006). Measurements of time-variable gravity shows mass loss in Antarctica, *Science*, 311, pp.1754-56.

Weir, J., *Global Warming,* Earth Observatory, 2002, April 8.

2

Climate Change and Green Economy with reference to India

Strategic Options at Different Scales

Ian Douglas

Introduction

With a rising global population and, perhaps more significantly, a rising relatively affluent middle class with high rates of per capita consumption, the emissions of green house gases continues to rise. These gases have so changed the atmosphere that weather systems have been perturbed and rainfall patterns altered. Such climate change is already posing problems and is leading to concerns about food security, adequacy of water resources and energy consumption. This chapter examines the issues that these changes may raise for India and suggests strategies that can be adopted for sustainable water, energy and food supplies and to help urban settlements adjust to climate change. It begins with a discussion of the ways climate change may affect India and then considers the nature of the green economy and how investment in the green economy may increase employment. It then examines the water, energy and food sectors, discussing how different actions can be taken at the community level, and the municipal and regional levels and at the national levels, emphasising the importancc of reliable infrastructure for distributing water and electricity. Multiple levels of responsibility are set out, showing how everyone from the individual householder to the national government can take appropriate actions to help adapt to climate change and develop the green economy. The notions of interdependence and collective responsibility are explored, with special reference to the goals of human ecology.

Climate Change

Nearly 700 million rural people in India are directly dependent on the climate for their livelihoods in agriculture, forestry and fisheries. They rely on natural resources (such as water, biodiversity, mangroves, coastal zones and grasslands) for their subsistence and economic activity. Their ability to adapt to climate change is generally low, but many are already being forced to change how they use these resources by increasingly unreliable rainfalls and rising sea levels. Climate change is expected to alter the climate of India in three main ways (Sathaye *et al.*, 2006):

- An annual mean surface temperature rise by 2100, ranging from 2.5° to 5° C, with warming more pronounced in northern India.
- A 20 per cent rise in all India summer monsoon rainfall and further rise in rainfall is projected over all States, except Punjab, Rajasthan and Tamil Nadu, which are forecast to have a slight decrease.
- Extreme maximum and minimum temperatures are also likely to increase. Extreme rainfall events are also likely to increase, particularly over the west coast of India and west central India. Such extremes will increase the risk of flooding, impacting rural and urban areas alike.

The impacts of climate change will be felt across most sections of the Indian economy. Droughts and floods could become more severe. Higher temperatures will mean higher evaporative losses at times of low river flows and the quantity of available runoff may decrease. Crop yields are likely to decrease with rising temperatures, however, higher atmospheric CO_2 may cause greater crop growth and offset some of the temperature effect. Forest character will change over about two-thirds of the country, with, if undisturbed by human action, forests shifting to wetter types in the north-east and drier types in the north-west. In the Bay of Bengal tropical cyclones and their associated damage are likely to become more frequent. Sea-level rise will force some coastal people to move out of low-lying areas, will inundate crop land and will extend coastal areas of saline soils. Increased rain and flooding may extend the range of malaria, especially in northern and western states. However, more severe droughts could lead to greater desertification is land resources are not carefully managed.

In addition to these direct impacts on rural livelihoods, there will be the disruptions to transport and urban life caused by flooding, water shortages and possible reductions in flows in streams used for hydropower generation. Food supply chains could become disrupted, increasing the difficulties faced by poor people in urban areas, by both rising food prices and unavailability of some foods at critical times. Climate change increases the uncertainty of life for those at the margins of society.

The Green Economy

The green economy involves investment in activities that promote sustainable living and the wise use of renewable natural resources. It includes investment in conservation (lowering per capita resource use), on- and off-shore wind power, hydroelectricity, solar power, bioenergy, waste energy recycling and smart grid electricity transmission systems (Pollin and Garrett-Peltier, 2009). Conservation includes building homes and offices that require less energy to cool or heat, collect their own water, and generally aim to be a close to zero-carbon emitting as possible. In higher latitudes where winter household heating is required, some 30 to 40 per cent of green house emissions are associated with static home energy use. The rising air conditioning provision in India increases the risk that the use of coal-generated electricity for this purpose will cause a growth in India's green house gas emissions.

Investment in the green economy leads to three types of job creation (Pollin and Garrett-Peltier, 2009):

1. *Direct Effects*—For example, work in retrofitting homes to enhance energy conservation or to build wind energy farms, or install solar panels.
2. *Indirect Effects*—The job associated with industries that supply intermediate goods for building retrofits, wind farms or solar panels, such as the timber, steel, glass and transportation industries.
3. *Induced Effects*—The expansion of employment in the regional or national economy that results when people who are paid in the construction or steel industries spend the money on other goods and services in the region or nation.

The impacts of the green economy on a given state or the entire nation will depend on the success of creating local manufacturing of the items required. Ideally, things like wind turbines and solar panels would be manufactured locally, but at present there is a highly active international supply market in which German, Japanese and Korean firms compete vigorously.

Investment in the green economy is highest in the big economies of the USA and China (about $ 45 to 48 billion each in 2011) but in terms of population, Germany and Italy have spent the most per capita (gross about $ 28 to 32 billion per country in 2011). Wind farms dot the German landscape and solar panels cover the roofs of many homes and other buildings. Countries like Britain, Japan, Spain and India only spent gross about $ 8 to10 billion per country on the green economy in 2011 (data from *The Economist*, 28th April, 2012).

Much of the Indian economy has traditionally been highly green, from the garbage collectors who recycle in cities, to rural farmers who use no fossil fuels at all. Modernisation has turned its back on the traditional green economy and India has become increasingly dependent on oil and coal, the causes of green house gas emissions and consequent climate change. Yet modernisation can use smart technologies to reduce energy use, and India has abundant solar energy that has supported the rural population for centuries, that could also be harnessed through the green economy, to provide the energy to power the lifestyles to which the new Indian middle class aspires.

Vulnerability to Climate Change

India's agricultural sector is particularly vulnerable to present day climatic variability, including multiple years of low and erratic rainfall (O'Brien *et al.*, 2004). This vulnerability will be increased by climate change. That increase is arriving at a time when a second form of vulnerability is being coming acute: vulnerability to global trends in food prices. With increasing liberalization of Indian trade in agricultural products, economic changes will bring uneven impacts. Some regions and farmers will benefit from market liberalization and from new inflows of investments and technology, but others may have difficulty in adjusting to a more open economy, particularly to the effects of increasing competition from agricultural imports. These potential impacts can be illustrated by a case study of a village in Rajasthan.

Jhalawar district in Rajasthan receives an average of 943 mm of rain per year, but is highly climate sensitive and has a low adaptive capacity (O'Brien *et al.*, 2004). Between 1993 and 2003, many farmers in Jhalawar moved from producing traditional crops such as sorghum and pearl millet to soybean cultivation for which they were paid more. However, because soybean prices fluctuate according to world market prices, the shift towards growing soybean left farmers in Jhalawar vulnerable to external market price shock. They are also highly vulnerable to climate change. In the years 2000-2004, drought reduced crop yields. Members of some farm households moved to Gujarat in search of work. Many farmers forced to take loans from money-lenders had increasing difficulty in making repayments. The economic difficulties give the farmers less ability to adapt to a changing climate. Any move towards the green economy have to be fitted into this context.

Acting to strengthen Resilience in Water Availability, Supply and Distribution

Water is an absolutely key resource that is under constant threat of over expoloitation, contamination and inaccessibility. It requires high quality infrastructure, sensible pricing policies and multiple sourcing (rain water, river water, groundwater, recycled water and sea water). India has excellent water policies and water engineers but has not yet managed to assure that all its citizens enjoy reliable supplies of clean safe water. On public health grounds alone, such safe supplies ought to be a priority.

In times of water shortage, all possible water sources have to maximised. This involves careful thinking at the household and enterprise level as to how we use the water that we have. New buildings could be equipped with dual water systems, using treated drinking water from the mains supply for cooking and eating and washing, but retaining "grey" water, such as rain water, for tasks such as vehicle cleaning, watering plants and flushing toilets. Although India is internationally at the forefront of rain water harvesting, dual supply systems should become the norm in all new developments. Uncontrolled groundwater pumping by urban property owners has created serious problems for the management of urban aquifers. The risks of groundwater contamination from badly managed waste disposal are still high. Together the pumping and the contamination are reducing the effectiveness of the groundwater resource. Future strategies to manage urban water will have to integrate supplies and endeavour to make use of every drop of rain. Finding ways of getting storm water to infiltrate to replenish depleted groundwater aquifers should be a priority. Where land is available, using sustainable urban drainage systems (SUDS) could be a most effective way of increasing groundwater availability. Conjunctive use schemes could be developed where water is taken out of rivers at high flows and inserted back down groundwater wells to recharge aquifers, which would then be able to produce a larger yield when pumped in drought periods. Groundwater should be regarded generally as a reserve resource, to be used when rain water harvesting and abstraction from rivers is not possible.

ENERGY : APPLYING THE GREEN ECONOMY

The energy sector offers abundant opportunities to change from heavy reliance on coal to renewable energy. Opportunities for hydropower development range from large dams to small microhydro installations. Villages in suitable locations could set up small hydropower schemes, as has been done in Nepal. Major dams could be controversial, but some electricity could be generated from existing irrigation reservoirs and from water supply reservoirs, using the energy of the water moving along the supply pipes to water treatment plants. The aim should be to get as many uses out of each drop of water as possible. Pumped storage schemes should be considered where there is electricity to drive the pumps available from a renewable source. A large solar panel array could supply enough power to pump water up to a storage reservoir when the sun is shining. When there is little or no sun, the water could be released to supply power. The water could also be released and times of peak electricity demand, and the solar-generated electricity used to pump when demand is low. Smart technologies could be important in managing such storages efficiently.

Solar Power

Solar power has a great future in India. The 200 MW solar panel array at Charanka, Gujarat is an example of what can be done. Its development was encouraged by the Gujarat government's decision to guarantee high prices of rupees 15 per kwh for the first 12 years of operation to solar power producers. Compared to the price paid by electricity distributors for conventional electricity of rupees 3 to 4 per kWh, solar operations appear to have a great advantage. The Indian national government has set a target of 20,000 MW of installed solar generation by 2022, but that would still be 5 per cent or less of the total installed power generation capacity in India. Just as European countries have had to with their solar electricity prices, India may have to pay less than the Gujarat rupees 15 per kWh in future. Rational energy pricing based on the long-run-marginal-cost principle would level the playing field for renewables, such as solar, increase the spread of energy-efficient technologies, and raise the economic viability of utility companies, ultimately leading to reduced green house gas emissions (Sathaye *et al.*, 2006). At the same time, the infrastructure of electricity distribution will have to be improved. Such action would greatly help increase India's resilience to climate change and create many jobs in the green economy.

Biofuels in India

About 220 million tonnes (MT) of firewood is used annually for cooking in rural areas and about 160 million tonnes of 'non-fodder' agricultural residues are produced every year in the country. Firewood consumption is increasing steadily (Kishore *et al.*, 2004). Another bio-resource in India is cattle dung. Nearly 600 million tonnes of wet dung is produced annually from a livestock population of about 270 million. This represents a sustainable resource, which if all converted into biogas would produce 36 billion cubic metres (bm^3) per year. If organic wastes such as sewage, municipal solid waste, distilleries, etc., can also be taken as feedstock for gas production, the total biogas potential would be 36.8 billion cubic metres per year. By comparison, 22.7 billion cubic metres of natural gas (excluding gas for LPG shrinkage) and 7.3 million tonnes of LPG were consumed in 2001/02. In terms of heat energy, this amounts to 1.08 exa joules (1 EJ=1018 J) per year, whereas the biogas production potential is 0.693 EJ/ year (Kishore *et al.*, 2004). This biogas potential, at nearly 70 per cent of the current gaseous fossil fuel consumption levels, represents a major asset that cannot be ignored.

Unfortunately, the present ways in which biomass fuels are used are inefficient and lead to green house gas emissions. The traditional devices burning biomass release products of incomplete combustion (PIC), chiefly carbon monoxide, methane, total non-methane organic compounds (TNMOC) and N_2O. These green house gases have higher global warming potentials (GWPs) than CO_2, their CO_2 equivalent contribution being nearly the same as the actual CO_2 emitted (Kishore *et al.*, 2004). Thus to gain benefits from biomass burning, new ways of using it have to be found.

The so far successful bagasse cogeneration power plants have an ultimate potential to generate 3500 MW. Rice husk based power plants producing 5 to 8 MW each in Andhra Pradesh and one based on mustard stalks in Haryana have been built and operated by private companies. Any future growth will depend largely on subsidies and the price per kWh. Biomass gasifier-based power plants, ranging from 5 to 500 kW, were established under subsidy schemes or demonstration projects but most appear to have closed, partly because of the difficulty of getting reliable supplies of biomass fuels (Kishore *et al.*, 2004). Many were run on wood chips, which may not be a sustainable fuel if it comes from new tree-felling.

Village biogas power generation schemes have not been as successful as expected, but the potential is still there. Technical improvements are necessary but there is also the possibly of funding such methods through the clean development mechanism (CDM) under the Kyoto Protocol (Kishore *et al.*, 2004). Advances elsewhere, including in the UK, suggest that biogas generation is possible in urban areas, particularly because the Indian urban waste stream has such a high percentage of organic matter. Solid waste disposal sites in the UK now have biodigesters, and sewage sludge is also used for cogeneration as part of its incineration process. All these opportunities using existing resources need to be fully explored.

Wind Power

From 32 MW in 1990, wind power installation in India rose rapidly over 1500 MW in 2001. However, following the introduction of a minimum alternate tax (MAT) and a general financial crisis in the industry, the rate of installation decreased after 1996 and the rate of expansion remains low (Kishore *et al.*, 2004). Offshore wind has made an impact in Europe. On one windy day, wind power supplies 20 per cent of Britain's electricity, but on average the proportion is much less. Wind generation complements solar power, often being readily available on cloudy days when there is less solar generation. The work in building wind turbines and erecting them uses the skills developed in ship building, aircraft and motor vehicle manufacture and is thus a valuable part of the green economy.

Energy for Transport

While electricity from renewables readily provides fixed pint needs in homes and businesses, it is less easily used for transport. However, the hybrid car and bus are already popular and are beginning to help reduce fossil fuel consumption. Taxation of vehicles by their CO_2 emissions helps to encourage people to buy lower carbon vehicles, but the savings are for too little to deter the really wealthy from buying large 4-wheel drive off-road vehicles with 3000 cc or more capacity engines. The prospect of electric vehicles capable of doing more than 150 km without recharging is still relatively far off.

The slowness with which electric private vehicles are emerging, suggests that in urban areas, both public transport and the way cars are managed need rethinking. Many cities have provided innovative express busways, interchange stations and rapid transit rail systems. Separate bus lanes are highly successful in making journeys by public transport more rapid. People will shift away from the private car, if the public system is frequent, comfortable and reliable. Many megacity commuter systems are overcrowded, but the speed of the rain system is so much greater than road traffic, that people choose to use it.

Within a city, traffic regulation can exclude private vehicles from certain streets, charge a congestion fee to enter the central business district (as in Singapore and London), make parking charges so high that people will prefer to park in the outskirts of a city and ride in on a bus (as with Park and Ride schemes at Oxford and Cambridge and many other UK cities), and tightly regulate on street parking. Increasing the traffic flow rate of the remaining vehicles moving in the city will greatly reduce the total emissions. There are considerable opportunities in the green economy for developing the control systems for congestion and parking fees, and considerable employment for traffic wardens and vehicle emission testers.

Food

Climate change will bring India more frequent floods and droughts due to increased variability of annual monsoon rains and rapid depletion of Himalayan glaciers. Whole overall food production may not be reduced, those least able to cope will likely bear the adverse impact of climatic changes. The socio-economic systems of countries like India are more vulnerable to climate change because of high reliance in natural resources, lower per capita GDP, high poverty and lack of safety nets. This negative impact of climate change on agriculture is likely to have a serious impact on poverty. A one percentage point decline in agricultural GDP growth could decrease the consumption of the poorest 30 per cent of the population by as much as 5 per cent. Higher temperatures will affect water availability for food crops by their impacts on evapo-transpiration. Monsoon season (*kharif*) crops may be little affected by moderate temperature and CO_2 rises, but winter (*rabi*) crops could suffer decreased yields, due to large temperature increases and greater uncertainty of rainfall. Many of the gains in yield that would be caused by a doubling of atmospheric CO_2 would be more than offset by reduction in yield resulting from temperature and moisture changes. During both seasons, crops could be increasingly disrupted by more frequent weather extremes, such as variations in monsoon onset, duration and frequency of drought and floods, and pest infestations. Although monsoon precipitation is likely to increase in many regions, winter rainfalls are likely to decrease in northern India affecting the output of winter wheat and mustard crops which could impact on food security. Lower rainfall in drier areas will make marginal agricultural areas less productive (Douglas, 2009).

Croplands presently liable to flooding are likely to become wetter for longer and additional land will be inundated in the worst floods due to increased river heights and rising sea levels. Crop damage and unemployment caused by floods will make an even larger section of population, particularly in deltaic areas, extremely vulnerable to starvation, malnutrition and even death. Unless action is taken in time, the higher future floods will disrupt existing transport and communications for longer periods and over a wider area. This will make the movement of food supplies more difficult. Longer wet periods could also affect the quality of food in storage, unless particular care is taken (Douglas, 2009). Ultimately, food supplies to major cities are disrupted.

ADAPTING CITIES TO CLIMATE CHANGE

Adaptation here means two approaches, reducing green house gas emissions on the one hand and making cities able to cope with future changes on the other. Much is said above about reducing emissions, especially in relation to transport. Coping with climate change involves trying to reduce the urban heat island effect; making buildings better insulated against heat to cut down on the use of air conditioning; protecting the urban environment against flooding through sustainable drainage and protection at the individual building level; and ensuring that essential services can operate in times of flood. Fire risk may increase with climate change and thus the ability of fire services to operate is essential.

Measures to modify the urban climate include increasing the tree cover to lower surface temperatures. Parks reduce the heat island effect in their vicinity, but green roofs and street trees also play their part. Where social conditions permit, vigorous tree planting and other urban greening schemes bring multiple benefits, in addition to assisting in mitigating climate change. Coping with flooding involves both hard and soft engineering. Following the principles of sustainable urban drainage, the same green spaces that reduce the urban heat island could be used, in appropriate locations, as temporary storm water storage basins to slow down the passage of rain water from hard surface to streams. This delay in the rate at which water enters rivers will lower the flood peaks and help protect people from the worst of the flood. At the individual building level, flood losses can be avoided by barriers to water entry, such as floodwalls at doorways and seals at other places where water could get under the floors of a building. At the community level, efforts can be made to keep drains clear and ensure that any flood basins are free of debris. At the city level, details mapping can help determine localities at risk from flood of particular frequency and thus warning systems can be established, so that if necessary residents can be evacuated. At the same time an urban flood management strategy can be developed coupled with an emergency planning scheme, to enable flood avoidance and flood rescue to operate.

Strategies for drought are also needed. Water supplies have to be maintained. One technique used in remote Malaysian villages is to have a rain water tank on the village hall or other public building that everyone respects as an emergency supply and which is kept strictly as a reserve for the worst situations. Similarly water rationing measures ought to be able to ensure that sufficient water is retained for the worst droughts. This will require water agencies to closely examine forecasts of future droughts and not to rely on patterns of droughts in the past as indicators of what might happen in the future. In today's climate change situation the future is uncertain.

Where the Responsibility lies

Everyone has to contribute to slowing climate change and reducing its impacts. People have to take individual responsibility for themselves and their families. They have to support the green economy, lower their own green house gas emissions, live sustainably, and adapt their way of life to the changing climate. So there a set of actions can be undertaken at the household level, from using the motorcar less, to rain water harvesting, growing food on what land there is around the home, if any, installing solar panels, burning any biomass fuels more efficiently and cooling the house without using excessive fossil fuel generated energy.

At the community level, efforts should be made to ensure green economy products are used in public buildings (such as solar power, lighting controls and rain water harvesting). Sharing transport and journeys to cut down vehicle use; small-scale hydro where relevant, biogas digestors, allotments for food production and community gardens that bring social and health, as well as green house gas and heat island reduction benefits are all possible.

At the municipal and higher levels, strategies for climate change adaptation are need that include municipal undertakings applying the green economy, using their buildings for rain water harvesting, solar power or green roofs, having adequate recycling facilities, and using hybrid or electric vehicles for their operations. Many municipalities worldwide do set examples in this way and some city halls are well ahead of their citizens generally in sustainability and green house gas emission reductions. Municipalities can run urban greening schemes, plan the recycling of waste water through high technology membrane filtration systems, require buildings to have dual water systems, require public transport operators to use low emission vehicles, and develop and enforce local building regulations that support the green economy. Their integrated, holistic approach to these issues can make their cities exemplars in preparing the world for the rest of the twentyfirst century.

National governments have particular responsibility to deliver laws and fiscal measures that will deter people for making excessive emissions and reward them if they support the green economy. Multinational corporations with interests in oil and natural gas are notorious for high publicity and political lobbying against large investment in alternative energy sources. Courageous politicians will resist this lobbying and take a wide future for all their people and the children and grand-children. Where politicians have espoused the ideals of human ecology and taken an integrated view of people and nature, they have usually argued for subsidies and payments to develop the green economy and to make non-fossil fuel energy a reality. All voters have to work to encourage more politicians to think that way, as well as changing their own lives to adapt to climate change.

NOTES

Douglas, I. 2009. Climate change, flooding and food security in south Asia. *Food Security*, 1,127-136.

Kishore, V.V.N.; Bhandari, P.M. and Gupta, P. 2004. Biomass energy technologies for rural infrastructure and village power: opportunities and challenges in the context of global climate change concerns. *Energy Policy*, 32, 801-810.

O'Brien, K.; Leichenko, R.; Kelkar, U.; Venema, H.; Aandahl, G.; Tompkins, H.; Javed, A.; Bhadwal, S.; Barg, S.; Nygaard, L. and West, J. 2004. Mapping vulnerability to multiple stressors: Climate change and globalization in India. *Global Environmental Change*, 14, 303-313.

Pollin, R. and Garrett-Peltier, H. 2009. *Building the Green Economy: Employment Effects of Green Energy Investments for Ontario*. Ottawa: Green Energy Alliance.

Sathaye, J.; Shukla, P.R.; and Ravindranath, N.H. 2006. Climate change, sustainable development and India: Global and national concerns. *Current Science*, 90 (3), 314-25.

3
Climate Change, Desertification and Water Resource Management in India

H.S. Sharma

Introduction

During last four decades it has been observed that global changes are increasingly putting stress on the Earth's biosphere, other less wholesale changes may have equally serious consequences for the viability of ecosystems. Loss and fragmentation of habitat, forest degradation and loss of wetlands all disturb and devastate the ecological niches occupied by species. Overexploitation of the natural world will lead to loss of renewable resources and biodiversity. Still more stress, and even a health hazard, are placed on populations by water and air pollution, either through catastrophic events such as oil-spills and explosions of chemical plants, or more insidious effects from the long-term use of insecticides, run-off of nitrogen based fertilisers, and air pollution in metropolitan areas. In addition to these threats to the natural world, managed systems are also subject to processes such as loss of fertility, desertification, water stress and erosion (ESA, Changing Earth 2006).

Anthropogenic Factor and No Analogue State of Earth

The studies conducted by the International Geosphere and Biosphere Programme (IGBP) under various projects reveal that the Planet Earth has always undergone major changes. The geometry of the Earth's orbit introduces regular changes in illumination conditions and thereby stimulates ice ages. Changes are a natural property of the Earth System, but there is mounting evidence that the changes which have been imposed on the Earth System during the last 150 years cannot be compared with any previous change. In the last century, humankind has driven the greenhouse gas concentrations on Earth far beyond the maxima reached during the last 1 million years, has become responsible for 70 per cent of the nitrogen and 95 per cent of the phosphorus cycle on

Earth, and has reduced tropical forest areas by 50 per cent. To determine whether these human-induced recent changes could ultimately destabilise the Earth System, both natural system variability and the consequences of human activities have to be fully understood and quantified. This represents the necessary scientific basis for sustainable future management of the Earth System as a whole (ESA, *The Changing Earth*, 2006).

Anthropogenic factors have become a most dominant environmental force, notably after the Industrial Revolution, that a new geological era, the **Anthropocene**, has been proposed to describe the last few hundred years. Increased population growth together with accelerated human activities and economic wealth over the past century have greatly increased the resource use, as reflected in agriculture, fisheries, forestry, industry, transport, energy and urbanisation. This has resulted in multiple and interacting global environmental impacts as seen by the current values of green house gases temperature. The Earth System has clearly moved far outside the range experienced over the last 700,000 years and hence is operating in a **"no-analogue"** state.

CLIMATE CHANGE

The earth's climate has changed frequently in the geological past at intervals of thousands of years. Sometimes it has been too hot or too cold. Closer to present day, in the late Pleistocene and Holocene, the climate change has been in a short interval of time. The last glacial period began about 110,000 years ago and ended between 10,000 and 15,000 before the present (around 12,500 years ago). Even during this period there were several changes with glacier advance and retreat. The maximum extent of glaciation [also called Last Glacial Maxima- (LGM)] was approximately 18,000 years ago. At present, we are in an interglacial period.

Global Warming

Beginning with the Industrial Revolution in the 1850s and accelerating ever since, the human consumption of fossil fuels has elevated CO_2 levels from a concentration of ~280 ppm to more than 380 ppm today. These increases are projected to reach more than 560 ppm before the end of the twentyfirst century. It is known that carbon dioxide levels are substantially higher now than at any time in the last 800,000 years. Along with rising methane levels, these changes are anticipated to cause an increase of 1.4°C to 5.6°C between 1990 and 2100.

For the last three decades, it has been observed that the global average atmospheric temperature is increasing at a high pace. Other than natural processes, human activities are responsible for accelerated rise in temperature. According to the report of the Intergovernmental Panel on Climate Change (IPCC, 2007), the global average surface temperature has increased by 0.6 ± 0.2°C since the late nineteenth century. Average surface temperature of the Earth is likely to increase by 1.1 to 6.4° C by the end of the twentyfirst century, relative to 1980-1990. It is now universally accepted that human induced green house gas emissions are responsible for global surface temperature.

Abrupt Climate Cooling in Near Future

Abrupt and drastic cooling in the climate can be the result of (a) Large scale melting of global ice due to warming, and/or (b) Prolonged sunspot minima

(a) Large Scale Melting of Global Ice due to Warming

As the global temperatures rise, ice and snow around the world are melting. Glaciers in the Garhwal region of the Himalayas are retreating so fast that researchers believe they will disappear by 2035. Arctic ice has thinned significantly over the past half century. Between 1979 and 2007, it has shrunk by 44 per cent, and it is predicted that a complete meltdown is possible by 2030 or sooner (Richard 2003).

The freshwater from the melting of ice in North America, Scandinavia and Europe is flowing into the North Atlantic Ocean. Researchers have shown that the Meridional Overturning Circulation (MOC) of the North Atlantic Ocean, a heat conveyor current in the oceans, is currently slowing down. This is likely to decrease the global surface temperatures over the next decade, and may temporarily offset anthropogenic warming. Such cooling events of Younger Dryas and 8,200 events are linked with the slowing down for the MOC for longer periods of time.

Scientists have predicted that the current slowing down of the MOC will last for only a decade. But if it prolongs further, then we are in for a longer cooling period. Its climate impact includes a severe alteration of rainfall patterns in the Indian subcontinent and Europe. The earlier short term colder events in the North Atlantic and European region had their effect on Indian region, though on a smaller scale. Therefore, it is desirable that whatever mitigation strategies are planned, they are for an eventuality of greater global warming as well as for an unexpected cooling event.

(b) Prolonged Sunspot Minima

There is a close correlation between variations in the sunspot cycle and Earth's climate. When the sunspots are few, the climate is cool and *vice versa*. Sunspot numbers follow a cyclicity of 11 years. However, at times, the cycle is delayed.

The most recent sunspot minimum was in March 2007. The new cycle was supposed to start soon after that, with a gradual build-up in sunspot numbers. It did not happen. This is the cause for our concern. Therefore, it is time to put aside the global warming dogma, to at least begin contingency planning about what to do if we are moving into another little ice age, similar to the one that lasted from 1100 to 1850 (*Joyce, 2008).*

The next little ice age would be much worse than the previous one and much more harmful than any warming may do. Global cooling would decrease agricultural output. Millions may starve if we do nothing to prepare for it (such as planning changes in agriculture to compensate), and millions more will die of cold related diseases.

INDIAN SCENARIO

With a rapidly expanding economy many changes are taking place in India today. Land use, cropping and water use patterns are changing, partly as responses to changing demographic and consumption patterns, and partly as responses to changing investment scenarios and economic growth. Some of these changes will have substantial implications on India's water future. Adverse impacts of global warming and resulting climate change will be superimposed on these changes and will exacerbate water shortages in many water-scarce areas of the country.

India has witnessed extreme weather events during last two decades which show that various parts of the country are already reeling under the impacts of climate change.

- Several studies by ISRO, GSI and other organizations indicate that the Northern Himalayan Glaciers are receding by about 16 m per year. The Gangotri glacier is retreating by about 28 m per year. The receding glaciers have major implications for water availability in the glacier fed rivers such as the Ganga and the Yamuna.
- Some studies show that the water level in almost all streams in the Kashmir Valley have fallen by two-third and groundwater level have dropped by more than one-third.
- In May 2002 severe heat wave conditions were reported in coastal Andhra Pradesh.
- In 2003, 22 out of 23 districts in Andhra Pradesh were in grip of a heat wave. In some years more than 80 per cent of the State faced a drought.
- In 1999, two cyclones hit the State of Orissa in quick succession.
- In the Sunderbans delta, 4 islands—Bedford, Lochahara, Superibhanga and Kabasgadi have submerged due to sea level rise.
- The desert of Rajasthan particularly Barmer was affected by floods in 2006 and subsequently in 2007 at Rawatsar in district Hanumangarh.
- Rajasthan was affected by severe drought in 2002. In 2009 also, as many as 26 districts are reeling under drought condition.
- In September 2009, bursts of sudden rainfall like the 600 per cent higher rain that caused deadly floods in southern States have gone upto 10-15 per cent across India.

Palaeoclimatic Changes in the Rajasthan Desert

The Rajasthan desert bears incomparable geomorphological, archaeological, palynological and pedological evidences of climate change. The landforms created by various geomorphic processes under different climatic phases, have been used to construct the chronological sequences of the climatic changes in different parts of the world during the Quaternary period. In Rajasthan especially in desert part, several climate-morphogenetic studies have been conducted by Singh, G (1971), Singh, *et al.* (1974), Singh (1977), Ghose (1964, 1965), Abhichandani and Ray (1966) and other workers from Central Arid Zone Research Institute, Jodhpur, Stein (1942), Raikes (1965), Sood *et al.* (1982), and Yash Pal *et al.* (1980), Allchin, Goudie and Hegde (1978), Meher-Homiji. (1973), Brysen and Barlis (1967), Ahmad (1969), Verstappan (1970), Agrawal, *et al.* (1975, 1976, 1980), Mishra and Rajaguru (1973, 1980), Whilhelmy (1969), Bakliwal (1999), Dhir (1995), Raghav K.S., (1992), Singhvi (1995), and Wadhawan (1995), Sharma, H.S. (1968, 1983, 1987).

The tentative chronology of the climatic changes in Rajasthan and adjoining parts has been presented in Table 3.1.

Table 3.1: Geomorphic Evidences of Desertification/Dedesertification during the late Quaternary Period in Western Rajasthan

Period in Years B.P. (After Singh et al. 1974)*	*Major Gemorphic events*	*Geomorphic processes*	*Climatic phases*	*Desertification/ dedesertification*
0- Early Centuries A.D.	Increasing Fluvial activities but increasing biotic stress on the land has resulted as salinity hazard along period drainage channels and more deflation and deposition of sediments in the sandy plains and duny tracts	Biotic activities	Slight amelior-ation of climate	Small scale amelioration of climate but biotic desertification is present
Early centuries A.D. 3,800	Formation of low active dunes like barchans. Shrubscoppice etc. reactivation of the flanks and crests of the stabilised dunes. Inland basins turned into saline depression	Aeolian	Arid	Climatic desertification
3,800-10,000	Major dunes stabilised, drainage channels revived, sandy inland basins filed with fresh water and fluvial sediments	Fluvial	Humid	Amelioration of climate leading to dedesertification
10,000 - ?	Formation of major dunes, disintegration of drainage systems of aeolian sand, creation of sandy inland basins.	Aeolian	Arid	Climatic desertifcation
?	Well integrated drainage systems, Vast alluvial plains	Fluvial	Humid	No evidence of desertification

Compiled by H.S. Sharma (1987).

*B.P. means before present.

A fairly large amount of work has been done at the Central Arid Zone Research Institute, Jodhpur, Cambridge-Baroda-Poona team of archaeologists and Scientists of Physical Research Laboratory, Ahmedabad, and geomorphologists on the changes in climate of Rajasthan desert in Quaternary and Recent times. While all agree that there has been a change from arid to semi-arid to sub-humid climate and a repetition of this cycle, contradictions are to be found in the duration assigned to these climatic changes. Broadly speaking, dry river beds are supposed to be indicative of aridity. There have also been attempts to correlate geomorphic features with the glacial and inter-glacial periods in the Himalayas. It is summarized that sand shields formed during the glacial period were washed out during the much longer fluvial (raining) inter-glacial periods. During these periods in the Rajasthan desert, the climate is supposed to have been arid and humid respectively fossil parabola seem to have shaped into longitudinal dunes and later berchans predominated. Major dunes are supposed to have been formed during the arid phase and stabilized during the humid phase. One interesting aspect of these studies is that a chronology could be established because of the occurrence of human implements with these landforms. Great emphasis has been laid on the degeneration of drainage in these parts though different authors attribute different reasons for this (tectonic, climate, river piracy, human interference, etc.). Now there is greater need of determination of TL ages of different representative samples from various parts of Rajasthan. However, a beginning has already been made by Singhvi (2005) and his team at RPL, Ahmedabad.

Changing Climate Scenario over Western Rajasthan

The studies indicate that the Indian summer monsoon has been stable as a large-scale system over the past 200 years. Most models simulate the enhanced monsoon precipitation in a global warming scenario (Kolli, 2006; Pant and Hingane, 1988). Barmer and Jaisalmer districts in particular are experiencing extreme weather conditions, such as the above average rainfall. For instance, Barmer has an average rainfall of 280 mm, but received 720 mm of rainfall in 1990, and 600 mm in 1994 and 2006. Such changes may also be the result of climate change in which the frequency of such extreme events is expected to increase.

Doi (2000) has also studied the rainfall trends in Rajasthan. The study reveals that the rainfall amount shows an increasing trend in western Rajasthan and a declining trend in eastern Rajasthan. Observing the rainfall data of last 16 years, it has been noted by Bharadwaj and Sharma (2008) that the desert part of Rajasthan gets more rainfall than the eastern part of the State. Studying the emerging trend of rainfall in the desert, they indicated that there were high rainfall events in the years 1968, 1979 1981, 1991, 1994 and 2006 followed by further high amount of rainfall in 2007. On the whole, it seems that climate extremes give way to a milder situation in desert.

Since Rajasthan is mostly arid, there is a need to channel excess precipitation to subsurface aquifers for storage, not just for future human use, but also for prevention of flooding. As an example, the 2006 deluge in Barmer caused flooding in Kawas, Malva and Uttarlai, but the subsurface layers of clay (bentonite and fuller's earth) of Tertiary and gypsite of Quaternary in the area (Laul and Laul, 2007) prevented the floodwaters from percolating down. Under the expected scenario of higher rainfall in the western Rajasthan, such heavy rainfalls may increase in frequency. It is required that geoscientists investigate and put forth some measures that would help policy makers in reaching a long term solution. One such investigation could be the **precise mapping of already known palaeochannels in western Rajasthan** (Sharma, H.S., 1987; Bakliwal and Grover, 1999; Sankaran, 1999), their continuity into the Rann of Kutch, and the determination of the nature of subsurface sediments. Geophysical investigations and deep drilling in selected locations would be ideal for discovering possible aquifers below the clay and other impervious strata. The palaeochannels and permeable sediments can be used for injection of excess surface water that will potentially fall in a short period of time, and reduce loss of life, property, agricultural products.

Climate Change and its Impact

Temperature and Rainfall

Table 3.2 shows selective studies by Hingani, L.S. *et al.* (1985); Rupakumar, K. *et al.* (1994); Thapliyal, V. *et al.* (1991); Rupakumar, K. (1992); Kriplani, R.H. *et al.* (1996), Koteshwaram, P. *et al.* (1973); and Singh, R.S. *et al.* (2001) observed changes in temperature and rainfall over India during the last century. These studies confirm the increase in temperature and change in rainfall pattern during the last century (Table 3.2).

Change in the River Course

This is an environmental problem of serious concern in the Indo-Gangetic Plain Region (IGPR). During different times in the past, different rivers changed their course a number of times. During the period 1731-1963, the course of the Kosi River (the Sorrow of Bihar) has shifted westward by about 125 km; courses of the Ganga, the Ghaghara and the Son at their confluence have shifted by 35 to 50 km since the epic period (Singh, R.L., 1971) (~1000 BC) and that of the Indus and its tributaries by 10-30 km in 1200 years in the same direction (Wilhilmy, H., 1967). Between 2500 BC and AD 500 the course of the Yamuna river shifted westward to join the Indus and then eastward to join the Ganga thrice (Raikes, R.L., 1968).

Sea Level Rise

Considerable amount of data has been generated during the last three decades for the Indian coastal margins as regards its tectonics and morphology. Attempts have also been made to generate regional sea level curves for the period encompassing the last glacial maxima about 18,000 years ago (Hashimi *et al.*, 1995; Kale and Rajaguru, 1985). These curves are based on limited data, and are a broad generalization for India. They have to be generated for different coastal segments that have behaved differently in response to neotectonic changes (Bhatt and Bhonde, 2006). For instance, the Saurashtra Peninsula of Gujarat is almost a neotectonically stable region, whereas the neighbouring Kutch is neotectonically unstable. Under the circumstances, only a detailed geoscientific investigation will tell us if the Rann of Kutch is going to be submerged under an anticipated sea level rise or it would continue to rise above the sea level, as has happened during the last 4,000 years. The available data from the western Saurashtra coast indicates that the sea level has been falling during the last 2,000 years (Mathur *et al.*, 2004) and may continue to do so in near future. If this holds good for the eastern Saurashtra coast as well, is yet to be determined. Under such circumstances, it is necessary that detailed geoscientific investigations are undertaken for different segments of the Indian coasts.

Das and Radhakrishnan (1991) reported a rising trend in the sea level at Mumbai (Bombay) during 1940-86 and Chennai (Madras) during 1910-33, based on the annual means of tide-gauge observations. Srivastava and Balakrishnan (1993) studied the atmospheric tide-gauge data and confirmed a rise in sea level by 8 cm. Hema Malini and Nageshwar Rao (2004) have also studied coastal erosion and habitat loss along the Godavari delta. The study indicates that decreased sediment supply through the river caused relative sea level rise under continued coastal subsidence along the delta front region. Recent studies conducted by Space Application Centre, Ahmedabad further justify the rise of sea level in the coastal areas. (Ajai, 2008)

Table 3.2: Observed Climate Change during the Past Century Over India

Region	*Temperature*	*Rainfall*	*Reference*
All-India	Increase in 0.4°C, 100 yrs in mean annual temperature		Hingane, L.S. *et al.*
All-India	Increase in 0.4°C, 100 yrs in mean annual temperature		Rupakumar, K. *et al.*
Western Himalayas	Winter season - Srinagar. Mussoorie and Mukteswar show increasing trend (0.5°C 100 yrs) Monsoon season - Srinagar. which is beyond the monsoon regime, shows significant increasing trend, whereas Mussoorie and Dehradun. which are at the foothills of the Himalayas. show decreasing trend	No increasing or decreasing trend for the last 100 years	Pant G.B. *et al.*
Indo-Gangetic Plain Region (IGPR)	Annual surface air temperature of the IGPR shows rising trend (0.53°C/100 yrs during 1875-1958).	Summer monsoon rainfall over western IGPR shows increasing trend (170 mm, 100 yrs) from 1900, while over central IGPR it shows	Singh N. *et al.*

(Contd.)

Table 3.2 *(Contd.)*

Region	*Temperature*	*Rainfall*	*Reference*
	Decreasing trend (–0.93°C 100 yrs during 1958-1997)	decreasing trend (5 mm 100 yrs) from 1939 and over eastern IGPR a decreasing trend (50 mm 100 yrs) during 1900-84 and increasing trend (480 mm 100 yrs) during 1984-99 Westward shift in rainfall activities over the IGPR	
All-India		Monsoon rainfall is trendless and is mainly random in nature over a long period	Moolay, D.A. *et al.* Thapliyal V. *et al.*
NE Peninsula. ME India. NW Peninsula, west coast and central peninsula		NE peninsula. NE India and NW peninsula show decreasing trend in the Indian summer monsoon rainfall (–6 to 8 per cent of normal 100 yrs), while increasing trend was noticed along the west coast and over central peninsula (–10 to 12 per cent of normal 100 yrs)	Rupakumar K. *et al.*
Western and eastern Himalayas		Western Himalayas gets more snowfall than eastern Himalayas during winter. More rainfall in the eastern Himalayas than in the western Himalayas during monsoon season	Kriplani R.H. *et al.*
All-India		Decadal departures in summer monsoon rainfall are found above and below the long-time average alternatively for three consecutive decades	Kothyari U.C. and Singh V.P.
Rajasthan desert In different locations		Slight increases in monsoon rainfall in spite of large inter-annual variations Pockets of significant long-term rainfall changes	Pant G.B. *et al.* Kotswaram P. *et al.* Raghavendra, V.K.
Luni River Basin (arid west Rajasthan)	Rising trend at Barmer, Jodhpur. Ajmer and Pali Decreasing trend at Udaipur and Jawaibandh	Annual rainfall indicating increasing tendency at 19 stations (around Ajmer in upper part of the Luni basin) Decreasing trend at the remaining nine stations in lower Luni basin, i.e., Barmer	Singh, R.S. *et al.* Doi, R.D. (2000) Bhardwaj, V. *et al.* (2008)

Source: Adopted from R.K. Mall *et al.*, *Current Science*, Vol. 90, No. 12, 25 June, 2006.

Various studies about climate changes using Global Climate Models (GCMs) and Regional Climate Models (RCMs) over India during the twentyfirst century (Bhaskaran, B. *et al.*, 1995 and Rupakumar *et al.*, 2003) indicate changing patterns in rainfall and an increase in temperature (Table 3.3). Utility of precipitation primarily depends upon its spatial as well as temporal distribution. Uniform precipitation over a larger area is more useful than its occurrence over a smaller region. Also, precipitation occurring over a larger time period would be more effectively utilized rather than when it occurs

Table 3.3: Projected Climate Change during Present Century over India

Region	*Temperature*	*Rainfall*	*Reference*
All-India	Increase in winter temperature by 1-4°C with increased CO_2 concentration	Precipitation increase of approximately 20 per cent Increase in heavy rainfall days during the summer monsoon period and an increased inter-annual variability	Bhaskaran, B. *et al.*
All-India	Average temperature change is predicted to be in the range of 2.33 to 4.78°C. with a doubling in CO_2 concentration	Increase in frequency of heavy rainfall events	Lonergan, S.
All-India	Area-averaged annual mean surface temperature rise is projected to range between 3.5 and 5.5°C by the end of the century. More warming in winter season	Increase of about 7 to 10 per cent in annual mean precipitation Decline of 5-25 per cent in winter precipitation Increase in monsoon precipitation is 10-15 per cent Monsoon season over northwest India —increase of 30 per cent or more in rainfall by 2050 Western semi-arid regions of India could receive higher than normal rainfall in a warmer atmosphere Decrease in winter precipitation between 10 and 20 per cent over central India by 2050	Lal M. *et al.*
All-India	Over the region south of 25°N (south of cities such as Udaipur. Khajuraho and Varanasi), maximum temperature will increase by 2-4°C during 2050s. In the northern region, increase in maximum temperature may exceed 4°C. A general increase in minimum temperature up to 4°C all over the country.	Decrease in number of rainy days over a major part of the country. This decrease is more in the western and central parts (by more than 15 days), while near the foothills of the Himalayas (Uttaranchal) and in northeast India, the number of rainy days may increase by 5-10 days. Increase in rainy days intensity by 1-4 mm/day, except for small areas in northwest India, where rainfall intensities may decrease by 1 mm/day.	Rupakumar K. *et al.*

Source: Adopted from R.K. Mall *et al.*, *Current Science*, Vol. 90, No. 12, 25 June, 2006.

within a short time-span. Therefore, projected changes in precipitation pattern over the Indian subcontinent in near future would result in deterioration of water resources. First, decrease in winter precipitation would reduce the total seasonal precipitation being received during December-February, implying greater water stress during the lean monsoon period. Secondly, intense rain occurring over fewer days, besides causing increased frequency of floods during the monsoon season, will also mean that much of the monsoon rain would be lost as direct run-off resulting in reduced groundwater recharging potential.

The rainfall data from 1951 to 2008 reveal that the frequency of sudden rainfall had increased by 10 per cent in India. Incidence of flash floods may increase in coming years because of the impact of climate on the monsoonal system. Rainfall patterns are changing faster than what we can realize. On the basis of various studies it may be concluded that extremes in rainfall and temperature are on the rise. Decrease in number of rainy days but increase in the intensity of rainfall on the day it rains has been witnessed during past decades. However no significant change in the date of onset of monsoon has been noticed but the variability of the date could be higher in future.

Mean sea level rise trends along the cost of India are about an average 1.30 mm per year. Estimates further indicate higher flood risks with storm surges along the southern part of the east coast.

Desertification

Climate change and desertification are closely linked. One of the impacts which global warming may have on the surface of the Earth is to aggravate the worldwide problem of desertification. Decrease in the total amount of rainfall in arid and semi-arid areas could increase the total area of dry lands worldwide and thus the total amount of land potentially at risk from desertification.

The 1977 UNCOD in Nairobi described desertification as the diminution or destruction of the biological potential of the land. In the recent Rio Summit in 1992 it has been defined as "Land degradation in arid, semi-arid and dry sub-humid areas resulting from various factors including climatic variations and human activities."

What causes Desertification?

- Apart from natural factors, overgrazing is the major cause of desertification worldwide. Plants of semi-arid areas are adapted to being eaten by sparsely scattered, large, grazing mammals which move in response to the patchy rainfall common to these regions. Early human pastoralists living in semi-arid areas copied this natural system. They moved their small groups of domestic animals in response to food and water availability. Such regular stock movement prevented overgrazing of the fragile plant cover.
- Cultivation of marginal lands, i.e., lands on which there is a high risk of crop failure and a very low economic return, for example, some parts of South Africa where maize is grown.
- Destruction of vegetation in arid regions, often for fuelwood.
- Poor grazing management after accidental burning of semi-arid vegetation.
- Incorrect irrigation practices in arid areas can cause salinization, (the build up of salts in the soil) which can prevent plant growth.

When the practices described above coincide with drought, the rate of desertification increases dramatically. Increasing human population and poverty contribute to desertification as poor people may be forced to overuse their environment in the short term, without the ability to plan for the long term effects of their actions. Where livestock has a social importance beyond food, people might be reluctant to reduce their stock numbers.

Status of Desertification in India

Recently in 2000, the Regional Action Programme (RAP) to combat desertification had been formulated by UN convention to combat desertification (UNCCD) and accordingly RAP for Asian region had been initiated as collective effort by the member countries. Under Regional Action Programme, Thematic Programme Network (TPN-1) is on Desertification Monitoring and Assessment. In India, Space Applications Centre (ISRO), Ahmedabad has been identified as the national focal institution to coordinate TPN-1 activities within the country and establish the national network for desertification monitoring and assessment.

To implement TPN-1 activities in India, a working group was constituted by Space Application Centre, Ahemadabad. The working group consisted of 17 institutions/organizations, which were given responsibility for desertification monitoring and assessment in the country. Under this project Status Mapping of Desertification was done at two scales, i.e., R.F. 1 : 50,000 and R.F. 1 : 500,000. The results are given in Table 3.2.

Table 3.4 indicates that the total area undergoing the process of land degradation in India is **105.48 million hectares**, which constitutes **32.07 per cent** of the total geographical area. There are three major processes of desertification in India. The water erosion is the major process of degradation in the country followed by **vegetal degradation and wind erosion** (Fig. 3.1).

Among the States, the NE-States have very high percentage of area under land degradation (79%), followed by Manipur (67%), Tripura (65%), Nagaland (64.27%). Outside NE-region, the other States having high per cent of desertification/degradation include Gujarat (68.27%), Rajasthan (67%) and J&K (60.70%). **In Rajasthan 67% of area is affected** due to desertification / land degradation. The most significant process of desertification in the State is wind erosion (44.42%), followed by water erosion (11.22%), vegetal degradation (6.25%) and salinization (1.70%).

Table 3.4: Process-wise Distribution of the Area under Land Degradation for India

Processes of Desertification/ land degradation	*Area covered*	
	Area (mha)	*% of Total Geog. Area*
Water Erosion	33.56	10.21
Vegetal Degradation	31.66	9.63
Wind/eolian Degradation	17.56	5.34
Frost Shattering	10.21	3.10
Salinity/ Alkalinity	5.26	1.60
Mass Movement	4.45	1.35
Water logging	0.98	0.30
Rocky areas/ Barren	1.65	0.50
Others (Man made, frost heaving, etc.)	0.15	0.04

Source: Desertification and Land Degradation Atlas of India, SAC, 2007.

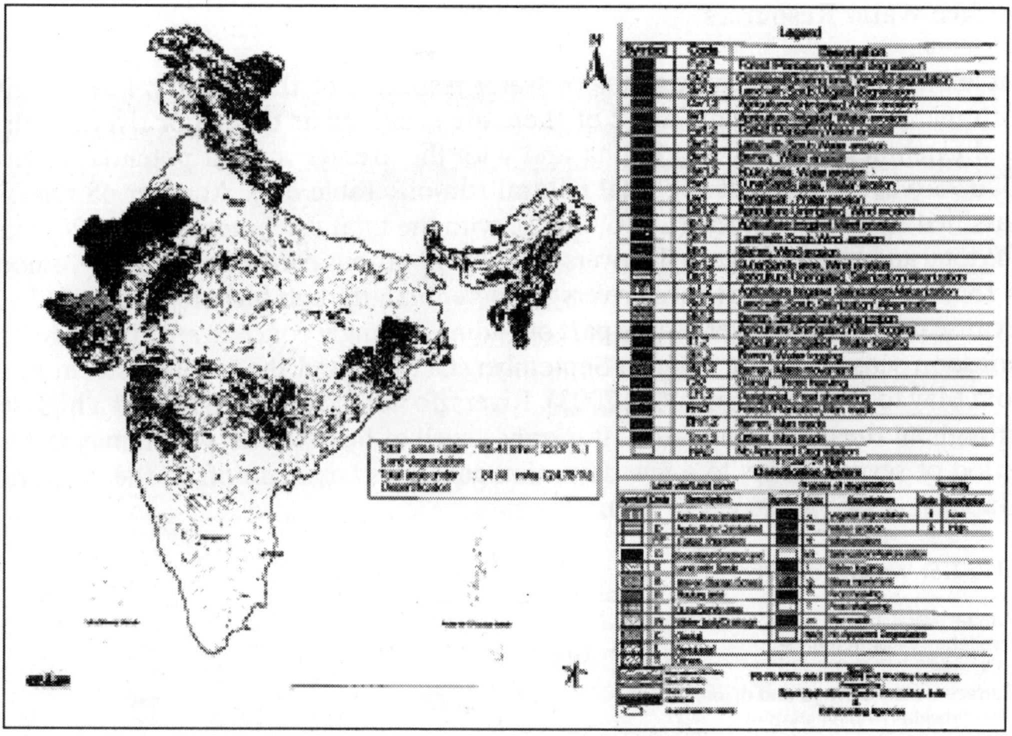

Fig. 3.1: Process of Degradation in the Country

Source: *Desertification and Land Degradation Atlas of India*, SAC, 2007

WATER RESOURCES

The climate changes are likely to affect water supply, quality and demand and may have ramifications for decision makers. In India, the major river systems of the sub-continent namely the Brahmaputra, the Ganga and the Indus which originate from the Himalayas are likely to be more vulnerable to climate change, because of substantial contribution from snow and glaciers in these rivers (Singh and Naresh Kumar, 1997).

Water resources will come under increasing pressure in the Indian subcontinent due to the changing climate. Presently, more than 45 per cent of the average annual rainfall, including snowfall in the country, is wasted by natural runoff to the sea. Rainwater-harvesting schemes are now being implemented in the country to minimize this run-off loss based on present rainfall scenarios over the country, to increase groundwater levels. However, for the success of these schemes it is necessary to focus on how the possible climate change will affect the intensity, spatial and temporal variability of the rainfall, evaporation rates and temperature in different agro-climatic regions and river basins of India.

Surface Water Resources

Table 3.5 shows the present status of water resources of the country. India is gifted with many rivers. As many as 12 of them are classified as the major rivers, with the total catchment area of 252.8 mha and with the average annual potential in rivers, 1570.98 b.cu.m. (85% of the total natural run-off; Table 3.3). Another 48 rivers are classified as medium rivers (CWC, 2002), with the total catchment area of 24.9 mha. The total annual discharge in the rivers that flow in various parts of the country, amounts to 1869 b.cu.m. Many of these rivers are perennial, though few are seasonal. This is because precipitation over a large part of India is strongly concentrated in the summer monsoon season during June to September/October and the tropical storm season from May to October (Kale, V.S., 2003). Rivers do not, however, remain at a high stage throughout the monsoon season. It is only a spell of heavy rains, which may last for a period of several hours to a few days, that generates large run-off in the catchments (Gupta A, 1995; Kale, V.S., 1998).

Table 3.5: Water Resources of India

Annual precipitation	4000 b.cu.m.
Available water resources	1869
Utilizable	1122
Surface water (storage and diversion)	690
Groundwater (replenishable)	432
Present utilization (Surface water 63%, groundwater 37%)	605
Irrigation	501
Domestic	30
Industry, energy and other uses	4

Source: CWC, 2002.

R.K. Mall, *et al.* (2006) have estimated that the average water yield per unit area of the Himalayan rivers is almost double of the south peninsular river systems, which indicates the importance of snow and glacier melt contribution from high mountains. The study further demonstrates that the rainfall contributions are greater in the eastern region, while the snow and glacier melt contributions are more important in the western and central Himalayan region. Apart from monsoon rains, contributions from snow and glacial melt are playing a significant role in water resources development of the country. In recent decades, the hydrological characteristics of the watersheds in the Himalayan region seem to have undergone substantial changes as a result of extensive land use (e.g., deforestation, agricultural practices and urbanization), leading to frequent hydrological disasters, enhanced variability in rainfall and run-off, extensive reservoir sedimentation and pollution of lakes (Ramakrishnan, P.S., 1999). Global warming and its impact on the hydrological cycle and nature of hydrological events have posed an additional threat to this mountainous region of the Indian subcontinent. Extreme precipitation events have geomorphologic significance in the Himalayas, where they may cause widespread landslides (Ives, J.D. and Messerli, B., 1989). The response of hydrological systems, erosion processes and sedimentation in this region could alter significantly due to climate change.

Groundwater Resources

The distribution of groundwater is not uniform in all regions of India due to spatio temporal variations in rainfall, geology and geomorphology. India is a vast country having diversified geological, climatological and topographic set-up, giving rise to divergent groundwater situation in different parts of the country. The prevalent rock formations, ranging in age from Archaean to Recent, which control occurrence and movement of groundwater, are widely varied in composition and structure. Similarly, not too insignificant are the variations of landforms from the rugged mountainous terrains of the Himalayas, Eastern and Western Ghats to the flat alluvial plains of the river valleys and coastal tracts, and the sandy deserts of Rajasthan. The rainfall patterns too show similar region-wise variations. The topography and rainfall virtually control run-off and groundwater recharge (CGWB, 2002). The available groundwater resources and replenishable resources (CGWB, 2006) are given in Table 3.6. The total annual replenishable groundwater resource is about 432.42 b.cu.m. The stage of groundwater development in India is 58 per cent (2004). The status of groundwater development is comparatively high in the State of Delhi, Haryana, Punjab, Rajasthan and Union Territories of Daman and Diu and Pondicherry where the development is more than 100 per cent, which implies that in these States, the average annual groundwater consumption is more than average annual water recharge.

Availability and Development of Groundwater Resources

The demand of water in agriculture, industrial and domestic sectors, has brought problems of over-exploitation of the groundwater resource, continuously declining groundwater levels, sea water ingress in coastal areas, and groundwater pollution in different parts of India. The falling groundwater levels in various parts of the country due to uncertainty of the monsoon system have threatened the sustainability of the groundwater resource, as water levels have gone deep beyond the economic lifts of pumping. The Central Groundwater Board has established more than **15,000 network-monitoring stations** in the country to monitor groundwater level and its quality. Water levels in major parts of the country generally do not show any significant rise/fall. However, significant decline in the level of groundwater has been observed in **certain pockets of 289 districts** in Andhra Pradesh, Assam, Bihar, Chhattisgarh, NCT Delhi, Gujarat, Haryana, Jharkhand, Karnataka, Kerala, Madhya Pradesh, Maharashtra, Orissa, Punjab, Rajasthan, Tamil Nadu, Tripura, Uttar Pradesh and West Bengal, Delhi, Haryana, Punjab, Rajasthan, Chandigarh and Diu have recorded the high level of groundwater development (more than 100%). In case of Rajasthan, low rainfall conditions have resulted in poor recharge to groundwater, whereas in case of Delhi,

Table 3.6: State-wise Groundwater Resource Availability, Utilization and Stage of development in India as on 31st March, 2004

State/Union Territory	*Annual Replenishable Groundwater Resource*	*Net Annual Groundwater Availability*	*Annual Groundwater Draft*			*Project Demand for Domestic and Industrial users upo 2025*	*Groundwater availability for future Irrigation*	*Stage of groundwater development %*
			Irrigation	*Domestic and industrial uses*	*Total*			
State								
Andhra Pradesh	36.50	32.95	13.88	1.02	14.90	2.67	17.65	45
Arunachal Pradesh	2.56	2.30	0.0008	0	0.0008	0.009	2.29	0.04
Assam	27.23	24.89	4.85	0.59	5.44	0.98	19.06	22
Bihar	29.19	27.42	9.39	1.37	10.77	2.14	15.89	39
Chhattisgarh	14.93	13.68	2.31	0.48	2.80	0.70	10.67	20
Delhi	0.30	0.28	0.20	0.28	0.48	0.57	0.00	170
Goa	0.28	0.27	0.04	0.03	0.07	0.04	0.18	27
Gujarat	15.81	15.02	10.49	0.99	11.49	1.48	3.05	76
Haryana	9.31	8.63	9.10	0.35	9.45	0.60	–1.07	109
Himachal Pradesh	0.43	0.39	0.09	0.02	0.12	0.04	0.25	30
Jammu and Kashmir	2.70	2.43	0.10	0.24	0.33	0.42	1.92	14
Jharkhand	5.58	5.25	0.70	0.38	1.09	0.56	3.99	21
Karnataka	15.93	15.30	9.75	0.97	10.71	1.41	6.48	70
Kerala	6.84	6.23	1.82	1.10	2.92	1.40	3.07	47
Madhya Pradesh	37.19	35.33	16.08	1.04	17.12	1.74	17.51	48
Maharashtra	32.96	31.21	14.24	0.85	15.09	1.52	16.10	48
Manipur	0.38	0.34	0.002	0.0005	0.002	0.02	0.31	0.65
Meghalaya	1.15	1.04	0.00	0.002	0.002	0.10	0.94	0.18
Mizoram*	0.04	0.04	0.00	0.004	0.0004	0.0008	0.04	0.90
Nagaland	0.36	0.32	0.00	0.009	0.009	0.03	0.30	3
Orissa	23.09	21.01	3.01	0.84	3.85	1.22	16.78	18
Punjab	23.78	21.44	30.34	0.83	31.16	1.00	–9.89	145
Rajasthan	11.56	10.38	11.60	1.39	12.99	2.72	–3.94	125
Sikkim*	0.08	0.08	0.00	0.01	0.01	0.02	0.05	16
Tamil Nadu	23.07	20.76	16.77	0.88	17.65	0.91	3.08	85
Tripura	2.19	1.97	0.08	0.09	0.17	0.20	1.69	90
Uttar Pradesh	76.35	70.18	45.36	3.42	48.78	5.30	19.52	70
Uttaranchal	2.27	2.10	1.34	0.05	1.39	0.08	0.68	66
West Bengal	30.36	27.46	10.84	0.81	11.65	1.24	15.32	42
Total States	432.42	398.70	212.38	18.04	230.44	29.12	161.92	58
Union Territories								
Andaman and Nicobar	0.330	0.320	0.000	0.010	0.010	0.008	0.303	4
Chandigarh	0.023	0.020	0.000	0.000	0.000	0.000	0.020	0
Dadar and Nagar Haveli	0.063	0.060	0.001	0.007	0.009	0.008	0.051	14
Daman and Diu	0.009	0.008	0.007	0.002	0.009	0.003	–0.002	107
Lakshadweep	0.012	0.004	0.000	0.002	0.002	—	—	63
Pondicherry	0.160	0.144	0.121	0.030	0.151	0.031	–0.008	105
Total Union Territories	0.597	0.556	0.129	0.051	0.181	0.050	0.365	33
Grand Total	433.02	399.25	212.51	18.09	230.62	29.17	162.29	58

* *Source*: CGWB, (2006).

Haryana and Punjab, the high rate of groundwater withdrawal is the main cause for high stage of groundwater development (Table 3.6). Gujarat (55%), Tamil Nadu (64%), Daman (70%) and Pondicherry also have significantly the high level of groundwater development. With rapid expansion in groundwater extraction, development related problems have started emerging. There is continuous growth in dark and over-exploited areas in the country. Out of **5,723 blocks / mandals / districts, in the country, 839 blocks are categorized as over-exploited**, whereas the stage of groundwater development exceeds the annual replenishment and **226 blocks are dark or critical, 4,078 units are safe and 30 units are saline** where groundwater development has reached a high level (Fig.3.2). Substantial decline of groundwater levels occurs even in blocks with sufficient groundwater resources due to climatic vicissitudes and localized development (Romani, S., 2004).

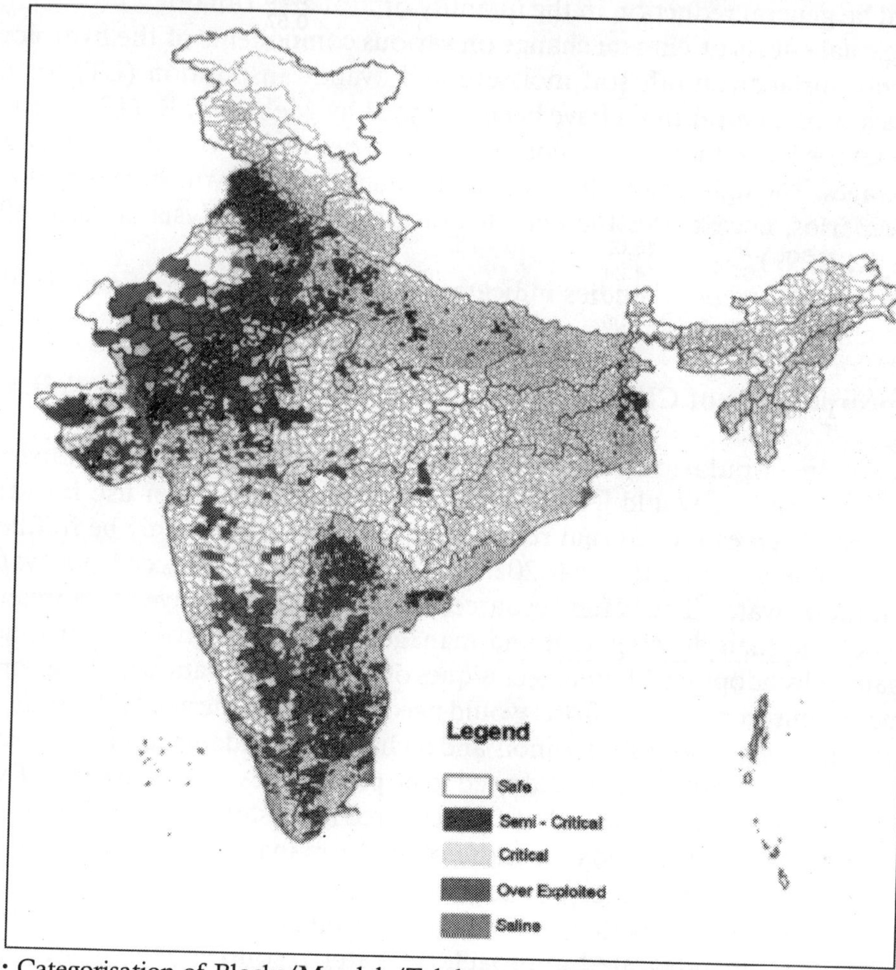

Fig. 3.2: Categorisation of Blocks/Mandals/Talukas as on March 2004

Source: CGWB, (2006).

Impact on Water Resources

As has been indicated earlier that several studies are available in India to show the likely impact of climate change on water resources. The increasing temperature over Indian subcontinent would result in an increase in pre-monsoonal and monsoonal rainfall and no change in winter rainfall over the central region. This would result in increase in evaporation and soil wetness during monsoon (Lal M. *et al.*, 1993). The studies further show 1 metre rise in sea level in Orissa and West Bangal and would inundate 1,700 km^2 on prime agricultural land in coastal area. The rise of Indian coastline is likely to affect a total area of 5,763 km^2 and put 7.1 million people at risk. Chattopadhyay and Hulme (1997) have estimated that there would be an increase in potential evaporation across India. Gosain and Rao (2003) have also indicated that there would be general reduction in the quantity of available run off.

The regional effects of climate change on various components of the hydrological cycle, namely surface run-off, soil moisture and evapotranspiration (ET) for three drainage basins of Central India have been analysed by Mehrotra, R. (1999). Results indicated that the basin located in a comparatively drier region was more sensitive to climatic changes. The significant effect of climate change on reservoir storage, especially for drier scenarios, necessitates the need for further critical analysis of these effects (Mehrotra, R., 1999).

The review of the above studies indicates that the Indian subcontinent is highly sensitive to climate change.

Need of Incorporation of Climate Change Impacts in Water Resource Policy

The increase in population, urbanisation, growth process and expansion of developmental activities would lead to increasing demand for water use for various purposes. It has been estimated that requirement of water would only be fulfilled by available utilizable water by the year 2025 but after 2025 it will exceed the available utilizable limit of water. The water resources would, therefore, have to be treated as national asset and their development and management would have to be planned in optimal manner by adopting different techniques of water conservation and appropriate water management techniques. Efforts would need to increase the available fresh water by harnessing the surplus monsoon runoff and recharging it underground and/recycle/ reuse of water. Special attention would need to be paid to the conjunctive use of water resources and use of saline/ brackish water resource in western parts of India. It has been agreed that water crisis today is more of water management than water itself. Unjudicious management, lack of political will, improper incentives and disparity in allocation of resources are threatening the available water resources. The real solution, therefore, requires an integrated approach of water resources management. The groundwater resource augmentation, conservation and protection should be the national policy in which there is an urgent need to adopt conservationists approach than exploitative approach for sustainable development of water resources.

An exploding population will continue to build pressure on land resource. This would further hasten the vegetal degradation by the ever increasing demand for firewood and other requirements. The future drought will have greater damaging effects in all walks of life particularly water resources of the country. Therefore, it is necessary to redress the issue of water resource management at three stages, i.e., **judicious use of available water resources, development of recharging methods and regulatory measures to utilize water resources.** The conservation of water should be integrated with land-use and land management in order to combat desertification.

The policy-makers must respond to the challenges posed by ensuing climate change to the water sector. Contemporary water-management decisions are made at a variety of space and time scales and are informed by assumptions about supplies, demands, weather, climate, and operational constraints at those scales.

Although a science based policy is an illusion. Political arguments are and will be more important in choosing between policy options than scientific arguments. However, decision making must be informed by purposeful knowledge and science is a must to illuminate the values, goals and instruments of policy making and to sensitize policy makers. Changes in planning and analysis that better accommodate non-stationarity will improve water management. These options would include the incorporation of lessons from responses to climate variability into longer term vulnerability-reduction efforts and within governance mechanisms from communities and watersheds to inter-State agreements.

In the end, it may be emphasised that research in Earth System Sciences must move beyond a disciplinary focus to address sustainability issues of Planet Earth in the framework of complex dynamical systems. Archives from the past ice cores, tree rings, archaeological and historical records must be studied more vigorously to provide trajectories of change, baseline conditions, insights into past societal resilience or fragility and perspectives on projections of future change. Earth System Sciences have already achieved much in support of sustainable development. The community of Earth System Sciences should, therefore, commit to help and build an improved, integrative, participatory and useable science that is applicable from local to global scales.

NOTES

Abhichandani, C.T. and Roy B. B. (1966), Rajasthan Desert—Its Origin and Amelioration, *Indian Geographical Journal*, Vol. 41 (3 and 4), pp. 35-43.

Adel, M. M., *International Journal of Climatology*, (2002), 22, 993-1016.

Afford, D., (1992) ICIMOD Occasional Paper No. 18, Kathmandu, p. 68.

Agrawal, D.P. and Kursumgar, S. (1975), Radio Carbon Dates of Some Late Quaternary Samples, *Current Science*. Vol. 44, pp. 149-51.

Agrawal, D.P.; Datta, P.S.; Husasain, Z.; *et al.* (1980), Palaeoclimate, Stratigraphy and Pre-history in North and West Rajasthan, Proceedings of the Indian Academy of Sciences, Vol.89 (1), pp. 51-66,

Agrawal, D.P.; Dhir R.P. and Krishnamurthy, R.V. (1976), The Relevance of Kankar Studies of the Archaeology and Palaeo-ecology of Rajasthan, *Annual Archaeological Congress*, University of Rajasthan, Jaipur.

Ahmed, E., (1969), Origin and Geomorphology of the Thar Desert, *Annals of Arid Zone*, pp.171-80.

Ajay (2007), Desertification and Land Degradation, *Atlas of India*, SAC, Ahmedabad.

Allchin, B., (1958), The Late Stone Age of Ceylon. *Journal of the Royal Anthropological Institute*, Vol. 88, pp, 179-201.

Allchin, B., (1959), The Indian Middle Stone Age. *Bulletin of the Institute of Archaeology*, London. Vol. 2, pp. 1-35. Allchin, B. and Goudie, A. 1971: Dunes, Aridity and Early Man in Gujarat, Western India, *Man*. Vol. 6, pp, 248-65

Allchin, B.; Goudie, A.; and Hegde, K., (1978), The Pre-history and Palaeogeography of the Great Indian Desert, Academic Press, London.

Allchin, B.; Hegde, K.T.M.; and Goudie, A., (1972), Pre-history and Environmental Change in Western India, A note on the Budha-Pushkar Basin, Rajasthan, *Man*, Vol. 74, pp. 542-64.

Allchin, Band Goudie A., (1974), Pushkar: Pre-history and Climatic Change in Western India, *World Archaeology*, Vol.5, pp. 358-68.

Arnell, N.W. (2006), Climate Change and Water Resources: A Global Perspective. In: H.J. Schellnhuber, W. Cramer, N. Nakicenovic, T.M.L. Wigley and G. Yohe (Editors), *Avoiding Dangerous Climate Change*. Cambridge University Press, Cambridge, pp. 167-175.

Arora, V.K. (2001) and Boer, G. J., *Journal of Geophysics, Research and Development*, 106, 3335-3348.

Bahadur, J., (1999), The Himalayan Environment (eds Dash, S.K. and Bahadur, J.), New Age International (P) Ltd., New Delhi, pp. 258-267.

Bakliwal, P.C. and Grover, A.K. (1999), Geol. Soc. India, *Mem.*, 42, 113-119.

Bhakar, R. (2004), Re-evaluating the sustainability of irrigated agriculture in arid zones- A case study of Indira Gandhi Nahar Pariyojna. *Indian Cartographer*, 24: 132-141.

Bhaskaran, B.; Mitchell, J.F.B., Lavery, J.R. and Lal, M., (1995), *Int. J. Climatol.*, 15, 873-892.

Bhardwaj V. & Geeta Sharma (2008), Flood Calamity: Challenge to Sustainable Development in the Thar Desert, *Geographer*, Vol. V, No. 1.

Biesbrouck, B.; G., W., Van Orschoven, J. and J., F., (2002), AVSWAT 2000. Laboratory for Soil and Water Management, Katholieke Universiteit Leuven, Leuven, Belgium.

Blaikie, P. and Brookfield, H. (1987), *Land Degradation and Society*. Methuen, London.

Bruce Battrick (2006), *The Changing Earth* (ed), ESA Publication Devision, The Nitherlands.

Bryson. R.A. and Barries., D.A. (1967), Possibilities of Major Climatic Modifications and their Implications: North West India. A Case for Study. *Bulletin of American Meteorological Society*. Vol. 48, pp. 136-42.

Chattopadhyaya, N. and Hulme, M. (1997) *Agric. for. Meteorol.*, 87, 55-73.

Dynamic Water Resource of India as on 31st March, 2004 (2006), *CGWB*, Govt. of India, New Delhi.

Das, P.K. and Radhakrishnan, M., (1991), *Proc. Indian Acad. Sci.* (Earth Planet. Sci.), 100, 177-194.

Doi, R.D. (2000), Rainfall Pattern and Agricultural Response to its occurrence in Rajasthan Studies in Geography, Vol. 18, pp. 56-64.

Freshwater Year (2003), Ministry of Water Resources, Government of India, New Delhi, 2003, p. 53.

Fushimi, H., (2000), In (eds Domoto, A. *et al.*), Tsukiji-Shokan Pub Co. Ltd., *Japan and The World Conservation Union*, Gland, Switzerland, 2000, pp. 42-45.

Ghose. B. and Singh. S. (1965), Geomorphological Studies in Rajasthan—Analysis of Landforms of Siwana Block from Aerial Photographs. *Annals of Arid Zone*. Vol. 4. (2), pp. 207-216.

Ghose. B., (1964), Geomorphological Aspects of the Formation of Salt Basin—The Lower Luni Basin. Proceedings of Symposium of Problems of India Arid Zone. Jaipur, pp. 169-78.

Gosain, A.K.; Rao, S. and Basuray, D., (2006), Climate Change Impact Assessment on Hydrology of Indian River Basins. *Current Science*, 90(3): 346-353.

Gosain, A.K. and Rao, S., (2003), Climate Change and India: Vulnerability Assessment and Adaptation (eds Shukla, P.R. *et al.*), Universities Press (India) Pvt. Ltd, Hyderabad, p. 462.

Gupta, A., (1995), In American Geophysical Union, *Geophysical Monograph* (eds Costa, J.E. *et.al.*), Vol. 89, pp. 125-136.

Hashimi, N. H.; Nigam, R.; Nair, R. R.; and Rajagopalan, G. (1995), Holocene Sea Level Fluctuations on Western India Continental Margin—An Update. *Jour. Geol. Soc. India*, 46, 157-162.

Hema Malini and Nageshwar Rao (2004), Coastal Erosion and Habitat Loss along the Godavari Delta Front—A Fallout of Dam Construction", *Current Science*, Vol. 87, No. 9, pp.1232-36.

Hingane, L. S.; Rupa Kumar, K.; and Ramana Murthy, (1985), Bh. V., *J. Climatol.*, 5, 521-528.

IPCC, (1992), Global Climate Change and the Rising Challenge of the Sea. Supporting document for the IPCC, WMO and UNEP, Geneva.

IPCC, (1998), Special Report of IPCC Working Group II (eds Watson, R.T. *et al.*), *Intergovernmental Panel on Climate Change*, Cambridge University Press, Cambridge, UK, p. 517.

IPCC (2007), *Climate Change*, Synthesis Report.

IPCC, (2007), Summary for Policymakers. Climate Change 2007: The Physical Science Basis. Contribution of Working Group I to the Fourth Assessment Report of the Intergovernmental Panel on Climate Change. Cambridge University Press, Cambridge.

Ives, J.D. and Messerli, B. (1989), The Himalayan Dilemma: Reconciling Development and Conservation, Routledge, London.

Joyce, Terrence (2008), Are We on the Brink of a 'New Little Ice Age? Woods Hole Oceanographic Institute website.

Kale, V.S. (ed.) (1998), *Flood Studies in India*, Geological Society of India, Bangalore, Memoir, 41, pp. 229-256.

Kale, V.S. and Rajaguru, S.N. (1985), Neogene and Quaternary transgressional and regressional history of the west coast of India. *Bull. Deccan Coll. Res. Inst.*, 44,153-165.

Kale, V. S., (2003), *Nat. Hazards*, 28, 65-84.

Kolli, Rupa Kumar (Nov. 2006), Modelling of Regional Climate in South Asia Region and Future Climatic Scenario. Global Change Conference, Islamabad, Pakistan, 16-17.

Koteswaram, P. and Alvi, S.M.A., (1969), *Curr. Sci.*, 38, 229- 231.

Kothari, U.C. and Singh, V. P., (1996), *Hydrol. Process.*, 10, 357-372.

Kripalani, R. H.; Inamdar, S.R.; and Sontakke, N. A. (1996), *Int. J. Climatol.*, 16, 689-703.

Kripalani, R. H.; Kulkarni, A.; Sabade, S.S.; and Khandekar, M.L. (2003), *Nat. Hazards*, 29, 189-206.

Kulkarni, A.V.; Rathore, B.P.; Mahajan, S.; and Mathur, P. (2005), Alarming Retreat of Parbati Glacier, Beas basin, Himachal Pradesh. *Current Science*, 88(11): 1844-1849.

Lal and Singh, S. (2001), *Mausam*, 52, 1.

Lal, M. and Chander (1993), S., JEH, 1, 3-13.

Lal, M. *et al.* (2001), *Curr. Sci.*, 81, 1196-1207.

Laul, V.P. and Vivek Laul (2007), Floods in Barmer and Other Tertiary Terrains in Rajasthan. *Current Science*, 92 (5), 558.

Leshnik, L.S. (1968) *Pre-Historic Explorations in North Gujarat and Parts of Rajasthan*. East and West (N.S.), Vol.1183 and 4, pp. 295-310.

Liverman, D. (Editor) (1998), *People and Pixels: Linking Remote Sensing and Social Science*. National Academy Press, Washington, DC.

Lonergan, S. (1998), Climate Warming and India (eds Dinar, *et al.*), *World Bank Technical Paper No. 402*, Washington DC.

Mathur, U. B.; Pandey, D.K.; and Tej Bahadur (2004), Falling Late Holocene sea-level along the Indian coast. *Current Science*, 87 (4), 439-440.

Mathur, U.B. (2008) Are We Prepared for Past Climate Change, Souvenir CHEC India, Jaipur, pp. 29-32.

Mall., R.K.; Akhilesh Gupta, Ranjeet Singh, R.S. Singh and L.S. Rathore (2006), Water Resources and Climate Change- An Indian Perspective, *Current Science*, Vol. 90, No. 12, pp 1610-1625.

Meher-Homji, V.M., (1973), Is the Sind-Rajasthan Desert the result of a recent climatic change, *Geogforum*, Vol. 15, pp. 47-57.

Mehrotra, R. (1999), *Climate Change*, 42, 725-757.

Mishra, V.N. (1961), Palaeolithic Cultures of Western Rajputana. Bulletin of the Deccan College, Research Institute, Vol. 21, pp. 85-156.

Mishra, V.N. (1973), Bagor—A Late Mesolithic Settlement in North West India, *WorldArchaeology*, Vol. 5, No.1, pp. 92-110.

Mishra, V.N.; S.N. Rajaguru; Agrawal, D.P.; Thomas, P.K.; Husain, Z. and Dutta, P.S. (1980), Pre-history and Palaeoenvironment of Jayal, Western Rajasthan. *Man and Environment*, Vol. IV, pp. 18-30.

Miyahara, Hiroko; Kimiaki Masuda, *et al.* (2004), Cyclicity of Solar Activity during the Maunder Minimum Deduced from Radiocarbon. *Solar Physics*, 224(1 & 2), Springer Netherlands.

Mooley, D.A. and Parthasarathy, B. (1984), *Climatic Change*, 6, 287-301.

Pant G.B. and Hingane, L.S. (1988), Climatic Change in and around the Rajasthan Desert during the 20th Century. *J. Climatol.*, 8, 391-401.

Pant, G.B. and Hingane, L. S., (1998), *J. Climatol.*, 8, 391-401.

Pant, G.B. and Rupakumar, K. (1997), *Climates of South Asia*, John Wiley, UK, p. 320.

Pant, G. B.; Rupakumar, K.; and Borgaonkar, H. P. (1999), *The Himalayan Environment* (eds Dash, S.K. and Bahadur, J.), New Age International (P) Ltd, New Delhi, pp. 172-184.

Petit, J.R. *et al.* (1999), *Nature*, Vol. 399, pp. 429-36

Raghav, K.S. (1992a), Geoenvironmental Appraisal of the Mendha and Bandi River Catchment Area of Jaipur upland, Jaipur District Record, GSI29 (7): 134-135.

Raikes, R. L., (1968), *Antiquity*, 63, 286-291.

Raikes, R.L. (1965), The Mohenjo-Daro Floods, *Antiquity*, Vol. 39, pp. 196-203.

Raikes, R.L. and Dyson, R.H. (1961), The Pre-historic Climate of Baluchistan and the Indus Valley. *American Anthropologist*, Vol. 63, pp. 265-282.

Ramakrishnan, P.S. (1999), *The Himalayan Environment* (eds Dash, S.K. and Bahadur, J.), New Age International (P) Ltd, New Delhi, pp. 213-225.

Report, National Commission for Integrated Water Resources Development, Ministry of Water Resources, September 1999, Vol. 1, p. 542.

Richard A. Wood, Michael Vellinga and Robert Thorpe (2003), *Global Warming and Thermohaline Circulation Stability*. Philosophical Transactions: Mathematical, Physical and Engineering Sciences, 361(1810).

Romani, S. (2004), *Groundwater Manage.*

Roy, B.B. and Pandey, S. (1970), Expansion or Contraction of the Great Indian Desert. Proceedings of the Indian National Science Academy, Vol. 36, B.No. 6, pp. 331-344.

Rupa Kumar, K. *et al.*, (2006), High-resolution climate change scenarios for India for the 21st Century. *Current Science*, 90(3): 334.

Rupa Kumar, K.; Krishna Kumar, K.; and Pant, G. B., (1994), *Geophys. Res. Lett.*, 21, 677-680.

Rupa Lumar, K.; Kumar, K.; Prasanna, V.; Kamala, K.; Deshpande, N. R.; Patwardhan, S.K.; and Pant, G.B. (2003), *Climate Change and India: Vulnerability Assessment and Adaptation* (eds Shukla, P.R. *et al.*), Universities Press (India) Pvt Ltd, Hyderabad, p. 462.

Rupa Kumar, K.; Pant; G. B.; Parthasarathy, B.; and Sontakke, N. A. (1992), *Int. J. Climatol.*, 12, 257-268.

Sankaran, A.V. (1999), Saraswati—The Ancient River Lost in the Desert. *Current Science*, 77(8).

Sharma, H.S., (1968), Genesis of Ravines of the Lower Chambal Valley, *India, Selected Papers*, 21st IGU, Vol. I, pp. 114-118.

Sharma, H.S. (1976), Morphology of Ravines of the Morel Basin, India, *International Geography*, Moscow, Vol. 1, pp. 219-223.

Sharma, H.S., (1983), *Inland Drainage of North-Eastern Rajasthan and Its Possible Link with the Ghaggar Yamuna System*. Transactions of Institute of Indian Geographers, Vol. 5, No.1, pp. 19-22.

Sharma, H.S. (1987), *Tropical Geomorphology*, Concept Publishing Company, New Delhi. 27.

Sharma, K. P.; Moore III, B. and Vorosmarty, C.J. (2000), *Climate Change*, 47, 141-165.

Sharma, K. P., Vorosmarty, C.J. and Moore III, B. (2000), *Climate Change*, 47, 117-139.

Sharma, V.K. (2007), Global Warming Its Impact on India Theme Paper (4), the Member Annual Conference, 118A, New Delhi

Singh and Kumar, J. (1997), *Hydrol.*, 193, 316-350.

Singh G.; Joshi, R.D; Chopra, S.K. and Singh, A.B. (1974), *Late Quaternary History of Vegetation and Climate of the Rajasthan Desert, India.* Philosophical Transactions and of the Royal Society, 267B, pp. 467-501.

Singh, G. (1971), The Indus Valley culture seen in the Context of Post-glacial Climate and Ecological Studies in North-west India. *Archaeology and Physical Anthropology in Oceania*, Vol. 42, pp. 178-89.

Singh, G. (1977), Stratigraphical and Palynological Evidence for Desertification the Great Indian Desert, *Annals of Arid Zone,* Vol.16, No.3, pp. 310-320.

Singh, N. and Sontakke, N. A., (2002), *Climatic Change*, 52, 287- 313.

Singh, P. and Jain, S.K., (2002), *Snow and Glacier Melt in the Satluj River.*

Singh, R. L., (1971), *India: A Regional Geography*, National Geographical Society of India, Varanasi, p. 992.

Singh, R. S.; Narain, P.; and Sharma, K.D. (2001), *Vayu Mandal*, 31.

Singh, S. (1977), Sand Dunes and Palaeoclimate in Jodhpur District, Western Rajasthan, *Man and Environment*, Vol, 1, pp. 7-15.

Singh, S. (1977), Geomorphology and Climatic Changes during Quaternary Period in Rajasthan Desert, *Recent Researches in Geology*, Vol. 9, pp. 121-130.

Singhvi, A.K. (1995), Luminiscence Dating: An Appraisal, *IGCP*, 349, I meeting at Jaipur, 8-11 April 1995.

Sood R, K.; and Pathan, S. K. (1982), Palaeoenvironment of Rajasthan—Observation from Landsat Imagery, Abstracts of paper Fourth Indian Geography Congress, Bombay,

Srivastava, H.N. and Balakrishnan (1993), Synthesis of meteorological and tide gauge observations on a decadal scale near the Indian coast. *TOGA Notes*, pp. 15-17.

Stein, (1942), A Survey of Ancient Cities along the 'Lost' Saraswati River, *Geographical Journal,* Vol. 99, pp, 1 73-82.

Thapliyal, V. and Kulshrestha, S. M., (1991), *Mausam*, 42, 333-338.

Theme Paper on Management of Floods and Droughts, Indian Water Resources Society, New Delhi, 22 March 2001, p. 53.

Upadhyay, D. S. (1995), *Cold Climate Hydrometeorology*, New Age International (P) Ltd, New Delhi, p. 345.

Verstappen, H. T. (1970), Aeolian Geomorphology of the Thar Desert and Palaeo-climates, Zeitschrift fur Geomorphische Abhandlungen, Vol. 60, pp., 341-353,

Verstappen, H. T.; Ghose, B., and Pandey, S. (1969), Landforms and Resources in Central Rajasthan (India) Result of Jalor Pilot Survey, Publi, Inst, *Aerial Survey and Earth Sc.*, Vol. 51, pp. 1-20,

Wadhawan, S. K. (1995), Quaternary Event Stratigraphy of Rajasthan Desert and its Palaeoenvironmental implications, *Abstract IGCP,* 349, I Meeting at Jaipur from 8-11 April 1995.

Wadhawan, S. K. (1995), Quaternary Event Stratigraphy of Rajasthan Desert and its Palaeoenvironmental Implications, *Abstract IGCP* 349, I meeting at Jaipur from 8-11 April, 1995.

Wilhelmy, H. (1969), Das Vrstromtal am Ostrand der Indusebene und dar Sarasvati-problem, Zeitochrift fur Geomorphologie, Supplement band Vol. 8, pp. 76-83,

World Meteorological Organization, Report, Geneva, Switzerland, 2002, p. 36.

Yash Pal; Baldev Sahai; Sood, R. K. and Agrawal, D.P. (1980), *Remote Sensing of the 'Lost'Sarasvati,* Space Application Centre (ISRO), Ahmedabad 53.

4

Changes in the Glacial Environment around the South-East—facing Slopes of Kanchenjunga in the Sikkim Himalaya

An Assessment of the Impact of Global Warming on the Glaciers

Guru Prasad Chattopadhyay

Introduction

The Sikkim Himalayan region, particularly the areas around Kanchenjunga summit complex, has drawn attention of many environmental scientists over the past centuries. Bose (*cf.* Mayewski and Jeschke, 1979), on his visit to this part in 1889, described the conditions of the glaciers nourished from the Kanchenjunga. Later Garwood (1902) managed to prepare a draft map of the position of Rathong, Onglaktang and other glaciers during his visit to this part of the Sikkim Himalaya in 1902. However, their reports were more of general documentations of the glacial environment of this part rather than pure geographical accounts. In the extensive tracts of high altitudes in this part, particularly above the timber-line, the variety of typical landforms bear with them certain geomorphological and geological mystery which is yet to be explored systematically. The wide variety of glacial, periglacial, nival and paraglacial features of this part reflect the pattern of typical Quaternary history. Such conditions and features have earlier been reported by a number of previous workers, *viz.,* from the Western Himalaya by de Terra (1936), from the Central (Nepal) Himalaya by Fuji and Higuchi (1976) and from the eastern Himalaya by Kar (1969). This chapter is a discussion based on the investigation by the author over the area around the Dzongri Bugiyal and the Rathong and Onglaktang glaciers at 4,200 to 5,200 m (approximately 14,000 ft to 16,500 ft) elevation on the south-east facing slopes of the Kanchenjunga summit complex in the western part of the Sikkim Himalaya during the long period from 1984 to 2010 (1985, 1998a, 1998b, 2008, 2010, 2011a, 2011b and 2012).

The major difficulty for a systematic geomorphological study of this part of Himalaya is the absence of any detailed geological and geographical report. As has been mentioned earlier the literature published so far have only passing mention of the

Quaternary environmental condition of this area which has further been complex and confusing. Another drawback is, for strategic reason, the Survey of India topographical sheets with detailed contour lines remain restricted and inaccessible.

Approach to the Area and its Geological Environment

Approach to the Area: Being situated in West Sikkim the Dzongri Bugiyal area, above the timber line (4,300 m) can be approached on foot from the last vehicle point of Yoksom village at a distance of about 30 km. The Dzongri Bugiyal is basically an alpine meadow of rugged terrain with occasional exposure of cliff walls and rocky ridges rising up to about 4,600 m The HMI base camp is situated at the south-east facing slope of Kanchanjunga summit complex in Sikkim Himalaya immediately below the Rathong Glacier valley which can be approached at a distance of 8 km from the Dzongri La (4,300 m). The area around the HMI Base Camp (4,600 m) is a rugged mountainous tract. It is surrounded by Singalila ridges, Koktang glacier, Rathong glacier, several snowcapped peaks like Kabru Dom (6,600 m) and Forked Peak (6,108 m), and the stream channel of Churang Chhu. The Churang Chhu is the main stream of this area, fed by Rathong glacier. From the Dzongri Bugiyal (Alpine meadow) to the east is the stream valley of Prek Chhu. This stream is fed by the Onglaktang Glacier lying on the eastern slope of the Kabru Dome summit of the Kanchenjunga.

Geological Environment of the Area: Geologically and structurally this part falls within the territory of "Darjeeling Gneiss". Gneissic rocks with occasional outcrops

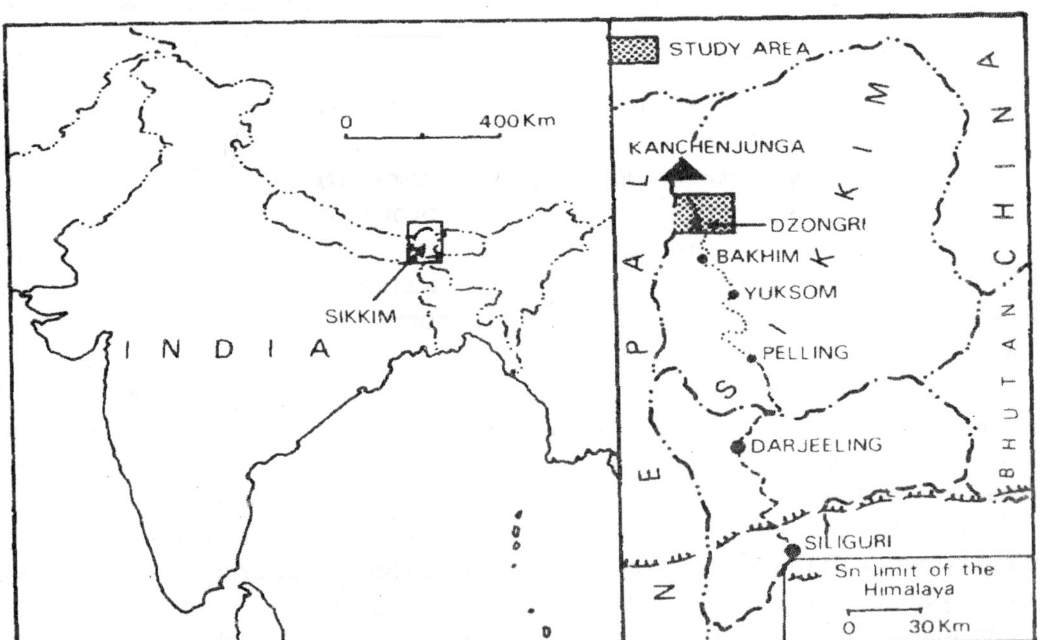

Fig. 4.1: Location of the Study Area in the Sikkim Himalaya

of mica schist and quartz veins predominantly form the mountain ridges. The limits of pre-existing glaciers in the form of lateral moraine and terminal moraine are conspicuous down the valley and from the existing glacier. The extensive frost shattered debris slope, block slope, talus slope, comprising massive gneissic boulders provide the evidence of periglacial condition of the region. The study and field investigation of landforms were concentrated upon the areas in this region above the timberline (4,300 m).

Investigation of Glacial Environment

The Physical Environment: From the timberline above, the first 200 m or so, the ground surface is characteristic of an Alpine meadow. On the bare mountain slopes frost-shattered (periglacial) detritus, comprising mostly massive gneissic boulders, occur extensively (Chattopadhyay, 1986). Numerous moraine ridges of pre-existing glaciers markedly traverse the main river valleys and the ridge-walls ascending up to the snowline. The two main valley glaciers, that exist today, are Onglaktang glacier (the source of Prek Chhu stream) and Rathong glacier (the source of Churang Chhu stream); both of these glaciers are fed by the ice from the main neveé field of the Kanchenjunga summit complex at its south face.

It has already been mentioned that the two main river valleys: Churang Chhu and Prek Chhu, and the ridge walls bounding them, are markedly traversed by numerous ridges of moraine loops of pre-existing glaciers. However, in order to explore the two main existing valley glaciers: Rathong glacier and Onglaktang glacier, of their maximum extension in the past, glacial and glacio-fluvial landforms, in the form of the successive loops of lateral and terminal moraines and other drift limits, were studied in detail. These moraine loops have been considered as the basis for reconstructing the glacial history of this region.

Characteristics of the Moraine Loops and their Interpretation: In both the cases of the Onglaktang and Rathong glaciers three sets of moraine loops through the valley, each comprising lateral and terminal moraines, extending down the valley from the active zone of the glaciers, have been clearly identified. These three sets of moraine loops around the two valley glaciers have some distinguished characteristics are demonstrated in Table 4.1.

From the data assembled in the Table 4.1 above it is obvious that the two glaciers experienced three distinct stages of advance and retreat. It is assumed that the energy of transportation and the subsequent deposition of moraine debris were maximum during Stage I when these glaciers extended downstream for another 6.0 to 6.5 km from their present point of termination. The terminal moraine limits corresponding to Stage II advance occurs at about 3 km from the existing snout position of both the glaciers. During Stages II and III advance the energy markedly diminished as is evidenced from the heights and extents of the terminal moraines and the associated drift limits. The fresh terminal moraine limits corresponding Stage III advances of the

Table 4.1: Characteristics of Moraine Loops and Drift Limits around the Onglaktang and Rathong Glaciers

Moraine loops	*Average distance of the terminal moraine from the snout*	*Altitude at which terminal moraine occurs*	*Maximum height of the drift limit from the valley bottom*	*Morphology and surface characteristics of the terminal moraine*
The farthest loop (Stage I)	6.0 to 6.5 km	4,000m	>500 m	Entirely vegetation-covered and traversed by melt-water channels
The intermediate loop (Stage II)	3.0 km	4,400-4,500 m	300 m	Partly vegetation-covered and traversed by melt-water channels
The nearest loop (Stage III)	1.5 km	4,550-4,600 m	200 m	Totally barren and composed of loose debris material

Source: Compiled by the author on the basis of field data.

two glaciers, corresponding to that of the Little Ice Age, occur at a distance of 1.0km to 1.5 km from the existing snout position.

Rathong Glacier: Rathong glacier, having its source from the neveé field of the Kabru Dome (6,600 m) summit area, descends over the west-facing slope of this summit. The source of this glacier lies at an elevation of 6,200 m (27°36′N and 88°06′E). With a total length and average width of about 3.5 km and 700 m respectively its lower part extends southward through the main trough between Koktang (6,147 m) and Kabru Dome (6,600 m) summits down to about 4,800 m level from where the headwater of the Churang Chhu stream rises.

The fresh moraine ridge, showing the evidence of its recent retreat following the Stage III (the Little Ice Age) advance, occurs at an approximate distance of 1.5 km from the snout. A reconnaissance survey of the present condition of Rathong Glacier was conducted traversing the area on foot from the H.M.I. Base Camp (at 4,500 m) upward over the fresh (Stage III) moronic ridges and the glacier body itself up to the corrie lake of Dudh Pokhri. With close observation and measurements with reference to the base map the existing average length, width and depth of this glacier have been recorded as 3,500 m, 700 m and 50 m respectively. This gives the calculated existing volume of the glacier as (3,500 m×700 m×50 m =) 122,500,000 m^3.

The extent of fresh lateral moraine loop of the Rathong glacier has been used as the denominator for calculating the former maximum volume. This calculation is based on the maximum height of the trim line of the fresh moraines. Here the average height of the fresh lateral moraine, corresponding Stage III, from the glacier surface is about 70 m and the average distance between the highest parts of the lateral moraine ridge on both sides of the glacier is about 1,500 m. On the basis of this observation the reconstructed former depth, width and length of the glacier would be (50 m + 70 m =) 120 m, 1,500 m and 5,200 m respectively. Hence, the calculated former volume of Rathong glacier would be (5,200 m×1,500 m×120 m =) 936,000,000 m^3, suggesting that with the recent retreat (following the Stage III advance) the volume of this glacier has reduced down to about 13.09 per cent (less than one-seventh) of its original size.

Most remarkable morphological features around the Rathong Glacier is the accumulation of fresh hummocky moraines mingled with detached parts of the glacier-ice, which cover extensive part of the recently deglaciated valley area in front of this glacier for a distance of about one kilometre. A number of large and small melt-water lakes also occur through the recently deglaciated valley floor. This makes a certain indication of how rapid is the existing rate of wearing of this glacier. A close field observation of its upper part also revealed that after the complete disappearance of the original neveé field around the Rathong Pass this glacier now owes its supply only from the neveé field on the southerly slope of the Kabru Dome summit. Fresh morphological evidences are preserved to show that due to growing water pressure dam burst of these melt-water lakes has taken place, particularly over the last two decades, causing flash flood down the Churang Chhu valley. Also as the deglaciation continues at a remarkably faster rate in the present day, many such flood or even more devastating ones would occur in this part in the coming years (Chattopadhyay, 2010).

Onglaktang Glacier: Onglaktang glacier, being situated in the north-eastern part of the study area, is also nourished from the same neveé field as the other around the Kabru Dome (6,600 m) summit and extends to the south-southeast direction (to the opposite side of the Rathong), through the valley for a distance of about 4 km with an average width of 800 m. The lower part of this glacier for a distance of about 3km is almost completely covered with debris of supra-glacial origin. The source of this glacier lies at an elevation of 6,000 m (27°36′N and 88°08′E). Average width of this glacier is about one kilometre and its snout part rests at about 4,700 m altitude from where the headwater of the Prek Chhu stream originates. Its two original tributary glaciers, one extending from the Forked Peak (6,108 m) to the west and the other from the Pandim (6,691 m) summit to the east, which used to feed its body, have now almost completely disappeared as a result of their apparent retreat.

The existence of fresh terminal moraine ridge, at a distance of about 3 km downstream from the present snout, clearly suggests that this glacier has assumed a significant recent retreat of about 3 km following the Stage III (the Little Ice Age) advance. In order to make a thorough investigation of the present condition and the stretch of this glacier the adjacent valley area beyond the Sumiti Lake right up to Jemathang for a distance of about 6 km, was traversed on foot and a GPS handset was used for recording altitudinal position of the successive parts of the glacier during the field investigation in 2010. With close observation and measurements in the field using the base map average width and depth of this glacier have been recorded as 800 m and 60 m respectively. Hence, considering the present approximate length (3,000 m), width (800 m) and depth (60 m) the approximate total volume of Onglaktang glacier would be (3,000 m×800 m×60 m =) 144,000 000 m^3.

The extent of fresh lateral moraine loop of the Onglaktang glacier has also been used as the denominator for calculating the former maximum volume. Here the average height of the fresh lateral moraine, corresponding Stage III, from the glacier surface is about 80 m and the average distance between the highest parts of the lateral moraine ridge on both sides of the glacier is about 1,600 m. On the basis of this observation the reconstructed former length, width and depth of the glacier would be 6,000 m, 1,600 m and (60 m+80 m =) 140 m respectively. Hence, the calculated former volume of Rathong glacier would be (6,000 m×1,600 m×140 m =) 1,344,000,000 m^3, suggesting that with the recent retreat (following the Stage III advance) the volume of this glacier has reduced to about 10.71 per cent (less than one-seventh) of its original size.

In common with the morphological features around the Rathong glacier this Onglaktang glacier also exhibit widespread accumulation of fresh hummocky moraines mingled with detached parts of the glacier-ice, which cover extensive part of the recently deglaciated valley area in front of this glacier for a distance of about 2 km. A number of large and small melt-water lakes also occur through the recently deglaciated valley floor indicating rapid rate of wearing of this glacier. This glacier is considered to have shrunk significantly due to the withdrawal of supply of ice by the two tributary glaciers as have been mentioned earlier. In common with Rathong glacier valley there are also physical evidences and records that, on a number of occasions particularly over the last two decades, fragile moraine dams across the melt-water lakes occurring through the recently deglaciated valley floor in front of the glacier collapsed causing destructive flash floods down the Prek Chhu valley.

Plate 4.1: Segments of glacier ice of the rapidly retreating Rathong glacier (Photograph of 1995)

Plate 4.2: The exposed valley-floor after the disappearance of the segments of the Rathong glacier ice (Photograph of 2009)

Plate 4.3: Fresh hummocky moraine and melt water lakes in front of the retreating Rathong Glacier (Photograph 2009)

Plate 4.4: Development of a large melt water lake in front of the rapidly retreating Rathong Glacier (Photograph 2009)

Plate 4.5: Expanding crevasses on the body of the Rathong Glacier (Photograph 2009)

Plate 4.6: The limit of the Stage III moraine across the Onglaktang glacier valley (Photograph 2010)

Plate 4.7: Accumulated fresh hummocky moraine through the recently retreated Onglaktang glacier valley (Photograph 2010)

Plate 4.8: Present form of the rapidly retreating Onglaktang glacier (Photograph 2010)

Mapping of Quaternary Geomorphologic Features and its Interpretation

Appropriate cartographic presentation of Quaternary geomorphologic features in this part of the Sikkim Himalaya is very rare. However, some attempts have been made in this respect in the previous decade (e.g., Chattopadhyay, 1998b). In this work Geomorphologic mapping of the distinguished and ostensible glacial and periglacial features has been done on the base map prepared upon 1:150,000 topo-sheet made available from the University of Bern, Switzerland on personal contact. The snow-covered area and the extension of glaciers were identified initially on the satellite Image (Google Earth Image) and placed accordingly on the map. This was further verified and necessary corrections have been made with ground truth verification during the field studies in 1995 and 2010. All periglacial landform features, however, were placed on the map entirely upon ground-truth verification. The Geomorphologic map, thus prepared, has been presented below.

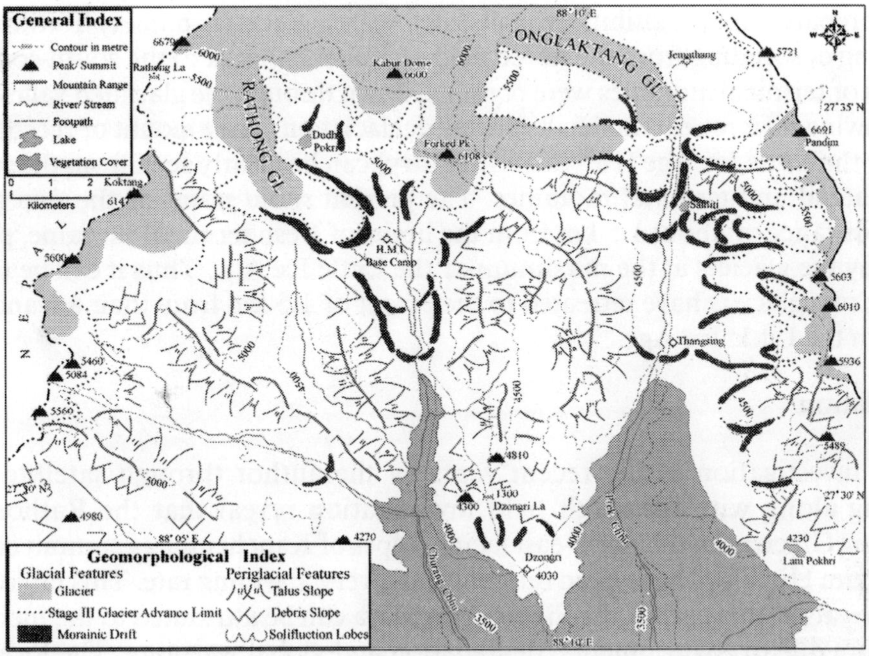

Fig. 4.2: Conditions of the glaciers on the south-east facing slopes of Kanchenjunga till 1986

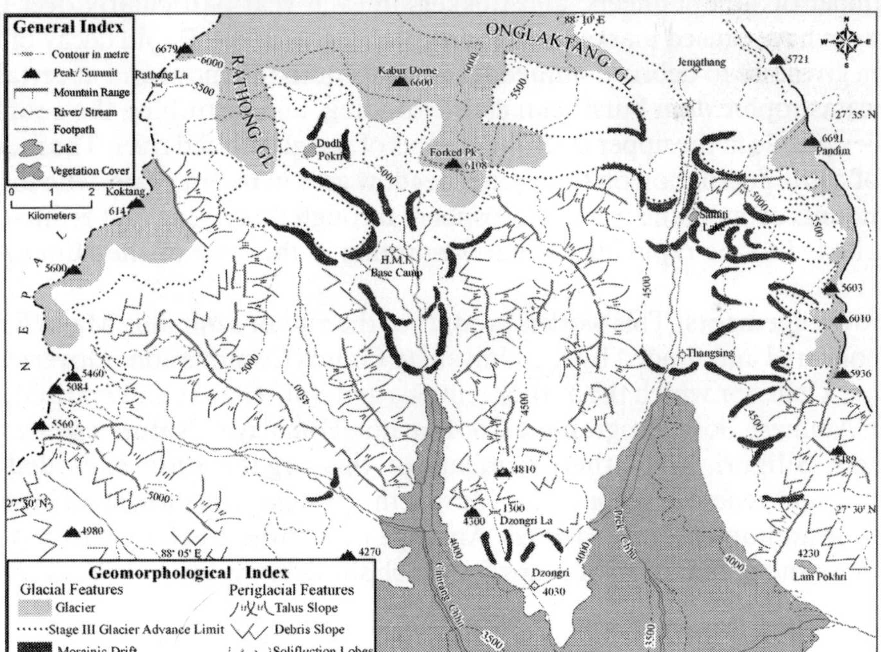

Fig. 4.3: Conditions of the glaciers on the south-east facing slopes of Kanchenjunga as updated on the basis of study conducted in 2009 and 2010

The prepared map exhibits two distinct valley glaciers, namely Rathong and Onglaktang descending from the neveé field of Kanchenjunga to the south-southeast. Three sets of terminal moraines were observed down through the glaciated valley before them, showing their retreats and halts during deglaciation. The amount of recent retreat following the Little Ice Age (16th to 19th century) can be clearly seen from the remains of fresh lateral and terminal moraines. The present snout point of the two glaciers rests at a distance of about 1.5 km from the limits of fresh terminal moraine, possibly dumped by the glaciers at the maximum of the Little Ice Age. Thus it can be referred that these two glaciers have retreated to the extent of 1.5 km from their advance limit attained in the Little Ice Age.

Consolidation

Detailed investigation in the recent years by the author through satellite image processing along with thorough field investigation reveal that the Rathong and Onglaktang Glaciers on the south-east facing slopes of Kanchenjunga summit complex in the Sikkim Himalaya have been retreating at a very alarming rate. This rapid rate of glacier retreats in this part of the Sikkim Himalaya can be attributed to the increase of temperature due to environmental degradation and global warming. The location of the HMI base camp in the immediate vicinity of the Rathong glacier and rapid increase in the number of mountaineers, and trekkers in each year particularly over the last three decades have caused marked environmental degradation. Rapid decay of glacier body often gives rise to debris-dammed lakes adjacent to the snout and at certain time intervals catastrophic dam bursts can cause flooding, and disturbing the equilibrium of the valley profile in the upper catchment areas of the glacier-fed river. The increasing pressure of cattle population on the alpine meadow area in recent years, who discharge higher amounts of methane gas in atmosphere through their dung and remnants, has also contributed to the rapid atmospheric warming in this part of the Himalaya to a significant extent.

Acknowledgements: This work is product of the author's ongoing Major Research Project sponsored and funded by the University Grants Commission (Government of India), New Delhi for which the author remains grateful. Thanks and gratitude is also due to Mr. Animesh Bose, Programme Coordinator, Himalayan Nature and Adventure Foundation, Siliguri, and other members of this organisation extended active participation and generous co-operation during the field investigation at different times for which the author remains indebted. Anasuya Das, Project Fellow, assisted in field investigations and preparation of maps for which she deserves sincere thanks from the author.

NOTES

Chattopadhyay, G.P. (1985), Nature and extent of periglacial phenomena in the Kanchenjunga area, Sikkim Himalaya. *Göttinger Geographische Abhandlungen, (Göttingen, Germany),* Vol. 81: 145-52.

Chattopadhyay, G.P. (1998a), Geomorphological processes and landforms in the glacial and periglacial environments. In *Environmental Issues in Geography*. edited by A.R. Ghosh, (University of Calcutta, Kolkata).

Chattopadhyay, G.P. (1998b), Nomenclature and mapping of glacial and periglacial phenomena in Arctic and Alpine Environments. *Indian Cartographer*, Vol. 18: p. 55-56.

Chattopadhyay, G.P. (2008), Recent Retreats of Glaciers on the Southeast-facing Slopes of the Kanchenjunga Summit Complex in the Sikkim Himalaya. Himalayan Geology, Vol. 29(2): 171-176.

Chattopadhyay, G.P. (2010), Glacial lake outbursts and related environmental hazards: Some observations in the Sikkim Himalaya. *Proceedings of International Conference on Global Climate Change (19-20 Feb 2010) eds. Balachandran, S. et al.* (Council for Scientific and Industrial Research and Visva-Bharati), p. 166-171.

Chattopadhyay, G.P. (2011a), Observations on the Quaternary landform features around the Dzongri-HMI Base Camp area on the Southeast-facing Slopes of Kanchenjunga in Sikkim Himalaya. *Advanced Journal of Geographical World,* Vol. I: p. 1-9.

Chattopadhyay, G.P. (2011b), Quaternary Environmental Changes around the Southeast-facing Slopes of Kanchenjunga: An Assessment with Geomorphological Mapping. In *Landforms Processes and Environmental Management;* edited by Bandyopadhyay, S. *et al.*, ISBN 81-87500-58-1; pp. 63-75.

Chattopadhyay, G.P. (2012), Glacier Retreats in the Himalaya and Dimension of Environmental Hazards. In *Emerging Issues in Geography*—Proceedings of the Eleventh Refresher Course in Geography, edited by Sukla Bhaduri. Academic Staff College and Department of Geography, University of Calcutta, Kolkata. p. 63-68.

De Terra, H. and Hutchinson, G.E. (1936), Data of post-glacial climatic change in northwestern India. *Current Science,* Vol. 1: 5-10.

Fuji, Y. and Higuchi, K. (1976), Ground temperature and its relation to permafrost occurrences in the Khumbu Himal and Hidden Valley, Nepal Himalayas. *Seppyo*, Vol. 38 (Special Issue): 125-28.

Garwood, E.J. (1902), Notes on a map of the glaciers of Kanchenjunga with remarks on some of the physical features of the district. *Geographical Journal,* 20(1): 13-24.

Kar, N.R. (1969), Some observations on periglacial features in the Sikkim and Darjeeling Himalaya. *Biuletyn Peryglacjalny*, (Lødz), Vol. 21(2).

Mayewski, P.A.; Jeschke, P.A. (1979), Himalayan and Trans-Himalayan Glacier fluctuation since A.D. 1812. *Arctic and Alpine Research*, Vol. 11(3):267-87.

5
Preparedness for Climate Change

T.I. Khan

The Present

Many natural environments on all continents and in most of the oceans are already affected by the changing climate. The snow and ice of Earth's cryosphere is melting causing unstable ground and changes in Arctic and Antarctic ecosystems. Rising water temperatures are the likely cause of changes in marine and freshwater ecosystems. Oceans have become more acidic as they take in more carbon dioxide from the atmosphere. On land, the plants, animals, and other living things of ecosystems have been affected by warming temperatures. Warmer temperatures have affected agriculture by changing planting dates and the impacts of fires and pests. Human health has been affected as warming increases dangerous heat waves and causes changes to the amount of pollen and the spread of diseases.

The Future

In the future, according to the report, we can expect that freshwater supplies will increase at high latitudes and in wet tropical areas. However freshwater supplies will become less available in areas where water is already in short supply. Areas that are affected by drought, such as southern and northern Africa, are expected to become more so. Flooding will likely become more common in areas that are already prone to flooding.

Within this century climate change and other global changes such as change in land-use and pollution will collectively be too much for many ecosystems to handle and they will not be able to adapt. Twenty to thirty per cent of plant and animal species will become extinct if global average temperatures increase by 1.5 - 2.5°C, which is within the range estimated by computer models for Twentyfirst Century.

Coastal communities, especially in low-lying regions, will be increasingly vulnerable to flooding as sea level rises, especially where tropical storm events are common. Plus, the frequencies and intensities of extreme weather events are very likely to increase. Small islands are especially vulnerable.

Responding to Climate Change

The number of communities that are taking steps to adapt to current warming and prepare for future warming is limited. The report urges that more action needs to be taken to as climate continues to warm in the future.

Climate and Weather

Weather in any particular location or region can change quickly from hour to hour, day to day, season to season and year to year, even within an unchanging climate. Such changes include shifts in temperature, snow and rainfall, winds, and clouds. They are caused by an interplay of a number of factors, including rapid shifts in air circulation, slower variations in ocean conditions, or seasonal changes in the amount of sunshine. The climate of a locality or region is calculated by averaging these weather conditions over an extended period of time, usually at least 30 years. The climate also describes how weather conditions can vary from these average values. Such variations are described in statistical terms such as standard deviations or frequency of occurrence.

CLIMATE CHANGE CONCEPT

Climate change occurs when the climate of a specific location, region or the entire planet is altered between two different periods of time. This usually occurs when something alters the total amount of the sun's energy absorbed by the earth's atmosphere and surface, or changes the amount of heat energy from the earth's surface and atmosphere that escapes to space over an extended period of time. Such changes can involve both changes in average weather conditions and changes in how much the weather varies about these averages. They can be caused by natural processes like volcanic eruptions, changes in the sun's intensity, or very slow changes in ocean circulation or land surfaces which occur on time scales of decades, centuries or longer. Alternatively, humans can also cause climates to change by releasing green house gases and aerosols into the atmosphere, by changing land surfaces, and by depleting the stratospheric ozone layer. Both natural and human factors that can cause climate change are called 'climate forcings', since they push, or 'force' the climate to shift to new values.

The "Greenhouse Effect" and its Effect on Climate

Earth is heated by sunlight. Although ozone in the stratosphere absorbs much of the harmful ultraviolet part of sunlight, most of the sun's energy passes through the atmosphere relatively unaffected by other gases in the atmosphere. About 31 per cent of the sunlight is reflected back to space by clouds and the earth's surface, but the remainder warms the earth's surface, oceans and atmosphere. However, in order to keep the atmosphere's energy budget in balance, the warmed earth also emits heat energy back to space as infrared radiation. As this energy radiates upward, most of it is absorbed by clouds and molecules of green house gases (including water vapour) in the lower atmosphere. These re-radiate the energy in all directions, some back towards the surface and some upward, where other molecules higher up can absorb the energy again. This process of absorption and re-emission is repeated until, finally, the energy does escape from the atmosphere to space. However, because much of the energy has been recycled downward, surface temperatures become much warmer then if it the green house gases were absent from the atmosphere. This natural process is known as the green house effect. Without green house gases, such as water vapour, carbon dioxide (CO_2), methane (CH_4) and nitrous oxide (N_2O), the Earth's average temperature would be –19°C instead of +14°C, or 33°C colder. Over the past 10,000 years, the amount of these green house gases in our atmosphere has been relatively stable. Then a few centuries ago, their concentrations began to increase due to the increasing demand for energy caused by industrialization and rising populations, and due to changing land-use and human settlement patterns.

Causes of Climate Change

Changes in the intensity of sunlight reaching the earth can cause cycles of warming and cooling that have been a regular feature of the Earth's climatic history. Some of these solar cycles, like the four large glacial-interglacial swings during the past 400,000 years, extend over very long time scales and can have large amplitudes of 5 to 6°C. For the past 10,000 years, the earth has been in the warm interglacial phase of such a cycle. Other solar cycles are much shorter, with the shortest being the 11 year sunspot cycle. However, the magnitudes of changes in climate for these shorter cycles are much smaller than those for the long cycles. Within the past 1000 years, for example, such changes have been within a range of about 1°C. Other natural causes of climate change include variations in ocean currents (which can alter the distribution of heat and precipitation) and large eruptions of volcanoes (which can sporadically increase the concentration of atmospheric particles, blocking out more sunlight). Most scientists are now convinced that human activities are also changing the climate. The main cause of such change is the increasing atmospheric concentration of green house gases. Particularly important is the increase in carbon dioxide, which is released through the burning of fossil fuels (coal, oil and natural gas) and through deforestation and land degradation. An increase in green house gases enhances the natural green house effect and leads to an increase in the Earth's average surface temperature. At the regional scale, emissions of other polluting gases and particles into the atmosphere can also have large effects, although some of these can have opposing impacts. Sooty aerosols, for example, tend to warm regional climates, while sulphate aerosols will cool it by reflecting more sunlight. While their direct effects will be felt primarily within the industrialized regions, these aerosols can also indirectly alter average global temperatures and wind currents. Finally, human induced depletion of ozone in the stratosphere also tends to cool the earth's surface, while land-use change can change the amount of sunlight reflected to space by the earth's surface and, hence, contribute to climate change.

The Impacts of Climate Change

We are already beginning to see the changes in biological and physical systems at the regional scale by the rapid increase in surface temperatures, and hence climate change. Climate change of only 1°C can significantly alter weather behaviour around the world. Climate change is projected to have both beneficial and adverse environmental and socio-economic effects, but the larger the change and rate of change, the more the adverse effects predominate. Adaptation has the potential to reduce adverse effects of climate change and can often produce immediate ancillary benefits, but will not prevent all damages (IPCC Synthesis Report, 2001).

The Kyoto Target

Recently the European Union and other industrialised countries have published new national emission inventories through 2005, which were officially submitted to the UN Climate Secretariat (UNFCCC). MNP has summarised the trend in historical emissions for the period 1990-2005 of the group of countries with an emission target under the Kyoto Protocol (so-called Annexure I countries to the Kyoto Protocol) and extrapolated the emissions trend of the last five years.

The group industrialised countries with a Kyoto target will probably meet their emission limitation requirements of about 4 per cent reduction. This target will even be met without accounting for the so-called CDM projects that aim at emission reductions in developing countries and fast growing economies and which reduction may be accounted for as part of the national Kyoto target (see Figure 4.1). When linearly extrapolating the trend of the last five years to 2010 (the average of the Kyoto target period 2008-2012), the emission reduction in 2010 of this group as countries will be close to 11 per cent. When including all 2800 CDM projects "in the pipeline", which account for approximately 500 megatonne (= million) CO_2 equivalent per year, the total reduction may even be as large as 15 per cent.

The collective reduction of 11 per cent, expected under the assumption listed above, is due to the limited increase in emissions in OECD countries (e.g., a stabilisation in the European Union for the last 10 years) and in particular due to the large reduction of about 40 per cent until 1999 in the Economies In Transition (EIT) (Russia, Ukraine and other Eastern European countries). In the EIT countries, as a result of their economic recovery, green house gas emissions started to increase again (at an average rate of 1 per cent per year), while the Kyoto target of the largest countries is +1 per cent and –8 per cent for the smaller countries. The latter also aids that the present EU-27 meets its collective Kyoto target.

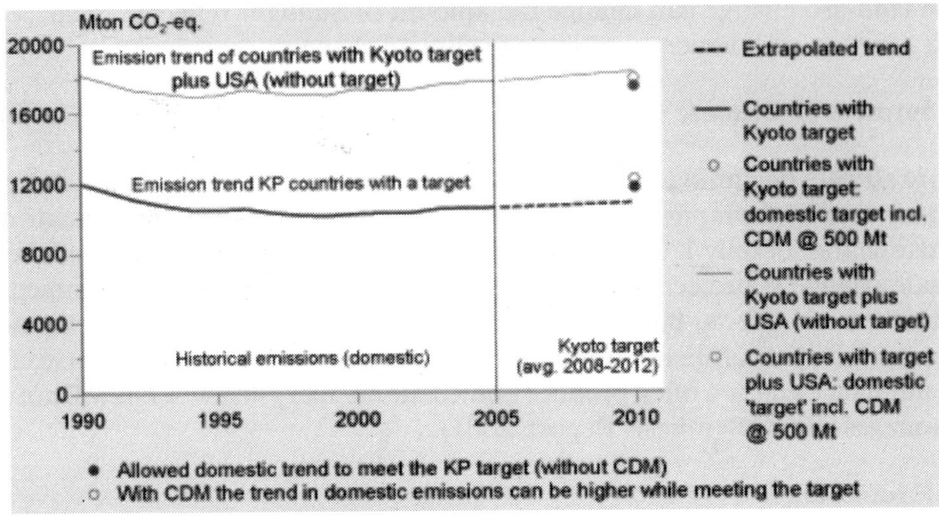

Fig. 5.1: Emission Trend of the Last Five Years

The Kyoto Protocol

International concern that increasing concentrations of green house gases (GHG) are contributing to climate change prompted more than 160 countries to gather in Kyoto, Japan in December 1997 where they agreed to set targets to reduce global GHG emissions. This agreement to set targets and the options available to countries to achieve them is known as the Kyoto Protocol. This legally binding agreement would commit industrialized countries to reduce their collective emissions of six GHG (carbon dioxide, methane, nitrous oxide, hydrofluorocarbons, perfluorocarbons, sulfur hexafluoride) by approximately 5 per cent of the levels emitted in 1990 by the year 2012. Rather than a specific target on each of the gases, the overall emissions targets for all six would be combined to give a single figure in "CO_2 equivalents". The burden of initial emission reduction rests on the industrialized countries, which have built high standards of living based on fossil fuel use, and which have produced most of the anthropogenic (human caused) input of green house gases in the Earth's atmosphere to date. Developing countries can be signatories to the Protocol with the understanding that they will not have any binding emissions targets until the next negotiation period of 2012-2017. Developing countries can participate in earlier emission reductions project, however, through the Protocol's Clean Development Mechanism. The Protocol becomes legally binding when it is ratified by at least 55 countries, emitting at least 55 per cent of the emissions addressed by the Protocol. With the approval of the Protocol by Russia in October 2004, the Kyoto Protocol became a binding international law for the ratifying parties on the 16th of February, 2005. The United States and Australia, however, have stated their unwillingness to ratify the treaty and remain outside the Kyoto framework at present.

The Intergovernmental Panel on Climate Change (IPCC)

The IPCC was formed jointly in 1988 by the United Nations Environment Programme (UNEP) and World Meteorological Organization (WMO). The IPCC brings together the world's top scientists in all relevant fields, synthesizes peer-reviewed scientific literature on global warming studies, and produces authoritative assessments of the current state of knowledge of climate change. The IPCC's Second Assessment (1996) and Third Assessment (2001) serve as the key references for knowledge on climate change by the global community.

The Top-20 CO_2 or GHG Emitting Countries

Below estimates for CO_2 emissions in 2006 from fossil fuel use and cement production have been ranked for the first 20 countries, including the EU-27 (the 27 member states that now form the European Union) as well as for total green house gas (GHG) emissions in 2005. To put these emissions into perspective, also the per capita emissions of these countries have been provided.

The Top-20 comprises 70 per cent of global total CO_2 emissions from fossil fuel use and çement production and 80 per cent of global total green house gas emissions. In 1990 the share of the countries with a Kyoto target was 30 per cent of the global green house gas emissions; in 2005 this share had decreased to about 20 per cent.

The Thermohaline Circulation

Thermohaline circulation is a global ocean circulation pattern that distributes water and heat both vertically, through the water column, and horizontally across the globe. As cold, salty water sinks at high latitudes, it pulls warmer water from lower latitudes to replace it. Water that sinks in the North Atlantic flows down to the southern hemisphere, skirts the Antarctic continent, where it is joined by more sinking water, and then crosses south of the Indian Ocean to enter the Pacific Ocean basin. There, the cold deep water rises to the surface, where heat from the tropical sun warms the water at the ocean's surface and drives evaporation, leaving behind saltier water. This warm, salty water flows northward to join the Gulf Stream, travelling up the Eastern coast of the United States and across the Atlantic Ocean into the North Atlantic region. There, heat is released to the atmosphere, warming parts of Western Europe. Once this warm, salty water reaches the North Atlantic and releases its heat, it again becomes very cold and dense, and sinks to the deep ocean.

The "Abrupt Climate Change"

The term "abrupt climate change" describes changes in climate that occur over the span of years to decades, compared to the human-caused changes in climate that are occurring over the time span of decades to centuries. From ice cores, ocean sediments, tree rings, and other records of Earth's past climate, scientists have found that changes in climate have occurred quickly in the past over the course of a decade. An example of an abrupt climate change event is the Younger Dryas (~12,000 years ago), a period of abrupt cooling that interrupted a general warming trend as Earth emerged from the last Ice Age. During the Younger Dryas period, average summer time temperatures in New England cooled by about 3-4°C. This and other abrupt events have been linked to changes in an ocean circulation pattern known as thermohaline circulation.

Cooling and Global Warming

As we rapidly increase Earth's average temperature, some regions, such as high latitudes, will experience greater warming than others, such as the tropics. As warming alters ocean and atmosphere circulation patterns, some regions could even experience cooling. Much of Western Europe is now warmed by ocean circulation as well as the atmosphere. Heat is transported to the region by a global ocean circulation pattern variably known as thermohaline circulation, the North Atlantic heat pump, or the "Great Ocean Conveyor Belt". This "heat pump" pulls warm salty water northward from the tropics into the North Atlantic, where heat is released, warming air temperatures over Europe. As Earth warms, melting of ice caps and glaciers, increased precipitation and other inflows of fresh water to the North Atlantic Ocean may weaken or shut down thermohaline circulation. This change in ocean circulation could disrupt the transfer of heat northward from the tropics, resulting in cooling in the North Atlantic region. Regional cooling of as much as 8 to 16°C has been seen in the past climate record. However, any regional cooling would be superimposed on the global warming that is already underway. Contrary to the "Day After Tomorrow" dramatization, abrupt climate change will not result in an ice age, because the cooling effects are regional and Earth is currently in an interglacial, or warm, period.

6

Rainfall Variability and Agricultural Productivity in Andhra Pradesh

A Spatio-Temporal Analysis

A.V. Shashikala and *S.Padmaja*

Introduction

The ever existing inter-relation between rainfall amount and agricultural productivity is an undisputable reality, the significance of which is ever increasing in the context of global climatic changes. In the same line the regional climatic variations in terms of variability of rainfall and increase in maximum and minimum temperatures, would have a telling impact on agricultural productivity. Even to day with rapid technological strides, good agricultural production is associated with good monsoon. There have been enumerous studies in this direction. However, in the case of Andhra Pradesh, like any other state is an agrarian state where majority of 75.7 million population are dependent on agriculture. It will be intriguing to analyze the correlation between agricultural productivity and rainfall variability at the end of twentieth century (1998-2000) and in the initial decade of twentyfirst century (2008-2010), which is witnessing not only climatic aberrations but is also coupled with various agrarian policies of the State government, like free power supply, setting of SEZ's NREGA, free seeds pesticides, easy loans WDP (watershed development programmes) and minimum prices for the crops.

In Andhra Pradesh, as two-thirds area comes under rain fed agriculture, the impact of rainfall on agriculture production is significant. Stability and sustainability of crop production can be achieved only if we can make efficient use of limited water, and this can be achieved only if dry land farming systems combine crops, pastures and fallow periods. Stability means the reliability or consistency of farm production. Sustainability is a multi-dimensional concept which deals with the maintenance of a given level of productivity over time. This chapter deals with rainfall variability in Andhra Pradesh and its impact on crop production in the State.

There are several factors that influence agricultural production, these factors could be natural or cultural. One of the most important natural factor that influences crop production is availability of adequate quantities of moisture for proper growth and sufficient yield of crops. In a State like Andhra Pradesh where climatic conditions vary to a large extent, calibrating the quantities of moisture existing at different areas becomes quite essential. It is also observed that the State contains a variety of geo-environmental conditions. There are arid and wet areas, plateau, *ghats* and coastal plains which respond differently in terms of agricultural production.

The primary cause of droughts and floods over a region is the variability in the rainfall received. The study of variability of rainfall during the monsoon and non-monsoon season and during the year is a useful indicator for studying implication of rainfall incidence on agriculture of a region, more so in case of dry land or rainfed agriculture in hot; subhumid, semiarid and arid parts of tropics and subtropics. The climate of AP does not differ from the general pattern of the climate of India. It can be classified as (i) Tropical rainy, and (ii) Hot steppe. The general characteristics of the first type are that the mean daily temperature for the coldest month is above 18°C with annual rainfall less than 250 cm, most of it occurring during the south-west monsoon. In the hot steppe region the mean annual temperature is above 18°C, summer is hot and dry followed by monsoon rain. In the case of Andhra Pradesh however considerable amount of rainfall occurs outside this stipulated period in the south coastal region. The annual rainfall varies from less than 60 cm in Western Rayalaseema to more than 140 cm in the Northern and North-Eastern part of the State. Except Telangana the State receives substantial amount of rainfall during the North-East monsoon period also. This is especially so in a Southern Coastal Belt which receives 50-75 cm from October to December. In Visakhapatnam district and its neighbourhood the unbroken range of Eastern Ghats lying to the West about 40 km inland and raising steeply exerts significant influence on rainfall and its distribution as they act as a barriers to the rain bearing Eastern winds blowing in association with depressions from Bay of Bengal during the South-West monsoon period. The rainfall here increases inland towards West from 100 cm near the coast to over 140 cm along the portions of Northern Boundary, while over the rest of the State it decreases inland. With this climatic backdrop, an attempt is made in this chapter to study the impact of climatic variation on agricultural production in the State of Andhra Pradesh between two time periods of 1998-2000 and 2008-2010, i.e., over a decade.

Objectives

1. To study the rainfall variability across Andhra Pradesh between 1998-2008 and 2008-2010.
2. To study the agriculture productivity using Jasbir Singh's Methodology in the same period.

3. To analyse the rainfall variability and agricultural productivity and correlate them accordingly.

Data Base and Methodology

In order to understand the spatial and temporal variations in rainfall and agricultural productivity, district-wise statistics of crop yields, area under various crops and rainfall data were collected for two sets of years (1998-2000 and 2008-10) from the Directorate of Economics and Statistics. The average of these sets of years were used to avoid the temporal fluctuations in the cropping pattern and rainfall. The level of agriculture productivity was measured and compared with the physiographic zones of the State.

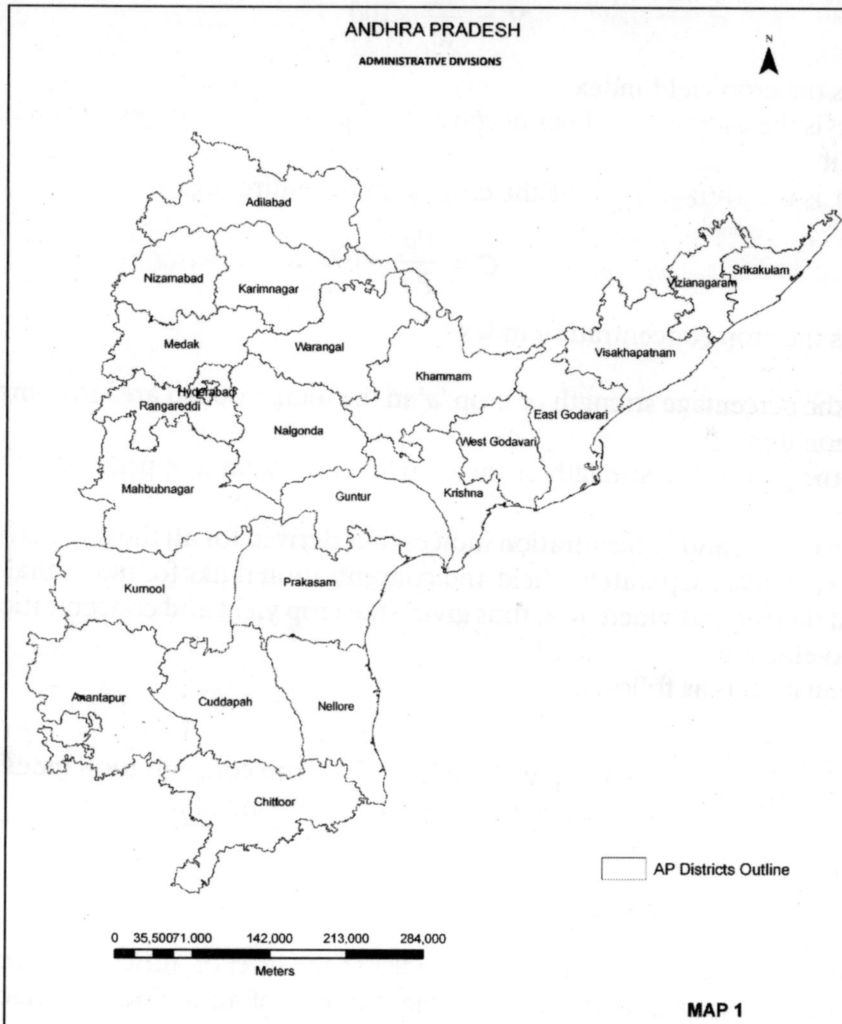

Map: 6.1: Administrative Divisions of Andhra Pradesh

A Number of geographers and agriculture scientists working on different aspects of agriculture have analyzed agriculture productivity in different ways. However, the work of Kendall (1939), Shafi (1974), Bhatia (1967), Singh Jasbir (1976) are noteworthy in connection with the development of various measurements in the study of agriculture productivity.

In the present study Jasbir Singh's method was used to measure the level of agriculture productivity. In this method, the relative crop yield and concentration indices are averaged in a ranking order and computed into an average ranking coefficient. This would give a measure, which may be called the crop yield and concentration indices ranking co-efficient for which the procedure is as follows:

$$Y_i = \frac{ya_c}{ya_r} \times 100$$

where Y_i is the crop yield index

Ya_c is the average yield per hectare of crop '*a*' in the component enumeration unit

Ya_r is the average yield of the crop '*a*' in the entire region

$$C_i = \frac{Pa_c}{Pa_r} \times 100$$

where C_i is the crop concentration index

Pa_c is the percentage strength of crop '*a*' in the total cropped area in component of enumeration unit.

Pa_r is the percentage strength of the crop '*a*' in the total cropped area in the entire region.

The crop yield and concentration indices thus derived for all the regional units and the crops are ranked separately. Yield and concentration ranks for individual crops are added and thereafter divided by 2, thus giving the crop yield and concentration indices ranking co-efficient.

The equation is as follows:

$$\text{Crop yield and Concentration indices ranking co-efficient for crop } a = \frac{\text{Crop yield index ranking of crop } a + \text{Crop concentration index ranking of a crop } a}{2}$$

The result thus derived will give us an idea of the level of agriculture productivity. The lower the ranking co-efficient, the higher the level of agricultural productivity and *vice versa*.

Agricultural productivity is the most important indicator to measure the level of agricultural development. The productivity index values have been arranged in the ascending order and mapped.

DISCUSSIONS AND ANALYSIS

Rainfall Variability in A.P.

At no point of time and over space we get uniform and same amount of rainfall due to the impact of physiographic factors associated with atmospheric components. Temperature and rainfall are integratedly interwoven and at any deviation in one parameter would automatically lead to a change in the other. In this section an attempt is made to study the rainfall distribution in two time periods namely 1998-2000 and 2008-2010 respectively (Figures 6.2 and 6.3)

When we study and analyse rainfall data for all the districts of Andhra Pradesh during 1998-2000 and 2008-2010, we get a clear indication that droughts and floods are fairly common in the State. While the districts of Rayalaseema experiences an equal number of droughts and floods, the districts of Telangana are prone to droughts and coastal Andhra is prone to heavy floods. The heavily drought prone districts of Andhra Pradesh are Mahaboobnagar, Nalgonda and Khammam of Telangana region and Kurnool, Anantapur and Cuddapah of Rayalaseema region.

A comparison between the two Figures 6.2 and 6.3 clearly highlights the fact that the districts receiving lower rainfall increased towards the western part of the State. Northernpart of the State became drier, whereas the southern part received lesser amount of rainfall as compared to the earlier times.

Agricultural Productivity

An examination of Figure 6.5 for the year 2008-2010 showing agricultural productivity indicates very high and high productivity in Srikakulam, Vizianagaram, Guntur, Krishna, West Godavari, East Godavari, Prakasham, and Nellore districts of coastal Andhra and Nizamabad, Khammam of Telengana region. The high productivity in Krishna Godavari delta region is due to high soil fertility of alluvial soils, intensive cultivation practices, use of hybrid seed varieties and assured water supply through canals. In this tract, Rice, Sugarcane and Pulses are predominant crops giving high yields. In the other districts cited above soil fertility and intensive cultivation practices are contributing factors. The pattern of agricultural productivity during 1998-2000 was different (Fig. 6.4). Very high and high productivity was noticed in almost all the districts of coastal Andhra and Rayalaseema region and also Nizamabad, Karimnagar, Warangal and Khammam districts of Telengana region.

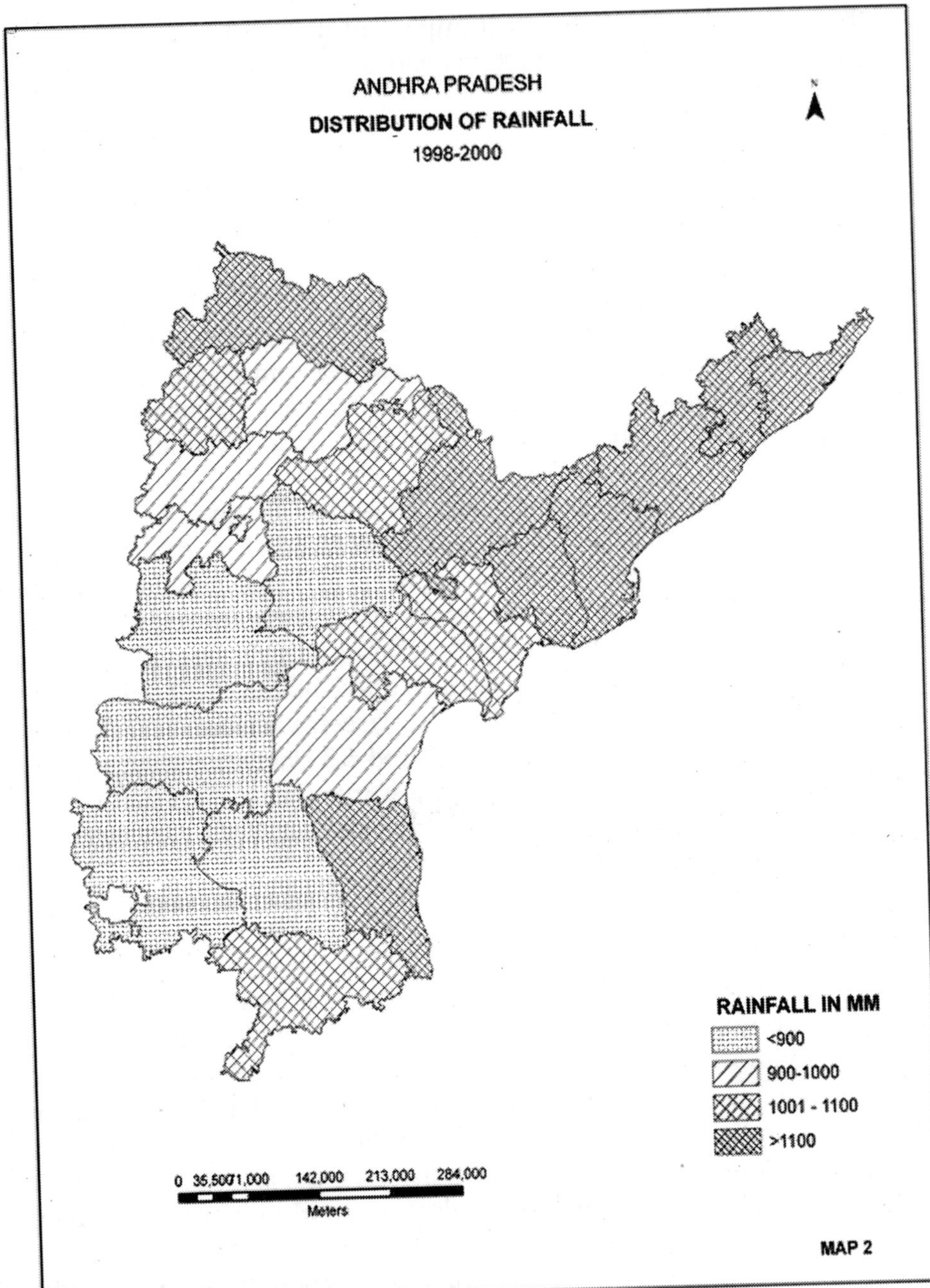

Map 6.2: Distribution of Rainfall during 1998-2000 in Andhra Pradesh

Moderate productivity is noticed during the period 2008-2010 in Visakhapatnam, Nalgonda, Mahbubnagar, Kadappa, and Chittoor disticts. The uneven topography and large forest area are the contributing factors for moderate productivity in Adilabad district, even though it gets a good amount of rainfall. During 1998-2000 moderate productivity is found in Nalgonda, Rangareddy and Adilabad districts.

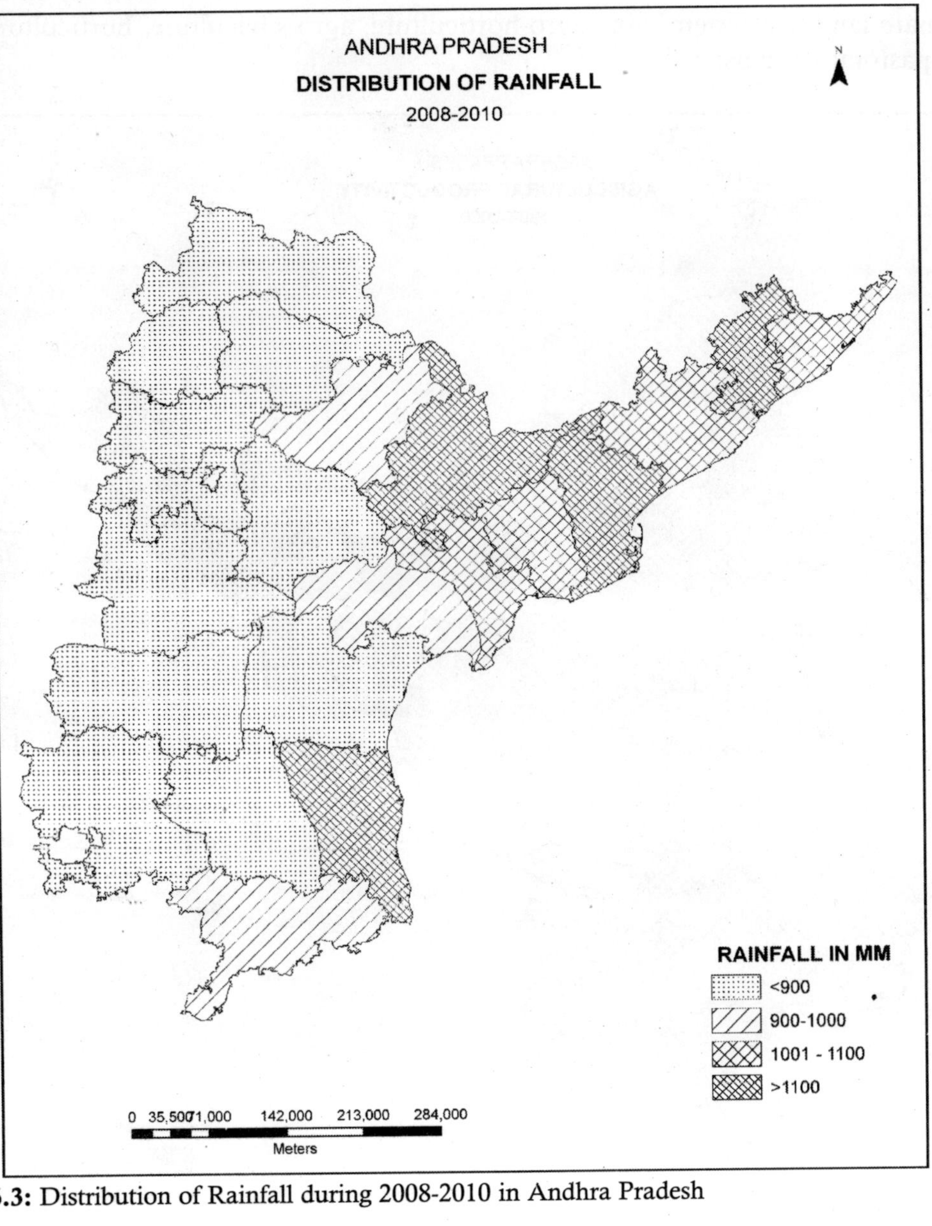

Map 6.3: Distribution of Rainfall during 2008-2010 in Andhra Pradesh

Low productivity is found in Ananthapur and Hyderabad districts of the State during 2008-2010 whereas during 1998-2000 (Fig.6.4) Mahbubnagar and Srikakulam showed low agricultural productivity. The productivity is low in Mahbubnagar district due to scanty rainfall and poor irrigation facilities, since this district is located in dry zone. Owing to these unfavourable conditions the yields of crops are low resulting in low agricultural productivity. As productivity is low in Ananthapur and Mahbubnagar These are most uneconomical and there may be an immediate need for shifting to alternate land use system, viz., agro-horticulture, agro-silviculture, horticultural and silvi-pastoral systems.

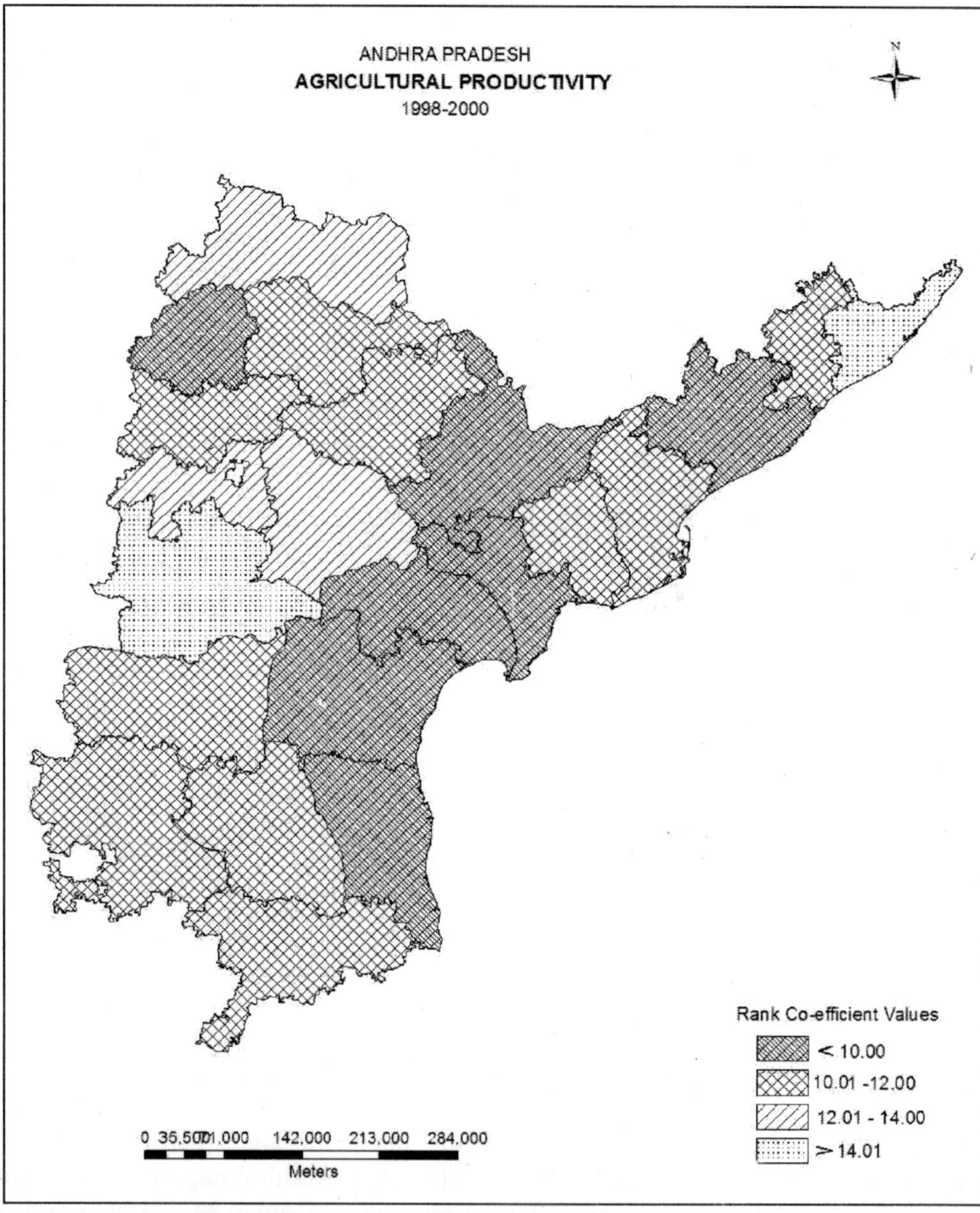

Map 6.4: Agricultural Productivity during 1998 to 2000 in Andhra Pradesh

It has been observed from the study that the spatial variation in agriculture productivity in the State is influenced by not only inherent soil fertility available, but also irrigation facility along with cultivation practices and cropped area. Although the yields are higher in case of certain crops in certain districts, the agricultural productivity is not very high due to low percentage of cropped area

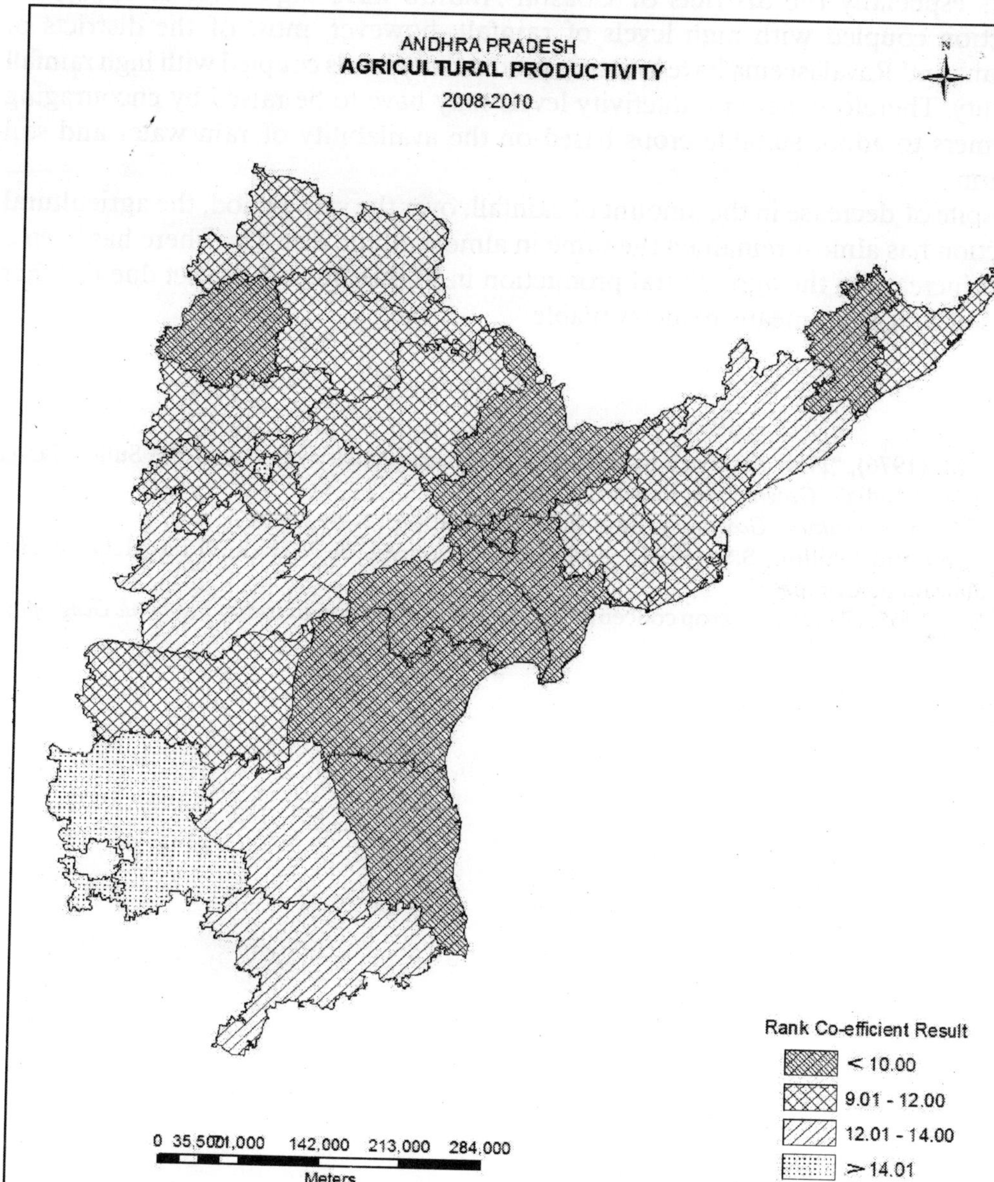

Map 6.5: Agriculural Productivity during 2008 to 2010 in Andhra Pradesh

Conclusion

The study clearly indicates that rainfall in Andhra Pradesh is highly variable. Further the agricultural systems within the State are affected by variability in rainfall because agriculture in the State is largely dependent on rainfall. A third of the total number of districts especially the districts of Coastal Andhra have high level of foodgrain production coupled with high levels of rainfall; however, most of the districts of Telangana and Rayalaseema have low crop productivity levels coupled with high rainfall variability. Therefore, here productivity levels may have to be raised by encouraging the farmers to adopt suitable crops based on the availability of rain water and soil condition.

In spite of decrease in the amount of rainfall, over the said period, the agricultural production has almost remained the same in almost all the districts. There has been a marked increase in the agricultural production in Mahbubnagar district due to other sources of irrigation means made available

NOTES

Hussain, M. (1976), "New approach to the agricultural productivity regions of the Sutlej-Ganga plains of India", *Geographical Review of India.*

Hussan, Majid, *Agricultural Geography*. Inter-India Publications. New Delhi.

Singh, Jasbir and Dhillon, S.S. (1994), *Agricultural Geography*, New Delhi, Tata McGraw Hill Publications Company.

Bhatia, S.S. (1965), Patterns of crop concentration and diversification in India, *Economic Geography*.

7
Chemistry of Stratospheric Ozone Depletion and its Possible Link with Climate Change

K.S. Gupta and *Yogpal Dhayal*

Introduction

Since last several decades, stratospheric ozone as a tracer for atmospheric circulation has been the subject of the study by physicists. British Antarctic Survey is engaged in routinely measuring the stratospheric ozone over Antarctic stations at Halley Bay (76° S 27°W) and at Argentine Islands (65°S 64°W). John Farman and his team in 1984 made a startling statement that value of total ozone had decreased dramatically for about six weeks in September-October (1980-84). The ozone layer exists at altitudes between about 10-40 km from Earth surface that is between 100 to 0.1 *mb* pressure altitude, just above the tropopause (Bridgman, 1990; Subbaraya and Lal, 1999). The natural seasonal and latitudinal variation in stratospheric ozone is now well established. The ozone distribution in stratosphere is strongly dependent upon stratospheric circulation patterns, which vary according to seasons, short term meteorological changes and photochemical processes of formation and destruction (Ramanathan and Kulkarni, 1960; Cunnold, 1995; McIntyre, 1980; Mahlman *et al.*, 1980).

The Importance of Stratospheric Ozone

The presence of ozone in stratosphere is responsible for absorbing UV-B radiations ($\lambda = 280–320$ *nm*). These radiations are absorbed only by O_3 and no other gas. It may be pointed that the UV-A ($\lambda = 320–400$ *nm*) radiations are not absorbed by O_3. On the other hand UV-C ($\lambda = 200–280$ *nm*) absorbed by other gases as well as O_3.

In fact ozone controls temperature profile of stratosphere. Any loss or depletion of ozone in stratosphere would lead to greater amount of UV-B radiations reaching the earth. This would create, among other problems an increase in melanoma (skin cancer) in humans as these damage surface cells of animals and plants. In such a scenario, the possibility of interference with the normal life cycles of animals and plants would become a reality. The UV radiations may, thus, interrupt lower level of food chains also.

These possibilities led to a keen public interest in the matter and brought the stratospheric chemistry and physics in the forefront.

Formation of Ozone in Stratosphere

In stratosphere, O_3 is produced by photodissociation of O_2 due to absorption of radiation of $\lambda < 242$ nm (1) followed by a thermolecular reaction of O-atoms with O_2 (2).

$$O_2 + hv\ (\lambda < 242\ \text{nm}) \rightarrow 2O \qquad ...(1)$$
$$O + O_2 + M \rightarrow O_3 + M \qquad ...(2)$$

O_3 concentration is limited by following two further reactions

$$O_3 + hv\ (\lambda \leq 1180 \rightarrow O + O_2 \qquad ...(3)$$
$$O + O_3 \rightarrow 2O_2 \qquad ...(4)$$

Reactions (1 to 4) constitute Chapman Mechanism (Chapman, 1930) for the creation/destruction of O_3. Reactions (2) and (3) are much faster than (1) and (4) and rapidly interconvert O and O_3 (Pilling and Seakins, 1995).

In atmospheric chemistry, a term odd-oxygen, which includes both O and O_3, is widely used.

$$[\text{Odd oxygen}] = [O + O_3] \qquad ...(5)$$

A simple relationship between the concentrations of ozone $[O_3]$ and oxygen $[O_2]$, may be derived as follows (Pilling and Seakins, 1995). Under steady–state regime,

$$d\ [O]/dt = 2J_1\ [O_2]–k_2[O]\ [O_2][M] + J_3\ [O_3]–k_4\ [O]\ [O_3] \qquad ...(6)$$

and

$$d\ [O_3]/dt = k_2[O]\ [O_2]\ [M]–J_3[O_3]–k_4[O]\ [O_3] \qquad ...(7)$$

On adding equations (6-7),

$$d[O]/dt + d\ [O_3]\ /dt = d\ [O + O_3]\ /\ dt = 2J_1\ [O_2]–2k_4\ [O]\ [O_3] \qquad ...(8)$$

Since under steady state regime

$$d\ [O + O_3]\ /\ dt = 0 \qquad ...(9)$$

from equation (8)

$$2J_1[O_2] = 2k_4[O][O_3] \quad ...(10)$$

The relative concentrations of O and O_3 may be found as follows. At steady state,

$$d[O_3] / dt = 0 \quad ...(11)$$

So from equation (7)

$$k_2[O][O_2][M]-J_3[O_3]-k_4[O][O_3] = 0 \quad ...(12)$$

As reaction (4) is much slower than reactions (2-3), ignoring term $k_4[O][O_3]$ in equation (12) we get

$$k_2[O][O_2][M] = J_3[O_3] \quad ...(13)$$

J_3 is a complex function of the rate constant of step (3), wavelength dependent solar light intensity and extinction coefficients.

From equation (13)

$$[O] / [O_3] = J_3 / k_2[O_2][M] \quad ...(14)$$

And $$[O] = [O_3]J_3 / k_2[O_2][M] \quad ...(15)$$

Substituting the value of [O] in equation (10)

$$2J_1[O_2] = 2k_4J_3[O_3]^2/k_2[O_2][M] \quad ...(16)$$

And

$$[O_3] = \{[J_1[O_2] / k_4]^{1/2}\} \{k_2[O_2][M]/J_3\}^{1/2}$$
$$= [O_2]\{J_1k_2M/k_4J_3\}^{1/2} \quad ...(17)$$

In the lower stratosphere short wavelength radiations for O_2 photolysis are highly diminished having been absorbed higher up, and in upper stratosphere, there is less O_2 because of decrease in O_2 as the altitude increases. Thus, there is layer type structure of O_3 in stratosphere, maximum ozone being around 30 km in the tropics and around 15-20 km over the polar region. Since the level of peak is considerably higher in altitude over the tropics than poles, the ozone is distributed pole-ward through the stratospheric circulation. Ozone accumulates over poles particularly in winter; some of it enters the troposphere over the poles (Bridgman, 1990).

Stratospheric Ozone Depletion

The problem of ozone depletion is an area of major concern in stratospheric chemistry and physics. As would be shown later the meteorology plays a major role. In the

Chapman mechanism, there is a serious imbalance in the rates of production and loss of oxygen (Dutsch, 1974). Step (18) accounts for only 20 per cent of O_3 loss. This necessitated the search of other pathways for O_3 loss to restore the equilibrium.

$$O + O_3 + M \rightarrow 2O_2 + M \quad ...(18)$$

A catalytic cycle can destroy O_3 as follows.

$$X + O_3 \rightarrow XO + O_2 \quad ...(19)$$

$$XO + O \rightarrow X + O_2 \quad ...(20)$$

$$\text{Net reaction: } O_3 + O \rightarrow 2O_2 \quad ...(21)$$

In equations (19-20), X is a radical which operates the catalytic reaction cycle, destroys ozone and is recovered at the end of reaction cycle. X can belong to the HO_x, NO_x or ClO_x families. Thus, X, which is a radical, can destroy thousands of O_3 molecules by chain reaction. The trace gases of importance in stratospheric ozone depletion are hydrogen species (OH, CH_4) nitrogen species (NO, N_2O, NO_2) and chlorine and bromine compounds. Recent studies have indicated that with decreasing use of CFCs and halons, bromine compounds, CH_3Br in particular, may become a major cause of ozone depletion in stratosphere (Lal *et al.*, 1994).

Role of NO, NO_2, N_2O

Crutzen (1970) was the first to suggest the ozone destruction through the involvement of NO_x. The latter were produced by the reaction of N_2O with O (^1D). N_2O, which has a long life time, is released in the troposphere and passes into the stratosphere. In the NO_x cycle, ozone is destroyed by NO which is formed through reactions (22) and (23). These reactions are the most important source of NO in the stratosphere.

$$N_2O + O(^1D) \rightarrow NO + NO \quad ...(22)$$

$$N_2O + hv \rightarrow NO + N \quad (\lambda < 400 \text{ nm}) \quad ...(23)$$

The catalytic cycle for ozone destruction is as follows:

$$NO + O_3 \rightarrow NO_2 + O_2 \quad ...(24)$$

$$NO_2 + O \rightarrow NO + O_2 \quad ...(25)$$

NO_x is the most important destroyer of ozone in 25-45 km altitude region.

Role of Chlorine Compounds

The destruction of ozone through involvement of chlorine was suggested by Stolarski and Cicerone (1974). In 1974, Molina and Rawland first suggested that the anthropogenic emissions of chorofluorocarbons. CFCs, might be depleting stratospheric ozone, through the removal of odd oxygen by chlorine atom. These compounds are man-made with a very long lifetime leading to increase in their abundance in atmosphere. Slowly these are transported in the stratosphere. Apart from CFC, methyl chloride is released naturally by oceans and biomass burning. Several other chlorine compounds are also present in stratosphere. The importance of bromine compounds in causing ozone depletion has also been established. The major sources of these compounds are oceans, biomass burning and industrial and other applications.

Chlorine atoms, released from CFCs/CCl_4/other active chlorine compounds during photodissociation, react with ozone to form ClO and O_2 (Molina and Molina, 1987; Molina *et al.*, 1987).

$$CF_2Cl_2 + h\nu \rightarrow Cl + CF_2Cl \qquad \text{...(26)}$$
$$Cl + O_3 \rightarrow ClO + O_2 \qquad \text{...(27)}$$
$$ClO + O \rightarrow Cl + O_2 \qquad \text{...(28)}$$

ClO can also react as follows:

$$ClO + NO \rightarrow Cl + NO_2 \qquad \text{...(29)}$$
$$ClO + h\nu \rightarrow Cl + O \qquad \text{...(30)}$$

ClO reacts about six times faster than any NO_x species. By far Cl is the most important destroyer of O_3. In the mechanism (26-28), destruction of an ozone molecule involves the consumption of an O atom, whose concentration is not high. To overcome this difficulty, Molina *et al.*, (1987) suggested the following modified mechanism.

$$ClO + O_3 \rightarrow ClO + O_2 \qquad \text{...(31)}$$
$$2ClO + M \rightarrow Cl_2O_2 + M \qquad \text{...(32)}$$
$$Cl_2O_2 + h\nu \rightarrow Cl + ClOO \qquad \text{...(33)}$$
$$ClOO + M \rightarrow Cl + O_2 + M \qquad \text{...(34)}$$

By a chain mechanism a single Cl atom thus destroys several thousands of O_3 molecules. The reactions of Cl atoms cease only when it is removed and placed as a reservoir species by reaction with CH_4 or NO_2.

The ozone depletion mechanism of bromine is similar.

Role of Hydrogen Species

The influence of HO, HO_2 and CH_4 is less important except in upper stratosphere. In the HO_x cycle, X is generally the hydroxyl radical, OH. It can react with O_3 as follows:

$$O_3 + OH \rightarrow O_2 + HO_2 \qquad \text{...(35)}$$
$$HO_2 + O \rightarrow O_2 + OH \qquad \text{...(36)}$$

At higher altitudes, photodissociation of water vapour, equation (33), is responsible for OH production in the atmosphere. However, at stratosphere altitudes, reaction of H_2O with O (^1D) produces OH.

$$H_2O + hv\ (\lambda < 240) \rightarrow OH + H \qquad ...(37)$$

The photodissociation of O_3 produces O (^1D). In addition, OH radicals are also produced in the stratosphere by the reaction of O (^1D) with methane.

$$O_3 + hv \rightarrow O_2 + O\ (^1D)\ (\lambda < 320) \qquad ...(38)$$
$$O\ (^1D) + H_2O \rightarrow OH + OH \qquad ...(39)$$
$$O^1D + CH_4 \rightarrow OH + CH_3 \qquad ...(40)$$

Other members of the HO_x family, which destroy O_3, are H and HO_2.

$$H + O_3 \rightarrow OH + O_2 \qquad ...(41)$$
$$OH + O \rightarrow H + O_2 \qquad ...(42)$$
$$HO_2 + O \rightarrow OH + O_2 \qquad ...(43)$$
$$OH + O_3 \rightarrow HO_2 + O_2 \qquad ...(44)$$

While the H/OH cycle, equation (40), is important in the upper stratosphere and mesosphere, the HO_2/OH cycle, equation (42), is important in the lower stratosphere and troposphere. Paul J. Crutzen, Mario J. Molina and F. Sherwood Rowland shared the 1995 Nobel Prize for their work on ozone chemistry.

Antarctic Ozone Hole

It was found that the development of the ozone hole started with the appearance of the sun over Antarctica after the long winter, and ozone reached a minimum around mid-October. The cause of the ozone depletion was a puzzle to the atmospheric scientists. Several theories sere propounded to explain this phenomenon. Through measurement of total columnar ozone with the help of Dobson spectrophotometer, Farman *et al.* (1985) were able to show that the total ozone concentrations over the Antarctic region had been depleting during September-October, with subsequent recovery during the November-December, and this has been going on since 1979. Although the decrease in O_3 varied from year to year, the results of the six year period firmly established the ozone loss. The evidence for this ozone loss was soon provided by other workers (Komhyer *et al.*, 1988). Media coined the phrase Antarctic *ozone hole* to describe this phenomenon. The results of ozone measurement over Antarctica appeared in a special issue of *Geophysical Research Letters, Vol. 13, Nov. 1986*. Subsequently, the results of a major international effort, the Airborne Antarctic Ozone Experiment (AAOE), on detailed trace gas, aerosol and meteorological measurements over the Antarctic region were published in *Geophysical Research Letters, Vol. 15, Jan. 1988* and *Journal of Geophysical Research, Vol., 94, 1989.*

Komhyr *et al.* (1988, 1989) carried out a detailed analysis of the vertical profiles to establish the extent of ozone loss over South Pole in 1986. It was found that the total ozone loss over Antarctica from mid August to early October was nearly 60 per cent, especially in the altitude region of 11-23 km. In 1987, ozone depletion was severe at about 75 per cent, and in a 4 km thick layer centered around 17 km, ozone was almost completely gone, creating ozone hole. The theory, which met the universal approval, is based on the chlorine chemistry occurring under unique meteorological conditions prevailing in stratosphere over Antarctic region.

The chemical theory requires an overabundance of ClO in the Antarctic stratosphere in the altitude region 12-25 km. AAOE results indicated that the concentrations of the NO_x species decreased toward the hole centre, and ClO concentrations were 100 to 500 times higher than observed outside the hole. In 1987, there was an increase in ClO concentration across a very sharp boundary layer, fluctuating between about 67 and 75°S. The evidence for the involvement of ClO was provided simultaneously monitoring ClO and O_3 during a flight through the hole. Spatial distribution of ClO and ozone showed a marked negative correlation inside the hole. This experiment is cited as the conclusive evidence for enhanced depletion of ozone by chlorine species. Farman *et al.* (1985) have also pointed out that the Antarctic ozone depletion began when the chlorine loading of the atmosphere exceeded a value of 2 ppbv. At the end of Southern Hemisphere winter, a circumpolar vortex of winds is at its strongest, extremely stable and durable at this time. The vortex blocks any incursion of warmer air from mid latitudes and allows an extensive drop in temperature. The temperature in the stratosphere falls to 193 K or less. The cold core is almost completely isolated from the rest of the atmosphere, and acts as a reaction vessel in which the constituents become chemically conditioned during the long polar night (Pilling and Seakins, 1995).

Although the stratosphere has little water vapour, the low temperature of the stratosphere creates thin polar stratospheric clouds (PSCs), which are located at 10-25 km altitude. The surface of these clouds provides a site for heterogeneous chemical reactions on the surface. The chlorine is present as inactive compounds, $ClNO_2$ and HCl, the so called reservoir species. When the air flows through the reservoir species interact with the frozen water. HCl dissolves in the frozen water and $ClONO_2$ remains in the gas form. The reservoir species are formed during winter and in the absence of light by the following reactions.

$$Cl + CH_4 \rightarrow HCl + CH_3 \quad \text{...(45)}$$
$$ClO + NO_2 + M \rightarrow ClONO_2 + M \quad \text{...(46)}$$

The following heterogeneous reactions convert the inactive HCl and $ClNO_2$ into active chlorine and HOCl.

$$ClONO_2\,(g) + HCl\,(s) \rightarrow Cl_2\,(g) + HNO_3\,(g) \quad \text{...(47)}$$
$$ClONO_2\,(g) + H_2O\,(s) \rightarrow HOCl\,(g) + HNO_3\,(s) \quad \text{...(48)}$$

Solid HNO_3 falls out of stratosphere into the troposphere as ice crystals. At the end of Southern Hemisphere winter, as the sun begins to appear the molecules containing chlorine begins to photodissociate in spring sunlight to begin the cycle of ozone destruction, equations (26-34).

The Cl cycle can be repeated much time before Cl atoms and ClO react with other species to generate the so called reservoir species, which hold the destructive chlorine atoms as unreactive species. It is possible to regenerate reactive Cl by reactions (46-47). Around late October, warming and eventual break down of strong spring vortex occurs. This is accompanied by the influx of mid-latitude ozone and disappearance of ozone hole. Thus the hole is caused by climatic and meteorological conditions coupled with the atmospheric chemistry.

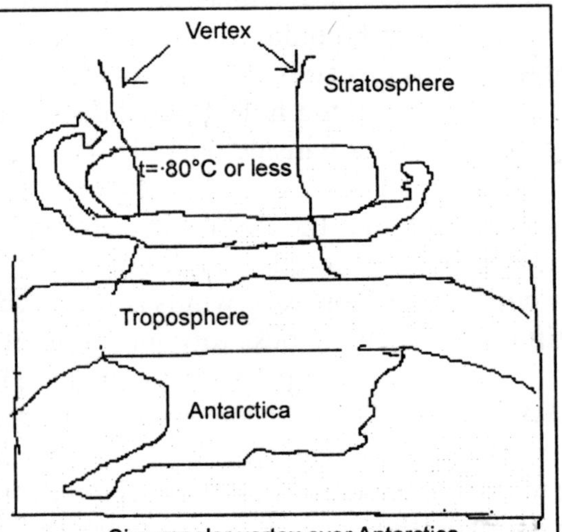

Circumpolar vortex over Antarctica

Stratospheric Ozone Depletion over the Arctic

In recent years significant ozone loss over Arctic region has also been observed but the extent of depletion is not as large as over Antarctica. This is due to the fact that the circumpolar vortex is neither as strong nor as stable as over Antarctica. Many times during winter, sudden stratospheric warming is brought about by air from lower latitudes which break through the vortex, bringing with it the mid-latitude ozone. The Arctic stratosphere is warmer and the prevalence of PSCs is lower than over Antarctica.

Control Strategies

Alarmed by the possible harmful effects of ozone depletion, at a historic meeting of several countries in Montreal, Canada in September 16, 1987, it was decided to phase out the use of CFCs and freeze the consumption of halons at 1986 level. Now the

CFCs are being replaced by HCFCs. In troposphere on attack by OH radicals, HCFCs are converted into H_2O and radical $CFCl_2$. The latter is oxidized and rained out.

$$HCFCl_2 + OH \rightarrow H_2O + CFCl_2 \qquad ...(49)$$

Ozone Depletion and its Link Climate Change

Ozone depletion and climate change has been the subject of intensive investigation owing to their influence on Earth. Both these phenomena modify global earth cycles, so one does not except these to be completely isolated from each other without any possible link. A deeper understanding of these phenomena is required to understand the possible link.

Ozone depleting substances such as chlorofluorocarbons (CFCs), hydrofluorocarbons (HCFCs), hydrofluorocarbons (HFCs) are all *greenhouse gases* like ozone itself. Among the several possible linkages, the most important one is how the ozone and the substances responsible for its depletion alter the radiation processes in the atmosphere resulting in global and stratospheric ozone simultaneously.

The ozone affects the climate primarily by changing temperature. In stratosphere, ozone generates heat both by absorbing solar ultraviolet radiation and by absorbing infrared radiations coming from troposphere. Obviously, more ozone in stratosphere means its higher temperature. For this reason a decrease in ozone concentration in stratosphere results in its cooling and its lower temperature. Mid to upper stratosphere (from 30-50 km above the earth surface) has been observed to cool by 1 to 6°C in recent times: incidentally, during this period the amounts of greenhouse gases in the troposphere have increased.

The strong influence of climate and ozone depletion on each other has been indicated (Shindell *et al.*, 1999). It is likely that the cooling in stratosphere might be rapid enough that even more ozone depletion takes place possibly due to feedback mechanism. Stratospheric low winter temperature over poles plays a big role in ozone depletion. When this temperature drops below –78°C, water, which is in very small quantity, is present as polar stratospheric clouds (PSCs). PSCs are actually ice crystals containing some nitric acid, and H_2SO_4. The chemical reactions on the surface of these PSCs begin with the onset of summer in Southern pole and chlorine radicals, responsible for ozone are released. Ozone loss lowers the stratospheric temperature and greater formation of PSCs, which in turn leads to faster loss of ozone. The ozone depletion in stratosphere would allow more UV radiation to reach troposphere and earth surface because ozone is an absorber of these radiation, particularly of UV-B type, in stratosphere.

The greenhouse effect on global warming is likely to cause changes in circulation pattern in the troposphere and that in turn may alter stratospheric circulation patterns. There is suspicion that these changes are aiding the cooling in the stratosphere over the poles.

As on today, our understanding of the link between ozone depletion and climate change is only cursory and much work and accumulation of data is necessary before a firm relation between the two could be even qualitatively be established.

NOTES

Beig G. and Mitra A.P. (1997). *J. Atmos. Solar Terr. Phys.*, 59, 1245.

Beig G., Walters S., and Brasseur G. (1993), *J. Geophys. Res.*, 98, 12767.

Brasseur G., Hitchman M.H. Walters S., Dymek M., Fatsie E., and Pure M. (1990), *J. Geophys. Res.*, 95, 5639.

Bridgman, H. A., (1990), *Global Air Pollution: Problems for 1990s*, Blhaven Press, London.

Chapman, S.A. (1930), *Mem. Roy. Met. Soc.*, 3, 103-125.

Cunnold, D. M.; Alyea, F.N.; Fhillips, N.; and Prinn R.G., (1975), *J. Atoms. Sci.*, 32, 170.

Dutsch, H.U., (1974), *Can. J. Chem.*, 52, 1491.

Garcia, R. R. and Soloman S. J., (1992), *J. Geophys. Res.*, 97, 12937.

Farman, J.C.; Gardner, B.G. and Shanklin, J. D., (1985), *Nature*, 315, 207.

Ko, M.K.W.; Schmider, H. R.; Shia, R.B.; Weisentein, D.K. and Sze N.D., (1993), *J. Geophys. Res.*, 98, 20429.

Lal, S.; Brouchers, R.; Fabiamn, P.; Patra, P.K. and Subbaraya, B. H., (1994), *Tellus,* 64b, 373.

Mahlman, J. D., Levy. H. and Moxin, W.J. (1980), *J. Atoms. Sci.,* 37, 655.

McLntyre, M.E., (1980), *Phil. Trans. Roy. Soc., London,* A 296, 129.

Molina, L.T. and Molina, M. J., (1987), *J. Phys. Chem.*, 91, 433.

Molina, M. J.; Tso. T.l.; Molina, L.T.; and Wang F.C.Y., (1987), *Science,* 238, 1253.

Molina, M.J.; and Rowland, F.S., (1974), *Nature*, 249, 810.

Ramanathan, K.R.; and Kulkarni, R. N., (1960), *Quart. J. Roy. Met. Soc.,* 86, 144.

Shindell, D.; Rind, R.; Balachandran, N.; Lean, J., and Llonergan, P., (1999), *Science*, 284, 305-308.

Subbaraya, B.H.; Lal, S. (1999), *Space Research in India: Accomplishments and Prospects–Stratospheric Chemistry and Ozone Depletion*, PkL Alumni Association, Physical Research Laboratory, Ahmedabad.

Subbaraya, B. H.; Appu, K. S.; Chatterjee, K. P.; Chizhov, A. F.; Grinchinkov, V. D.; Jayaraman, A.; Khokia, G. A.; Kononlov, V. A.; Moshnikov, I. S.; Sreedharan, C.R. and Zalpuri, K. S., (1987), *Proceedings Indian Academy of Science (Earth and Planetary Science)*, 96, 25-40.

8

The Impact of Coal Mining on Climate Change

Retrospect and Prospect in Indian Context

S.Padmaja, R. Pavanaguru and *Ch. Ravi*

Introduction

Coal plays a pivotal role in sustainable development. It is the most widely used energy source for electricity generation and an essential input to most steel production (Upadhyay, R.P. 2012, Annual Report MOM 2004-2005). As estimated by the World Coal Institute, coal currently fuels 40 per cent of the world electricity and this proportion is set to remain as static over the next 30 years. About 70 per cent of the world's steel production is based on coal. As per Integrated Energy Policy Committee of Planning Commission, coal will remain India's most important energy source till 2031-32 and beyond. About 77 per cent coal output in India is consumed in power generation. Further, industries like cement, fertilizer, chemical, paper and thousands of medium and small-scale industries consume coal for their process and energy requirements.

Global commodity demand is growing and is forecasted to rapidly increase in many developing nations. Demand for coal in India by 2006-07 is 448 MT, whereas the supply will be 405 MT and by 2011-12 the demand of coal rises to 775 MT, and the supply will be only 672 MT. There is a definite shortfall in the supply of the coal and the gap in the coal demand and supply is to be met either by increasing the production through public private participation or through coal imports. This scenario brings about the new challenges in the deregulated regime.

The past statistics reveal that the production of coal at 430.8 million tonnes in 2006-07 increased by 6.1 per cent to 457 million tonnes in 2007-08. Lignite production in 2007-08 at 34 million tonnes increased by 8.6 per cent from that in the previous year. India ranks third in world coal production. It is the eleventh largest energy producer (2.4% of world energy production) and the sixth largest energy consumer (3.7% of world energy consumption) (Sastry, K.R.S. 2012).

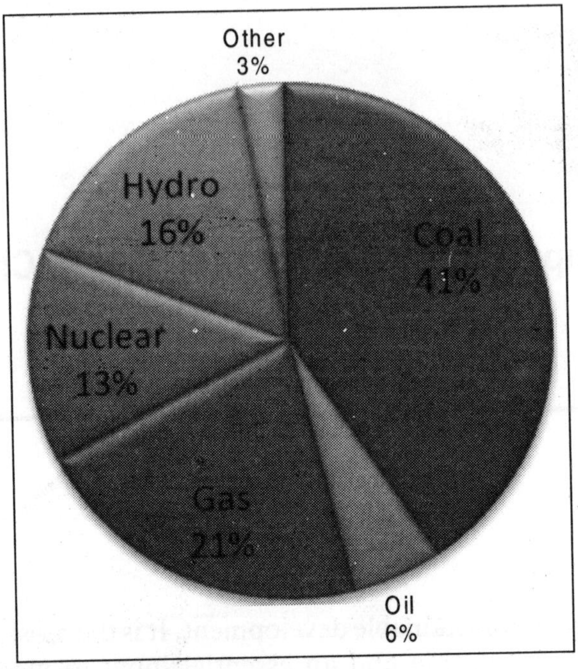

Fig. 8.1: World Electricity Generation

Inventory of Geological Resources of Coal in India

India possesses vast resources of coal spread in Godwana and Tertiary sequences of India. The exploration carried out up to the maximum depth of 1200 m by the GSI, CMPDI, SCCL and MECL, etc., indicated a cumulative total of 285862.21 Million Tonnes of Geological Resources of Coal in the country as on 1.4.2011. The details of State-wise geological resources of coal are given as under:

(A) Gondwana Coalfields

State	*Geological Resources of Coal (in Million Tonnes)*			
	Proved	*Indicated*	*Inferred*	*Total*
Andhra Pradesh	9296.85	9728.37	3029.36	22054.58
Assam	0	2.79	0	2.79
Bihar	0	0	160	160
Chhattisgarh	12878.99	32390.38	4010.88	49280.25
Jharkhand	39760.73	32591.56	6583.69	78935.98
Madhya Pradesh	8871.31	12191.72	2062.70	23125.73
Maharashtra	5489.61	3094.29	1949.51	10533.41
Orissa	24491.71	33986.96	10680.21	69158.88
Sikkim	0	58.25	42.98	101.23
Uttar Pradesh	866.05	195.75	0	1061.80
West Bengal	11752.54	13131.69	5070.69	29954.92
Total	**113407.79**	**137371.76**	**33590.02**	**284369.57**

(B) Tertiary Coalfields

State	*Geological Resources of Coal (in Million Tonnes)*				
	Proved	*Indicated*	*Inferred (Exploration)*	*Inferred (Mapping)*	*Total*
Arunachal Pradesh	31.23	40.11	12.89	6.00	90.23
Assam	464.78	42.72	0.50	2.52	510.52
Meghalaya	89.04	16.51	27.58	443.35	576.48
Nagaland	8.76	0	8.60	298.05	315.41
Total	**593.81**	**99.34**	**49.57**	**749.92**	**1492.64**

Source: Geological Survey of India.

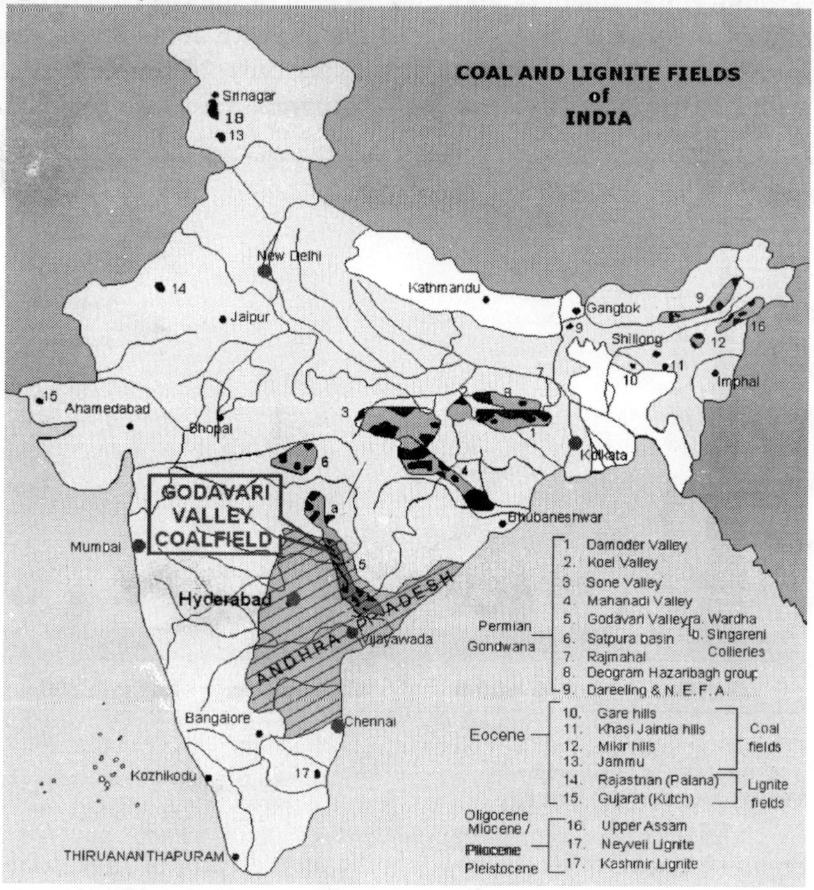

Fig. 8.2: Coal and Lignite Fields of India

The Formation-wise and Category-wise coal resources of India as on 1.4.2010 are given in Table 8.1;

Table 8.1: Formation-wise and Category-wise of Coal Resources of India

Formation	*Proved*	*Indicated*	*Inferred*	*Total (in Million Tonnes)*
Gondwana Coals	113407.79	137371.76	33590.02	284369.57
Tertiary Coals	593.81	99.34	799.49	1492.64
Total	**114001.60**	**137471.10**	**34389.51**	**285862.21**

Category-wise Distribution of Coal Resources in India

In India, out of about 595 operating mines, 385 are at present underground mines. With below ground employment of about 54 per cent of the workforce, Bord & Pillar method continues to dominate in Indian coal mining with about 93 per cent of total underground production. However, it contributes only 20 per cent of total coal production with Low productivity at about 0.87 tonnes/manshift (India 2020).

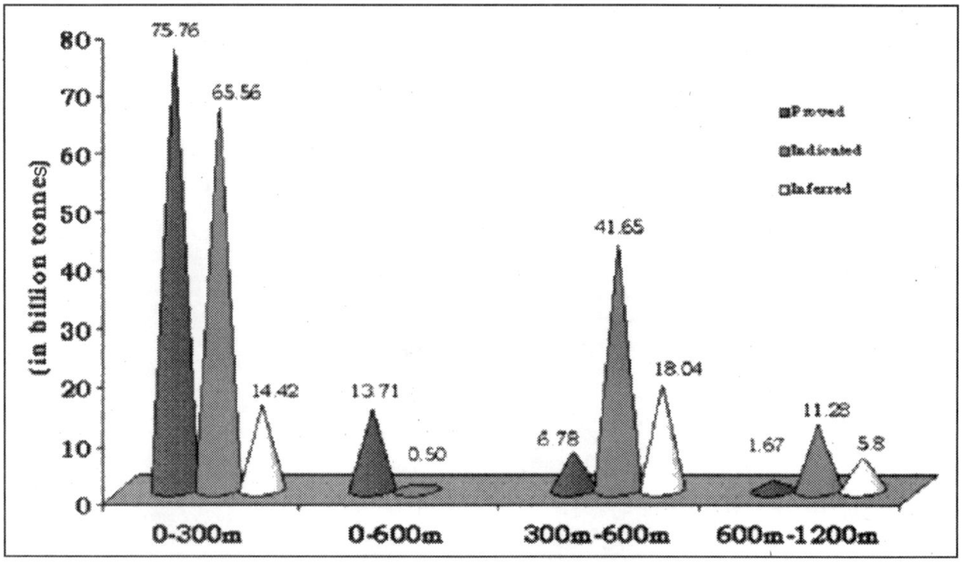

Fig. 8.3:

Coal — A Major Source of Energy

Growth of Energy especially electricity is significant indicator of Industrial growth of a nation. The per capita consumption of electricity in India is 612 kWh compared to 1500 kWh in China. Energy demand of India is on rise. Coal is predominantly used as a cheap source of energy since times immemorial. India has one of the biggest coal reserves in the world. As on 1.04.2011 India has around 286 BT of Coal inventories, out of which 114 BT are in proved category and 160 BT are in indicated category (Annual reports of MOM 2011), 60 per cent of these coal are deposited up to 0-300 metres depth and the rest is more than 300 metres, which is difficult to mine and beneficiate. 87 per cent of the total is non-cooking coal. The inventory of lignite sources stands at 40.90 BT as on 1.04.2011 with 6.14 in proved category.

Coal and Indian Economy

The Indian economic growth is largely dependent on industrial performance which in turn is dependent on energy resources (energy sector). In the Five Year Plan programmes of national planning, the industrial sector received a major boost. A slow, steady and consistent growth is designed and recorded, which today stands at 5.5 per cent (though projected at 8 per cent it is plunged on account of huge power shortage.

It is a well known fact that every industrial activity needs energy inputs. During 1990-2005, the Indian GDP has grown by 148 per cent, commercial energy supply by 118 per cent, thermal electricity generation by 170 per cent and total CO_2 emissions by 100 per cent. In India industrial growth is highly dependent on thermal energy, which is causing alarm with regard to climate change. It is observed that 70 MT of CO_2 per day is added to the atmosphere by coal utilization. Indian CO_2 emissions were 1229 MT in 2005. Presently coal contributes more than 72 per cent of all India CO_2 emissions, overwhelmingly from power and steel plants.

The world population is estimated to grow at 8.7 billion by 2050, which demand 1.7 per cent increase in energy per year to meet the requirement. Accordingly this would enhance emission of GHG due to over dependence on fossil fuels.

The total commercial energy consumption in India is estimated to increase by 7.5 times over 30 years from a level of 285 MTE (million tonnes of oil equivalent) in 2001 to 2123 MTE by 2031. If it follows the energy development path. For a large country like India with its over one billion population (likely to grow to 1.5 billion in 2050) and expected rapid economic growth rate, no single energy resource or technology constitutes a solution to address all issues related to availability of fuel supplies, and environmental impact, climate change and health issues. This becomes a separate gamut of climate change which warrants immediate attention.

Coal and Environment

Coal being 'location specific', there is clearly a need to review this approach. A Group of Ministers is working on resolution of these issues and it is essential that an appropriate balance be struck between the need to protect the environment and the need for energy security.

Underground coal mining has the potential of greatly reducing the disturbance caused to the environment. However, current output levels from underground mining at 60 MT are very low and these mines are mostly old. There has been very little fresh investment in underground mining. It may be necessary to sharply increase the scope and share of underground coal mining and this will involve much greater mechanization and investment by private players.

M/s Coal India Ltd. along with its 8 subsidiaries contributes a lion's share of 81.1 per cent of indigenous production of coal in India and also it is the world's largest producer of Coal with an output of 431.32 MT in the year 2010-11. CIL operates through 81 mining areas spread over 8 provincial States of India. CIL has 471 mines (i.e., 273 underground, 163 open cast and 35 mixed mines). SCCL with its headquarters at Kothagudem in Andhra Pradesh is a State run public sector undertaking involved in the mining operations and production of coal in Godavari Valley Coal Belt. Currently it produces about 51 million tonnes of coal. However, the demand is so high that it is not in a position to supply completely and satisfy the utilization agencies.

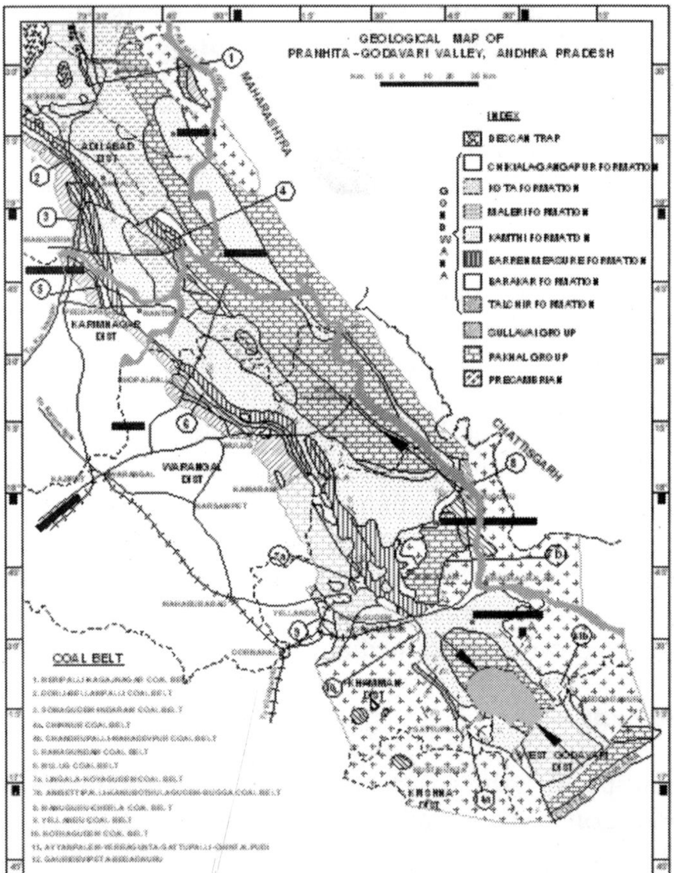

Fig. 8.4: Geological Map of Pranhita-Godavari Valley, Andhra Pradesh

The Godavari valley coal fields are located strategically catering to the energy needs of the southern part of our country (King, W. 1881, Fox, CS. 1934, Krishnan, M.S. 1964, Nargundkar, S.K. 1962). Since coal accounts for more than 40 per cent of energy requirements, the demand for coal is ever increasing logarithmically with enlarged mining activities contemplated for the exploitation of shallow deposits posing a lot of threat to the environment, so it is warranted to develop an Environmental Information System, which facilitate the planners to evolve mitigative measures at right earnest. Coal Supplies are augmented through modern and cutting edge Technologies and Automation. The production targets have enhanced manifolds since last few decades. (5 million to 51 million tonnes in Godavari valley coal fields). It has caused impetus on the mining activity, causing geogenic contamination in the exogenic environment. The increase in OCPs and the associated mechanization have cut down the gestation period of coal production particularly in the recently developed mines.

Coal Characteristics of GVC in Andhra Pradesh

The concise account of the coal characteristics ascertain that the coal in GVC is Non-coking, Sub-bituminous, Low to medium rank, ungraded and banded with high ash ranging from 25 to 40 per cent. Mineral matter in the coal is quite considerable which can be classified into—(i) Inherent and (ii) Extraneous mineral matter (Razvi, S.K. and Ramana Rao, N. 1969). The specific gravity ranges between 1.4 to 1.65. However, consistent dilution is observed in the coal produced from open cast mines. Sulphur content is about 3 per cent (average 2%) and the presence of pyrites is conspicuous. P ranges between 0.001 to 0.01 per cent and N 1.6 to 2.2 per cent. In all the coal fields dull coal predominates—Semi bright and bright follows. Petrographically duroclarain predominates at Yellandu, clarodurain at Kothagudem and clairovitrain at Manuguru (Chandra D. 1965; Ramana Rao, N. 1968; Subbaramaiah, K.V. 1979; Pavanaguru and Malleswara Rao V. 1992)

CFRI Grading of coal from GVCF

GRADE	% ASH
A	0.796
B	2.600
C	21.000
D	27.200 *
E	30.400 *
F	32.600 *
G	35.300 *

*Ungraded

Note: UHV varies from 1200 to 6900 K.Cal

The provenance of the lithounits suggests that the basement has contributed to occurrence of trace element content of the overlying coal bearing horizons. In all probability the basement rocks of Gondwana group consistently exhibits the potentiality of trace metals (Borreswara Rao and Venkateswara Rao 1964, Torrey 1978, Mukherjee *et al.*, 1982, Sahu, K.C. 1982) and their migration due to the anthropogenic activity has been highlighted while dealing their incidence on the environment (Pavanaguru

1990, 1991). Due to the enlarged mining activity to meet the demand deeper seams in close proximity to the basement are also mined which favour their migration on to the surface in acid mine waters as majority of elements are soluble in water with dissolved oxygen. Based on the degree of solubility of different elements in the exogenic environment most of the metals are migrated into the environment and their levels of migration depends upon the content of these elements available in the lithounits and in coal. Further migration is effected by the coal utilizing industries. The trace element content of different lithological units of lower Gondwana horizons in Godavari Valley Coal-fields is given in Table 8.2. The incidences of these elements have produced different types of irregularities on the exogenic environment.

Table 8.1: Trace Element Analyses of the Litho Units in Kothagudem Area (in ppm)

Rock Type	*Co*	*Cr*	*Mn*	*Ni*	*V*	*Zr*	*Cu*	*Pb*	*Ba*	*Sr*	*Ga*
Sandstone	15	150	700	60	110	300	70	15	50	60	20
Sandy Shale (bottom seam)	15	350	600	55	120	250	70	50	60	50	25
Talchir Shale	10	200	800	80	100	170	60	30	100	230	10
Carbo. Shale (Pariting)	15	120	800	80	130	200	80	20	60	80	25
Carbo. Sandstone	15	100	750	70	120	250	60	25	50	70	20
Coal (Semi Bright)	40	100	500	80	100	170	100	25	220	80	10
Caib.sandstone (parting)	15	210	700	50	100	250	75	25	220	30	30
Carb. Shale	20	120	700	100	160	225	120	25	110	350	30
Coal (Bright)	50	120	600	120	110	170	90	35	60	80	10

Impact of Coal and Climate Change

Coal mining, which is of two types. Underground and Open Cast Methods of mining cause environmental degradation of air, water, and land. The effects of scarification are consistently observed in the open cast mines from where immediate increases of production have been noticed. The mining activity does not identify the quality of coal and the associated intercalated zones thus enhancing the dilution of the grade of coal. The amount of fines generated (Gurudeep Singh and Puri S.K. 2004, Pavanaguru *et al.* 2000)) during the course of activity directly depends upon the non coal intercalations. This not only hampers the coal quality and has impinging effects on the air borne particulate matter. The finer the grains the greater the area it covers from the core zone of mining to the buffer zones (Banerjee, S.P. 2004). Thus the SPM ratio in the ambient environment has a telling effect on the overall difference in the areas. Experimental studies conducted at Kothagudem Coal fields with the help of respirable dust samples proved a point in this regard. The mineralogical constitution of the SPM of above 10 microns was subjected the x-ray diffraction studies to understand the mineralogical composition of the air particulates and the SEM studies of the finer particulates of –4 microns substantiated the presence of large amount of quartz which ultimately becomes a causative feature for the incidence of silicosis in this type of coal field.

Table 8.2: Showing SPM in Manuguru Mine Area of Kothagudem Coal Fields (GVCF)

Air Monitoring at Manuguru OC - II						
PRE- MONSOON						
Sample Location (summer)	*I. R.*	*F. R.*	*Difference*	*Time*	*Flow rate*	*SPM U gm*
Mine site office	2.78	2.91	0.13	24 hrs	1.00-.84	2462
Coal haul road	2.78	2.79	0.01	1.5 hrs	1	1136
OB KCSP Drill	2.78	2.81	0.03	11.1 hrs	0.96-1.02	1136
Kondapuram RLSP	2.78	2.81	0.03	7.5 hrs	1.06	1136
Coal Handling plant (building)	2.78	2.86	0.09	1.3 hrs	1.15-0.98	3409
ICC No. 6	2.78	2.83	0.05	6.0 hrs	1.06	1893
Substation West of OC - II Vagu	2.78	2.88	0.1	2.1 hrs	1.15-1.00	3787
View point OC -II	2.78	2.92	0.14	24 hrs	1.08-0.83	5303
West of east substation	2.78	2.97	0.19	25.30 hrs	1.05-0.90	7197
East substation	2.78	2.85	0.07	21.25 hrs	1.10-1.04	2651

Table 8.3: Showing X-Ray Diffraction Mineralogy of SPM

Sl. No.	*Location*	*Mineralogy*
10	E.Sub-station	Quartz, Na-illite
9	West of E. Sub-station	Quartz, Na-illite
1	M.S. Office	Quartz, Kaolinise
3	KCSP	Quartz, Kaolinic, montimorillonite
7	Sub-station West of OC II	Quartz, Na-illite, Kaolinite

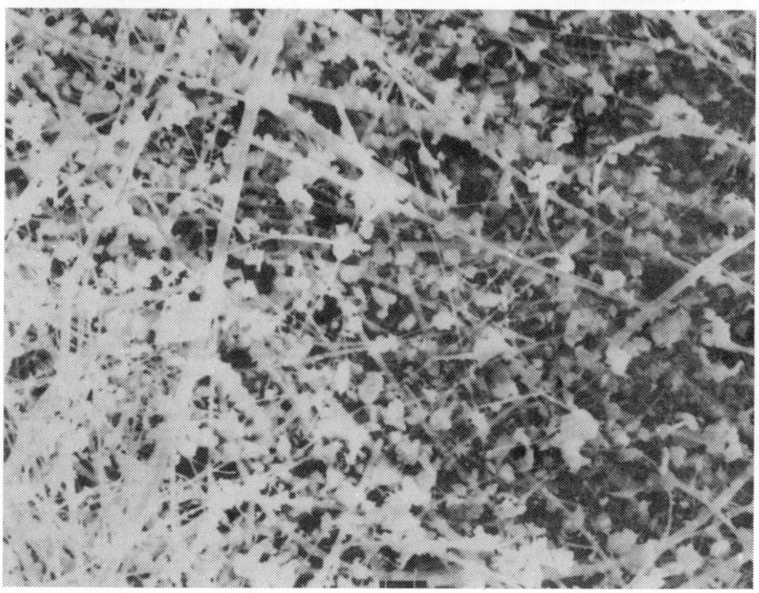

Fig. 8.5: S.E.M. Photograph of S.P.M. (4μ)

Greenhouse Gases (GHGs)

Gases in the atmosphere affect the earth's temperature by trapping radiation from the sun—this is refered as the green house effect. Greenhouse gases (GHGs) include carbon dioxide (CO_2), methane (CH_4), nitrous oxide (N_2O) and water vapour (H_2O), amongst others, and occur naturally. CO_2 makes up almost 80 per cent of anthropogenic GHG emissions. Over the last century, the amount of CO_2 in the atmosphere has risen, in large part driven by fossil fuel use (coal and oil) but also because of other factors such as land-use change and deforestation. (http:/www.indiaenergyportal.org/overview-detail.ptp)

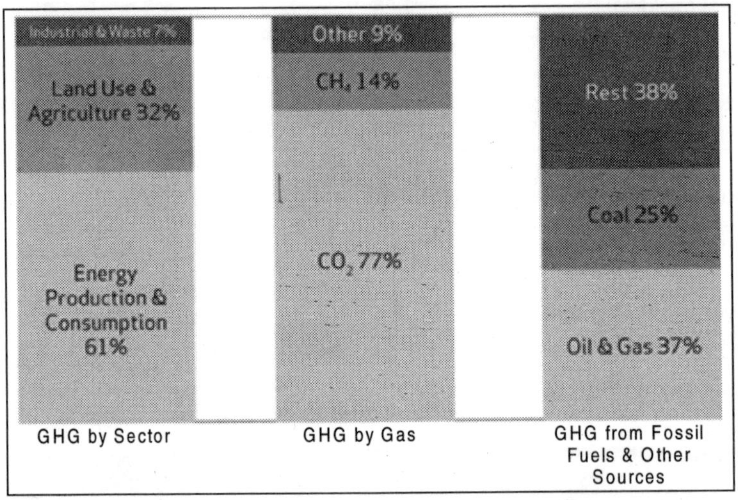

Fig. 8.6: Global Greenhouse Gas Emissions Sources and Activities

India is committed to minimize its GHG emissions, research has found that replacing older coal-fired power stations with larger more efficient plants could reduce global GHG emissions by 5.5 per cent. Emissions from all coal mining related activities—extraction, transport and storage—accounted for 7.4 per cent of total global anthropogenic methane emissions in 2005.

Table 8.4: Projections of percentage contributions from major LPSs to all-India CO_2 emissions (%)

Sector	2000	2010	2020	2030
Power	50.1	52.5	50.5	48.2
Pulverized Coal (sub, super and ultra critical)	49	50.4	45.9	42.8
Integrated Gasification Combined Cycle	—	0.4	1.3	2.2
Gas	1	1.6	3.2	3.1
Oil	0.1	0.1	0.1	0.1
Steel	8.9	9	8.5	7.4
Cement	7.5	8.6	7.8	7.1
Petroleum refining	0.8	1.2	2.4	3.1
Fertilizer	2.2	0.3	0.2	0.2
Paper	0.4	0.5	0.5	0.5
Aluminum	0.1	0.2	0.2	0.3
Petrochemicals	0.2	0.3	0.3	0.5
Hydrogen plants	—	—	0.2	0.3

The global warming potential (GWP) of methane is 23 times greater than that of CO_2. Methane remains in the atmosphere for a period of approximately 12 years after it has been emitted. In comparison, CO_2 is estimated to have an atmospheric lifetime of tens of 50-200 years.

The main green house gas released by Coal Mining—(Methane and CO_2) is Methane and is released in Underground Coal Mining through ventilation and by degassing process. Surface Mining release methane by blasting. Surface mine coal are shallow and low ranks and hence contain less CH_4

Table 8.5: Depth-wise Distribution of CH_4

Depth (m)	Mean CH_4 content (m^3 /t of coal)
100	0.02
500	0.99
1000	3.73
1500	4.89
2000	7.09

Coal contributed 72 per cent of all-India CO_2 emissions, overwhelmingly from power and steel plants. The top three sector contributors were power plants (51.9%), steel plants (8.4%) and cement plants (8%).The IPCC global average method uses emission factors of 10-25 m^3/tonne of coal mined for underground coal mines and 0.3 to 2 m^3/tonne of coal mined for Surface coal mine.

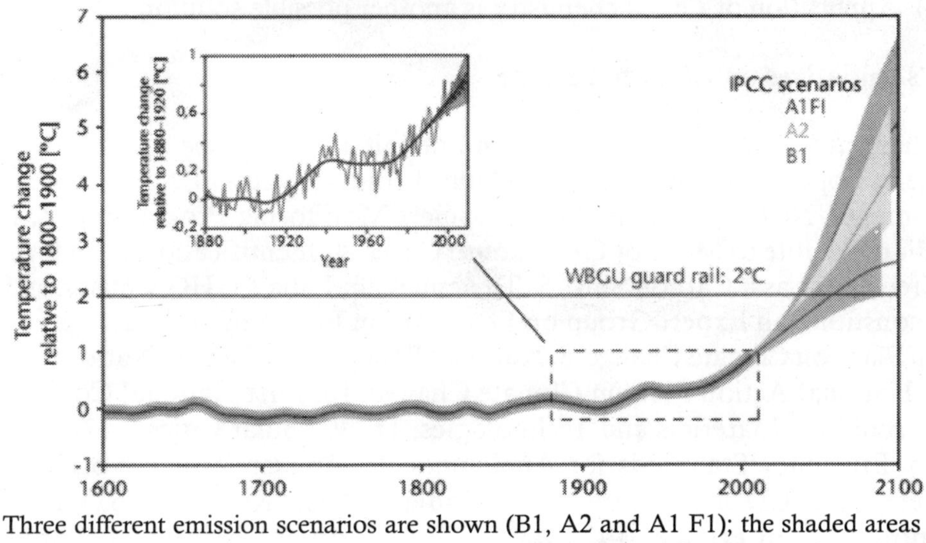

Note: Three different emission scenarios are shown (B1, A2 and A1 F1); the shaded areas are the associated climatological uncertainty ranges. Without successful climate change mitigation, the 2°C guard rail will be exceeded even under the most optimistic emission scenario (B1), Inset: Comparison of observed temperatures with projections of 2001.

Source: WBGU, modifed after Rahmstorf *et al.* 2007

Fig. 8.7: IPCC Scenarios

Mitigation Measures are as follows

*The process of mining in India has witnessed large scale conversion of forest land. The associated anthropogenic activities further escalated the deforestation activities to support industrial estates and SEZs and in SAT regions and enacted several policy initiatives by both C&S Governments. It is generally observed that 70 per cent of mining activities (*Krishnamurthy, K.V. 2004, *Singh A.N. et al. 2002, Gurudeep Singh 2012) have affected the green lands which have created inequilibrium in the ecosystem. The recent verdict of the Supreme Court promulgating the ban of mining in Kudremukh Iron Ore Project in Karnataka State is an ample testimony to this effect. Hence, warrant immediate generalized mitigative measures such as aforestation programmes, groundwater and surface water conservation bodies, protection of Reserve forests (wildlife habitats), disaster management efforts, viz., abatement of floods and landslides, desiltation of lake and river beds, etc.* Based on the above observations it has been proposed to have the following mitigative measures of Natural resource management in the form of:

(i) LPG replacing free coal supply;
(ii) Non–mining activities in forest areas if not possible alternatively;
(iii) Extraction of balance reserves from closed mines;
(iv) Eucalyptus plantations for use as timber;
(v) Technology upgradation to reduce timber consumptions;
(vi) Green Belt development advanced aforestation programmes;
(vii) Fly ash utilization for making bricks, etc.; and
(viii) Application of Green chemistry is another possible solution.

India's Initiatives on Climate Change

India has put forward the following frontline initiatives to cope up with the climate change at Copenhagen meet. They include: 1. Indian Network for Climate Change Assessment (INCCA), 2. Himalayan Glaciers Monitoring Programme, 3. Launch of Indian Satellite to Monitor Greenhouse Gases, 4. Identification of India's Forest and Tree Cover as a Carbon Sink, 5. To demarcate India's GHG Emissions Profile, 6. To constitute an Expert Group on Low Carbon Economy, 7. Drawing up a State Action Plans on Climate Change, 8. National Policy on Biofuels, 9. National Missions under National Action Plan on Climate Change, 10. First National Conference on Green Building–Materials and Technologies, 11. 30 "Solar Cities", 12. To identify Energy Efficiency Standards for Appliances, 13. To standardize Norms for Fuel Efficiency, 14. CDM Programme, 15. India to host 'Rio+20', 16. UN Climate Technology Conference, 17. SAARC Environment Ministers Conference, 18. Periodical submissions on climate change of the country to UNFCCC, 19. State of Forests Report 2009, 20. Launch of CAMPA, 21. Green India Mission, 22. Capacity Building in Forestry Scheme, 23. Intensification of Forest Management, 24. Inclusion of Forestry within NREGA.

Coal mining is an important anthropogenic source of methane emission. (that generate GHG emissions). There is an urgent need to envisage proper mitigative measure to combat this change.

Mitigation measures: For sustainment of fast growing economy and reducing the emission of anthropogenic GHG, a continuous and secured environmental friendly energy supply is needed, therefore, in view of it is proposed to support carbon sequestrian programmes with the application of different methodologies, viz., aforestation programmes, and identification of carbon sink techniques.

Renewable energy holds the key for sustainable development. Therefore a policy needs to be formulated for a transition from hydrocarbons to renewable energy. Nanotechnology can combat climate change. It has potential to removc the finest contaminants in the environment. Further nano materials are very much helpful in energy production by creating clean, less expensive sources enabled by novel nano materials and processes improved solar cells.

SCCL from Godavari Valley Coal fields, with only 7 per cent of nations coal reserves SCCL contributes to 9 per cent of nations energy needs, is implementing several measures for control of GHG emissions. (Mohan Rao. S. 2011)

Conclusion

- Economic growth, energy supply, thermal electricity generation and national CO_2 emissions have profound linkages on climatic changes in India.
- The increased rate of exploitation and production of Coal, to meet the energy demand in India is causing mechanisation and automation, which is associated with complex issues concerning environment in the country.
- Coal contributed 72 per cent of all-India CO_2 emissions, overwhelmingly from power and steel plants. The top three sector contributors were power plants (51.9%), steel plants (8.4%) and cement plants (8%).
- The Gondwana coals exhibit heterogeneity in lithology and coal mining contribute considerable SPM. Nature and mineralogy of SPM in air by SEM and X–ray studies in GVC has indicated the nature and composition of SPM in ambient air.
- The inherent and extraneous mineral matter contributes cumulatively to ash content.
- The coal and coal bearing horizons are repository of trace elements in the Godavari Valley Coal Belt and their incidence need to be further investigated.
- The ultimate objective of the UNFCCC is the: "*stabilization of green house gas concentrations in the atmosphere at a level that would prevent dangerous anthropogenic interference with the climate system. Such a level should be achieved within a timeframe sufficient to allow ecosystems to adapt naturally to climate change, to ensure that food production is not threatened and to enable economic development to proceed in a sustainable manner*".

- In view of the enlarged mining activity and for sustainable development it is essential to develop an Environmental information system which facilitate proper planning and development of the area.
- India's frontline initiatives to cope up with the climate change need to be implemented with letter and spirit so as to achieve the above objective.

NOTES

Annual Report, Ministry of Mines 2004-2005, 2011.

Banerjee, S.P., (2004), Mineral Availability and Environmental Challenges—A Vision 2020 Statement. Prof. S.K. Bose Memorial Lecture, ISM, Dhanbad.

Borreswar Rao and Venkateswar Rao (1964), Minor Elements in Indian Coals, *Jour. Ind. Acad. Geoscience*, Vol. 4.

Chandra, D. (1965), Application of coal microscopy. *Jour. Min. Met. Fuels*, Vol. 12, pp.335-343.

Chandra, D. (1991), Trace element content of Indian coals and their impact on the environment. International Conference on Environmental Impact of Coal Utilization, Bombay.

Fox, C.S. (1934), The Lower Gondwana Coal Fields of India. *Mem. Geol. Sur. Ind.*, Vol. 69.

Gurdeep Singh, and Puri, S.K. (2004), Air quality assessment in Korba Coalfield, *Indian Journal of Air Pollution Control*, IV (2) 31-41.

Gurudeep Singh (2012), Environmental issues with best management practice of Mining in India. National Conference on "India 2020: Challenges, Opportunities and Growth Strategies in Coal & Lignite sector in India. Hyd.

King. W. (1881), The geology of Pranhita Godavari Valley, *Mem. GSI.*, Vol. 18, Pt. III

Krishnan, M.S. (1964), *Geology of India and Burma*. Higginbothoms Ltd. Madras.

Krishnamurthy, K.V. (2004), Environmental impacts of coal mining in India, Proceedings of the National Seminar on Environmental Engineering with special emphasis on Mining Environment (NSEEME), Dhanbad.

Mohan Rao. S., "Mitigation measures in Singareni Collieries Company Limited", in Climate Change Perspectives and Projections: A Systems Approach edited by K. Janardhan Reddy, Sashidhar Rao Beedu, R. Pavanaguru, Macmillan Publishers, pp. 314-323.

Mukherjee *et al.* (1982), Trace element studies in coal fields of India. *Bull. GSI*, No. 49.

Nargundkar, S.K. (1962), Coal in Andhra Pradesh. Proc. Symp. On Mineral Industries in Andhra Pradesh, Feb. 1962.

Pavanaguru, R. and Malleswar Rao, V. (1982), Petrographic Nature of Coal from Godavari Valley and its significance in coal preparation. National Seminar on Appropriate Technology for Indian Coal Mining by 2000, Kothagudem. A.P.

Pavanaguru R. (1990), Geochemical significance of some major and minor elements and their incidence on the environmental conditions of Godavari Valley coal fields–A Case Study, *Ind. Jour. Envt. Protection,* Vol. 10, No. 3.

Pavanaguru, R. (1991), Geochemical significance of some trace elements in coal bearing horizons of Kothagudem area, Godavari Valley Coal Belt. A.P.—A Case Study. *International conference on Environmental Impact of Coal Utilization*. IIT Bombay.

Pavanaguru, R.; Narasing Rao A.; J. Vijaya Kumar and M.N. Reddy (2000), Geoscientific considerations and its impact on the environment—Kothagudem area, A Case Study. *Journal of Environment and Pollution* 7(3), pp. 189–195.

Ramana Rao, N. (1968), A Petrographic study of coals from Kothagudem coal field, Godavari Valley, A.P. India. Proc. Res. Sem. Coal—A Petrographic approach—Nagpur.

Razvi, S.K. & Ramana Rao, N. (1969), Nature of inorganic mineral matter in coal from Kothgudem coal fields. *Jour. Ind. Geo. Sci. Assn.*, No. 10.

Sahu, K.C. (1982), Environmental impact of coal utilization. *Jour. GSI*, Vol. 30, No. 5

Sastry, K.R.S. (2012), Background paper on the National Conference on "India 2020: Challenges, Opportunities and Growth Strategies in Coal & Lignite sector in India. Hyderabad.

Singh, J.J. & Deepak Adarsh (1979), Environmental and climatic impact of coal utilization. Proc. of Symposium, Academy Press, p. 655.

Singh, A.N.; Raghubanshi, A.S.; and Singh, J.S. (2002), Plantations as a tool for mine spoil restoration, *Current Science*, 82:12.

Subbaramaiah, K.V. (1979), Petrographic and chemical characteristics of banded ingredients of coal from Kothagudem coal field, A.P. India. Unpublished Ph.D. thesis. O.U.

Torey (1978), *Trace Contaminants from Coal in Pollution Technology Review*. Noyer Data Corporation U.S.A.

Upadhyay, R.P. (2012), Coal a major source of Energy. National Conference "India 2020: on Coal and Lignite sector, 2012, Hyderabad

Twelfth Five Year Plan approach paper on the environment constraints in Coal and Lignite sector. National Conference on Coal and Lignite sector, 2012, Hyderabad.

http:/www.indiaenergyportal.org/overview–detail.ptp

9
Indian Desert
Problem of Conservation and Management of Biodiversity

R.P. Misra

India is on the move. After several centuries it is in the news for good reasons. It is the second fastest growing economy of the world. Its snail's pace GDP growth sarcastically referred to as Hindu rate of growth, is now referred to as Indian miracle. As in 2009, it is a trillion dollar economy; by 2050 it would be a 27 trillion dollar economy. The only country ahead would be China.

But behind the glitter of economic growth emanating from the concrete jungles of our mega cities is a lengthening trail of natural and cultural decay. The development model we have adopted speeds up national economic growth but at a huge human and ecological cost. Growing poverty in the midst of plenty and increasing destruction of the very nature that is the source of all that is alive is the order of the day. Unless we change our thinking, style of living, and development models wherein human values respect for nature Indian culture stood for, we are destined to jeopardize all life, human included, on planet earth.

CHEC India has been engaged in discussions, dialogues and research on relationships between number and distributions of human beings are causally linked with the biotic and abiotic environment, and how the environment influences human beings and how the human being influence the environment. The domestication of fire was the first major step to change vegetation structure and influence populations of food plants and animals. Vegetation burning in the tropical regions especially is the continuation of our ancestors began in the early stages of their cultural development. But it was not devastating. The emergence of human beings with sophisticated tools and techniques triggered off the second phase of destruction of nature. The extinction of large vertebrates was the gift of this era. The third major phase of human modification of the natural environment began with the development of agriculture. Today no less than 40 per cent of terrestrial primary production is in the hands of people, and the percentage is rising. If we look at the tropical areas, the percentage is much higher to at times touch 70 per cent. CHEC India has very rightly taken up the theme of biodiversity of Indian desert for it is here that the preservation of biodiversity is an uphill task.

Ecological Setting of Indian Desert

Located in western Rajasthan covering around 70 per cent of the land of Jaisalmer, Barmer, Bikaner and Jodhpur districts it is a part of the Thar Desert spread over 446,000 sq km on both sides of Indo-Pak border covering the southern part of Haryana, Punjab and northern part of Gujarat State of India and Sind province of Pakistan. Bounded by River Sindh in the west, River Sutlej in the north-west, the Aravali Mountains in the east, and the salty marshes of the Rann of Kutch in the south, it extends over 208,110 sq km.

Climatically, it is hot and dry; rainfall is scanty. Its physical build is, however, not so uniform. There are sand dunes, plains, hills, salty marshes, and a few oases here and there. Luni is the only river that meanders through the desert and reaches the Arabian Sea through the Rann of Kutch; it is said to be the remnant of River Saraswati along which Vedas, the first written books of the world were composed. Apparently, what is scrubland today was full of lush vegetation once upon a time. 300 millions years ago roamed in this part of India the dinosaurs and their ascendants.

Earth scientists are not yet sure of the origin of the Thar Desert. Some consider it 4000 to a million years old while others think that aridity started here much earlier. Some western Indologists think that the area turned arid just around 2000 to 1500 B.C. because it was during this period that Ghaggar which was once Saraswati started drying up. Many scholars do not share this view. They postulate that the Sutlej flowed into Ghaggar as did Jamuna and due to tectonic movements Sutlej course moved westward and that of the Yamuna eastwards. An archaelogical study of Kalibangan, a dead city in the desert by Robert Raikes clearly indicates that the town was abandoned because of the shifting of the river course. It was abandoned around 2000-1900 B.C. on account of the drying up of the Saraswati now known as Ghaggar says B.B. Lal, former Director General of ASI. This conclusion has been backed by other studies by archaeologists and hydrologists.

Thar Desert in Indian History and Culture

As stated earlier, Thar region has influenced Indian culture and history from earliest times. River Saraswati is one of the main *Rig* Vedic rivers. According to the Nadistuti hymn of the *Rigveda*, River Saraswati was a major river flowing between Yamuna in the east and Sutlej in the west. The later Vedic texts, however, mention it as a dried up river. *Rigveda* also refers to River *Aúvanvatî* along with river Drishadvati. Whether they were the same with different names or two separate rivers, is not clear. Similarly, River Helmand in south-western Afghanistan is at times referred to as Saraswati. Whether Helmand got the name Saraswati because some people moved westward or the people moving to the east called Saraswati as Helmand, is not clear. What is clear is that the human habitations on the banks of rivers Saraswati and Drishadvati had shifted to the east and south directions prior to *Mahabharat* period. In those days the present day Bikaner was known as Kurujangala and Jodhpur as Madrajangala. Apparently, the the region that is desert today was foested then.

Ramayana refers to this region as *Lavanasagara* (Salt-ocean). When Rama moves his *Vanara* army to attack Lanka, he used his *agneyashtra-amogha* to dry up the *drumakulya sea* situated on north of *Lavanasagara*. Thus was born Pushkar, a fresh water lake. *Mahabharat* mentions the Kamyaka forest located on the west of the Kuru kingdom's Kurukshetra plain on the banks of River Saraswati. The forest had a lake called Kamyaka. While on exile, the Pandavas went towards Kurukshetra, crossed rivers Yamuna and Drishadvati, and reached the banks of River Saraswati. When they reached the Kamyaka forest they found that it was the abode of many ascetics. It abounded in wild life especially the birds and deer. It took three days by their chariot to reach the forest from Hastinapura.

What made this land of rivers and forests a desert is long drawn story. To the west every continental mass of the world is a desert; Indian subcontinent is no exception. As at present it is marked with extremes of climate. The summer temperatures touching 49° C is not uncommon. The scorching heat of June and July parches everything. The winters limited to the months of December and January are fairly cold (less than 20°C) especially when the winds passing through the Himalayan ice cover descend and chill the desiccated land surface. Temperature may touch the freezing point (0°C). Diurnal variations are sharp: days are hot while the nights are cold. Annual rainfall rarely exceeds 25 cm.

How well the people have adapted themselves and their activities to the demands of nature, have to be seen in this desert to be believed. There is water shortage but the ingenious ways people have evolved to conserve water keeps the life of the desert people moving smoothly.

Biodiversity of Indian Desert

Flora and Fauna

Indian desert is not all sand; there are hillocks and sandy and gravel plains too. This diversity in habitat has given rise to more diversity in vegetation, animal life and human culture in comparison to the other desert regions of the world. Trees are few; but thorny bushes and shrubs in small patches are scattered all over the region. The main tree species found here are: *Acacia nilotica*, *Tamrix aphylla*, *Prosopis cineraria* (Khejri). The dominant scrubs are *Calligonum polygonoides*, *Crotalaria* spp, and *Haloxylon recurvum*. Among the *xerophilious* grasses of the region are *Aristida ädscensionis*, *Cenchrus biflorus*; and *Lasiurus scindicus*.

The desert of Rajasthan contains 25 species of serpents and 23 species of lizards. The endangered Great Indian Bustard (*Ardeotis nigriceps*), the Blackbuck (*Antilope *cervicapra*), Indian Wild Ass (*Equus hemionus khur*) and the Indian Gazelle (*Gazella bennettii*) are found here. The Desert National Park in Jaisalmer houses the Bengal fox, chinkara, desert fox, wolf, desert cat, the black buck, a plethora of migratory birds and seashells. The gigantic fossilized tree trunks found here bear witness to the presence of water and vegetation in the past.

Some of the wildlife species vanishing in other parts of India are well preserved in Thar desert. Important among them are: the Great Indian bustard, *(Ardeotis nigriceps)*, the Blackbuck *(Antilope cervicapra)*, the Indian Gazelle *(Gazella bennettii)* and the Indian Wild Ass *(Equus hemionus khur)* in the Rann of Kutch. Two factors have contributed immensely to their survival. First, due to lack of water, human encroachment of grasslands for tillage is minimal; and two, the Bishnois, a local community protects them as a part of their religious duty. Another factor that has helped them survive is their small size in comparison to other animals, and they are nocturnal.

Conserving Biodiversity

Biodiversity can be conserved in two ways: *ex situ* (i.e., out of the natural habitat) and *in situ* (within the natural habitat). The latter is habitat specific while the former is species specific.

In situ conservation of species maintains the genetic diversity of the species, while at the same time helps the species adapt to the changing environment caused by nature or anthropogenic activities. It also helps in preservation of other related species of the habitat. For this type of conservation technique certain areas are designated as protected sites. It is being promoted by the Man and Biosphere (MAB) Programme of the United Nations Educational, Scientific and Cultural Organization (UNESCO). It serves two purposes: it helps conserves biodiversity and biological resources while at the same time, allows sustainable use of natural resources for people. The programme started in 1976 has now spread to nearly 100 countries. By 2001, their number was 393 spread over in 94 countries.

There are a number of biodiversity conservation sites in Indian desert. The most important and by far the largest among them is Desert National Park, Jaisalmer. Spread over 3,162 km², it is an excellent example of the ecosystem of the Thar Desert, and its diverse fauna. Then there are two relatively small reserves: Talchhapar Sanctuary in Churu District some 210 km from Jaipur in the Shekhavat region; and Jalore Wildlife Sanctuary in Jalore district (130 km from Jodhpur). It has a sizeable population of rare and endangered wildlife including the Asian-Steppe Wildcat (Ornata), leopard, zird, birds, lizards, snakes, insects, desert fox, and Indian Gazelle.

The *ex situ* methods biodiversity conservation include creation of zoos where captive breeding programmes are carried out; development of aquaria for research, public information and education; and plant collections through seed storage and breeding. Zoos are not just public display facilities and for educating people about wild animals, but also for captive breeding specially of the vertebrates such as panda and dormouse that are facing extinction. Breeding of species in the zoos and their introduction in the wild is not without problems. Constant monitoring is needed to ensure that the reintroduction does not spread diseases. Genetic changes in the process of breeding also pose some problems as do the behavioural adjustments once the animals bred in captivity are reintroduced in the original habitat. Aquaria too are used for breeding of fresh water as well as marine species.

The *ex situ* methods of conservation of fauna are more expensive and should be regarded as complementary to *in situ* conservation methods. They should be used only when *in situ* conservation is no longer possible. However, it is much easier to maintain the populations of plant species artificially. They can better adjust to minor changes in the habitat and need less care. It is also far easier to breed and propagate plant species in captivity.

Among the measures being adopted to conserve and preserve the plant life in Indian desert is the greening of the desert. The soil of the Thar Desert remains dry for much of the year and is prone to wind erosion. High velocity winds blow soil from the desert, deposit some on neighbo8ring fertile lands, and cause shifting sand dunes within the desert, which bury fences and block roads and railway tracks. Permanent solution to this problem of shifting sand dunes can be provided by fixation of the shifting sand dunes with suitable plant species and planting windbreaks and shelterbelts. They also provide protection from hot or cold and desiccating winds and the invasion of sand.

The local species of trees are slow growing and hence exotic species of Eucalyptus, Acacia, Cassia and other genera from Israel, Australia, USA, Russia, Southern Rhodesia, Chile, Peru and Sudan have been tried in recent years. Among the important plants selected for plantations to check the shifting of sand dunes in Indian desert are *Acacia tortilis* and *jojoba*. The contribution of Rajasthan Canal in reclaiming land and checking the spread of the desert is great indeed.

Concluding Remarks

The pressure of population on the land and biosphere is getting unbearable all over the world; Indian desert is no exception to this. It is not only the number of people which matters, the consumerist culture being promoted matters far more. Since 1951, Population of Rajasthan has increased from 16 million in 1951 to 56.50 million in 2001. If the growth rate is similar to the one during 1991-2001, Rajasthan touched 68.62 million people in 2011. The villages, towns and cities will grow in size and encroach upon the open land removing whatever green there still is and leaving lesser area for the wild life.

The present policies on biodiversity conservation are not working well. Many of the protected species of wild life are on the verge of extinction. Forests are decreasing in area as the pressure of population on land increases. The stage has come, when each village and city should be asked to reserve at least 20 per cent of its land for forests. It may be that several villages can join hands and have joint forest reserves along the rivers, on the hills, and other areas not used for agriculture. Some of the less productive areas can be devoted to forestry. This may not be a very feasible proposition in States like Rajasthan when climatic restrictions come in the way of forestry.

Clash between humans and beasts in and around the biodiversity reserves has become quite common. Wild animals attack people; and people then attack the animals. The ideal situation will be one in which the bio-reserves are fenced so that the animals do not go out. Yet another problem is the deliberate killing of animals for pecuniary benefits as the bone, horns, and skins of the animals are highly valued in the market. The poachers have no religion except money. They often work in collusion of the people looking after the reserves. More stringent measures are needed to stop this menace.

A very serious threat to wildlife has been posed by chemical fertilizers, insecticides and pesticides used by the farmers. Many species of birds like sparrow which once adorned the trees with their beautifully constructed nests hanging through the branches of the trees especially near the tanks are on the verge of disappearing. The vultures that used to reach spot where an animal died within a couple of hours, are no longer seen even in the villages. Crows too are disappearing as are the other common birds. Many other wild species like snails, frogs, condors, eagles, falcons, hawks, owls, squirrels, foxes, and jackals are becoming scarce. This is more so in the desert areas where survival opportunities are lesser.

Apart from the measures being taken by the national as well as State governments to organize *in situ* as well as *ex situ* conservation, the need for educating the children and youth about the contribution that flora and fauna make to human life both culturally and economically, cannot be over-emphasized. The concept of *vasudhaiva kutumbakam* has to be revived, and inculcated. The children have to be told that this earth of ours was once inhabited by various species of plants and animals alone. Human beings came later took control of it. They cut the forests and killed the animals but left enough land for them to grow and prosper. Now they want to lead a luxurious life and for that they need more space. Not even the areas legally reserved for them are safe for them. Once upon a time, man was afraid of the wild animals; now the wild animals are afraid of the man. What a progress we have made !

10
Biodiversity Conservation in Traditions and Rituals of Rajasthan

P.C. Trivedi

Rajasthan is the largest State of India occupying an area of 342,274 square kilometres. It covers nearly 11 per cent of the area of India lying between 23°3′ and 30°2′ North latitudes and 69°3′ and 78°7′ East longitudes. In shape it is almost rhomboidal. It is surrounded by Pakistan in the West and the States of Punjab, Haryana, U.P., M.P., and Gujarat in the north, east and south. The presence of the Great Indian Thar desert in its western portion makes Rajasthan a unique State of India. The Thar forms the eastern extremity of the desert belt of the earth. The Aravallis, forming barriers to the desert in the east, are the oldest folded mountain chain of the world. This range not only divides the State geographically into two portions but also climatically different regions, thus affecting the vegetation, mode of life, occupation and culture of the people. The soils of the desert are typical grey brown desert soil containing 90-95 per cent sand and 5-10 per cent clay, high percentage of soluble salts and high pH value. The climate of the desert zone is characterized by extremes of temperatures, low rainfall, low relative humidity and high wind velocity. The important topographical feature of the desert is the presence of salt lakes in Sambhar, Degana, Kuchaman and Didwana which narrate the past geological history of the area. The Luni is the only seasonal watercourse in this region.

The ancient Aravallis having undergone various stages of erosion, span a long distance of 692 km in the form of alternating hills and valleys, diagonally, across the entire State with Delhi in the north–east and Champaner in the south-west being its two ends. The two extreme points within Rajasthan are Khetri (north–east) and Khedbrahma (south-west) separated by 550 km. Detached from the main mountain range, exists a hill tract in Sirohi where rises Mount Abu with Guru–Shikhar, the highest peak of the Aravallis (1,727 metres above M.S.L.).Between the Himalayas in the north of the country and the Nilgiris in the south, this is the highest mountain.

The rivers in the west and south of the Aravallis are not very large. They extend up to the Arabian sea through the Rann of Kutch. The most important rivers of the State are the Chambal, Mahi, Luni, Banas, Sabarmati and Banganga. In Rajasthan, no natural fresh water lakes are found. The Sambhar Lake, though natural, is a salt lake.

Plant Life in Rajasthan

Of the nine recognized botanical regions of India, three fall in the territory of Rajasthan—the Western Dry Region, the Gangetic plains and the Deccan Plateau. The State of Rajasthan is rich in floral diversity as 1,911 wild species belonging to 780 genera and 154 families grow here. Gymnosperms are represented by single species only (*Ephedra foliata*). Rainfall and altitude decide the type of vegetation in the different regions of the State ranging from extremely xerophytic scant vegetation of the desert region to the lush sub-tropical evergreen forests of the Abu Hills.

Tribes in Rajasthan

The State of Rajasthan is inhabited by several tribes, namely Bhil, Bhil–Meena, Damor, Dhanaka, Garasia, Kathodia, Koli, Mina, Nayaka, Patelia and Sahariya. Besides these, there are some nomadic, Semi-nomadic tribes and denotified communities also. Nomadic tribes are Banjara, Gadia-Lohar and Kalbelia, whereas semi-nomadic tribes are Rebari, Jogi and Masani. Among denotified communities, Bori, Kanjar, Sansi, Babri, Jat and Bhat are included.

Traditions, Rituals and Practices in Rajasthan

The folk traditions of Rajasthan are many centuries old. They have respected nature and made sincere efforts to keep nature in its pristine glory. Unfortunately, many of us under the materialistic glamour of Western culture, have started to forget such conservation concepts. Rajasthan has many emulative traditions and rituals to remedy this malady to a considerable extent. We must, therefore, re-realise and revive those practices in Rajasthan to be able to live harmoniously with the environment.

Tree/Forest Protection

The Rajasthanis still abide by some traditions, rituals and practices which help to conserve and protect trees and forests of which a few are indicated as under:

(i) **Kesar chhanta:** In southern Rajasthan, when villagers decide to provide protection to a certain patch of forests, they collect Kesar (*Crocus sativus*) from a temple and on a prefixed date, the village community announces with drumbeat that they gather at a definite point and then proceed to the forest area proposed to be safeguarded against felling and other adverse biotic factors inimical to the perpetuation of the forest vegetation. After reaching the pre-determined spot, they sprinkle Kesar on the trees standing on the outer periphery of the desired forest land. They, thus impose a self-restricting ban on cutting of trees and grazing, etc.

(ii) ***Khadana or Saka (Sacrificing Life):*** Western Rajasthan is prone to famine and has been so for centuries. This forced the cattle owners to migrate to green pastures in adjoining States each year. Guru Jambheshwarji (Jambhoji) considered the matter seriously and felt that lack of green trees contributed considerably to this distress. He enunciated 29 (20+9 *bees* + *nau*) principles and those who followed these principles were declared *Bis–noi*. One of the most important principals that these Bisnois still sincerely practice is not to cut green trees nor allow them to be cut by others in their areas. *Khejadi* or *Shami* tree is one such tree which is preserved by them at all costs and there are instances when Vishnois laid down their lives to save these trees. Such examples where they offered their life for saving the *khejadi* tree are known as *Khadana* or *Saka*.

In a village, Ramasadi (Jodhpur), two Vishnoi ladies named Karman and Gauran are on record that they volunteered to be killed to save the Khejadi tree on Saturday in *Sanwat* 1661. It is the first known case where people sacrificed their lives in an effort to save trees– an incomparable component of nature. Many Sakas followed and the largest and most prominent of these *Khadanas* is the one that took place in village Khejadali (Jodhpur) on 12 September,1730 (*Bhadwa Sudi* 10, *Sanwat* 1787). The Maharaja of Jodhpur wanted large sized trees for some construction work but they were available only in Vishnoi areas. The village Khejadali, not too far from Jodhpur, had many huge sized khejadi trees and the Maharaja's men went there to cut them. The Vishnois of this village offered strong resistance and did not permit the green trees to be felled. A Vishnoi lady Amritanji came forward and hugged the Khejadi tree and sacrificed Women which included newly married girls. In all 363 Vishnois (289 males and 74 females) laid down their lives but did not allow the green Khejadi trees to be felled. The last words of Amritanji were "*sar sante runk ruke to bhi sasto jan*" (it is a good bargain if a tree is saved even by giving your head).

This instance is enough to prove that our faith and practices upheld living harmoniously with nature. In fact, this is the first **Chipko Movement** (hugging trees to save them from being felled). To commemorate this sacrifice, an annual fair is also held at Khejadali where thousands of people come to pay homage to these great martyrs who gave their lives to save the trees.

(iii) Oran culture: Oran is the distorted form of Aranya which denotes forests. The protection of forests in the underlying concept of this religious faith. In western Rajasthan practically every village had a piece of land of natural vegetation attached to a temple or any religious endowment of a local deity. The cutting of trees and grazing of animals in them was prohibited and they served as a *sanctum sanctorum* on account of religious considerations.

(iv) Pavitra Kunjs (Sacred Groves): These *kunjs* (tree groves) are observed practically over the whole of the Rajasthan with a preponderance in the Aravalli Mountains. There are three categories of these groves. First category are those sacred groves which are in the tribal village, often near the water source. These forest-like areas are often preserved on hill tops or on hillocks devoted to a local deity, e.g., Bherunji, which are dedicated to Lord Mahadev and Shiva. These aim at the protection of watershed areas. A few instances are Gautemeshwarji, Tameshwarji, Jhameshwarji, etc. In the third category fall such groves in which only one species constitute the grove. In Kotra range of Udaipur division, there are many *bargad* (*vatvriksha*) tree with their aerial roots forming props on the ground and looking like big groves. Such sacred groves while providing welcome greenery and comfortable climate, also serves as shelters for wildlife and birds.

Veneration of Trees

Trees and plants have been taken as our benefactors and well-wishers for ages. The traditions go on to say that various god and goddesses have their preference for their stay in various trees.

Tree		*Associated God*
Peepal	–	Vishnu, Laxmi, Pitradev
Aawala	–	Laxmi
Gular	–	Vishnu, Rudra
Kadam	–	Krishna
Vat	–	Brahma, Kuber
Palas	–	Gandharva, Brahma
Karpur	–	Chandra
Vilva	–	Shiv, Souls
Neem	–	Shitala, Manasa, local deity
Ashoka	–	Indra, Buddha

In order to please lord *Brahaspati,* the banana plant is worshipped. *Aak, nariyal, mahuwa* and *semal* are worshipped as symbols of enhanced fertility. *Tulsi* is grown in a specially constructed pot in the courtyard of the home and worshipped. *Tulsi* is also wedded to *Shaligramji* with religious fervour.

Kalpavriksha Fair

In Rajasthan, the *Kalpavriksha* growing in Balund (near Deoli in Tonk District) and Banswara are adored but the importance of three *Kalpavriksha (Baobao tree)* along the Ajmer–Beawar National Highway at Mangaliawas has no parallel in Rajasthan, India or the World. A big fair is held on *Hariyali Mawas of Shrawan* (rainy season) to worship these trees.

Shami (Khejadi) Poojan (Worship)

On Dussehra day, there are many recorded instances from former princely states of state worshipping the *Khejadi* tree. In Thar Desert, *Khejadi* is considered as *Kalpataru* because of its multifarious utilities. *Khejadi* has been credited with the qualities of blessings and infertile women with son and this belief has existed for a considerable time.

Gotra Namkaran (Adopting Surname after a Tree)

Some tribal particularly Bhils feel honoured in adopting names after a particular tree. Some of the surnames are *Salaria* (after salar), *Semalia* (after semal), *Palasia* (after Palas), *Jamania* (after jamun) and *Anwalia* (after anwala).

Gram Namkaran (Naming the Village after a Tree)

Some places are named after the preponderance of a particular tree or some incident related with a tree. In Rajasthan, Banswara (after *bans*, bamboo), Kherwada (after *Khair*), Sangwada (after *Sangwan*), Semalwara (after *Semal*), Sariska (after *Siris*), Neemada (after *Neem*) and Peepalu (after *Peepal*) are some examples. Khejadali which is now internationally famous for the *Khandana* of 1730 A.D. has its name after the *Khejadi* tree.

Not Accepting Flowers

It seems strange to many that Jain monks do not accept flowers. This Jain community believes in living in harmony with nature and its various creations. When we pluck out the flowers, the availability of seeds is reduced and there is obvious interference in getting adequate natural regeneration of the plants which results in disturbing the ecological balance.

Recognition of Trees by Government

Governments have also appreciated the traditions of the people and many plants have been given the status of State's trees. In Rajasthan, the *Khejadi (Prosopis cineraria)* tree has been declared the State Tree while the *rohida (Tecomella undulata)* flower is the State flower. Similarly national tree is "Banyan tree" (*Ficus benghalensis)* and national flower is lotus.

WILDLIFE PRESERVATION

There are a number of instances of our traditions and practices which contribute to the conservation of wildlife of the State. Some of them are:

Vishnois: The Wildlife Protectors

Following the sermon of not cutting the green trees, this Vishnoi community is an ardent supporter of Wildlife conservation. Because of this *Chinkara* and *Kalahiran* (Indian gazelle and blackbuck) roam freely in Vishnoi village in Rajasthan.

Worshipping Animals

Our forefathers imbibed in us the cause of maintaining biodiversity for environmental congeniality. It is on this age-old faith that we worship the venomous Cobra (*Nag*) on *Nagpanchami.* The appearance of *dumuhi* (Earth boa) snake in the cereal godown is taken as the arrival of goddess of wealth, Laxmi (the Earth boa feeds on rats which consume large amounts of foodgrains).On *Goverdhan* day, cows and bulls are honoured while on *Shiv Ratri,* (Nandi bullock) is venerated. On Dussehra day, the *ashwa* (horse) is worshipped.

Vehicles of Gods

To save animals from unnecessary annihilation by men, our ancestors named various animals as the means of God's transport and thus they enjoyed immunity on religious grounds. Some examples are:

Gods		*Vehicles*
Vishnu	–	Garud (eagle)
Mahadev	–	Nandi (bullock)
Ganesh	–	Musak (mouse)
Kartikeya		Mayur (peacock)
Bhairav	–	Shwan (dog)
Vayu	–	Hiran (antelope)
Yama	–	Bhainsa (buffalo)
Goddesses	–	
Laxmi	–	Veloo (owl)
Durga	–	Sinha (lion)
Katyayani	–	Bagh (tiger)
Shitala	–	Gardabh (ass)
Saraswati	–	Hans (swan)

State Animals

In Rajasthan the *Chinkara* is the State animal while the Great Indian Bustard has been given the honour of State bird. The National Animal is the tiger while the National Bird is the peacock.

EFFICIENT USE OF WATER

Our traditions treat water as god varun. In fact, water is the elixir of life. Notorious for frequent famines, villagers in the Thar desert use each drop of water extremely efficiently. Some of the water conserving devices practiced are:

Tanka Construction

Water is a very scarce commodity in Western Rajasthan. Most of the people make a *tanka* for their family. This *tanka* is a pit like well whose steining is stone construction in lime mortar and its catchment is artificially designed in a small sloping area towards the well. This slope is plastered with kankar paste so that the water does not percolate in the soil but travels down to the well through the holes on the steining (wall of the well). The top of the *tanka* is covered with a lid. The water is taken out only for drinking purposes.

Khadin Construction

In scarce rainfall areas as in the Thar desert, Khadins are made in depressions like small dams. Water as run-off from the rains, collects in them and is utilized for irrigation. In case these beds dry up, crops are grown on it and good harvest is obtained on account of the better moisture regime.

Collecting Water on Roofs

Every drop of rain is useful in the desert region. The roofs of houses are so constructed that the meagre rains that take place are collected on the roofs and from there the water is stored and utilized.

ENVIRONMENT-FRIENDLY FESTIVALS AND RITUALS

There are many festivals in Rajasthan which are common with the rest of the country. These festivals, fairs and rituals, along with local customs, make the pleasure of life worthwhile. In reality, our traditional practices and customary festivals are related intimately to nature and is an integration of these nature oriented and eco-friendly activities.

Gangaur

Gangaur is the festival for praying to *Isar* (Shivji) and *Gaur* or *Gawar* (Parvati) so that virgins get a good husband. It is also the practice of ladies and girls to go to gardens woodlots for picnics. The girls carry *kalash* (metal pots for carrying water) with *durva* (green blades of grass), etc., in a decorative way. This is a tradition to enjoy nature and keep greenery. In Jaipur, the *Gangaur Teej* fair is a big attraction for rural as well as urban people and of late foreign tourists are also invited.

Shitalastami

On *Shitalastami* we observe fast, change the *matkas* (earthen pots for cooling water), start *basoda* that is eat stale and cold food and also look after the most ignored and innocent animal or donkey; for it is regarded as the vehicle of *Shitalamata.* This tradition indicates that the Food composition must change to suit the ensuing hot season. The deity is supposed to ward of measles (*bodri mata*), chicken pox (*mata*) and small pox (*badi mata* or *chechak*).

Akha Teej (Akshyaya Teej)

This is the day when marriages are performed in large numbers and belief is that girls married on this day will have a long wedded life and that they will die as *Suhagins* (death when husband is still alive). In many parts, this is the time when *Kanji-bada* (ground moong or urad fried in oil and kept in earthen pot with water which is fermented with powdered mustard) is prepared. This preparation is an antidote for sunstroke.

Other major festivals, viz., Makar Sakranti, Diwali, Holi, Shivratri and Sharad Purnima are common with the rest of the country.

OTHER ENVIRONMENT CONSCIOUS TRADITIONS

We have always been deriving inspiration from nature and discouraged indiscriminate exploitation of natural resources. Harmless harvesting of natural products has been our motto and aim so that sustained yield is ensured for future generations. Some of them are:

Using Biodegradable Products in Functions

In our social functions like marriages and other large-scale party gatherings, we use leaf plates and cups (*pattal dauna)* and earthen Pots (*shakora* or *khullads* for water) instead of plastic articles which do not decompose while the leaf products and earthen utensils get degraded biologically and provide nutrients to the soil.

Swallowing Basil (*Tulsi*) Leaves

In our temples and other religious places, the *Prasadans* given is of *tulsi* leaves along with *charanamrit* (holy aromatic water) and we swallow then rather than chew them. Many disregard this as a superstitious act but doctors have established that these *tulsi* leaves contain mercury which, if chewed, erodes the enamel of the teeth and makes them sensitive to cold, hot and sweet liquids and solids whereas this chemical in the body provides strength.

Animal given as Dowry

It is often practice in some communities like Gurjars, Jats to present a cow or an ox as dowry in weddings. Some villagers also follow this tradition. In the nomadic communities like *Kalbelia* (snake charmers) a nag (cobra) and in jogis, a donkey in dowry is given and this is indicative of social status. In the Rebaris (camel breeders), the dowry invariably includes a camel. The Marwadas (nomadic cattle owners) give goats and sheep in dowry.

Yagyas (Holy Fires)

It may be a nicknamed as religious activity but research has established beyond doubt that it purifies the atmosphere.

Miscellaneous Practices

There are many other practices in vogue which are beneficial for us like occasionally walking bare footed on gravelly surface (gives the advantage of acu-pressure, for most of the acu-pressure points lie in the soles of the feet), wearing wooden slippers (*Khadaun* or *chattis*) which helps in retaining the body energy in body as wood is a bad conductor of energy (sages and saints often use them), one usual custom is to keep *joneyu* (sacred thread) on the body and put around the ear while urinating which is said to complete the flow of waste products of the body, that is, urine and keeping kidneys free from unnecessary pressure. In north Rajasthan, there is the practice to the feet on green branches of *Khejadi* while leaving the cremation ground so that the germs of the dead person are got rid off. The antiseptic properties of *neem* are also taken advantage of. They are hung on the sill of the door of the room in which mother with a new born baby lives and even when there is eye-sore or small-pox eruption. We ornament our newly constructed houses and all the places where auspicious ceremonies for the family are performed with green leaves of mango or *Ashok* tree, etc.

Conclusion

I have given some examples of Rajasthani traditions, practices, rituals, customs and folklore which are based on long experience and are in perfect balance with the local factors and locality. Most of them are intimately integrated with the sustainable environment not only congenial to the generations of those times, but even to the following generations. The need of the hour is to take out those traditions from our senior citizens and make sincere efforts to gauge their usefulness in the present context rather than simply discarding them as fancies of early backward generations.

11

Tiger Habitat Suitability and Potential Relationship with Forest Biomass in Sariska Wildlife Reserve, India

A Geospatial Perspective

M.S.Nathawat, L.K.Sharma and *Suman Sinha*

1. Introduction

The world is experiencing a rapid loss of biodiversity leading to extinction crisis and this loss is likely to hasten the climate changes and *vice versa*. Climate change effect on wildlife is prominent at local, regional and global levels. Impact of climate change on forests will fluctuate based on the region accounting to the degree of local change. The United Nations Framework Convention on Climate Change (UNFCCC) has supported for economic incentives to help facilitate REDD (Reducing Emissions from Deforestation and Degradation) in forest-rich developing countries (Gibbs *et al.*, 2007). The advent of recent climate policies like REDD has escalated the necessity of accurate monitoring of the forest (van Leeuwen *et al.*, 2011). Forest biomass is an important component for studying the carbon cycle dynamics as potential carbon emission can be released to the atmosphere due to deforestation (Lu, 2006) and forest fires (Kanga *et al.*, 2010) and is one of key indicators of forest health. Amongst the effects already reported, enhanced atmospheric carbon dioxide (CO_2) levels are considered to be enhancing growth and increasing the forest carbon sequestration rate in areas with adequate rainfall (DeLucia *et al.*, 1999). The impact on associated wildlife due to extreme changes in the structure and function of forest ecosystem can result in the extinction of sensitive and specialized species. Apart from natural causes, there are anthropogenic activities that can trigger species extinction (Sinha *et al.*, 2011a). However, its far more difficult to pose a control over natural causes. Surface temperature of forest and forest biomass can be related to forest quality, which in turn is a vital parameter for wildlife habitat management, as the habitat depends on forest stratum (Sinha *et al.*, 2011b). Tiger habitat can also alter due to dynamic nature of land-use/ land-cover features (Sinha *et al.*, 2011a, Sharma *et al.*, 2012). The present analysis comprises of spatial modelling for estimating forest biomass and conducive tiger habitats by integrated geospatial approach. This analysis can quench the requirements of targetting ecosystem services and policies for monitoring forests for tiger habitat management and conservation, however, the method also bears potential application towards the conservation and management of any wildlife in general.

Methodology

Study Area and Datasets

The study area of Sariska Wildlife Reserve (Figure 11.1), located among the Aravalli hill ranges in Alwar district of Rajasthan, India with spatial extent of 27°13′to 27°31′N latitude and 76°15′ to 76°33′E longitude. The park covers an area of 800 km^2 and the study area comprises of nearly 1183 km^2. 50 per cent of the area is covered under forests (Sinha *et al.*, 2011b). From a view point of the succession, and concept of continuum of vegetation, the large-scale formation types in the area are *Acacia catechu* and *Anogeissus pendula* vegetation types (Puri *et al.*, 1982). According to Champion and Seth (1968), two major vegetation types namely "Tropical Dry Deciduous Forest" and "Tropical Thorn Forest" are present in the area based on the structural attributes. Satellite data of Landsat ETM+ was obtained online free from GLCF (*http://glcf.umiacs.umd.edu*) and Survey of India toposheets were used as reference for image interpretation along with extensive ground survey. Thermal band (band 6) of Landsat ETM+ has a spatial resolution of 60 m.

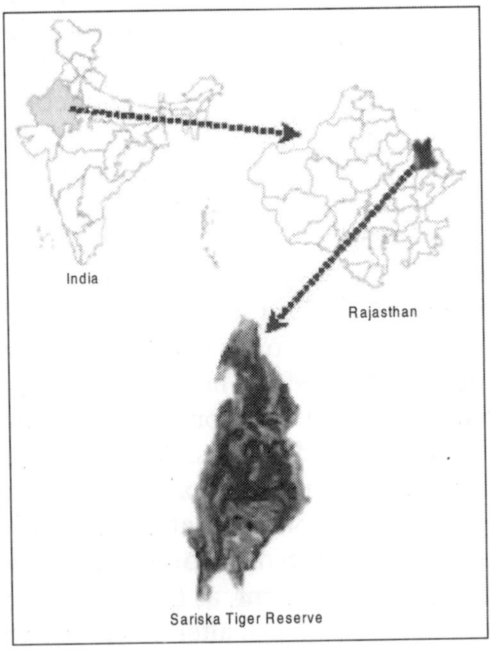

Fig. 11.1: Location of the Study Area

Image Interpretation

Satellite data of Landsat ETM+ was co-registered and geo-rectified taking SOI toposheets of scale 1:50,000 as reference. The image was clarified with the prominent identifiable ground features, the information of which was gathered during extensive field survey. Spectral signatures for different land-use and land cover types were identified and False-Colour Composite (FCC) was interpreted based on visual image interpretation techniques. Ground truth was collected from study area to identify different elements. The image was processed for classification of the different features on the ground. Maximum Likelihood Classification (MLC) was used in supervised classification where training sets were selected from FCC imagery with three bands of green (2), red (3) and near-infrared (4) based on the collected sample points for respective LULC classes (Sinha *et al.,* 2011a, b).

Retrieval of Spectral-based Forest Biomass

Forest strata maps based on visual interpretation of the image and NDVI values and validated with ground truth were generated. Sample plots of 0.1 ha were then laid in order to collect DBH and height of trees which would give the tree volume by general volumetric equations and then multiplying it with specific gravity would finally give plot-wise above-ground bole biomass (Equation 1). Plot-wise NDVI is extracted and regressed with estimated plot-wise biomass, which gives the biomass model (Equation 2). The predicted and estimated biomass had an r^2 correlation value of 0.7.

Volume (stem volume) = f (DBH, height)
Biomass = Volume (stem volume) × Specific gravity ...(Eq. 1)
$AGB = -4.5713 + 231.3735 * \text{NDVI}$ ...(Eq. 2)

Retrieval of HSI-based Suitable Tiger Habitats

Major parameters governing tiger habitat suitability are determined from literature survey, expert knowledge, reconnaissance information and field study. Parameters considered are land-use/land-cover (forest type and density, waterbodies, settlements, road), slope of the terrain and anthropogenic intervention in the form of settlements. All these factors are represented in the form of thematic maps prepared with the help of satellite image, toposheets and field survey in GIS domain. Proximity analysis for each parameter is carried out and for every thematic layer; appropriate weights are assigned in GIS-MCDA using AHP. Habitat Suitability Index (HSI) map is generated using equation 3 (Sinha *et al.*, 2011a, b). GPS data for tiger locations are obtained from PCCF, Jaipur, Rajasthan, which in turn can validate the HSI map output.

$$HSI = \frac{1}{(1+e^{-x})} \qquad \text{...(Eq. 3)}$$

where, *x* refers to the summation of the overall weighted parameters.

RESULTS AND DISCUSSIONS

3.1 Retrieval of Spectral-based Forest Biomass

Field based AGB was assessed from the data collected in the field from the sampling plots. Among the sampled plots, Anogeissus forest with 40-70 per cent density showed the maximum AGB of 135.22 t; while minimum for Boswellia of 6.36 t. However, overall it was found that *Acacia zizyphus* contributes to the maximum AGB of 164.2 t among the sample plots. When AGB was expressed in tonnes per hectare of the area, it was observed that Boswellia forests put in the maximum of 37 per cent of the total. The entire plot-wise AGB of the area is estimated to be 795 t/ha. The modelled biomass map shows the maximum value of 130 tonnes. The range of biomass that occurs the maximum in the area is 40-60 tonnes and then within the range of 60-80 tonnes. Figure 11.2 shows the spatial distribution of forest biomass with GPS data for tiger locations obtained.

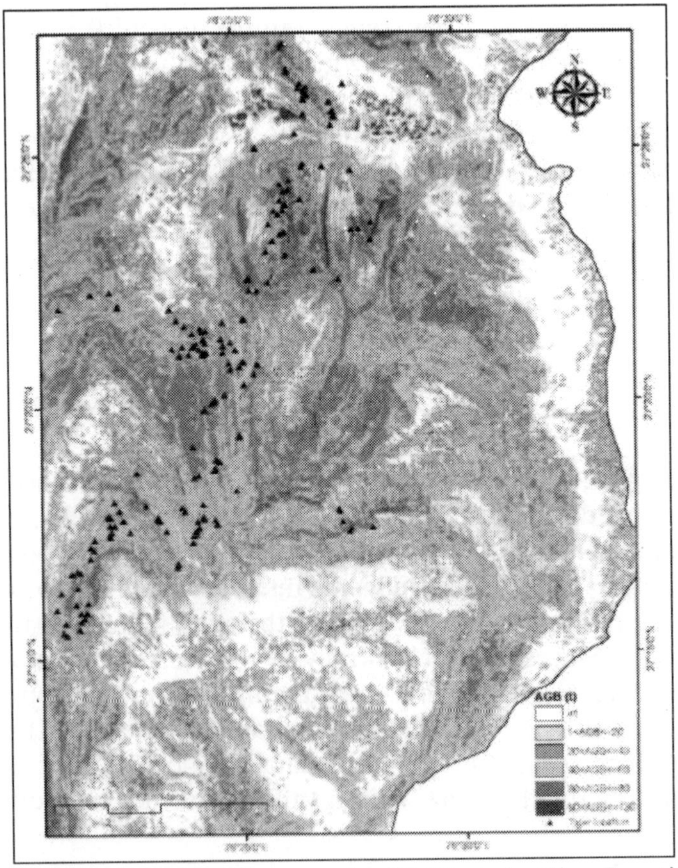

Fig. 11.2: Spatial Distribution of First Biomass with GPS Data for Tiger Location

Retrieval of HSI-based Suitable Tiger Habitats

Open forest covers most of the area (468.247 km^2), where *Anogeissus pendula* covers nearly 25 per cent of the total forest area, while bulk of the forest of nearly 34 per cent is covered with *Acacia-zizyphus* mixed forest. Approximately 47 per cent (269.288 km^2) and 34 per cent (198.959 km^2) of the forest lies in 40-70 per cent and 10-40 per cent forest canopy density classes respectively. Forest area with >70 per cent canopy density comprises less than 3 per cent (16.381 km^2) of the forest area. Dispersed heterogeneous settlements are scattered throughout the area within the main core zone covering approximately 4.17 km^2. Distance maps of road and waterbodies are prepared which is a critical step in the habitat suitability and is important for planning management strategies. Along with these slope which conveys information regarding the terrain is considered an important factor in tiger habitat modelling. Table 11.1 gives the weights associated with the parameters derived through crisp AHP λ max method. HIS is found to be 0.87 ± 0.06. Habitat suitability map of tiger (Figure 11.3) is derived through GIS-MCDA HSI modelled approach using equation 3, the output is validated by the GPS data for tiger locations obtained.

Table 11.1: Weightage assigned by Crisp AHP λ max method

Sl. No.	*Parameters*	*Weights*
1.	Forest type	0.332
2.	Forest density	0.092
3.	Settlement	0.122
4.	Waterholes	0.187
5.	Drainage	0.187
6.	Road	0.054
7.	Slope	0.026

Conclusions

The research offer improved management strategies for wildlife, especially for spatially determining suitable sites for habitat conservation, relocation and evaluation. The value for habitat suitability indicates that the area is an admirable destiny for conservation and relocation of tigers. Figures 11.2 and 11.3 show the spatial distribution of biomass and tiger HSI respectively with GPS data for tiger locations. The figures show apparent spatial relation or link among the parameters. This reveals the existence of forest biomass as vital parameters for habitat suitability of Sariskan tigers. This study did not consider these parameters for tiger habitat suitability. If considered, the habitat suitability of tigers could increase. So, a potential relationship exists between forest biomass and tiger habitat suitability in Sariska Wildlife Reserve, India. The study recommends Sariska to have great potential for quality tiger habitats, provided proper effective conservation and management strategies are implemented.

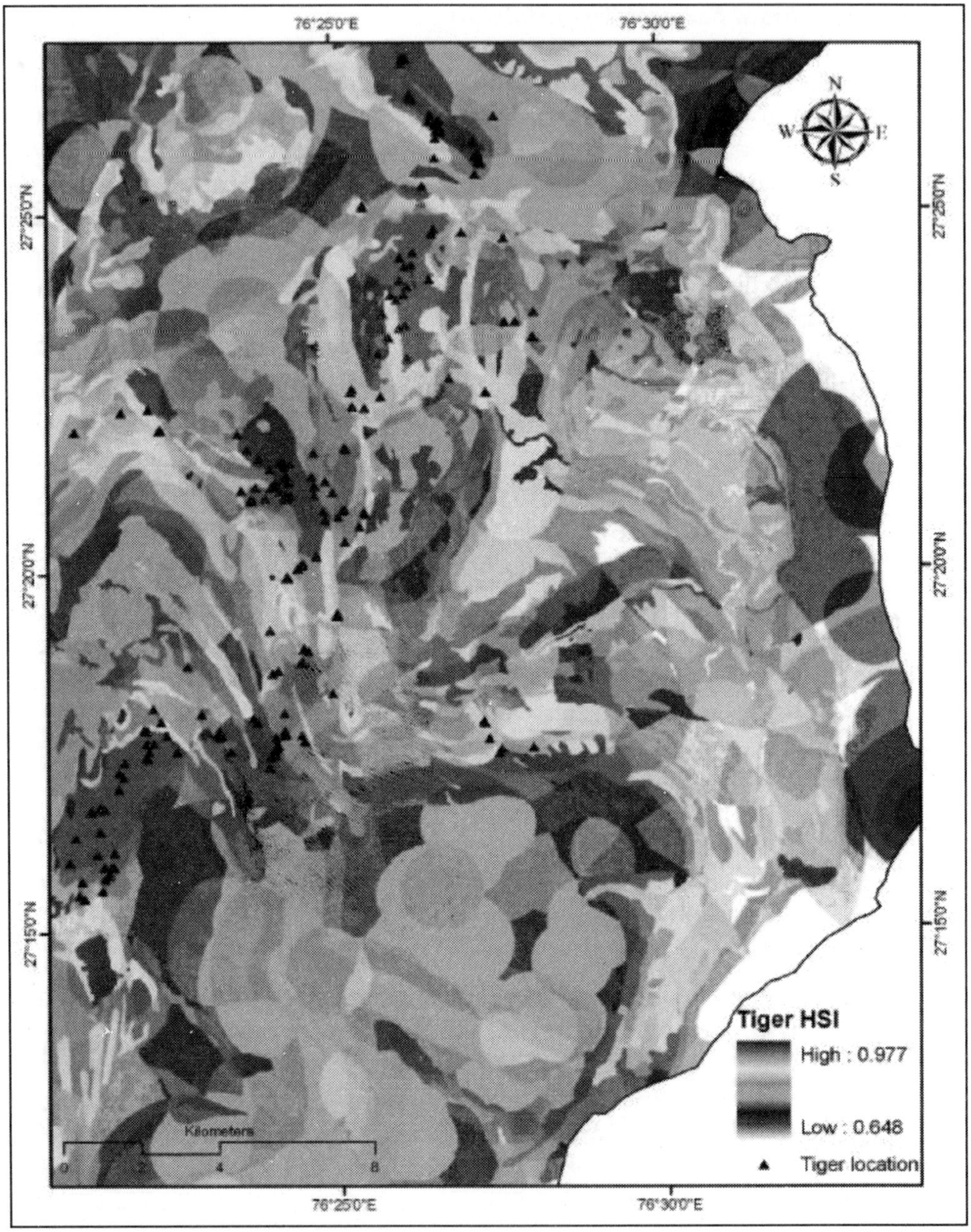

Fig. 11.3: Tiger habitat suitability Index Map derived through GIS-MCDA HSI modelled approach and validated by the GPS Data for Tiger locations obtained.

Acknowledgements

Authors express sincere gratitude to the Department of Science and Technology (DST),

Government of India for providing funds under DST-INSPIRE Programme (AORC scheme) to carry out the research to Mr. Suman Sinha. PCCF (Jaipur, Rajasthan) and officials of Project Tiger, Sariska and Sariska Forest Division are acknowledged for their support.

NOTES

Champion, H.G. & Seth, S.K. (1968). *A Revised Survey of the Forest Types of India,* Manager of Publications, Govt. of India, New Delhi, pp135-139.

DeLucia, E.H.; Hamilton, J.G.; Naidu, S.L.; Thomas, R.B.; Andrews, J.A.; Finzi, A.; Lavine, M.; Matamala, R.; Mohan, J.E.; Hendrey, G.R. and Schlesinger, W.H. (1999). Net primary production of a forest ecosystem with experimental CO_2 enrichment. *Science*, 284(5417): 1177-1179.

Gibbs, H. K.; Brown, S.; Niles, J.O. and Foley, J.A. (2007). Monitoring and estimating tropical forest carbon stocks: making REDD a reality. *Environmental Research Letters, 2.* (DOI:10.1088/1748-9326/2/4/045023) online access: erl.iop.org.

Kanga, S.; Sharma, L.K.; Pandey, P.C.; Nathawat, M.S. and Sinha, S. (2011). Forest fire susceptibility mapping with GIS modelling. *International Journal of Earth Science and Engineering*, 4, 80-87.

Lu, D. (2006).The potential and challenge of remote sensing-based biomass estimation. *International Journal of Remote Sensing, 27,* 1297-1328.

Puri, G.S.; Meher-Homji, V.M.; Gupta, R.K. and Puri, S. (1982). Forest Ecology. In Phylogeography and Forest Conservation, 2nd edn, Vol. 1. IBH Publishing, New Delhi.

Sharma, L.K.; Pandey, P.C. and Nathawat, M.S. (2012). Assessment of land consumption rate with urban dynamics change using geospatial techniques. *Journal of Land Use Science*, 7, 135-148.

Sinha, S.; Sharma, L.K.; Pandey, P.C.; Nathawat, M.S. and Kanga, S. (2011a). Impact of human intrusion on tiger habitat and conservation using integrated geospatial techniques. *International Journal of Earth Science and Engineering*, 4, 39-45.

Sinha, S.; Sharma, L.K. and Nathawat, M.S. (2011b). Retrieving tiger habitats: Conserving wildlife geospatially. *Applied Remote Sensing Journal*, 2, 1-5.

van Leeuwen, T.T.; Frank, A.J.; Jin, Y.; Smyth, P.; Goulden, M.L.; van der Werf, G.R.; Randerson, J.T. (2011). Optimal use of land surface temperature data to detect changes in tropical forest cover. *Journal of Geophysical Research*, 116, G02002 (DOI: 10.1029/ 2010JG001488).

12
Some Major Threats to Herpetofauna of Central Aravalli Ranges and Thar Desert of Rajasthan, India

Sharma, Krishan Kumar; Sharma, Vivek; Kumawat Rakesh and *Sharma Neha*

India is one of the 12 mega-diversity countries of world with high endemism in which herpetofaunal species endemism is highest among vertebrates (Daniels, 1996, 1997). Rajasthan is largest provenance of India having arid and harsh type of climatic conditions. It comprised by most diversified types of habitats: Thar desert, Aravalli mountain ranges, Eastern Plains and Southern Plateau. Due to general believe of arid conditions and hot climate, naturalists supposed to be very poor herpetofaunal diversity in this part of globe. This leads to neglected studies in concern to herpetofaunal diversity from this region so there are no major previous studies which conclude effect of anthropogenic activities mainly urbanization and their associated activities on the herpetofauna.

Indian desert or thar desert extends over about 0.32 million km^2 forming approximately 10 per cent of the total geographic area of India. Aridity results from evaporation being higher than precipitation thus, comprises this part of globe as arid climatic zone. More than 60 per cent of the desert lies in the State of Rajasthan followed by 20 per cent in Gujarat (Krishnan, 1977). It is one of the smallest deserts of the world but exhibits a wide variety of habitats and high biodiversity in general and herpetofauna in concern, due to juxtaposition of palaeoarctic, oriental and Saharan elements (Blandford, 1901; Pocock, 1939 and 1941; Prakash, 1963).

Aravalli mountain ranges are one of oldest mountain ranges of India. This also provides a unique habitat to many floral and faunal species. It plays an important role as a natural barrier between great Indian desert (Thar desert) and eastern plains. It also has great geomorphological values as it shows varied altitudinal variation throughout its expansions. Taragarh (855 m) is the highest peak of central parts of these mountain ranges and it is typical ensialic mountain range of olden proterozoic rocks having age between 2500 and 850 million years.

Herpetofaunal Diversity of the Thar Desert and Central Aravalli Foothills of Rajasthan

Amphibians and Reptiles are collectively designated as Herpetofauna.Living amphibians and reptiles are representatives of a small number of the many historical tetrapod radiations Amphibians and reptiles (collectively, herps) are not each other's closest relatives evolutionarily, yet they have traditionally been treated as though they are related (e.g., "herpetology" does not include birds and mammals). These are often the most abundant terrestrial vertebrates at any locality in the warmer parts of the world. Thar desert of Rajasthan and varieties of habitats of central Aravallis provide abundant food, shelter and breeding grounds to the herpetofauna of western India. Due to their high species diversity and abundance, their impact on other animal and plant species is very significant. Amphibians and reptiles have played prominent roles in ecosystems.

Some major threats include, habitat loss and defragmentation due to natural disasters and developmental activities such as Mining, Deforestation and Urbanization etc. Some other natural reasons of habitat loss includes landslides, soil erosion (By Air and Water), reduction in groundwater level and desiccation of open water sources such as pond, lakes and seasonal rivers or streams. Many anthropogenic activities. such as indiscriminate use of pesticides, road accidents, poaching, illegal trades, Secondary threats includes such as these factors have resulted in population decline of herpetofauna in several parts of the world including central Aravalli ranges and Thar desert of Rajasthan. It has been pointed by many naturalists that some species have undergone extinction and there is no record available due to lack of documentation. We do not know how many species have undergone extinction. Due to this reason the Zoology Department of MDS University Ajmer conducted thorough survey of herpetofauna of Rajasthan particularly the Thar desert and the central Arravalis along with assessment of major threats. Primary data base of herpetofauna from this part is now available (Table 12.1). A total of 59 herpetofaunal species were documented from the study area. Out of these 9 species represented 3 families of Amphibians, 25 species represented by 7 families if lizard group (Sauria), 2 individual species represented turtle/tortoises and 23 species of snakes represented by 6 families (Table 12.1).

For tourist purposes, most attention on biodiversity conservation has been paid on big mammals. Very little attention has been paid for conservation of non-mammalian species particularly amphibians and reptiles which are facing great threats of local extinction. Herpetofaunal species not only inhabits in forest areas but also migrates in urban localities because of deforestation and mining activities in areas nearby forest. Therefore, there is a need to document biodiversity in areas other then protected areas for meaningful assessment of population status of herpetofaunal species. And to prepare a proper conservation action plans.

Major Threats to Biodiversity of Herpetofauna

In recent past biodiversity is facing natural as well as man-made threats on large scale. Natural threats include climate change mainly whereas man-made threats include urbanization, mining, habitat destruction, poaching and road accidents, etc. Both threats have either resulted in species loss or shift to any unknown sites which may or may not provide suitable feeding and breeding grounds to new inhabitants, thus local extinction of valuable herpetofaunal species takes place. So any unwanted anthropogenic activities at any place of such type of biodiversity rich areas leads to population decline of such ecobiologically important species.

Table 12.1: Observed Herpetofaunal Species of Western India

S. No.	*Amphibians*	
	Common Name	*Scientific Name*
		Bufonidae Family
1.	Marbled Toad	*Duttaphrynus stomaticus*
2.	Common Asian Toad	*Duttaphrynus melanostictus*
		Dicroglossidae Family
3.	Indian Skipper Frog	*Euphlyctis cyanophlyctis*
4.	Indian Cricket Frog	*Fejervarya limnocharis*
5.	Indian Bull Frog	*Hoplobatrachus tigerinus*
6.	Short-headed burrowing Frog	*Sphaerotheca breviceps*
7.	Indian Burrowing Frog	*Sphaerotheca rolandae*
		Microhylidae Family
8.	Ornate Narrow mouthed Frog	*Microhyla ornata*
9.	Marbled Balloon Frog	*Uperodon systoma*
		Reptiles / Lizards
		Agamidae Family
10.	Garden Lizard	*Calotes versicolor*
11.	Fan Throated Lizard	*Sitana ponticeriana*
12.	Burrowing Lizard	*Agama agilis*
13.	Spiny tailed Lizard	*Saara hardwicki*
14.	Toad headed lizard	*Bufoniceps laungwalansis*
15.	Sand agama	*Phrynocephalus euptilopus*
		Chamaeleonidae Family
16.	Indian Chameleon	*Chamaeleo zeylanicus*
		Gekkonidae Family
17.	Keeled Rock Gecko	*Cyrtodactylus scaber*
18.	Brook's House Gecko	*Hemidactylus brooki*
19.	House Gecko	*Hemidactylus flaviviridis*
20.	Bark Gecko	*Hemidactylus leschenaultia*
21.	Termite Hill Gecko	*Hemidactylus triedrus*
22.	Sindh Sand Gecko	*Stenodactylus orientalis*
		Eublepharidae Family
23.	Common Leopard Gecko	*Eublepharis macularius*
		Lacertidae Family
24.	Long Tailed Lacerated	*Eremias guttulata watsonana*
25.	Snake eyed Lacerated	*Ophisops jerdoni*
26.	Large Snake eyed Lacerated	*Ophisops microlepis*
27.	Fringe Toed Lizard	*Acanthodactylus c. cantoris*
		Scincidae Family
28.	Yellow Bellied mole Skink	*Eumeces taeniolatus*
29.	Bronze Skink	*Mabuya macularia*
30.	Common Indian Skink	*Mabuya carinata*
31.	Common Spotted Skink	*Riopa punctata*
32.	Three Stripe-tad Skink	*Mabuya dissimilis*
		Varanidae Family
33.	Indian Monitor Lizard	*Varanus bengalensis*

(Contd.)

Table 12.1: (*Contd.*)

S. No.	Amphibians	
	Common Name	Scientific Name
34.	Desert Monitor Lizard	*Varanus griseus*
	Reptiles / Turtles and Tortoises	
35.	Indian Star Tortoise	*Geochelon elegans*
36.	Indian Flap Shell Turtles	*Lissemys punctata punctata*
	Reptiles / Snakes	
	Typhlopidae Family	
37.	Brahminy Worm Snake	*Ramphotyphlops braminus*
	Pythonidae Family	
38.	Indian Rock Python	*Python molurus molurus*
	Boidae Family	
39.	Common Sand Boa	*Gongylophis conicus*
40.	Red Sand Boa	*Eryx johnii*
	Colubridae Family	
41.	Trinket Snake	*Coelognathus helena helena*
42.	Indian Rat Snake	*Ptyas mucosa*
43.	Glossy Bellied Racer	*Platyceps ventromaculatus*
44.	Black Headed Royal Snake	*Spalerosophis atriceps*
45.	Red spotted Royal Snake	*Spalerosophis diadema*
46.	Striped Keelback	*Amphiesma stolatum*
47.	Sind Awl Headed Snake	*Lytorhynchus paradoxus*
48.	Russell's Kukri Snake	*Oligodon taeniolatus*
49.	Common Kukri Snake	*Oligodon arnensis*
50.	Common Wolf Snake	*Lycodon aulicus*
51.	Barred Wolf Snake	*Lycodon striatus*
52.	Checkered Keelback	*Xenochrophis piscator*
53.	Common Cat Snake	*Boiga trigonata*
54.	Afro-Asian Sand Snake	*Psammophis schokari*
55.	Leith's Sand Snake	*Psammophis leithii*
	Elapidae Family	
56.	Common Krait	*Bungarus caeruleus*
57.	Spectacled Cobra	*Naja naja*
	Viperidae Family	
58.	Russell's Viper	*Daboia russelli*
59.	Saw-scaled Viper	*Echis carinatus*

Mining, pollution, urbanization, land encroachment, habitat loss and more other anthropogenic activities have increased along with explosion of human population. So there is an urgent need to re-evaluate population status and distribution including impact of recent threats to various aspects of herpetofaunal biodiversity. We have been exploring various aspects related to herpetofaunal biodiversity including population status, distribution pattern and recent threats facing by these species of interest and conservational aspects too of Herpetofauna of this part of India from nearly two decades (Sharma and Mehra, 2007, 2009; Sharma *et al.,* 2009; Sharma and Sharma, 2009a, b; Sharma *et al.,* 2010, 2011 and 2012).

Mining

Nature has endowed Rajasthan with several rich mineral deposits. Perusal of available records shows that sixty-four kinds of metallic and non-metallic minerals, ores and deposits are exploited in the State of Rajasthan. A wide spectrum of non-metallic compounds like gypsum bentonite, marble, fuller's earth, siliceous earth, white clay, silica-rich sands, limestone, yellow ochre, selenite, jasper, lignite, asbestos, granite, sandstone, etc., account for a major share of State's mineral resources. State also has rich deposits of metallic ores like tungsten, zinc and copper in some regions. Guidelines under Central and State enactment, call for a sketch of mining area when a mining lease is applied for. It was found by *Lokayukta* that sometimes actual mining areas are not related to sketch given with applications, without officials crosschecking them. Further mining applicants falsely claim a prohibited forest area as a revenue area. Often actual area of mine is much bigger than the claimed area. Indian Bureau of Mines rules which control the type of mining, allow a maximum mining depth six metres to prevent environmental degradation. But miners have flouted this rule to over-extract iron ore. Illegal mining in India is widespread in various ore-rich states of India, and has generated controversy, which spans encroachment of forest areas, underpayment of government royalties and conflict with tribes regarding land-rights. After raising a very serious talk by various government and non-government organization, Ministry of Environment and Forest release a notification dated 7th May, 1992 regarding restriction of certain activities in specified area of Aravalli Range, which are causing Environmental Degradation. These activities are restricted by legislations of Government of India under Sections 3 (1) and 3 (2) (v) of Environment Protection Act 1986 and Rule 5 (3) (d) of Environment Protection Rules, 1986.

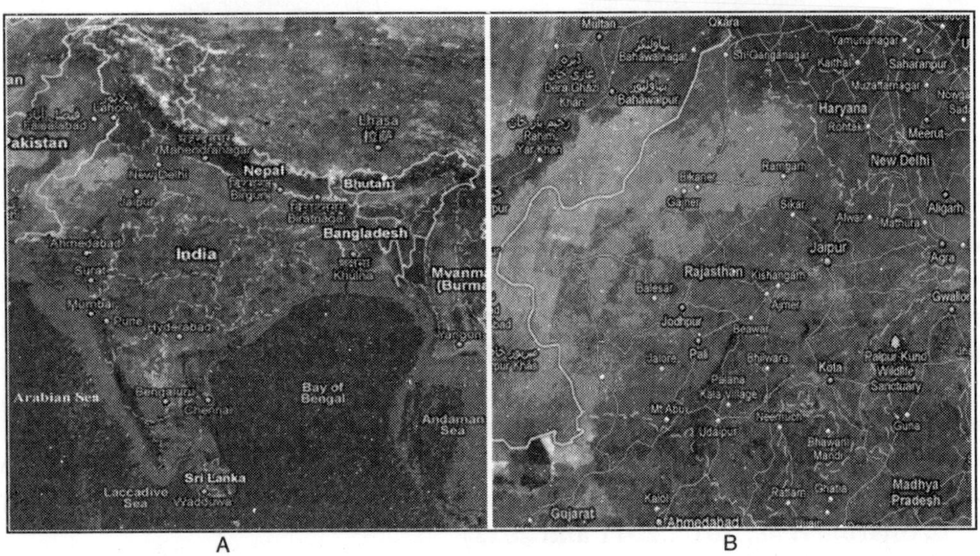

Fig. 12.1: (A) Map of India; (B) Study Area in Rajasthan State

Due to illegal mining and Urbanization habitat composition was completely changed at central Aravalli and Thar Desert areas that may lead to severe impact on various floral and faunal species. These facts have been forced us for the idea of present study. Present article is an approach to fulfil these research gaps as a record of distribution, population trend of Herpetofauna from Central Aravalli and Thar Desert areas of western India and other associated recent major threats facing by population of these fascinating herpetofaunal species at study area.

Now-a-days Aravalli mountain ranges are facing serious mining activities which lead to habitat loss for many herpetofaunal species. There were 360 lease mines in 2009 which increased to 830 lease mines in 2012 at central Rajasthan out of these lease mines 16 per cent mine's lease are timed out but they are still continuing and damaging this known oldest natural habitat of various herpetofaunal species severely. All observed study sites shows significant decline in population trend of Herpetofauna in subsequent years from 2009 to 2012. These mining activities at study area of Aravalli mountain ranges have caused almost complete removal of certain species and leads to local extinction in wild and simultaneously pressurized rest of individuals to move to new microhabitats that causes range extension (Sharma *et al.,* 2011).

Fig. 12.2: (A) Central Aravalli Ranges prior to 2009 (B) and after the mining activities 2011

Mining activities also lead to habitat loss at semi-arid zone to arid zone (Fig. 12.3 and 12.4) which introduced the certain invasive species those shows competition for food and habitat and generated the adverse effects on population of residential herpetofaunal species.

Urbanization and Habitat Alteration

Major threats facing by several species of Herpetofauna are increasing rate of urbanization which leads to habitat alteration, habitat fragmentation and finally habitat loss for species of interest. Land encroachment for agricultural and other practices also include in this category. Due to introduction of the Indira Gandhi Nahar Project (IGNP) Sri Ganganagar district of Rajasthan, land characteristics have been drastically changed particularly near the IGNP canal system. In these areas irrigation is possible throughout year for farmers and field workers who have occupied most of lands for

Fig. 12.3: Habitat Fragmentation and Habitat Alteration (A sand patch clearly seen in typical Aravalli Ranges due to the mining)

Fig. 12.4: Secondary effects includes urbanization and land encroachments which directly causes the habitat loss of herpetofauna

farming and agricultural practices. These areas previously has been used as feeding and breeding grounds of lizard *Varanus griseus*, and other reptile species but now these days population of *Varanus griseus* sharply declines.

Man-Wildlife Conflicts

Whenever the man-wildlife conflict takes place in observed areas in most of time animal losses its life in these areas because people usually kill monitor lizards, snakes and anurans due to fear or some myths associated with its coloration, bifid tongue and ability to make loud hissing when they encountered. Another threat to these little tiny creatures is road accidents since various herpetofaunal species often meet road accidents while crossing main roads situated in between microhabitats where they inhabited.

Hunting Pressure

Certain local tribes and labours at brick manufacturing units who reside in nearby microhabitats of *G. elegans, V. bengalensis, S. hardwickii*, etc., hunt these species for food and other uses.

Pet Practices and Orthodox Issues

It is believed in Chinese fengshui and Indian Vastushastra, that rearing of turtles and tortoise in house helps increments in financial status and social status. These types of orthodox issues have increased pet practices of *G. elegans* in study area. One more myth is that tortoise is representative of goddess Laxmi (in Indian mythology goddess of wealth) and rearing of *G. elegans* brings good health-wealth to family, such practices cause removal of breeding individuals from their natural habitats. When they go for breeding in domestic environment with hard floor it often results in failure of hatching and egg mortality.

Illegal Trade

Illegal trade of wild animals is a worldwide issue and *G. elegans* is not exception (Jenkins, 1995; Sekhar *et al.*, 2004; Shephard *et al.*, 2004; Vyas, 2009). Several records show that the *G. elegans* and other turtles and tortoises are facing serious problems for their survival due to illegal trade for food, medicinal uses and pet practices all over the world.

Recommendations

Regular monitoring of *G. elegans* during the study period has demonstrated that the population is facing serious threats in the study area of Central Aravalli mountain ranges. The threats to the population of *G. elegans* were mainly destruction of habitats, expansion of urbanization in the habitat resulting in habitat fragmentation, deforestation, mining activities, pet practices, orthodox issues, illegal trade for food and medicinal uses, etc.

The primary cause which is responsible for declining the population of *Hemidactylus triedrus* is identified as mining and habitat fragmentation. And many other causes, such as urbanization, deforestation and climate change that might causes disturbance in the normal life cycle or lifestyle of this ecologically important lizard species. However, during the observations it was noticed that large areas of hills slope are dug out by local inhabitants to collect stones, plants and for their domestic use.

Other three species of lizards *Cyrtodactylus madarensis, Cyrtodactylus fedtschenkoi* and *Mabuya dissimilis* was not sighted in the period of 2000-2010 (Sharma *et al.*, 2010). This indicates the decline status of population of the mentioned saurian species. All of these species observed having very short range of activities at central Aravalli but due to the increased rate of mining, urbanization and land encroachment the residential habitats of these lizards along with *H. triedrus* have lost so the populations of these species are facing severe threats. It has been also observed that the species of nocturnal habit are affected more severely as compared to the diurnal ones (Sharma *et al.*, 2010). These anthropogenic activities severely affect this lizard because the niche and preferred habitat is highly specific. The rate of declining the population is very fast and if a check on mining and land encroachment in the central Aravalli range is not put at the earliest, this lizard may vanish completely from this part of globe. Therefore, the species needs serious consideration for its conservation.

Suggestions for Conservation: It is suggested that efforts should be made to protect the natural habitats of this lizard by awareness and by signboard mounted near the road sides where this lizard often crosses the roads for feeding and breeding. The myths about the danger of this lizard for human beings may be removed by education and awareness.

Conservation Strategies and Recommendations: *G. elegans* has been kept in Schedule IV of Wildlife Protection Act 1972; Appendix II in CITES and Vulnerable in IUCN Category, but this is not sufficient for its conservation. Hardly any action has been evidenced by the forest authorities or district administration against illegal pet practices, hunting and trade of this species. Further its distribution and population status is needed to be properly documented that would clarify the picture of its occurrence at protected areas and outside the protected areas. This will help in proper conservation plan and management. There is a need to develop awareness among the local co-inhabitants about its ecological importance. So the trends of pet practices are reduced. It is also required to remove myths and orthodox issues related to this species. Most people are not aware about the legal issues related to Wild Life Protection Act 1972. It is necessary to develop awareness about this and other acts protecting the environment and animal welfare. Habitat restoration, *in situ* and *ex situ* breeding and helping the individuals crossing the highway shell definitely increase the population of *G. elegans* in the central Aravalli region of our planet (Sharma, 2003; Rao, 2005).

Acknowledgements: We are thankful to the DST, Govt. of India, for financial support. The effort of corresponding author (Prof. K.K. Sharma) for the innovative idea of Live Zoology in the curriculum of Master Degree course so that such a study could be made is highly acknowledged. We are also thankful to the Save Environment and Welfare of Animals Society (SEWA), Jaipur, Rajasthan for the technical support and motivation.

NOTES

Blanford, W.T. (1901): *The Distribution of Vertebrate Animals in India, Ceylon and Burma*. Philos. Trans. R. Soc. Lond. 194: 335-436.

Boulenger, G.A. (1890): *The Fauna of British India including Ceylon and Burma: Reptilia and Batrachia*. Taylor and Francis, London. xviii + 541 pp.

Chanda, S.K. (2002): *Handbook of Indian Amphibians*. Zoological Survey of India, Kolkata. viii + 335pp.

Daniel, J.C. (2002): *The Book of Indian Reptiles and Amphibians*. Bombay Natural History Society and Oxford University Press, Mumbai.

Daniels, R.J.R. (1996): Biogeographic and taxonomic uncertainties as illustrated by Indian amphibians. *Cobra* 24: 2-6.

Daniels, R.J.R. (1997): *A Field Guide to the Frogs and Toads of the Western Ghats, India*: Part I. Cobra 27: 1-25.

Daniels, R.J.R. (2005): *Amphibians of Peninsular India*. Indian Academy of Sciences and Universities Press, Hyderabad.

Dutta, S.K. (1992): *Amphibians of India and Sri Lanka* (Checklist and Bibliography). Odyssey Publishing House, Bhubaneswar, India xiii + 342 pp.

Envis. (2000): Bulletin, *Wildlife and Protected Area*, 3, No.1.

Heron, A.M. (1953): *Geology of Central Rajasthan*. Mem. Geol. Surv. India, 79:339.

Heyer, W. R.; M.A. Donnelly; R.W. McDiarmid; L.A.C. Hayek and M.S. Foster (1994): *Measuring and monitoring biological diversity: Standard methods for amphibians*. Washington, Smithsonian Institution Press.

Jenkins, M.D. (1995): *Tortoises and Freshwater Turtles: The Trade in South Asia*. TRAFFIC International, U. K.

Krishnan, A. (1977): A climatic analysis of the arid zone of North-western India. In: *Desertification and its control*. ICAR, New Delhi. pp. 69-76

Pocock, R.I. (1939) and (1941): The fauna of British India, *Mammalia*. Vols I & II. Franas and Taylor, London.

Prakash, I. (1963): Zoo-geography and evolution of the mammalian fauna of Rajasthan desert. *Mammalia* 35: 384-423.

Rao, R.J. (2005): *Captive Breeding of Star Tortoise (Geochelone elegans) in the Deer Park*, NFCL, Kakinada, Andhra Pradesh. Reptile Rap. 7, 07-09 pp.

Roy, A.B. and Jakhar, S.R. (2002): *Geology of Rajasthan (Northwest India) Precambrian to Recent. Jodhpur*, Scientific Publishers. 421 pp.

Sekhar, A. C.; Gurunathan, N. and Anandhan, G. (2004): *Star tortoises—A victim of the exotic pet trade*. Tiger Paper 31 (1): 4-6.

Sharma Vivek; Meena Dinesh; Kumawat Rakesh and Sharma, K.K. (2012): Assessment of faunal diversity in and around certain wetlands of Central Aravalli foothills of Rajasthan. *International Journal of Environment & Animal Conservation*, Vol. 1 (1), 34-37.

Sharma, K.K. and S.P. Mehra (2009): The Thar of Rajasthan (India): Ecology and conservation of a desert ecosystem. In *Faunal ecology and conservation of the Great Indian Desert*. Eds. Sivaperuman, C.; Baquri, Q.H.; Ramaswamy, G. and Naseema, M., Springer Germany, pp. 1-10.

Sharma, K.K. and Vivek Sharma (2009a): *Records of two Microhylids; Microhyla ornata (Ornate Narrow Mouthed Frog)* and *Uperodon systoma (Mrbaled Balloon Frog)* from Ajmer district, central Rajasthan (India). Cobra, Vol 3 (III). 11-15 pp.

Sharma, K.K. and S.P. Mehra (2007): Need of anuran studies in habitats of southern Rajasthan, India. *Frog Leg* 13: 12-16.

Sharma, K.K. and Sharma Vivek. (2009b): Records of New locality of royal snake species in central semiarid region of Rajasthan (Ophidia: Colubridae; Colubrinae; Spalerosophis diadema atriceps). *Cobra*, Vol 3 (III). 16-18 pp.

Sharma, K. K.; M. Sathyanarayana; M.A. Akbarsha; Vivek Sharma and Neha Sharma (2011): Role of Amphibians in the Ecosystem, The Current Threats and Conservation Strategies. International Conference on Ecosystem Conservation and Sustainable Development. 10-12 February, 2011. Organized by Department of Biology, Ambo University, Ethopia. (Proceedings- pp- 163-164, Abstract).

Sharma, K. K., Sharma Neha and Sharma Vivek (2010): Current status and distribution of anurans from foothills of Aravalli ranges at central Rajasthan, India. *Flora and Fauna*. Vol. 16 (2), pp. 244-248.

Sharma, R.C. (2002): *The Fauna of India and the Adjacent Countries-Reptilia (Sauria)*, Volume II. Director, Zoological Survey of India, Kolkata.

Sharma, S.K. (2003): Is a wrong rescue practice of Testudian animals responsible for dwindling of their numbers in Rajasthan? Cobra (52): 13-14.

Shepherd, C. R.; Burgess, E.A. and Loo, M. (2004): *Demand Driven: The Trade of Indian Star Tortoises Geochelone elegans in Peninsular Malaysia*. TRAFFIC Southeast Asia.

Smith M.A. (1943), The Fauna of British India, including Ceylon and Burma. *Reptilia and Amphibia*. Vol. III. Serpentes, Taylor and Francis, London. Xii + 583 pp.

Vyas, R. (2009): The status of North-Western population of Star tortoise (*Geochelone elegans*). *In*: Vasudeven, K. (Ed.) *Freshwater Turtles and Tortoises of India*. ENVIS Bulletin: Wildlife and Protected areas, Wildlife Institute of India, Dehradun, India, Vol. 12 (1), 95-104 pp.

Whitaker R. and Captain A. (2004), *Snakes of India : The Field Guide*, Draco Books, Chennai, 481 pp.

13 Medicinal Plants–Biodiversity, Conservation and Bioprospecting

Raka Kamal

The age old traditional values attached with the various forest types and the varieties of forest products (i.e., medicinal plants) have gained tremendous importance in the present century. Furthermore, the cosmetic industries are increasingly using natural ingredients in their products, and these natural ingredients include extracts of several medicinal plants (Kit, 2003). India and China are two of the largest countries in Asia, which have the richest arrays of registered and relatively well-known medicinal plants for the use of medicinal plants as raw materials in the production of new drugs is ever increasing because of their potentials in combating the problem of drug resistance in microorganisms.

Interest and support for the conservation and development of medicinal plants is increasing in all parts of the world. This is due, in part, to a growing recognition given to the role of medicinal plants in the provision of culturally relevant and affordable health care in creating sustainable livelihoods and in the vital conservation of biodiversity. This has also drawn the attention of the world community towards the need for creating mechanisms to ensure sustained development of the sector and to allow sharing of information between countries, organizations and agencies.

Table 13.1: Floral Diversity in India

Numbers	Species
15,000 - 18,000	Flowering plants
23,000	Fungi
25,000	Algae
1,600	Lichens
1,800	Bryophytes
30 million	Microorganisms

Rising Demand

In India, Medicinal plants sector has traditionally occupied an important position in the socio cultural, spiritual and medicinal arena of rural and tribal lives. Recognizing its importance, the Government of India established the Department of Indian System of Medicine and Homeopathy, and more recently the Medicinal Plants Board to develop, promote and regulate the sector for maximizing the benefits to the people as well as to ensure sustainable growth. Medicinal plants have been identified as one of the thrust areas by the Ministry and different programmes have been initiated for conservation of medicinal plants found in the forests and protected areas as well as cultivation of these plants in the degraded forest areas.

World Scenario

Generally the terms "medicinal plants" and "herbals" are used interchangeably. The traditional or popular name of medicinal plants varies from country to region depending on the historical and cultural aspects.

The World Health Organization (WHO) has estimated the present demand for medicinal plants is approximately US $ 14 billion per year (Sharma, 2004). The demand for medicinal plant based raw materials is growing at the rate of 15 to 25 per cent annually, and according to an estimate of WHO, the demand for medicinal plants is likely to increase more than US $ 5 trillion in 2050. In India, the medicinal plant related trade is estimated to be approximately US $1 billion per year (Joshi *et al.*, 2004). According to an estimate, the quantity of export of Ayurvedic products produced in India has tripled between last two financial years (2001-2002 and 2002-2003.Demand for medicinal plants is increasing in both developing and developed countries. Research on medicinal plants is one of the leading areas of research globally. However, there is a need to pay closer attention to the issue of bioactivity-safety evaluation and conservation of medicinal plants. Although some of the screening tests on medicinal plants are performed *in vitro*, the fact still remains that the ultimate aim of the researcher is to use the medicinal plants to treat diseases in humans and animals, who has to take the product orally or through other means into the system. There is also the need to conserve the medicinal plants to prevent their total extinction from the natural flora. This is because the expanding trade in medicinal plants has serious implications on the survival of several plant species, with many under serious threat to become extinct. The aim of this review is to justify and emphasize the need for the bioactivity-safety evaluation and conservation of the medicinal plants.

In India, of the 17,000 species of higher plants, 7500 are known for medicinal uses (Shiva, 1996). This proportion of medicinal plants is the highest proportion of plants known for their medical purposes in any country of the world for the existing flora of that respective country. Ayurveda, the oldest medical system in Indian sub-continent, has alone reported approximately 2000 medicinal plant species, followed by Siddha and Unani. The Charak Samhita, an age-old written document on herbal therapy, reports on the production of 340 herbal drugs and their indigenous uses (Prajapati *et al.*, 2003). Currently, approximately 25 per cent of drugs are derived from plants, and many others are synthetic analogues built on prototype compounds isolated from plant species in modern pharmacopoeia (Rao *et al.*, 2003).

Medicinal plants have strong acceptance in religious activities of north Indian native communities, who worship the plants in the form of various gods, goddesses and minor deities (Silori & Badola, 2000; Dhyani, 2000). *Origanum vulgare, Saussurea obvallata, Ocimum sanctum, Cedrus deodara, Cynodon dactylon, Aegle marmelos, Juniperus communis, Musa paradissica, Nardostachys grandiflora, Zanthoxylum armatum, Ficus benghalensis,* and *Ficus religiosa* are examples of the medicinal plants highly used for medicinal as well as a religious purposes by the Hindus in northern India. The Buddhist community regards *Terminalia chebula* as an important medicine as well as sacred fruit. It has been stated long ago that the therapeutic potency of medicinal plants is more effective and better suited to a person of a particular region or culture in which the plant is naturally growing (Nadkarni & Nakarni, 1908). This idea has given a way to the development of a new drug for heart patients of specific ethnic groups in African countries (Pearson, 2004).

Medicinal plants are plants containing inherent active ingredients used to cure disease or relieve pain (Okigbo *et al.*, 2008). The use of traditional medicines and medicinal plants in most developing countries as therapeutic agents for the maintenance of good health has been widely observed (UNESCO, 1996). The World Health Organization estimated that 80 per cent of the populations of developing countries rely on traditional medicines, mostly plant drugs, for their primary health care needs (Schmincke, 2003). Medicinal plants represent a consistent part of the natural biodiversity endowment of many countries in Africa (Okigbo *et al.*, 2008). Also, modern pharmacopoeia still contains at least 25 per cent drugs derived from plants and many others which are synthetic analogues built on prototype compounds isolated from plants. Interest in medicinal plants as a re-emerging health aid has been fuelled by the rising costs of prescription drugs in the maintenance of personal health and well-being and the bioprospecting of new plant-derived drugs (Lucy & Edgar, 1999). Furthermore, an increasing reliance on the use of medicinal plants in the industrialized societies has been traced to the extraction and development of several drugs and chemotherapeutics from these plants as well as from traditionally used herbal remedies (UNESCO, 1998). The medicinal properties of plants could be based on the antioxidant, antimicrobial antipyretic effects of the phytochemicals in them (Cowman, 1999; Adesokan *et al.*, 2008).

Medicinal Plants Resource Base

Current practices of harvesting are unsustainable, resulting in depletion of resource base. Medicinal plants based industries in India also lack a proactive and socially responsible approach, and are, therefore, partly responsible for inefficient, imperfect, informal and opportunistic marketing of medicinal plants.

The collections in most cases are done by villagers and tribals, without paying attention to the stage of maturity, dried haphazardly and stored for long periods under unsuitable conditions. The quality of collected material, as such is often degraded.

Several medicinal plants have been assessed as endangered, vulnerable and threatened due to overharvesting or unskilful harvesting in the wild. The Government of India has put 29 species in the negative list of export.

The incidence of renal disease is increasing dramatically and has become a significant public health problem both economically and medically (Salako, 2005). Several studies (Bwititi *et al.*, 2000; Ijeh & Agbo, 2006) have indicated the possibility that the use of plant extracts in high doses could lead to toxic injury to the kidneys which interfere with renal tubular functions and induce acute renal failure. Folkloric history has indicated attempts made by inhabitants using plant-derived recipes in parts of Nigeria to treat what they described as "fever of crisis", shifting joint pains, exacerbations especially during rainy seasons and "constant abnormality of the blood" though relatively few have been validated scientifically (Egunyomi *et al.*, 2009). This calls for a need to further investigate safer concentration of ethnomedicinal preparations in view of the increasing reports of acute renal failure (Ijeh & Ukweni, 2007).

Department of Indian System of Medicine and Homoeopathy (ISM&H)

The Government of India established the ISM&H, under the Ministry of Health and Family Welfare in 1995, with the objective of rendering simple and effective remedies to people, particularly in the area of preventive health, maintaining healthy life style, alleviating discomfort caused by specific diseases and treating chronic diseases.

Activities of the ICM&H include:

- Upgrading educational standards in ISM&H;
- Strengthening existing research and ensuring time bound research programmes in ISM&H;
- Drawing up and administering schemes for promotion, cultivation and regeneration of medicinal plants; and
- Evolving pharmacological standards for ISM& H.

Pharmacopoeia Standards

The Ministry of Health and Family Welfare has taken up the task of developing Pharmacopoeia standards, both for single as well as compound drugs, through Pharmacopoeia Committees. Pharmacopoeia standards are mandatory for the drug testing provisions under the Drugs and Cosmetic Act 1940. These standards are also used for checking samples of drugs available in the market for their safety and efficacy.

Quality Standards for Medicinal Plants

The Medicinal Plants Unit of Indian Council for Medical Research has announced that it would be shortly releasing Quality Standards for 32 Indian Medicinal Plants, based on research conducted at four national research institutions. These standards will provide guidelines for herbal drugs manufacturers, practitioners, academicians, researchers as well as regulatory bodies.

Good Manufacturing Practices (GMPs)

The Indian System of Medicine has adopted the GMPs with effect from June 2000 for herbal products manufacturing units, with the objective of ensuring quality assurance covering all aspects of the manufacturing system. GMPs were formulated initially in a WHO report on specifications for pharmaceutical preparations. These have been expanded/elaborated from time to time. In the present form, the GMPs lay down specific requirements for various aspects such as location, surroundings, factory premises, buildings, water supply, disposal of waste, storage area, machinery and equipment, health and hygiene of workers, documentation and records, and quality control system. The manufacturing unit is issued a GMP certificate, valid for 2 years, after proper inspection of the premises

International Standards and Regulations

The legal situation regarding herbal preparations varies from country to country. In some countries, phytomedicines are well established, whereas in others they are regarded as food, and therapeutic claims are not allowed. Developing countries, however, have a large number of traditionally used herbal medicines and tremendous folk knowledge about them, but no significant steps have been taken to establish these traditionally used herbal medicines as part of the drug legislation.

The Need for Conservation of Medicinal Plants

Demand for medicinal plants is increasing in both developing and developed countries and surprisingly, the bulk of the material traded is still from wild harvested sources on forest lands and only a very small number of species are cultivated (Schmincke, 2003). The expanding trade in medicinal plants has serious implications on the survival of several plant species, with many under threat to become extinct. Okigbo *et al.,* (2008) reported the need for effective conservation strategies for medicinal plants. They stated that saving Africa's medicinal plant resources from extinction calls for intensive management and conservation through more research and increased level of public awareness about these vanishing heritage.

Many medicinal plants are fast disappearing and some are in danger of extinction. There is a great need to conserve medicinal plants because they contain highly bioactive components which can be developed into pharmacologically active agents. Some of the pharmacological and other beneficial effects of anti-nutritional factors in plants have been reviewed by Soetan, (2008). Many signs reveal that medicinal plants are gradually facing extinction. They are (i) People walk long distances to collect them. (ii) Some medicinal plants are no longer found. (iii) What used to be a thick forest of diverse plant species is reduced to bush and areas that have diverse flora are fast disappearing; and (iv) Many medicinal plants are not maturing and seeding because the young plants are being harvested before they mature (ITDG and IIRR, 1996). The causes have been attributed to a number of factors like rising number of human and animal population which cause pressure on plant survival. Human agricultural activities result in clearing of natural habitats for farming and grazing, trees are felled for timber, charcoal and other commercial uses and lands are used in a way that are not sustainable. Others are inappropriate ways of harvesting medicinal plants, like removing all the bark or uprooting the whole plant without leaving part to regrow, bush burning and commercialization of plant sources, lack of awareness that plants are sources of the conventional medicine, poverty in arid or semi-arid areas and religious factors. For example, some western religions view the use of traditional medicine as a form of evil worship or witchcraft (ITDG and IIRR, 1996).

As a result of the numerous potentials of medicinal plants in the combating of drug resistance by micro-organisms, strict conservation measures should be put in place to prevent the total extinction of the medicinal plants from the natural flora. They are medicinal plants should be harvested in the proper way to avoid serious damage and proper control of overgrazing and deforestation. This could be done by practicing rotational grazing so as to allow plants to rejuvenate/regrow. The Government can reserve some areas strictly for medicinal plants and prevent encroachment by undesirable plant species. Individual small-scale cultivation of medicinal plants should also be encouraged as is done for tree planting that is, medicinal plants should be incorporated in agroforestry and reafforestation programmes. Mini-forests on individual farms should be maintained to increase biodiversity. Governments should also have reserve stocks of medicinal plants and encourage community actions to collect, retrieve and plant seeds of medicinal plants.

The removal and export of rare and scarce medicinal plants should be discouraged/ prohibited by legislation. The intellectual property rights of practitioners with great discoveries in medicinal plants should be protected so as to encourage them. Scientists and toxicologists should investigate the active components of certain medicinal plants *viz-à-viz* their toxic potentials so as to determine their safe level of consumption. Public education on medicinal plants as a potential source of modern medicine should be promoted in schools and tertiary institutions of learning.

Replantation of Medicinal Plants

The Forests Department of the State government should be entrusted with this very important task for conservation of medicinal plants. Based on earlier experience, about 2 lakh plants per district per year should be made available to the Forest department for this purpose.

Development of Data Bank and Information Centre

Managing and dissemination of market information is very critical for success of the export promotion initiative. Hence, it is proposed to set up a state of art Data Bank and Information Centre, to be maintained by the Extension Department of the State Government. The centre will maintain detailed records on medicinal plants production and development in India as a whole. The centre also provide information on various international practices/researches on growing, harvesting, transportation, etc., and latest R&D initiatives will be made available to farmers, pack houses and exporters.

Bioprospecting the Medicinal Plants

Bioprospecting of plants entails the search for economically valuable biochemical resources from the floral wealth of a country. Such initiatives hold the promise of new medicines as well as biodegradable pesticides that can be a source of income for developing countries, thus providing incentives to conserve biodiversity. The bioprospecting practices of the institute focused on six main areas: (1) identifying bioactive constituents, (2) producing plant-based safer pesticides, (3) antifertility agents (3) studying the efficacy and safety of plant-based composite medicines, (4) producing plant-based edible oil, (5) cultivating herbals having high value chemicals, e.g., *Catharanthus roseus* yielding vincristine and vinblastine alkaloids, and (6) developing programmes to increase the value of raw plant materials.

Today, adverse health effects and pest-resistance are two of the most serious problems associated with the use of synthetic pesticides. Thus, there is growing interest in the discovery of natural pesticides that are affordable, biodegradable and host-specific. India has a rich tradition in the use of plant materials as pesticides. Based on the folk use, researchers at the institute selected several abundantly available plants and screened them for pesticidal properties (Kamal, 1985, Kamal & Mangla,1991; Kamal & Mehra, 1991; Kamal & Mathur 1991,1992).

From this effort, two plant-based pesticide formulations were developed. Initial field trials on cotton and horticulture crops have shown extremely promising results. Both formulations were found to be active against white fly, an insect considered one of the most destructive agriculture pests in India. Patents on both formulations have been filed (Kamal,1976). Population control is another major problem of the developing countries and role of phytobased antifertility agents have been reviewed by Kamal *et al.* (2003).

The medicines for internal use prepared in the traditional manner involve simple methods such as hot- or cold-water extraction, extraction of juice after crushing, powdering of dried material, formulation of powder into pastes using water, oil or honey, and even fermentation after adding a sugar source. Traditionally, the herbal medicines were produced manually in small quantities, by the practitioner himself/herself who was able to identify the correct plant species. However, with the introduction of commercial production of herbal drugs, while the processing efficiencies have improved considerably, it has also introduced an element of doubt about the authenticity and quality of plant material used. Therefore, there is a need to select proper and appropriate technologies for the industrial production of traditional medicines such that the effectiveness of the preparation is maintained. The production of garlic-based edible oil has been another focus of the institute. Garlic (*Allium satirvm*) is a common herb used extensively all over the world. Both the folk use and scientific literature confirm its efficacy against several health disorders. Ajoene, a constituent of garlic, is reported to have clinically proven cholesterol-lowering effect in human patients. Researchers have developed a process for isolating and chemically enriching the ajoene. This additive, when blended with the edible oil, can reduce blood-cholesterol levels substantially with moderate use. *Catharanthus roseus* is one of the most important medicinal plants in the world. The plant contains two binary indole alkaloids, vinblastine and vincristine that are drugs of choice for pediatric leukemia and Hodgkin disease. Because of its importance, the institute has cultivated *Catharanthus roseus* and produced vincristine and vinblastine at its plant section for bioprocessing. These medicines are produced from natural sources due to very high synthetic cost and poor yields. A large-scale isolation process was developed that gives 30-fold higher yields (up to 60 grams per tonne of leaves) of vinblastine than the previous process. Bioprocessing also involves adding value to the large quantities of plant materials that are exported in crude forms from India. This includes *Glycyrrliza glabra, Hypercum performatus, Dioscorea* sps, *Digitalis* sps, *Commiphora mukul, Carum copticum, Aloe vera, Asparagus racemosus, Azadirachta indica, Bacopa monniera, Bauhinia purpurea, Barleria* sps, *Brassica campastris, Butea monosperma, Calotropis* sps, *Camellia sinensis, Capparis decidua, Capsicum annum, Cassia* sps., *Centella asiatica, Chlorophytum arundinaceum, Cinnamomum verum, Citrullus sps,* Coleus sps, *Commiphora mukul, C. whightii, Convolvulus* sps, *Crocus sativus, Crotolaria* sps., *Curcuma longa, Cymbopogan citrates, Datura* sps, *Delbergia sissoo, Eclipta alba, Momordica charantia, Moringa oleifera, Morus alba, Murraya koenigii, Nigella sativa, Ocimum sanctum, Parkinsonia* sps, *Pedalium murex, Phyllanthus niruri, Prosopis* sps., *Plumbago zeylanica, Psoralea corylifolia, Punica granatum, Rubia cordifolia, Sesamum indicum, Sida cordifolia, Solanum* sps, *Syzigium cumini, Tephrosia* sps., *Terminalia arjuna, T. bellarica, Tinospora cordifolia, Tribulus terrestris, Trigonella foenum-graecum, T. polycerata, Withania somnifera and Zingiber officinalis, Zizyphus* sps etc.Triphala, Churnas for stomach disorders, Goggul, Senna, Rotenoids, Pyrethrins, Homeopathic tinctures, Ayurvedic preparations, etc. These plants are collected from wild sources and sold to local traders who, through a chain of middlemen, are able to reach international traders. In this chain of profit makers, the plant gatherers receive virtually nothing. The logical solution to this problem would be to eliminate the middlemen and add value to the herbal material.

Conclusions

Determining the biological (activity) properties of plants used in traditional medicines is helpful to the rural communities and informal settlements. Several authors are currently being undertaken to isolate the active compound(s) by bioassay-guided fractionation from the species that showed high biological activity during screening. Therefore, these scientific investigations may be utilized to develop drugs for these diseases. Further research is deserved to isolate the compounds responsible for the observed biological activity.

NOTES

Adesokan, A.A.; Yakubu, M.T.; Owoyele, B.V.; Akanji, M.A.; Soladoye, A. and Lawal O.K. (2008). Effect of administration of aqueous and ethanolic extracts of *Enantia chlorantha* stem bark on brewer's yeast-induced pyresis in rats. *Afri. J. Biochem. Res.* 2(7): 165-169.

Bwititi, P.; Musabayane, C.T. and Nhachi, C.F. (2000). Effects of *Opuntia megacantha* on blood glucose and kidney function in streptozotocin diabetic rats. *J. Ethnopharmacol.* 69 (3): 247-252.

Cowman M.M. (1999). Plant products as antimicrobial agents. *Clin. Microbiol. Rev.* 12: 561-582.

Dhyani P.P. (2000). Common plant species have potential for economic upliftment of rural populace-Bantulsi a case in point. *Hima-Paryavaran.* 12:11-13.

Egunyomi, A., Moody, J.O. and Eletu, O.M. (2009). Antisickling activities of two ethnomedicinal plant recipes used for the management of sickle cell anaemia in Ibadan, Nigeria. *Afri. J. Biotechnol.* 8(1): 020-025.

Ijeh, I.I. and Agbo, C.A. (2006). Body organ weight changes following the administration of aqueous extracts of Ficus exasperate. Vahl. J. Anim. and Vet. Adv. 5: 277-279Anim. and Vet. Adv. 5: 277-279.

Ijeh, I.I. and Ukweni A.I. (2007). Acute effect of administration of ethanol extracts of *Ficus exasperata* vahl on kidney function in albino rats, *J. Med. Plants Res.* 1(2): 027-029.

I.T.D.G. and I.I.R.R. (1996). Ethnoveterinary medicine in Kenya: A field manual of traditional animal health care practices. Intermediate Technology Development Group and International Institute of Rural Reconstruction, Nairobi, Kenya.

Joshi, K.; Chavan, P.; Warude, D. and Patwardhan, B. (2004). Molecular markers in herbal drug technology. *Curr. Sci.* 87:159-165.

Kamal, R. (1985). *Naturally Occurring Insecticides from Plant Tissue Culture.* A review. 10th Plant Tissue Culture Association Meet (India) pp. 11-12.

Kamal, R.; R.S.Gupta and N.K. Lohiya. (2003). *Plants for Male Fertility regulation.* Phytotherapy Research, UK, 17 (6): 579-590.

Kamal, R. (1976). Govt. of India Patent on the process of 'Isolation of Pyrethrins from *Tagetes erecta* Indian Patent *No. 142643, 1976.*

Kamal, R. and M. Mangla, (1991). The compatibility of Rotenoids and Piperonyl butoxide with Pyrethrins from *Tagetes minuta*, against mosquito larvae. Pyrethrum Post 18(2): 65-69.

Kamal, R. and P. Mehra, (1991). Efficacy of Pyrethrins extracted from *Dysodia tenuifolius* and *Ageratum conizoides* against larvae of *Anopheles stephensi.* Pyrethrum Post 18(2): 70-73.

Kamal, R. and N. Mathur, (1991). *In vitro* amoebicidal activity of World Standard Pyrethrins against *Entamoeba hystolytica.* Pyrethrum Post, 18(3): 92-93.

Kamal, R. and M. Mangla, (1992). A phytochemical investigation of few plant species. *Vegetos* 5(1&2): 41-45.

KIT. (2003). *Cultivating a Healthy Enterprise.* In Bulletin 350 Royal Tropical Institute, Amsterdam, The Netherlands

Lucy, H. and Edgarm, J.D. (1999). Medicinal Plants: A reemerging Health aid. *Electronic. J. Biotechnol.* 2(2): 1-15.

Nadkarni, K.M. and Nadkarni, A.K. (1908). *Indian Materia Medica.* Bombay: Popular Prakashan.

Okigbo, R.N.; Eme, U.E. and Ogbogu S. (2008). Biodiversity and conservation of medicinal and aromatic plants in Africa. *Biotechnol. Mol. Biol. Rev.* 3(6): 127-134.

Pearson, H. (2000). First 'black drug' nears approval. *Nature*

Prajapati, N.D.; Purohit, S.S., Sharma, A.K. and Kumar, T. (2003). *A Handbook of Medicinal Plants.* Jodhpur: Agrobios. India.

Rao, M.R.; Palada, M.C. and Becker, B.N. (2004). Medicinal and aromatic plants in agro-forestry systems. *Agroforestry Systems.*, 61:107-122.

Silori, C.S. and Badola, R. (2000). Medicinal plants cultivation and sustainable development: A case study in buffer zone of the Nanda Devi Biosphere Reserve, Western Himalaya, India. *Mountain Research and Development.* 20:272-279.

Salako, B.L. (2005). The Relationship between kidney and hypertension: A Review. *Afr. J. Med. Med. Sci.* 34: 335-340.

Schmincke, K.H. (2003). *Medicinal Plants for Forest Conservation and Healthcare. Non-Wood Forest Products* 11, Food and Agriculture Organization of the United Nations

Sharma, A.B.: Global Medicinal Plants Demand May Touch $5 Trillion By 2050. *Indian Express,* 2004.

Shiva, M.P. (1996). *Inventory of Forestry Resources for Sustainable Management and Biodiversity Conservation.* New Delhi, India. Indus Publishing Company.

Soetan, K.O. (2008). Pharmacological and other beneficial effects of Antinutritional factors in Plants—A Review. *Afr. J. Biotechnol.* 7(25): 4713-4721.

UNESCO. (1996). *Culture and Health,* Orientation texts—World Decade for Cultural Development Documents CLT/DEC. PRO-1996, Paris, France, p. 29.

UNESCO. (1998). *Terminal Report: Promotion of Ethnobotany and the sustainable use of Plant Resources in Africa.* p. 60.

14

Environmental Uranium and Human Health

A Review from an Epidemiological and Biochemical Perspective

K. Vali Pasha, R. Pavanaguru and *B. Sashidhar Rao*

Introduction

Fossil fuels are lasting assets and are dwindling at faster rate warranting need to identify alternate energy sources. Current energy resources of petroleum are not sustainable for a long term. Energy from nuclear fission of uranium appears to be more viable alternative that could sustain economic and human development indefinitely. However, processing technology and extraction of uranium has significant adverse effects on the environment if it is not properly identified and managed. Therefore, efforts should be undertaken to understand and minimize the potential health risks to the mine workers and also the environment (Cuttler, 2009). The geological resource potential of uranium ore, for instance, at Thummalapalli area in A.P., India, a recent find under extraction is placed at 170,000 tonnes, making it as the world's largest strata-bound uranium deposit. The Deptartment of Atomic Energy recently discovered that the upcoming mine in Thummalapalli has close to 49,000 tonnes uranium reserve and there were indications that the total quantum of uranium could go up to 1.7 lakh tonnes. This is a new area where characterization of different effects of incidence and biochemical remediation can be identified and put to practice.

Uranium occurs as essential constituent in more than one hundred minerals. The most important ore minerals are uraninite and the weakly crystallized variety, pitchblende, coffinite, branmerite, davididte, iranothorite and uranothorianicnite. Uranium, the heaviest of the naturally occurring elements, is widely present in the environment as a result of leaching from natural deposits, released in mill tailings, emissions from the nuclear industry, the combustion of coal and other fuels, and the use of phosphate fertilizers and weapons that contain uranium. Uranium is found in various chemical forms and different levels in all soils, rocks, sea and bed rock (Bosshard *et al.*, 1992, Kurttio *et al.*, 2002, Moss *et al.*, 1983).

Extraction of Uranium ore from earths' crust, milling and chemical processing are producing large amount of residues from which heavy metals are leachable by rain. Uranium and heavy metals are hazardous and may found their way to the groundwater in concentrations exceeding groundwater and drinking water standard limits (Salem *et al.*, 2000). Leaching and erosion of mill tailings and mine drainage not only can cause contamination of groundwater but also it can pollute the nearby rivers and lakes. This can lead to the increase in levels of contamination affecting fish, or crops and even plants if water is used for irrigation. Air pollution may result from wind erosion (Veiga *et al.*, 1998). Inappropriate conditioning and disposal of tailings waste permit the contaminants to spread into air, soil, sediment, surface water and groundwater, if not properly disposed of. Uranium mill tailings constitute a potential hazard to public health for very long period of time (Cothem, 1983).

The environmental effect of uranium is concerned with the air, water and land. The distribution and migration of uranium from the primary set-up into the exogenic environment is observed at various levels. Hence, the identification of these migrants and characterization of aerosols in the core zone attract immediate attention for necessary mitigation.

Effect of Uranium on Health

In the extraction of uranium ore, and its chemical processing, mill tailings contain radioactive elements such as U, Th, Ra, Rn, and non-radioactive heavy metals in low concentrations. Inappropriate conditioning and disposal of tailings waste permit the contaminants to spread into air, soil, sediment, surface water, and groundwater. Uranium deposits weather naturally by oxidizing groundwater flowing through fractured rock and infiltrating from the surface (Otton *et al.*, 1989; Yanase *et al.*, 1995). Uranium that goes into solution can migrate over long distances and can be concentrated both in surface and groundwater. Such concentration is enhanced by the ability of uranium to form complexes, which are most stable in solutions where the pH is greater than 7.5. Generally, levels of more than 4 ppb U are considered to be anomalous. However, in arid continental conditions contents of several hundred ppb U have been recorded, and in the vicinity of U deposits it would reach 2000 ppm U or more (Thornton, 1983).

Increased uranium levels in groundwater are associated with uranium-rich ores and high solubility of uranium under oxidizing conditions in soft and bicarbonate rich waters (Salonen, 1994). Consequently, exceptionally high uranium concentrations have been found in private drilled wells located mostly in the southern part of Finland. The chief raw material of uranium is pitchblende or carnotite ore. This element is largely limited to use as a nuclear fuel. Table 14.1 shows the various epidemiological studies of uranium contamination and its impact.

Table 14.1: Epidemiological Studies and its Health Impacts

	Study	Major findings
1.	Fernald uranium processing plant (Ohio)	Study finds decreases in white blood cell counts and alterations in systolic blood pressure among residents in the vicinity
2.	Uranium mines (New Mexico)	Increased likelihood of kidney disease and diabetes among people who live close to abandoned uranium mines
3.	Jadugoda uranium mine (Jharkhand, India)	Survey finds excess deformities and cancer
4.	Spanish uranium mills	Increase in leukemia, lung and renal cancers
5.	Karnes County, Texas	Chromosome aberrations with residents of uranium mining area
6.	Shiprock, New Mexico	Increase in birth defects from mothers living near uranium tailings
7.	New Jersey Department of Health, New Jersey	Lung cancer cases
8.	109 Municipalities in Sweden	Elevated Radon concentrations
9.	Pershagen and co-workers	586 women and 774 men with lung cancer diagnosed
10.	Kazakh Uranium mines	Excess chromosomal aberrations
11.	French uranium miner cohort	Association between cerebrovascular diseases mortality and cumulative Radon exposure
12.	German Uranium miner study	Evidence of increased risk of extra-pulmonary cancers associated with Radon
13.	Wismut Uranium miners	Lung cancer/Leukaemia
14.	Hungarian Uranium miners	Long term persistence of chromosomal aberrations
15.	Namibian uranium miners	Chromosomal aberrations
16.	Native American uranium miners	Higher risk of non-malignant respiratory disease

Uranium and Human Health

The uranyl ion is rapidly absorbed from gastrointestinal tract. About 60 per cent is carried as a soluble bicarbonate complex, while the remainder is bound to plasma protein. 60 per cent is excreted in the urine within 24 hours about 25 per cent may be fixed in the bone. Following inhalation of the insoluble salts, retention by the lungs is prolonged. Uranium tetrafluoride and uranyl fluoride can produce a typical toxicity because of hydrolysis to HF. Skin contact (burned skin) uranyl nitrare has resulted in nephritis.

The soluble uranium compound and those that solubilize in the body by the formation of a bicarbonate complex produce systemic toxicity in the form of acute renal damage and renal failure, which may be fatal. However, if exposure is not severe enough, the renal tubular epithelium is regenerated and recovery occurs. A study of uranium mill workers suggested that workers long-term long level exposure is associated with β_2-microglobulinuria and aminoaciduria. Table 14.2 gives the various biological effects of uranium.

Normally 10-35 per cent and up to 70 per cent of the depleted uranium is estimated to be aerosolized on impact or when DU catches fire. Most of the dust particles are less than 4 μm to be inhaled or ingested by humans causing respiratory and other problems.

Table 14.2: Biological Effects of Uranium

Biological Effect
Decreased in white blood cell counts and alterations in systolic blood pressure
Chromosome aberrations
Increase in birth defects
Fibrosis
Deformities
Lung cancer
Aplastic anemia
Small cell carcinoma
$Beta_2$ -microglobinuria
Aminoaciduria
Albuminuria
Toxicity because of HF
Elevated blood urea
Lysosomes accumulation
Osteogenic sarcoma
Loss of cellular immunity
Autoimmune disease
Joint disease
Kidney and circulatory system disease
Nervous system effects
Alterations in respiratory enzymes
Decline in functional mitochondria
Changes in male reproductive capacity
Diabetes
Renal cancer

Depleted Uranium and its Effects

Depleted uranium is the by-product in the process of converting natural uranium for use as nuclear fuel. Several studies have reported on the health effects of exposure to depleted uranium. Immediate health risks associated with exposure to depleted uranium include kidney and respiratory problems, kidney stones, chronic cough and severe dermatitis. Long term risks include lung and bone cancer. Several published reports implicated exposure to depleted uranium in kidney damage, mutagenicity, cancer inhibition of bone, neurologic deficits, adverse effects on the reproductive and central nervous system (Arzuga *et al.*, 2010). After an accidental absorption of Uranium compounds, the soluble forms of uranium is carried in the blood, filtered by the glomerules and partly excreted in urine. Uranyl nitrate induced nephropathy has been extensively studied in animal models.

Some of the environmental effects of uranium on human health has been reviewed by Taylor *et al.*,1997; Brugge *et al.*, 2005. Many uranium mine workers have been disabled and died of pulmonary fibrosis that was not recognized as an occupational disease (Archer *et al.*, 1998). Chromosomal aberration in lympochytes of Namibian uranium miners were also reported (Loyal *et al.*, 2001). A study undertaken in New Mexico uranium mines showed that babies from mothers who lived near the tailings dump, suffered a significant increase in birth defects (Schields *et al.*, 1992).

Renal Effects

The first study on the effects of chronic Uranium ingestion from drinking water showed impaired renal function, with the involvement of proximal tubule. (Zamora *et al.*, 1998). A much larger study on exposure of humans to Uranium in drinking water revealed nephrotoxic effects over at low concentration without a clear threshold (Kurttio *et al.*, 2002).

The kidney has been considered the main target organ of chemical toxicity of uranium in humans, but effects in other tissues or organs remain poorly unknown (Paivikurttio *et al.*, 2005). After an accidental absorption of Uranium compounds, the soluble forms of uranium is carried in the blood, filtered by the glomerules and partly excreted in urine. Uranyl nitrate induced nephropathy has been extensively studied in animal models. Nephrotoxicity was also observed in young children exposed to uranium (Magdo *et al.*, 2007).

Various methods to prevent the contamination of aquifers with uranium and halt the expansion of uranium contaminated groundwater plumes were attempted and ways to immobilize uranium in a geochemically inert form *in situ* were looked into (Ewing, 1999). Many isolated studies conducted on the mechanisms for the toxic effects of uranium at moderate to high acute doses on experimental animals have shown that the major health effect of uranium in chemical kidney toxicity rather than a radiation hazard (Miller *et al.*, 1998). The intake of natural uranium through drinking water is associated with increased functional excretion of calcium and phosphate in urine (Kurttio *et al.*, 2002).

Renal toxicity with the classic signs of albuminuria, elevated blood urea nitrogen, and loss of weight, is brought about by filtration of the bicarbonate complex through the glomerulus, reabsorption by the proximaltubule, liberation of uranyl ion, and subsequent damages to the proximal tubular cells. Uranyl ion is most likely concentrated inter-cellularly in lysosomes. Studies also demonstrated kidney as the organ affected from depleted uranium toxicity. In general, renal injury occurs within days of exposure and manifests itself as a change in the proximal convoluted tubules, resulting in increased urinary levels and enzymes excretion. Depleted uranium-induced kidney damage involved combined uranium with bicarbonate, citrate or plasma protein with blood at low pH these complexes split and the resulting uranyl ion may combine with proteins on the tubular wall to cause renal damage.

In addition, only a few studies have been attempted to characterize the effects of chronic exposure to uranium through drinking water (Zamora *et al.*, 1998). Chronic exposure to uranium in humans has been clearly associated with increasing urinary glucose, alkaline phosphatase and β_2–microglobulin, suggesting the involvement of proximal tubule alterations (Kurttio *et al.*, 2002; Zamora *et al.*, 1998). Although both functional and histological damage to the proximal tubules resulting from acute uranium exposure has been clearly demonstrated, little is known about the long term environmental uranium exposure in both humans and animals (Sun *et al.*, 2002, Mao *et al.*, 1995, Zamora *et al.*, 1998). A study of uranium mill workers suggested that workers long term low level exposure is associated with β-2 microglobulinuria and aminoaciduria (Magali Taulen *et al.*, 2004, Thun, *et al.*, 1985). Several studies have reported on the health effects of exposure to depleted uranium. Immediate health risks associated with exposure to depleted uranium include kidney and respiratory problems, kidney stones, chronic cough and severe dermatitis (Aqel *et al.*, 2002, Legget, 1989).

Carcinogenic Effects

Underground mine workers are at risk of developing lung cancer where the levels of radiation are above permissible limits. Radon levels in homes that are comparable to those in mines surely confer risks to the residents. The mining epidemiology indicates that short exposure to high levels of radon and daughter products produces a clear excess of lung cancer (NIH, 1994). Using a new bio-kinetic model, the body retention of uranium in an average Indian adult were predicted and it almost agrees with the measured value (Singh *et al.*, 2008). Although the risk of the developing lung cancer from radon exposure among underground mine workers is firmly documented and quantitative risk estimates are available, the current interest lies in whether this risk carries over into environmental situations.

In the 11 large studies conducted, underground miners exposed to high concentrations of radon and radon daughters, and the documentation of excess lung cancer is convincing (NIH, 1994). The carcinogen in the case of radon is actually the alpha-emitting short-lived daughters of radon, ^{218}Po and ^{214}Po. As the airway lining (bronchial epithelium) is only 40μm thick, the alpha particles emitted are able to reach and transfer a significant amount of energy to the entire cell implicated in lung cancer induction.

Reports have tried to link exposure to depleted uranium with increased cancer incidences in various parts in the world. An example is the recent illness and death of NATO veterans in European countries from leukemia after serving in the Kosovo war, which was confirmed by experiments. Miller *et al.*, (1998) reported that *in vitro* incubation of human osteoblast cells with depleted uranium-uranyl chloride transformed the cells to tumorigenic. Introduction of enriched and depleted uranium into the rat lungs caused carcinomas at the site of deposition, whereas no lung tumours occurred in the rats with depleted uranium oxide (Bachelor *et al.*, 1980).

Effects on Bone

The early deposition of uranium in the skeleton is similar to that of calcium (Legget, 1994). Acute or continuous exposure to uranium may lead to decreased bone formation rate and also increased bone resorption (Guglielmotti *et al.*, 1984; Diaz Syslester *et al.* 2002; Ubios *et al.*, 1991) in rats. The above studies suggesting in addition to kidneys, bone may be another target for uranium toxicity. Inhibition of bone formation has been reported following dermal exposure of rats to depleted uranium. Uranyl ion accumulates in bone and could induce bone cancer. Long term risks include lung and bone cancer. Gugliemotti *et al.* (1984) reported that rats injected i.p. with uranyl nitrate caused alteration of bone formation. Long term risks include lung and bone cancer.

Effects on Nervous System

Exposure to depleted uranium can result in adverse effects on the nervous system as shown in several studies. Six months after implantation of depleted uranium pellets, electro physiological changes in hippocampus slices isolated from treated rat brain that could contribute to neurological deficit were found. Pellmar *et al.* (1999) found that depleted uranium concentration accumulating in the rat brain in a dose-dependent manner. Several published reports implicated exposure to depleted uranium in kidney damage, mutagenicity, cancer inhibition of bone, neurologic deficits, adverse effects on the reproductive and central nervous system (Arzuga, *et al.*, 2010).

Birth Defects

Several studies from Iraqi and other areas where depleted uranium has been used or left behind after the gulf war suggest a significant increase in birth defects and mutant crops. Treatment of male and female mice with an oral dose of uranyl acetate resulted in a lower growth rate of offspring compared with control animals. The number of dead young per litter increased significantly at birth and at day 4 of lactation.

Biochemical Assessment

Heavy metal exposure (including uranium) can cause loss of cell immunity, autoimmune diseases, joint diseases such as rheumatoid arthritis, and diseases of the kidney, circulatory system, and nervous system. Heavy metals supplant the normal calcium and other minerals in enzymes, and cause these molecules to lose their important functions in the body. Decline in functional mitochondria is most damaging to those organs that have the highest energy demands per gram of tissue, namely, the heart, kidney, brain, liver, and skeletal muscle, in that order. These organs become poorly protected against circulating uranium particles, as well as various other pathogens.

The kidney has been considered the main target organ of chemical toxicity of uranium in humans but effects on other tissues or organs remain poorly unknown. There is a need to the development of new biomarkers to evaluate uranium health hazards in mining workers.

With the above perspective in view, a detailed investigation is necessity for monitoring mine workers and evaluating work related pathologies by using biochemical techniques. Uranium concentration in the areas of mining activity and its surroundings must be determined followed by the assessment of secondary migration in the food chain and uranium levels in biological fluids. The characterization of aerosols in the core mining area has also to be undertaken. Further, the biochemical assessment of health of mine-workers is to be undertaken by looking at major organ function tests and hormonal status. The already established biochemical effects of Uranium published on the mine workers shall be reinvestigated along with the normal population. The results generated in such studies will be useful to evaluate likely potential health risks among mine workers.

For any uranium mining activities it is essential to consider the remediation and mitigation process for the safety of human health. A number of better, cheaper, and faster methods have been suggested for the removal of uranium from contaminated and mining areas using bioremediation technologies. Evaluation of human health status has to be assessed by looking at various biochemical parameters related to health and also on some of the already established health effects of uranium need to be looked into based on the levels of radiation and its likely incidence on the environment. Uranium toxicity in humans is observed in kidney and bone while its effect on other major vital organs remains poorly understood warranting, a detailed investigation which is necessary for monitoring mineworkers and evaluating work related pathologies by using biochemical assessment (Sujatha Naik, 2010). The methodology includes (i) statistically sound sampling design; (ii) Background and threshold values of uranium in soils, ores and mine waste; (iii) Bio-geo-chemical studies in core zones; (iv) biochemical analysis of body fluids; and (v) organ function tests of the population and the mine workers.

Conclusions

1. Environmental impacts have been identified with uranium extraction and processing.
2. The incidence of its effects depends upon the degree of intensity of radiation, etc.
3. Bio-remediation is one of the best methods of mitigation of environmental imbalances
4. The technology is to be implemented taking into consideration of the threshold limits of the uranium and selection of geo-bio-chemical methods of remediation.
5. Unnecessary propaganda is to be avoided without understanding the nature of vulnerability of the impacts at a given place as the effects vary from place to place.
6. Consortium approach leads to the development of a data base for all futuristic studies.

NOTES

Aqrel, W.; Abu-Qare and M.B. Abou, Donia (2002), Depleted Uranium—the growing concern, *Journal of Appl. Toxicology,* 22, 149-152.

Archer, VE *et al.*, (1998), chromic diffuse interstitial fibrosis of the lung in uranium miners, *Journal of Occupational and Environmental Medicine*, 40 (5), 460-474.

Arzuga, X.; Rieth, S.H.; Bathija, A.; Cooper, G.S. (2010), Renal effects of exposure to natural and depleted uranium; a renew of the epidemiological and experimental data. *J. Toxical. Environ. Health B. Crit. Rev*, 13(7), 527-45.

Batchelor, A.L.; Buckley, P.; Gore, D.J.; Jenner, T.J.; Major, I.R.; (1980), The carcinogenic effect of localized fission fragment irradiation of rat lung, *Int. J. Radiat. Biol.*, 37, 249-266.

Boddhard, E.; Zimmerli, B.; Schalatter, C.; (1992), Uranium in diet: risk assessment of its nephro and radiotoxicity. *Chemosphere* 24(3), 309-321

Brugge, D.; de Lamos J.L.; Oldmixon, B.; (2005), Exposure pathways and health effects associated with chemical and radiological toxicity of natural uranium: a review, Rev. Environ. *Helath*, 20(3), 177-93.

Cothem, C.R.; Lapperbusch, W.L. (1983), Occurrence of Uranium in drinking water in the United States, *Health Phys.* 45: 89-99.

Diaz Sylvester P.L.; Lopez, R.; Ubios, A.M.; Cabrini, R.L.; (2002), Exposure to subcutaneously implemented Uraniumdioxide impairs bone formation. Arch. Environ. *Health* 57, 320-325.

Ewing, R.C.; (1999), Radioactivity and the 20th century, In Uranium Minerology, Geochemistry and the Environment, 38 Edt. By F.R. Burns, P.C. Minerological Society of Americas, Washington DC, 1-21.

Salem, H.M. (2000), Uranium ores and the environmental impact on human health risks, *ICEHM 2000*, Cairo Iniversity, 580-585.

Cuttler, J.M. and Myron, Pollycove (2009), Nuclear Energy and Health, Dose Response, I, 52-89.

Kurttio, P.; Aurinen, A.; Salomen, L.; Sala, H.; Pekkanen, M.A.; Makeleinen, I., *et al.*, (2002), Renal effects of Uranium in drinking water, *Environ. Health Physics*, 110(4), 337-342

Legget, R.W. (1989), The behaviour and chemical toxicity of Uranium in the Kidney: a reassessment, *Health Phys.* 57, 365-383.

Legget, R.W.; Pellmar, T.C. (2003), The biokinetics of Uranium migrating from embedded DU fragments *J. Environ. Radioact.* 64: 205-225.

Legget, R.W. (1994), Basin for the ICRP's age-specific biokinetic model for uranium, *Helath Phys.* 67, 589-610.

Loyal, D.C.; Lucer, J.N.; Edwards, A.A.; Deng, W.; Valente, E.; Hone, P.A.; and Moquet, J.E. (2001) "A study to verify a reported excess of chromosomal aberration in Blood lymphocytes of Namibian Uranium Mines" Radiation Research 155(6), 807-817.

Mao, Y.; Desmeules, M.; Schaubel, D.; Berude, D.; Dyck, R.; Brule, D. *et al.*, (1995), Inorganic components of drinking water and microalbuminuria, *Environ-Res.* 71(2), 135-140.

Magali Taulan, Francor's Paquet, Christophe Maubert, Olivia delissen, Jacques Demaelle and Marie Catherine Romey (2004), Renal Toxicogenomic response to chronic Uranylnitrate insult in Mice. Environmental Health perspection 112 (16),1628-1635.

Magdo, H.S.; Forman; J., Graba, N.; Newman, B.; Klein, K.; Satlin, L.; Amler, R.W.; Winston, J.A.; Landrigan, P.J. (2007), Grand rounds, Nephrotixicity in a young child exposed to uranium form contaminated well water. *Environ. Health Perspect*, 115, 1237-41.

Moss, M.A.; McCurdy, R.F.; Dooley, K.C.; Girner, M.L.; Dymond, L.C.; Stayter, J.M. *et al.*, (1983). Uranium in drinking water—report on clinical chemistry of Metals (Broom S.S, Savory J.eds) London- Academic Press, 149-152.

Miller, A.C.; Fuciarelli, A.F.; Jackson, W.E.; Ejnik, E.J.; Hogan, J.; Stracko, S.; Emond, C.; Page, N.; Pellmar, T.C. (1998), Urinary and Serum mutagenicity studies with rats implanted with depleted uranium or tantalum pellets, *Mutgagenesis*, 13(6), 643-648.

NIH: (1994), Radon and lung cancer risk: A joint analysis of 11 underground miner studies: US Dept. of Health & human services, National Institute of Health, *NIH publications* 94-3644, Betheska.

Otton, J.K.; Zielinsky, R.A.; Beem, J.M. (1989), Uranium in Holocene valley-fill sediments and uranium, radom and helium in waters, lake Tahoe-Carson Range area, Nevada and California, USA, *Environ Geol. Water Sci.*, 13, 15-28.

Paivi Kurttio, Hannu Komulainen, Aila Leino, Laria Saloren, Anssi Auvinen and Haikki Saha (2005), Bone as a possible target of chemical toxicity of Natural uranium in drinking water, *Environmentall Health Perspectives*, 113(1), 68-72

Pellmar, T.C.; Fucianalli, A.F.; Ejnik, J.W.; Hamilton, M.; Hogan, J.; Strocko, S.; Emond, C.; Mortaz, H.M.; Landauer, M.R. (1999), Distribution of Uranium in rats implanted with depleted Uranium Pellets, *Toxicol. Sci.*, 49, 29-39.

Salonen, L., (1994), 238U series radio nuclides as a source of increased radio activity in groundwater originating from Finnish bedrock. In: Future groundwater resources at risk (Soveri, J., Suokko, T. eds) IAHS Publication 222, Oxford Press, 71-84.

Singh, I.S.; Jaiswal, D.D.; Nair, S.; Vijayagopal, P.; Bhati, S.; Garg, S.P. (2008), Comparison of observed body retention of uranium in natural condition in an average Indian adult with the values predicted by the ICRP biokinetic model, *Radiate. Prot. Dosimetry,* 131(4), 425-30.

Sun, D.F.; Fujugaki, Y.; Fujimato, T.; Yanemura, K.; Hishida, A. (2000), Possible involvement of myofibriblasts in cellular recovery of Uranyl acetate induced acute renal failure in rats. *Am. J. Pathorol* 157(4), 1321-1325.

Sujatha, Nayak, R.B., Sashidhar, (2010), Metabolic intervention of aflatoxin B, toxicity by curcumin, *J. Ethnopharmacol.*, 127, 641-644.

Shields, L.M.; Wiese, W.H.; Skipper, B. J.; *et al.* (1992), Navajobirth outcomes in the ship rock uranium mining areas, in: *Health Physics*, 63(5), 542-551.

Taylor, D.H. and Taylor, S.K. (1997), Environmental Uranium and human health, *Rev. Environ. Health* 12(3), 147-57.

Thorton, I., (1983), Applied Environmental Geochemistry, *Academic Press*, 501 p.

Thunm, M.J., Baku, D.B.; Steenland, K., (1985), Renal toxicity in Uranium mill workers, Scand J. Work. *Environ Health* 11, 83-90.

Ubios, A.M.; Gughiel, Motti, M. B.; Staimetz, T., Cabrini, R.L., (1991), Uranium inhibits bone formation in physiologic alveolar bone modelling and remodelling, *Environ Res.* 54, 17-22.

Veiga, L.H.S.; Amaral, E.C.S.; Fernandes, H.M. (1998); Human health risk of radioactive & Non-radioactive contaminents due to Uranium industry: *J. Environ, Radioactivity* 39(1), 69-85.

Yanase, N.; Payne, T.E.; Sekine, K. (1995), Groundwater geochemistry in the Koongarra ore deposit, Australia (1) Implications for uranium migration, *Geochem. J.*, 29, 1-29.

Zamora, M.L.; Tracy, B.L.; Zieluslia, J.M.; Mayahof, D.P.; Moss, M.A. (1998), Chronic ingestion of Uranium in drinking water: a study of kidney bio-effects in humans. *Toxicol. Sci*: 43(1), 68-77.

15

Limits to Green Economy, Limited Potential and Many Conflict Situations

K.R. Dikshit

The Green Economy Initiative launched by the United Nations Environmental Programme aims at encouraging an economic development that is ecologically risk free, economically productive and socially inclusive. This is an extension of the programme for preserving the natural environment and overcoming indirectly the impact of climate change on the abiotic and biotic features of the earth. These may include submergence of low lying coastal land following the melting of glaciers and continental ice, change in the rhythm of the hydrological cycle, plant resources of the world, changes in the biodiversity and areal spread of forests and grasslands and above all agricultural productivity following a water balance at variance with the present. A rise in temperature may change the seasonal rhythm that may not suit the existing cropping pattern and agricultural systems, requiring perhaps a greater intensity of irrigation or re-engineering of crops genetically to suit the changed conditions.

While the nations of the World are geared up to find ways to cut down the emission of the GHG gases, the only or at least the main culprit for rise of temperature, green economy is another slogan where there is emphasis on economic practices which minimize the use of fossil-fuel based generation of energy. By implication it includes use of renewable sources of energy like solar, wind or hydroelectric power to avoid burning fossil fuels.

Any such practice that eliminates the risk of pollution of air, soil and water and leads to the protection of green cover could be considered a step towards green economy. But saving energy by minimizing the use of power by industries, and using nonpolluting alternative sources of energy could all lead to green economy. Organic farming, Social forestry, Watershed Management, water conservation, and above all protection of green cover are good initiatives towards green economy. Indian programme of MGNREGA (Mahatma Gandhi National Rural Employment Guarantee Action) is also counted as an effort towards Green economy.

The Crux of the Initiative: *Minimize the use of power, avoid using fossil fuels for* *generating energy, refrain from polluting natural resources like water, soil and air, preserve green cover and protect wild life.*

The Fallacy in Assumptions: One of the points made in GEI (Green Economy Initiative) is that such an initiative is socially inclusive. This carries a moralistic tone, and deep social concern is the preservation of the nature. This implies that the poor and the marginalized will benefit greatly if nature is preserved. The assumption is that the poor people, especially the tribals living in the forested areas depend on the forest and other natural resources of the region and the degradation, destruction or pollution of any of the environmental resources will deprive them of their means of livelihood. It is frequently observed that the deprived groups existing on the margin of the society whose case is advocated through Green Economy Initiative show minimal interest in preserving these invaluable environmental resources. Many of these tribal communities have not changed their outlook to the system of exploitation of environmental resources and remorselessly destroy them in the name of their traditions.

Destruction of Forests and Biodiversity: Shifting cultivation, still practiced in India, is a case in point. The country has almost one million hectares of land under shifting cultivation, annually. Of this, roughly 387,000 hectares are in the seven States of North-East India., where roughly 450,000 families are engaged in shifting cultivation. *jhuming.* Contrast this with the agenda of IUCN (International Union for the Conservation of Nature) which delineates North-East region of India, especially the part extending from Arunachal to Manipur. The enormous destruction of forests and, in turn, biodiversity, is effected by the marginalized group who would like to guard their traditional rights for an immediate gain without thinking of the long term implications.

Is organic Farming a Viable Option in India?: Organic Farming, another programme included in GEI, has many aspects. It prohibits the use of chemical fertilizers, insecticides or herbicide. The use of chemical fertilizers is believed to destroy the natural character of the soil as a large residual of salts remain in the soil which in due course make the soil highly alkaline causing harm to the growth. It is known that extremes of pH values of soil cause physiological drought under which the plants are unable to take their sustenance from the soil. This may result from excessive use of chemical fertilizers, especially under arid or semiarid conditions. In rainy climate, much of this residual salts are likely to be flushed away. There are areas where farmers have been using chemical fertilizers for over fifty years without experiencing much adverse impact. But, there are areas in semi-arid regions where the salts remain impregnated in the soil, and appear on the surface through capillary action as encrustation.

The second question is, despite the merits of organic farming, doubts are expressed about the yield and scale of this kind of farming. Products of organic farming are sold at a price, double or triple the market rate. Only the well–to-do will afford such products, and with all the good intentions organic farming cannot meet the demand for food, fodder or fibre because it has low yield and may be even labour intensive.

By itself, organic farming is a healthy practice but for a country with 1.2 billion population, organic farming is not an option. The poor yield of this mode of farming cannot meet even half the requirement of agricultural and horticultural products in the country.

The question is, do we have a substitute for chemical fertilisers?. The environmentalists come with an answer in the form of green manure. Assuming that this could be a good substitute, it requires two things: a good rainy season for the green plants to decay and change into green manure over a period of five to six months. But more serious is the problem of the land being occupied for the preparation of green manure, limiting the possibility of multiple cropping. In an overall assessment the total yearly yield from a plot of land would be far less than what the farmers get otherwise.

Green Energy: Green energy implies pollution free energy, and this is the lynchpin of the Green Economy Initiative programme. All fossil fuels pollute and generate GHGs, though natural gas pollutes much less than petrol and the latter less than crude. No doubt, coal is the most polluting fossil fuel producing large amount of carbon dioxide.

The alternatives available are 1. Solar energy; 2. Wind Energy; 3. Hydroelectric energy; 4. Tidal Wave energy; and 5. Nuclear energy. All these pollution free sources of energy can substantially promote green economy, though the last one carries considerable risk as any malfunctioning or failure of the nuclear plants is likely to cause nuclear radiation affecting the health of all living entities in the neighbourhood. The potential of each one of these sources needs to be tapped. That would mean that much less use of polluting fuels.

The Indian Power Generation Scenario: The total installed capacity for power generation in India as on 30th April 2012 is as follows

Fuel	*MW*	*%ag*
COAL	113,782.38	56.42
GAS	18,381.05	9.11
OIL	1,199.75	0.59
Total Thermal	133,363.18	66.14
Hydro	38,990.4	19.33
Nuclear	4,780.00	2.37
Renewable Energy Resource	24,503.45	12.15
TOTAL	201,637.03	99.99

As can be seen, two-thirds of India's installed capacity for power generation is based on coal, gas and oil, these are non-green polluting sources of energy. Among the cleaner sources opf energy are Hydel power which accounts for 20 per cent of the total installed capacity.

Development of hydroelectric power as well as nuclear power meet stiff societal resistance for one reason or the other. To take an example, Lower Subansiri Hydro

Power in Arunachal has a 2000 MW potential, but it is slowed down and is meeting stiff resistance from people in Assam. As for the nuclear power, there is a genuine fever of radiation, and a slow movement in this direction is understandable. Despite the fact that Arunachal has a potential of 40 to 50,000 MW hydro power, resistance is met at every stage. It is also to be born in mind that the energy requirement is increasing progressively and to generate power to the required level is not going to be easy. Granted that the entire water power potential is utilised, still the country will have to depend for long on fossil fuels.

Solid waste Disposal: This is an area where much can be achieved with a bit of enterprise and imaginative planning. This is an achievable task. Many western countries have taken the initiative to dispose of the Municipal Solid Waste (MSW) by grading and separating combustible material from the recyclable like metals, glass and plastic. The combustible materials used as fuel for generating power which is used for heating water or other purposes. Many million cities in India have such plans, but few if any have achieved it.

Recycling of Water and Water Harvesting: This is a very controversial issue. Conservation of water is intervention in the hydrological cycle. Building a dam or restraining surface water is to prevent the runoff. Recycling the water implies multiple uses and minimizes the volume of water required. Conservation of water means greater availability of water for generating green cover and other uses.

Of the total utilizable water in India, about 90 per cent is used for irrigation; domestic and industrial consumptions use 5 per cent of utilizable water each. Thus there is much greater scope to rationalize and conserve water used for irrigation. In India, there is always a talk of water harvesting and conservation in urban areas. Not so much emphasis is given in minimizing the use of water in irrigation without compromising irrigation requirements. The critical component is the method of irrigation as the most common method is flood irrigation. There is much greater scope for bringing far greater acreage under irrigation by adopting suitable method and saving water from excessive evaporation.

Limits imposed by the Use of Fossil Fuels: Most industries in India or for that matter anywhere else, with the exception of a few like IT industries, are polluting industries. They pollute the air, water and soil. The availability of carbon credits should induce them to minimize the production of GHG which certainly they must be doing. Most industries don't allow any waste of energy and use it to the last bit and some of them recycle water. But there is a limit beyond which they cannot curtail the use of fuels that add to the GHG. Similarly, the paddy fields will certainly add some amount of methane in the atmosphere. Burning of coal/kerosene is a common practice in poorer families and we cannot whisk it away easily.

People in India are aware of the harmful effects of using fuels that add CO_2 to the atmosphere, but there are still limited options. Greater research in the generation and storage of solar power may offer a partial but substantial solution. Equally wind energy is not harnessed to the full. India is a large country and some part or the other of the country is always sunny. It may be feasible to think of a device by which generators in the sunny regions are switched on and off. How much will the system contribute to the generation of energy is uncertain but worth an effort.

Is a substantial generation and use of nuclear energy as in France a solution.

Symbolic Gestures by Groups of People: Recently certain media houses have started programmes of using non-polluting source of energy for lighting; and the most common example is giving solar lanterns to adopted villagers. This is laudable yet it remains symbolic. A greenathon, implying long distance running to induce awareness about nature conservation, by some youths may heighten awareness of nature conservation. The intention of those participating in a 1500 km run from Delhi to Mumbai is certainly noble and has to be admired and appreciated, but the results of such efforts have to be judged by the impact visible on the ground and not on tele-screen.

The key to GREEN ECONOMY lies in preserving the greenery we have, multiplying it, but above all in conserving energy and finding alternative sources of nonpolluting energy resources.

16

Bioremediation of Toxic Metal Ions polluting the Environment

Green Approach based on Utility of a Natural Biopolymer (Gum Kondagogu)

B. Sashidhar Rao, and *V.T.P. Vinod*

Introduction

Heavy metal cations such as lead, cadmium, nickel, mercury, zinc and copper are highly hazardous to the aquatic ecosystem and pose possible human health risk. The presence of these toxic metal ions in effluents resulting from rapid industrialization and technological advances is a worldwide environmental problem. Recently, considerable attention has been focused on the utilization of eco-friendly materials for the removal of heavy metal cations from polluting aqueous discharges. In this perspective ***green approach*** has attracted the attention of environmental scientists as it is based on the principles of sustainable technology. Biosorption of heavy metal ions using natural products has become an attractive alternative to traditional methods of industrial waste-water treatment, such as precipitation, adsorption, ion exchange, reverse osmosis, evaporation and membrane filtration.

Gum kondagogu [*Cochlospermum gossypium* (L.) DC.] is an Indian tree gum derived as an exudate and belongs to the family *Bixaceae*. In the recent past, gum kondagogu has been established as a biosorbent to remove toxic metals like lead, cadmium, nickel and total chromium from aqueous solutions. The morphological, physico-chemical, structural and rheological properties of gum kondagogu have been comprehensively studied. The primary structure of this biopolymer contains sugars, such as, arabinose, rhamnose, glucose, galactose, mannose, glucuronic acid and galacturonic acid. Based on the spectroscopic categorization, the probable structural feature consigned to gum kondagogu is (1→2) β-D-Gal *p*, (1→6), β-D-Gal *p*, (1→4) β-D-Glc *p* A, 4-0-Me-α-D-Glc *p* A, (1→2) α-L-Rha and (1→4) α-D-Gal *p* A. Gum kondagogu is an acidic gum with high content of uronic acid and the major functional groups identified in the gum are hydroxyl, acetyl, carbonyl and carboxylic groups. Gum kondagogu fulfils the requirements as a biosorbent as it contains many of the functional groups recognized earlier of having involvement in metal binding. Thus, this biopolymer has an impending application as a matrix for toxic metal bioremediation from industrial effluents for mitigating environmental pollution.

Gums are hydrocolloids, which are hydrophilic in nature and are found in almost every biosphere on earth; in plants, animals and many bacteria. They contain a large number of hydroxyl groups arranged usually in a fairly regular manner along the backbone of the molecule, which allows for the chelation of mono- and divalent cations, thereby cross-linking the hydrocolloid chains together and forming complex macrostructures. Hydrocolloids are some of the most well known polysaccharides with complex structures with glycosidic bonding. Most gums are heterogeneous polysaccharides with complicated structures and extremely high molecular masses. The simplest interactions of hydrocolloids are with water and it is this interaction that is key to their use in foodstuffs. The gum hydrocolloids are effective water adsorbents and to a greater and lesser degree will be solubilized by water. Owing to the high number of hydroxyl groups, water is held within the molecular structure by hydrogen bonding and also within the voids created by the complex molecular configuration (Hall, 2009).

Exudate gums are amongst the oldest natural products used by mankind. They were being used as thickening and stabilizing agents for centuries because they provide excellent water binding capacity. Some gums possess a natural biological activity related to the glyco-proteins they contain, which results in a particular role of gums in the pharmaceutical industry. The three major exudates are gum Arabic, gum tragacanth, and gum karaya. These gums have been used in food, pharmaceutical, adhesive, paper, textile and other industries (Verbeken *et al.*, 2003).

Tree gums are obtained as the natural exudates of different tree species and exhibit unique properties with a wide variety of applications. Commercially, gums have a separate niche in the world market as a commodity of international trade. Most gums are heterogeneous polysaccharides with complicated structures and extremely high molecular masses. They contain a large number of hydroxyl groups arranged usually in a fairly regular manner along the backbone of the molecule, which allows for the chelation of mono-and divalent cations, thereby cross-linking the polymer chains together and forming complex macrostructures. The three major exudates gums namely, gum Arabic, gum tragacanth, and gum karaya possess a unique range of functionalities (Phillips and Williams, 2001). Biochemically, more than 90 per cent of the gum biomass is in the form of carbohydrate polymers (polysaccharides). Since these biopolymers are abundant, come from renewable sources, are relatively inexpensive, are non-toxic, and are amenable to both chemical and biochemical modification, it is not surprising that they find widespread and extensive use. They find important applications in food, pharmaceutical, adhesive, paper, textile, beverage and other industries (Verbekan *et al.*, 2003). Many trees and shrubs produce exudate gums as a natural defence mechanism, particularly in semiarid regions. When the plant's bark is injured, an aqueous gum solution is exuded to seal the wound, preventing infection and dehydration of the plant. The solution dries in contact with air and sunlight, to form hard, glass-like lumps, which can easily be collected.

Indian forests are a major source of large number of non-wood forest products. One such product is an exudate tree gum, regionally called as gum kondagogu (*Cochlospermum gossypium*) (**Fig. 16.1**), belonging to the family *Bixaceae*. This gum is an important non-wood forest natural product collected by tribals in the State of Andhra Pradesh and marketed by Girijan Cooperative Society, Andhra Pradesh, India (**Table 16.1**). Three grades of gum are marketed as Grade I, II and III. The production in the State of Andhra Pradesh is in the range of 0.5 to 1.5 tonnes of gum per season. This gum is also produced and marketed by other Indian States, namely Rajasthan, Madhya Pradesh and Chhattisgarh.

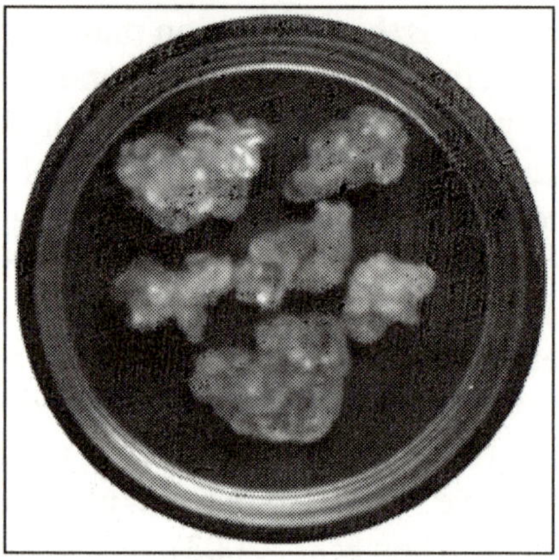

Fig. 16.1: Gum Kondagogu Specimen (Grade–I)

Table 16.1: Gum Kondagogu Collection in Andhra Pradesh and its Export Sale Value

Year	*Quantity in Quintals*	*Value in Rs. Lakhs*
2000-01	887	84.37
2001-02	1321	143.16
2002-03	1150	116.07
2003-04	1222	136.06
2004-05	1136	126.69
2005-06	1158	135.67
2006-07	1165	145.89
2007-08	1175	155.09

Source: Girijan Co-operative Corporation, Andhra Pradesh.

Experimental work carried out earlier in our laboratory on Gum kondagogu had resulted in assigning a separate identity to this gum as compared to the well established and commercially exploited gum karaya. Gum kondagogu has unique physico-chemical properties as compared to other tree gums. Proximate analysis of gum indicates that it has high volatile acidity and water binding capacity (hydrogel property). Janaki and Sashidhar (1998) pioneered the studies on gum kondagogu in relation to its physico-chemical properties and as a potential new food additive. Their investigation provided a separate identity to this gum as compared to other established tree gum such as gum

Arabic, gum tragacanth, and gum karaya. The physico-chemical analysis revealed that gum kondagogu differs from gum karaya with respect to fibre, protein, tannin, minerals, sugar compositions, intrinsic viscosity, water binding capacity and pH (Janaki and Sashidhar, 1998). The toxicological evaluation of gum kondagogu established that this gum to be non-toxic (Janaki and Sashidhar, 2000). Even though, gum kondagogu is an important forest produce, its commercial exploitation was limited due to non-availability of scientific information, especially in relation to its morphological, physico-chemical, structural, rheological, solutions, and conformational properties.

Physico-Chemical and Structural Characterization of Gum Kondagogu

Morphological Studies

Morphological analysis of gum kondagogu was carried out using scanning electron microscopy (SEM), atomic force microscopy (AFM) and transmission electron microscopy (TEM). Microscopy (SEM, TEM and AFM) was used to provide visual evidence for the formation of multi layer interfaces around colloidal particles. Scanning electron microscopy (SEM) of native and deacetylated gum showed that native form of the gum had irregular particle size. Deacetylated gum was found to be fibrilar, indicating the loss in particulate morphology that was observed in the native form of the gum, suggesting that acetyl groups are essential in the structural integrity of the gum for particulate appearance in the native form. Topographic image of gum kondagogu based on AFM analysis suggests an irregularly shaped network structure, indicating the involvement of inter- and/or intra-molecular aggregations. TEM analysis of gum kondagogu showed elongated chain structures with extensive branching (Vinod and Sashidhar, 2010a) (Fig. 16.22 to 16.4).

A B

Fig. 16.2: (A) Scanning electron microscopy (SEM) of native gum shows that irregular particle size and (B) Deacetylated gum was found to be fibrilar, indicating the loss in particulate morphology that was observed in the native form of the gum, suggesting that acetyl groups are essential in the structural integrity of the gum for particulate appearance in the native form.

Fig. 16.3: Topographical phase of native gum kondagogu determined by AFM (phase image) analysis suggesting an irregular network shaped structure for gum kondagogu.

Fig. 16.4: TEM analysis of native gum kondagogu showing backbone and branched network structures. **Scale bar 100 μm**

Compositional Analysis

Composition analysis of gum kondagogu was based on HPLC, LC-MS and GC-MS analysis. The analytical data on sugar composition suggests that gum kondagogu contains very high concentration of uronic acids [D-glucuronic acid (19.26 mol %), β-D-galacturonic acid (13.22 mol %) and α-D-galacturonic acid (11.22 mol %)] (Vinod *et al.*, 2008a & b) (Fig. 16.5). The amino acids present in gum kondagogu includes, alanine (32.25±1.44 μg^{-1}), glycine (5.05±0.55 μg^{-1}), valine (7.24±0.60 μg^{-1}), leucine (3.85±0.22 μg^{-1}), proline (42.45±2.56 μg^{-1}), methionine (44.24±2.25 μg^{-1}), aspartic acid (72.85±3.45μg^{-1}), threonine (30.45±1.54 μg^{-1}), glutamic acid (34.25±1.56 μg^{-1}),

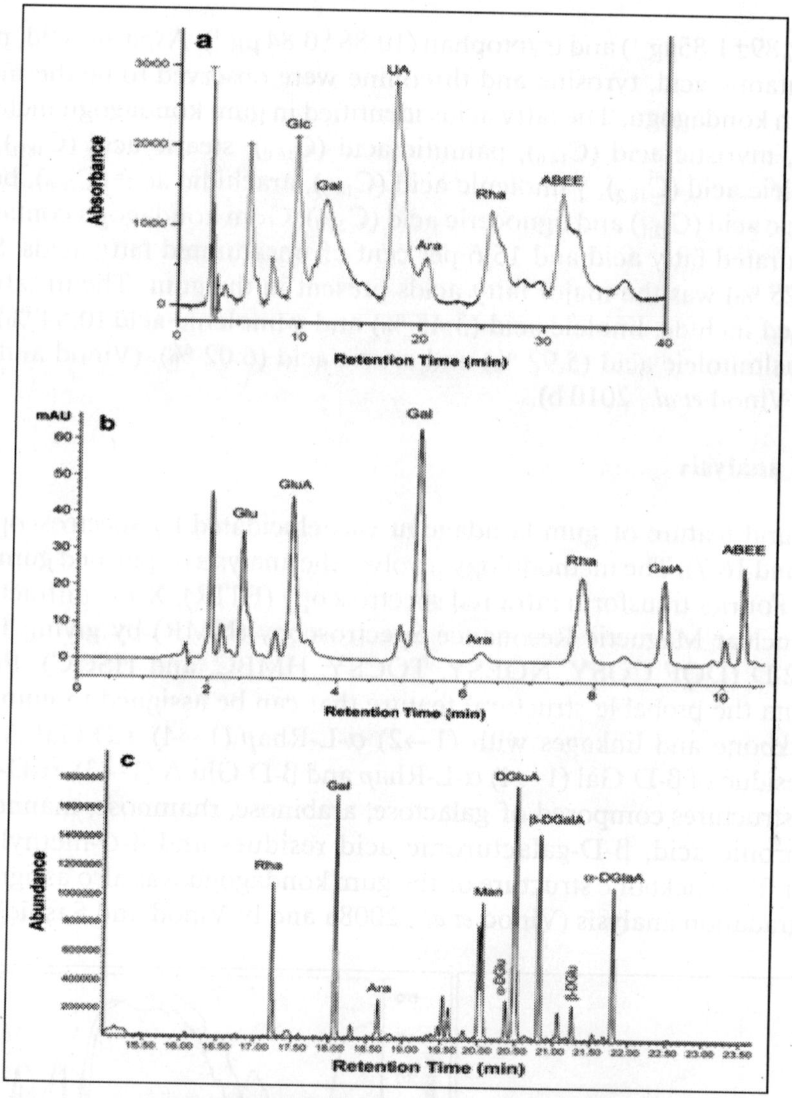

Sugar residue	Composition (mol %)
Rhamnose	12.85
Galactose	18.95
Arabinose	2.52
Mannose	8.30
α-D-glucose	2.48
D-glucuronic acid	**19.26**
β-D-galacturonic acid	**13.22**
β-D-glucose	2.52
α-D-galacturonic acid	**11.02**

Fig. 16.5: Compositional Analysis of Gum Kondagogu: Sugar Profile of Gum Kondagogu as determined by (a) HPLC, (b) LC-MS and (c) GC-MS Analysis.

tyrosine (32.89±1.85μg^{-1}) and tryptophan (10.88±0.84 μg^{-1}). Aspartic acid, methionine, proline, glutamic acid, tyrosine and threonine were observed to be the major amino acids in gum kondagogu. The fatty acids identified in gum kondagogu included: lauric acid ($C_{12:0}$), myristic acid ($C_{14:0}$), palmitic acid ($C_{16:0}$), stearic acid ($C_{18:0}$), oleic acid ($C_{18:1}$), linoleic acid ($C_{18:2}$), γ linolenic acid ($C_{18:3}$), arachidic acid ($C_{20:0}$), behenic acid ($C_{22:0}$), erucic acid ($C_{22:1}$) and lignoceric acid ($C_{24:0}$). Gum kondagogu contains 84.4 per cent of saturated fatty acid and 15.6 per cent of unsaturated fatty acids. Stearic acid ($C_{18:0}$) (37.25 %) was the major fatty acids present in the gum. The unsaturated fatty acid detected include, linoleic acid (3.45 %) and γ-linolenic acid (0.54 %), oleic acid (0.50 %), palmitoleic acid (5.92 %) and erucic acid (6.02 %). (Vinod and Sashidhar, 2010a and Vinod *et al.*, 2010 b).

Structural analysis

The structural feature of gum kondagogu was elucidated by spectroscopic methods (Fig. 16.6 and 16.7). The methodology involves the analysis of purified gum kondagogu sample by Fourier transform infra-red spectroscopy (FTIR), X-ray diffraction analysis (XRD), Nuclear Magnetic Resonance Spectroscopy (NMR) by giving 1-D (^{1}H and ^{13}C) and 2-D (DQF COSY, NOESY, TOCSY, HMBC, and HSQC). Based on the spectral data the probable structural feature that can be assigned to gum kondagogu was a backbone and linkages with (1→2) α-L-Rha*p* (1→4) a-D-Gal A (1→2) and pendent residue of β-D-Gal (1→4) α-L-Rha*p* and β-D-Glu A (1→3)–α-D-Gal A, with branched structures composed of galactose, arabinose, rhamnose, mannose, glucose, β-D-glucuronic acid, β-D-galacturonic acid residues and 4-*O*-methyl derivatives (Fig. 16.7). The backbone structure of the gum kondagogu was also assigned based on Smith degradation analysis (Vinod *et al.*, 2008a and b; Vinod and Sashidhar, 2010a).

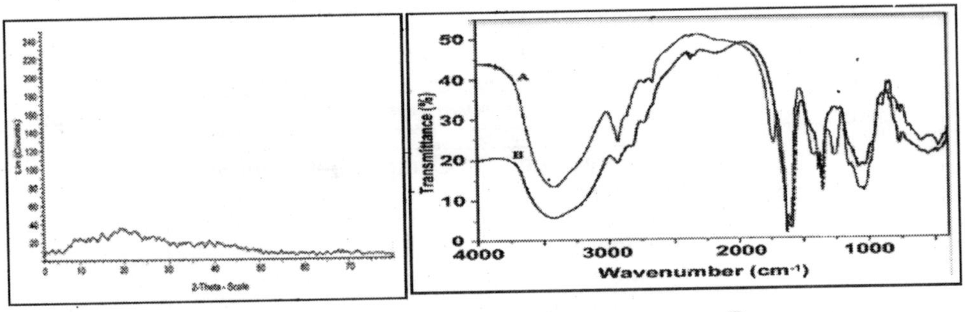

A B

Fig. 16.6 (A):The XRD pattern of the native gum indicates a completely amorphous structure. (B) FT-IR analysis of native and deacetylated gum kondagogu. The major functional identified from FT-IR spectrum include, 3431 cm^{-1}(–OH), 1731 cm^{-1} (CH_3CO^-), 1632 cm^{-1}–COO^-), 1429 cm^{-1} (–COO), 1249 cm^{-1} (–CH_3CO). Absence of peaks at 1736 cm^{-1} and 1227 cm^{-1} were attributed to acetyl groups; these groups were absent in deacetylated gum.

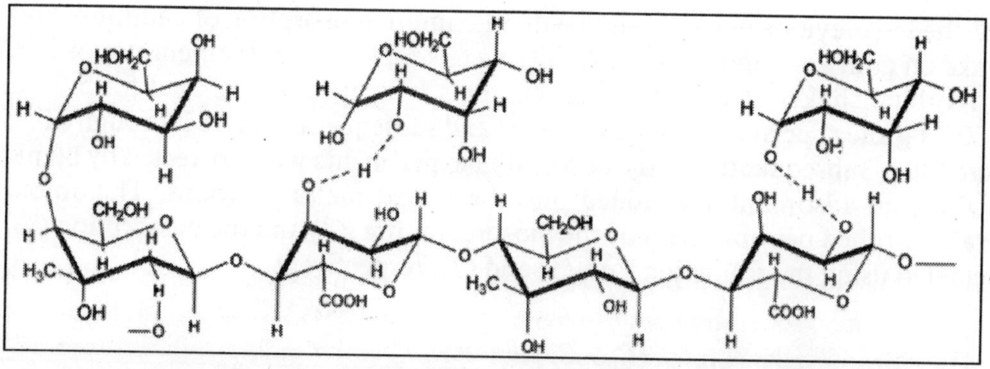

→2)α-L-Rha*p*-(1→4)-α-D-Gal*p*A-(1→4)-α-L-Rha*p*-(1→4)-β-D-Gal*p*(1→

4 ↑ 1 β-D-Gal*p* (at α-L-Rha*p*); 3 ↑ 1 β-D-Glc*p*A (at α-D-Gal*p*A); 2 ↑ 1 α-D-Gal*p* (at β-D-Gal*p*)

Fig. 16.7: The assigned structure of the gum kondagogu by NMR and GC-MS analysis. Structural assignment was carried out using acid hydrolysis, Smith degradation and NMR studies [^{1}H, ^{13}C, 2-D NMR (TOCSY and NOESY)

Bioremediation of Toxic Metal Ions by Gum Kondagogu

Elucidation of structural and compositional properties of gum suggested its impending applications for food and non-food uses. Potential application of gum kondagogu as a novel biosorbent for removal of the industrially relevant toxic heavy metals form effluents was explored (Vinod *et al.*, 2010c, Vinod and Sashidhar, 2011).

Preparation Biosorbent: Gum kondagogu was powdered in a high-speed mechanical blender, (Philips, Mumbai, India) and later sieved using a bin (mesh size–250 mm), so as to obtain a fine and uniform sample. Gum kondagogu powder (2 g) was accurately weighed, and dispensed into a clean glass beaker containing one litre of de-ionized water. The whole gum solution was kept on a magnetic stirrer at room temperature and gently stirred over night. Later, the gum solution was allowed to stand at room temperature (30° C) for 12 h, so as to separate any suspended matter. The gum solution was filtered through a sintered glass funnel (#G-2 followed by #G-4). The clear solution so obtained was freeze-dried and stored, until further use.

Langmuir isotherm The isotherms were derived at the optimum pH 5, kept in an orbital shaker at 200 rpm for 2 h at a temperature of 25° C. The gum kondagogu (1 g) were mixed with varying concentration of individual metal ion (Cd^{2+}, Cu^{2+}, Fe^{2+}, Pb^{2+}, Ni^{2+}, Zn^{2+}, Hg^{2+} and Cr between 25 to 300 mg L^{-1}), in a volume of 1L.

Biosorption Studies. In a typical experiment, 0.1 g of biosorbent was added into 100 mL of wastewater solution and pH (5.0±0.1) was adjusted previously by 0.1 M HNO_3 and 0.1 M NaOH. The pH (5.0±0.1) value was selected based on the results published in previous works dealing with maximum biosorption of cadmium and lead uptake on gum kondagogu. The samples were kept for 2 h at temperature of 25±2° C in an orbital shaker at 200 rpm. The contents were then centrifuged for 5.0 min. at 10,000 × g and supernatants were collected and subsequently filtered through a 0.45 µm micro filter. Subsequently, all the adsorption experiments were corrected by blank tests in which no adsorbent was added into the mixed metal solutions. The amount of metals adsorbed per unit mass of the biosorbent (mg g^{-1}) and the biosorption (%) was calculated using the following Eqs. (1) and (2), respectively.

$$q\left(mg\ g^{-1}\right) = \frac{(C_o - C_e)}{W} \times V \quad \ldots(1)$$

$$\text{Biosorption (\%)} = \frac{C_i - C_o}{C} \times 100 \quad \ldots(2)$$

Where, q is the adsorption capacity in mg g^{-1}, C_i, C_0 and C_e are the initial, final and equilibrium concentration of metals in mg L^{-1}, V is the volume of metal solution in L and W is the total amount of gum kondagogu in g.

***Collection of Industrial Effluents and Elemental Analysis*:** The industrial effluent samples were collected from five different industrial locations (Nacharam, Cherlapally, Tank bund, Balanagar and Kattedan) in city zone of Hyderabad (Andhra Pradesh, India). Three independent samples were collected (1L) at each location and later transferred into clean polypropylene containers and transported to laboratory for analysis.

Instrumental Methods: Biosorption of toxic metals were quantitatively determined by ICP-MS (Inductively coupled plasma mass spectrometry) (ELAN DRC 11, Perkin-Elmer SCIEX, Toronto, Canada). FT-IR spectra were obtained using an IR spectrophotometer (420, Jasco, Tokyo, Japan). IR spectra of gum kondagogu and gum kondagogu—metal complexes were obtained in the spectral region of 4000–400 cm^{-1} using resolutions of 4 cm^{-1} and 64 co-added scans.

The Langmuir isotherms of Cd^{2+}, Cu^{2+}, Fe^{2+}, Pb^{2+}, Ni^{2+}, Zn^{2+}, Hg^{2+} and total Cr ions on the gum kondagogu are comparatively illustrated in Fig. 16.8. The amount of metal ions adsorbed at equilibrium per gram of gum kondagogu (q) is represented as a function of the equilibrium metal ion concentration (C_e) as shown in Fig. 16.8. The uptake of metal ions by gum kondagogu varies in the range of $Cd^{2+} > Cu^{2+} > Fe^{2+} > Pb^{2+} > Hg^{2+} >$ total Cr $> Ni^{2+} > Zn^{2+}$ over the whole range of metal initial concentrations examined.

Five effluent samples collected from different industrial areas of Hyderabad city zone were analyzed for the bioremediation property of gum kondagogu for removal of toxic metal ions contaminants. Gum kondagogu was capable of competitively biosorbed eight metal ions from the samples of industrial effluents tested, which was observed to be in the following order; $Cd^{2+} > Cu^{2+} > Fe^{2+} > Pb^{2+} > Hg^{2+} >$ total Cr $> Ni^{2+}$

>Zn^{2+}. The adsorption capacity (*q*) of metals by gum kondagogu varied in the range of 31 to 37 mg g^{-1} for Fe^{2+} and a minimum of 5.5 to 9.3 mg g^{-1} for Hg^{2+}, in the tested effluent samples. The competitive biosorption capacity of gum kondagogu for removal of metal ions was best effective at pH 5. **(Table 16.2)**

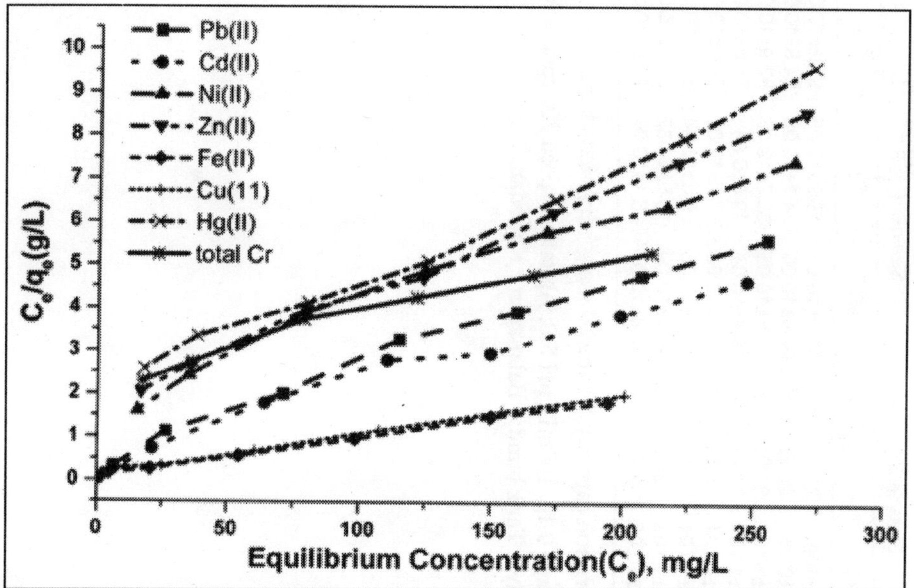

Fig. 16.8: Langmuir plots for metal ions (Cd^{2+}, $Cu^{2+,}$ Fe^{2+}, Pb^{2+}, Ni^{2+}, Zn^{2+}, $Hg^{2+}$and total Cr) biosorpion by gum kondagogu. Conditions: initial metal ions concentrations (25-300 mg L^{-1}); gum kondagogu dose (1 mg mL^{-1}); pH- 5; Temperature; 25° C; Time of contact 2 h.

The interactions between metal ions (Cd^{2+}, $Cu^{2+,}$ Fe^{2+}, Pb^{2+}, Ni^{2+}, $Zn^{2+,}$ Hg^{2+} and total Cr) and functional groups on the gum kondagogu were confirmed by FTIR analysis (Fig. 16.9). The FT-IR result indicates that bioremediation of metal ions by gum kondagogu occurred as a result of electrostatic interaction as well as complexation with hydroxyl, carboxylic, carbonyl, alcoholic, ester, amino, acetyl groups and high amount of uronic acid of the biopolymer. Gum kondagogu biopolymer possesses functional groups, which can adsorb heavy metals. The major functional groups of the type include, carboxylate anion (–COO), hydroxyl (–OH), and others (–C–N, –C–O, –C–H, –C=O, –NH). The wide band with maxima at 3405 cm^{-1} was assigned to the stretching of O–H group of macromolecular association, the band at 2925 cm^{-1} was assigned to stretching of CH_2 bond of methylene groups, and the sharp band observed at 1708 cm^{-1} was assigned to a C–O bond of carboxylic acid. The strong peak that appears at 1622 cm^{-1} can be attributed to C–O stretching vibration of a carboxylic acid that exists because of intermolecular hydrogen bond. The bands at 1125 cm^{-1} and 1127 cm^{-1} were due to the –C–O stretching of ether groups. The bands at 1072 cm^{-1} and 1062 cm^{-1} were assigned to the–C–O stretching of alcoholic groups. These band changes result in the absorbance

Table 16.2: Bioremediation of Toxic Metal Ions from Effluents by gum kondagogu. Metal Biosorption (%) or Adsorption Capacity (mg g^{-1}) by gum kondagogu as determined by ICP-MS in Industrial Effluent Samples.

Metal ions	*Sample 1*[a]		*Sample 2*[b]		*Sample 3*[c]		*Sample 4*[d]		*Sample 5*[e]	
	% B[f,g]	*g*[f,h]	*% B*[f,g]	*g*[f,h]	*%B*[f,g]	*g*[f,h]	*%B*[f,g]	*g*[f,h]	*%B*[f,g]	*g*[f,h]
Cd^{2+}	97.48±0.34	17.7±10.02	97.9±10.12	15.4±10.02	97.0±109	17.7±10.02	97.2±10.0	19.9±10.05	97.1±10.08	23.6±10.04
Cu^{2+}	93.7±0.08	19.3±10.05	94.8±10.09	21.1±10.02	94.0±0.08	25.8±10.06	94.4±10.08	24.1±10.08	94.8±10.05	27.5±10.04
Fe^{2+}	87.1+0.09	35.4±10.10	87.9±10.08	37.3±10.08	87.8±10.02	31.2±10.06	87.6±10.05	33.7±10.08	87.9±10.02	35.9±10.09
Pb^{2+}	75.1±0.22	15.4±10.02	74.8±10.08	16.9±10.04	75.5±10.04	18.6±10.04	75.4±10.04	19.6±0.04	75.6±10.04	25.1±10.05
Hg^{2+}	71.7±0.08	6.1±10.02	70.8±10.06	6.3±10.02	70.6±10.04	5.5±10.02	71.2±10.08	7.7±10.02	70.4±10.05	8.1±10.02
Cr (Total)	52.0±10.08	6.5±106	51.9±10.04	6.8±10.02	51.3±10.04	7.6±10.03	51.6±10.08	9.3±10.04	51.9±10.05	10.5±10.03
Ni^{2+}	39.1±10.08	6.6±10.02	40.0±10.05	7.5±10.03	40.2±10.03	8.2±10.03	39.4±10.09	8.5±10.02	39.8±10.05	10.2±10.06
Zn^{2+}	34.1±10.08	11.0±10.09	34.2±10.08	8.8±10.05	34.8±10.05	7.7±10.04	34.8±10.04	8.1±10.04	34.9±10.03	9.3±0.03

[f]Values = Mean ± S.D; n = 3: [g] (%B) = Biosorption (%); [h]q_e = Adsorption capacity (mg metal ion g^{-1} dry biosorbent).

Conditions: Metal ions concentration (5-35 mgL^{-1}; biosorbent dosage 0.1g/100mL; pH 5.0; Agitation speed 200 rpm; temperature 25 °C. Samples collected from Industrial area of, [a] Nacharam; [b] Cherlapally; [c] Tank bund; [d] Balanagar; [e] Kattedan.

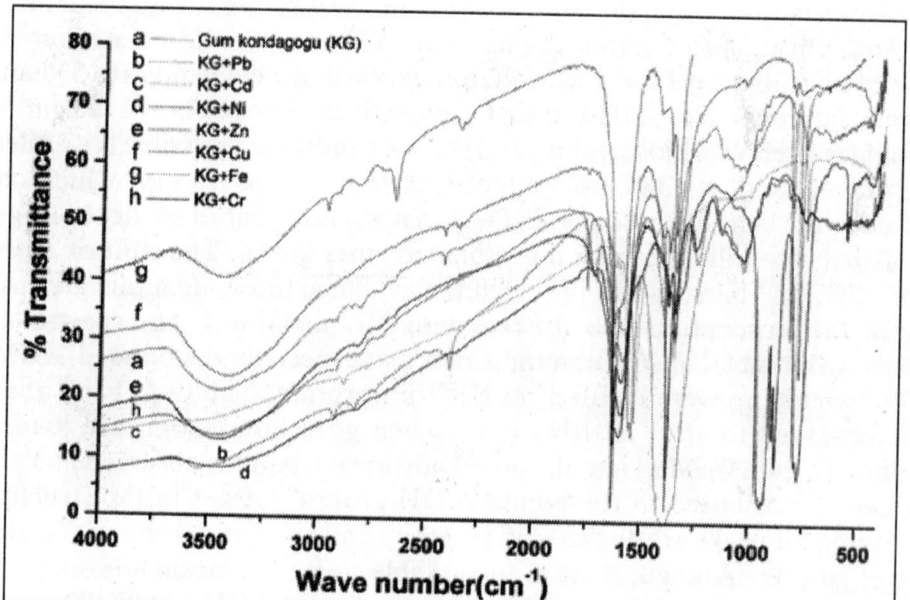

Fig. 16.9: FT-IR spectra of (a) gum kondagogu before metal bio sorption, (b) Pb^{2+} loaded, (c) Cd^{2+} loaded, (d) Ni^{2+} loaded, (e) Zn^{2+} loaded, (f) Cu^{2+} loaded, (g) Fe^{2+} loaded, (h) Cr loaded.

band at 1380 cm^{-1} became stronger. The C=C stretching in aromatic ring at 1610 cm^{-1} shifts to1601 cm^{-1} after Cd^{2+} biosorption. The –OH and (–COO) groups in gum kondagogu were involved in Cd^{2+} biosorption. The FTIR spectra of unloaded and lead loaded gum kondagogu in the range of 400 to 4000 cm^{-1} were taken to find out which functional groups are responsible for the biosorption. The FTIR spectrum of lead loaded gum kondagogu indicated that the peaks around 1455 cm^{-1} and 1280 cm^{-1} gradually disappeared with increase of Pb^{2+} ions concentration in solution, while the peak around 1380 cm^{-1} was found to increase. The shift of peak at 1455 cm^{-1} transposing to lower frequencies up to disappearance is due to the complexation of amino and hydroxyl groups with Pb^{2+}. Another shift was observed at 1280 cm^{-1} to higher frequencies corresponding to complexation of oxygen from the carbonyl–C–O bond. The functional groups involved in bivalent metal biosorption included hydroxyl, carboxyl, ether and alcoholic groups. A comparison of the spectra for native gum kondagogu with that of lead-loaded gum reveals characteristic changes of the hydroxyl groups and the acetyl groups, which shift from 3415 cm^{-1} (–OH), amide (1654 cm^{-1}) before lead adsorption to 3427 cm^{-1} (–OH) and amide (1649 cm^{-1}) after lead adsorption. Lead-gum kondagogu interaction is more evident at 3415, 1654, 1550, 1250 cm^{-1} because of –OH, –C=O, –COOH and –C–C vibrations. The broad stretching absorption band at 3407 cm^{-1} shifted to 3414 cm^{-1} after Ni^{2+} biosorption. The band intensity at 1725 cm^{-1} clearly decreased after Ni^{2+} biosorption, i.e., there would be an interaction of Ni^{2+} with carboxylate groups. The band at 1384 cm^{-1} appears only in the spectra at 1082 cm^{-1}, which could be assigned to C=O from polysaccharide, while the band at 1787 cm^{-1} was found to disappear after metal interaction. On the other hand, the band at 1488 cm^{-1} would be attributed to the C–H bending shifted to 1468 cm^{-1}. The appearance of new bands at 1535 and 1240 cm^{-1} could be assigned to –NH and C=O, respectively. The peaks at 1725 cm^{-1} turned to a shoulder after Ni^{2+} biosorption, and this may indicate a possible Ni^{2+} interaction with C=O. A band at 1385 cm^{-1} appears after Ni^{2+} biosorption, which could be attributed to the bending mode of C–O–H that would occur in an alcoholic group or a protonated alcoholic group or a protonated ether group. The shift of absorbance peak of after Ni^{2+} biosorption provided the evidence those alcoholic groups would be one of the biosorption sites for removing Ni^{2+}metal ion. The spectral analysis before and after metal adsorption indicated that especially the bonded –OH groups and C–O stretching were involved in Ni^{2+} biosorption. The C–O band absorption peak is observed to shift to 1047 cm^{-1} when gum kondagogu are loaded with chromium. In gum kondagogu, the broad adsorption band is observed at 3307 cm^{-1}, which can be attributed to the bonded –OH groups present in the structure. The other prominent peaks are due to $-OCH_3$ group and aromatic compounds. However, in case of gum kondagogu, there is remarkable shift in position and shape of the –OH group indicative of chromium binding mostly with –OH groups. These changes in FT-IR spectra confirm the band shifts indicating the bonded –OH groups and /or –NH stretching and carboxyl groups (C=O) may especially play a major role in chromium biosorption on gum kondagogu. Biosorption of metals ions (Cd^{2+}, Cu^{2+}, Fe^{2+}, Pb^{2+}, Ni^{2+}, Zn^{2+}, Hg^{2+} and total Cr) by gum kondagogu occurs as a result of electrostatic interaction and complexation with hydroxyl, carboxylic, carbonyl, alcoholic, ester, amino, uronic acid and acetyl groups present in the biopolymer, that are amenable for the interaction with metal ions at pH 5. (Vinod *et al.*, 2010c, Vinod and Sashidhar, 2011)

Discussion

Water is one of the vital natural resource supporting human activities and ecosystems. The hydrologic cycle is a fundamental component of climate and the anthropogenic activity and rapid urbanization has impacted in more than one way the water quality due to dumping of industrial effluents and chemicals from farm run offs waste into river waters without being treated, severely polluting the usable water supply (Bates *et al.*, 2008). Safe and effective disposal of wastewater containing heavy metals is always a challenging task for industrialists and environmentalists, as cost-effective treatment alternatives are not available. One of the approaches to mitigate the environmental water pollution is by employing novel bioremediation technologies that are based on the principles of green and clean chemistry. Conventional technologies for the removal of heavy metals such as chemical precipitation, ion exchange, or electrochemical processes are often neither effective nor economical, especially when used for the reduction of heavy metal ions at low concentrations. Biosorption technology based on the ability of certain biomasses to remove metallic ions from aqueous solutions and its potential for wastewater treatment has received increasing awareness

One of the methods of decontamination includes green approach, which utilizes the available natural products that have low operating cost, improved selectivity for specific metals of interest, removal of heavy metals from effluent Gum kondagogu—a natural biopolymer has been exploited as an efficient, environmental friendly matrix for removal of toxic metal pollutants from industrial effluent discharges. The results of our experiments demonstrate that the biosorption of metals from waste waters using biomass byproducts is a viable and cost-effective technology that should be included in process evaluations. The combination of biodegradability, high metal uptake capacity and high uronic acid content contributes to make gum kondagogu a suitable material for toxic, heavy metal remediation by the biosorption process. The result indicates that adsorption metal ions by gum kondagogu occurs as a result of electrostatic interaction as well as complexation with hydroxyl, carboxylic, carbonyl, alcoholic, ester, amino, acetyl functional groups and high content of uronic acid of the biopolymer. Gum kondagogu being poly-anionic polysaccharides was successfully employed as metal binders for nucleation and precipitation of heavy metals present in effluents. A schematic diagram showing the interaction of metal cation with gum kondagogu is presented in Fig 16.10.

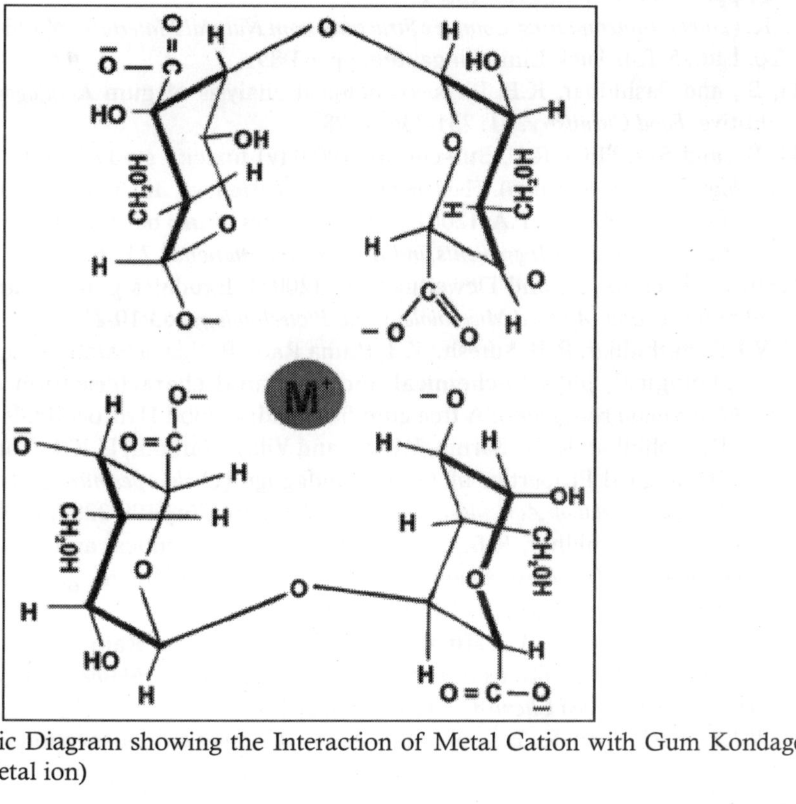

Fig. 16.10: Schematic Diagram showing the Interaction of Metal Cation with Gum Kondagogu (M^+= Metal ion)

Future Outlook and Perspectives

Biopolymers (exudate gums) possess a unique combination of functionalities and properties that can never be matched by any other alternative synthetic polymers, which makes their complete substitution impossible. Importantly, these biopolymers are eco-friendly as they are biodeteriorable. One of the important outcomes of the present investigation is the interaction of metal ions with gum kondagogu, other being its potential applications in food and pharmaceutical industry. Gum kondagogu can be used as a nontoxic reductant for metal-nanoparticles constructs which offers numerous benefits ranging from environmental safety to ready integration of these nano-materials in biologically relevant systems, catalysis, antimicrobial agents, and biosensors, with biomedical applications. Additionally, appropriate chemical modification of the functional groups present in gum kondagogu may lead to the development of novel pH responsive hydrogels for slow drug delivery systems.

NOTES

Bates, B.C.; Z.W. Kundzewicz; S. Wu and J.P. Palutikof, Eds., (2008). *Climate Change and Water. Technical Paper of the Inter-governmental Panel on Climate Change*, IPCC Secretariat, Geneva, 210 pp. ISBN: 978-92-9169-123-4.

Hall, S.R. (2009). *Biotemplating-Complex Structures from Natural Materials*. World Scientific Publishing Co. Ltd., 5 Toh Tuck Link, Singapore, pp. 63-89.

Janaki, B., and Sashidhar, R.B. Physico-chemical analysis of gum *Kondagogu:* A potential food additive. *Food Chemistry*. **61**: 231-236, 1998.

Janaki, B., and Sashidhar R.B. Sub-chronic (90-day) toxicity study in rats fed gum kondagogu (*Cochlospermum gossypium*). *Food and Chemical Toxicology*, **38**: 523-534, 2000.

Phillips, G.O., and Williams, P.A. (2001), Tree exudates gums: natural and versatile food additives and ingredients. *Food Ingredients and Analysis International,* **23**: 26-28.

Verbeken, D.; Dierchx, S.; and Dewettinck. K. (2003). Exudates gums: Occurrence, production, and applications. *Applied Microbiology and Biotechnology*, **63**:10-21.

Vinod, V.T.P.; Sashidhar; R.B. Suresh.; K.I. Rama Rao.; B. Vijaya Saradhi.; U.V.R. Prabkar Rao; T. Morphological, physico-chemical and structural characterization of gum kondagogu (*Cochlospermum gossypium*): A tree gum from India. Food Hydrocolloids, **22**: 899-915, 2008a.

Vinod, V.T.P.; Sashidhar, R. B.; Sarma, V.U.M. and Vijaya Saradhi, U.V.R. Compositional Analysis and Rheological Properties of Gum Kondagogu (*Cochlospermum gossypium*): A Tree Gum from India. *Journal of Agriculture and Food Chemistry*, 56: 2199-2207, 2008b.

Vinod, V.T.P.; and Sashidhar, R.B.; Surface morphology, chemical and structural assignment of gum kondagogu (*Cochlospermum gossypium* DC): An exudate tree gum of India. *Indian Journal of Natural Product Resource*, 1: 181-192, 2010a.

Vinod, V.T.P.; Sashidhar, R. B.; Sarma, V.U.M.; Satyanarayana, Raju; S. Comparative amino acid and fatty acid compositions of edible gum kondagogu (*Cochlospermum gossypium*) and karaya (*Sterculia urens*). *Food Chemistry*, 123:57-63, 2010b.

Vinod, V.T.P.; Sashidhar, R. B.; Sukumar, A. A.; Competitive adsorption of toxic heavy metal contaminants by gum kondagogu (*Cochlospermum gossypium*): A natural hydrocolloid. Colloids and Surfaces B: *Biointerfaces* 75: 490-495, 2010c.

Vinod, V.T.P.; Sashidhar, R.B.; Bioremediation of industrial toxic metals by gum kondagogu (*Cochlospermum gossypium*): A natural carbohydrate polymer. *Indian Journal of Biotechnology*, 10: 113-120, 2011.

17
Can Ecotourism Practices Support Green Economy? A Conceptual Framework

Smriti Ashok and *Anshul Ashok*

Introduction

Over the last quarter of a century, the world economy has quadrupled, benefiting hundreds of millions of people. In contrast, however, 60 per cent of the world's major ecosystem goods and services that underpin livelihoods have been degraded or used unsustainably. The cost of this loss is not directly computed as we do not have a practice of representing our priceless and extremely limited natural resources in monetary terms. Thus, came, the concept of Green Economy into existence, which advocates attaching economic value to our natural capital.

In this chapter an attempt has been made to explore the economic aspects of Ecotourism and examine its relation with Green Economy. Authors have tried to identify the underlying synergy between these two concepts and they have been able to conclude that Ecotourism practices would definitely support the cause of Green Economy. It can be said that after the process of greening the economy has attached economic values to the natural capital and people start to recognize its true potential and value, they would start actively participating in the conservation drive. In second stage with development of Ecotourism in true manner and its implementation it would provide the desired livelihood to the local and marginal community, by protecting the natural and cultural resources and attracting the tourists.

Green Economy (GE) : The Concept

UNEP defines a green economy as one that results in improved human well-being and social equity, while significantly reducing environmental risks and ecological scarcities. In its simplest expression, a green economy can be thought of as one which is low carbon, resource efficient and socially inclusive. In a green economy, growth in income and employment should be driven by public and private investments that reduce carbon emissions and pollution, enhance energy and resource efficiency, and prevent the loss of biodiversity and ecosystem services (UNEP, 2012).

The Global Citizens Centre, led by Kevin Danaher, defines green economy differently from the use of pricing mechanisms for protecting nature, by coining a term triple bottom line. That is an economy, which is Environmentally Sustainable, Socially Just and Locally Rooted. (Dual Citizen Inc., 2011).

The Global Green Economy Index, published annually by consultancy Dual Citizen Inc., measures and ranks the perception and performance of 27 national green economies based on 4 primary dimensions defining a national green economy like (Dual Citizen Inc., 2011):

1. Leadership and the extent to which national leaders are champions for green issues on the local and international stage.
2. Domestic policies and the success of policy frameworks to successfully promote renewable energy and green growth in home market.
3. Cleantech Investment and the perceived opportunities and cleantech investment climate in each country.
4. Green tourism and the level of commitment to promoting sustainable tourism through government.

This index ranked European Economies as first which dominate the top of our Leadership, Cleantech Investment and Domestic Policies segments with respect to implementing Green Economy while New Zealand is the global best practice case for promoting Green Tourism.

UNEP report, Towards A Green Economy, identified 10 key economic sectors which can contribute towards transition of Brown Economy to a Green Economy, resulting into increasing human well-being and social equity, and reducing environmental risks and ecological scarcities. These sectors are Agriculture, Construction, Renewable Energy, Fisheries, Forestry, Industry, Tourism, Transport, Waste Management and Water Conservation. Comprehensively Renewable Energy Sector, Sustainable Forms of Agriculture, Sustainable Urbanization, Solid Waste Management, and Ecotourism, etc., can contribute towards Greening the economy (UNEP, 2011).

Among above mentioned economic activities Ecotourism can be considered one of the most promising activities, which can support Green Economy—as it depends upon the principles of sustainability similar to the Green Economy concept. Achievement of the goal of the Green Economy is only possible when such economy follow the concepts of Ecological Green Economy, which is an economic development model based on Sustainable Development and knowledge of Ecological Economics. (*Wikipedia*, 2012)

What is Ecotourism?

The concept of ecotourism came into being in the early 60s, it gained prominence during 1980s, when Ceballos-Lascurian defined ecotourism as "an experience of travelling to relatively undisturbed areas with specific objective of studying and admiring the scenery, its wild plants and animals as well as any other cultural manifestations (both past and present) found in these areas" (Page and Dowling, 2002; Ross and Wall, 1999). Likewise it has been defined by international organizations as follows:

The Ecotourism Society defined Ecotourism as "Responsible travel to natural areas that conserves the environment and improves the well-being of local people". (TIES, 1990)

UNWTO (2001) defines ecotourism as:

- Nature-based forms of tourism in which the main motivation of the tourists is the observation and appreciation of nature as well as the traditional cultures prevailing in their natural surroundings.
- Containing educational and interpretation features.
- Generally, but not exclusively, organized for small groups by specialized and small locally-owned businesses. Foreign operators of varying sizes also organize, operate and/or market ecotourism tours, generally for small groups.
- Minimizing negative impacts on the natural and socio-cultural environment.

Supporting the Protection of Natural Areas

On the basis of above definition ecotourism can be defined as a form of tourism where tourists visit to fragile, pristine, and usually protected areas. It is intended as a low-impact and often small scale alternative to standard commercial (mass) tourism. Its purpose may be to educate the traveller, to provide funds for ecological conservation, to directly benefit the economic development and political empowerment of local communities, or to foster respect for different cultures and for human rights.

Gradually, it has been adopted as a strategy for reconciling conservation with development in ecologically rich areas by many international organizations—World Tourism Organization (WTO), World Wide Fund for Nature (WWF), International Union for Conservation of Nature and Natural Resources (IUCN) and The International Ecotourism Society (TIES). UN declared 2002 as the "International Year of Ecotourism" with the aim to encourage the cooperative efforts of Governments, International, National and Regional Organizations as well as NGOs to fulfil the objectives of Agenda 21—a document submitted after Earth Summit, 1992 at Rio de Janeiro, to achieve sustainable development for promotion of environment. (Gray, 2003)

Relationship between Green Economy and Ecotourism

The relationship between Green Economy and Ecotourism can be judged under following themes and sub-themes.

Value of Natural Capital

Green Economy and Ecotourism both recognize the need and importance of protection of biodiversity—the life support system of this live planet—Earth. Green Economy recommends that the development path should maintain, enhance and where necessary, rebuild natural capital as a critical economic asset and as a source of public benefits, especially for poor people whose livelihoods and security depend on nature. In this context Biodiversity, (the living fabric of this planet includes life at all levels: Genes, Species and Ecosystems) contributes to human wellbeing by providing four types of services: (i) Provisioning Services, (ii) Regulating Services, (iii) Cultural Services, and (iv) Supporting Services.

These services are mostly in the nature of public goods and services whose economic invisibility has thus far been a major cause of their undervaluation, mismanagement and ultimately resulting in loss. In this context, Ecotourism can play an important role, as it takes care of all four services provided by ecosystem. In fact in its purest form, it is found to be on the philosophy of sustainable development as well as sustainable tourism. It supports conservation of biodiversity and natural resources, creates environmental awareness and ensures equitable distribution of the socio-economic benefit to the local community while offering optimum satisfaction to the tourists (Balmey, 2001; Buchsbaum, 2004; Lascurain, 2001; Newsome, *et al.*, 2002).

Sustainability: A Common Goal of Green Economy and Ecotourism

Sustainable development emphasizes a holistic, equitable and far-sighted approach towards decision-making, at all levels. It emphasizes not just strong economic performance but intra-generational and inter-generational equity. It rests on integration and a balanced consideration of social, economic and environmental goals and objectives in both public and private decision-making (Agenda 21, 1992).

The basic idea behind adopting green economy is that it can contribute to achieving the goal of sustainable development by generating economic sustainable wealth, facilitating jobs and using natural resources in an optimal way. This is only possible when such economy follow the concepts of Ecological Green Economy, which is an economic development model based on Sustainable Development and knowledge of Ecological Economics. (*Wikipedia*, 2012)

Thus we can infer that the concept of a "green economy" does not replace sustainable development. In fact to achieve Sustainability principles of Green Economy is required to be followed. There is now a growing recognition that achieving sustainability rests almost entirely on getting the economy right. Decades of creating new wealth through a "brown economy" model have not substantially addressed social marginalization and resource depletion, and we are still far from delivering to the Millennium Development Goals. Sustainability is still a vital long-term goal, but we must work on greening the economy to get us there. (UNEP, 2011)

On the other hand "Achieving Sustainability is the goal of Ecotourism. Ecotourism is considered as part of Sustainable tourism which is envisaged as management of all resources in such a way that economic, social and aesthetic needs can be fulfilled while maintaining cultural integrity, essential ecological processes, biological diversity and life support systems" (UNWTO, 1995). It has been advocated by many scholar that Ecotourism, to be successful, must promote sustainable development by establishing a durable productive base that allows local inhabitants and ecotourist service providers to enjoy rising standards of living. (Lascurain, 2001; Newsome, *et al.*, 2002)

Principles of Green Economy and Ecotourism

Principles of Green Economy

Principles of Green is based on three factors such as:

(a) Environmentally Sustainable;
(b) Socially Just; and
(c) Locally Rooted.

Principle of Ecotourism

The underlining sustainability principles of ecotourism are reflected in definitions given by many scholars like Hetzer (1965), Ceballous Lascurain (1987), Martha Honey (1999), TIES, etc., and on the basis of above definition and other literature review the principles of ecotourism has been comprehended by Kumari (2008) in her Ph.D. thesis.

(a) *Ecologically Sustainable*: Ecotourism is ecologically sustainable tourism undertaken in a natural setting. Its primary goal is to foster sustainable use through resource conservation, cultural revival, and economic development and diversification. This principle advocates the protection of natural and cultural resources, which extend the life support system for the survival of human beings.
(b) *Generation of Socio-Economic Benefit to the Local Community*: Improvement in the socio-economic condition of the local people is one of the major constituents of the concept of ecotourism. Without including this component any tourism cannot be defined. Ecotourism, as a concept, focuses on travel for education of tourists and on indigenous culture as an attraction. It should encourage all people to value their cultural heritage.

(c) *Generation of Environmental Awareness and Capacity Building:* Environmental education and interpretation is the central tenet of ecotourism and this helps in understanding the complexity of nature (Kumari, 2008). It is well researched that "a proper understanding of biological or more specifically ecological factors can significantly reduce the scale of environmental damage associated with recreational and tourism development" (Newsome *et al.*, 2002). Ecotourism can be an effective means to achieve conservation of biological diversity, both for the benefit of the tourism industry and for the intrinsic and economic values of biodiversity.

(d) *Optimization of Touristic Aspirations*: Satisfaction of visitor's experience is essential for long-term viability of ecotourism destinations (Page and Dowling, 2002). Ecotourists in general, are highly educated, have high levels of income, and have high level of environmental awareness. They demand high levels of ecological and cultural information (Cater, 1993). The quality of the environment and the visibility of both flora and fauna and indigenous culture are essential features of their experience. This demands the preservation and maintenance of ecosystem (Kumari, 2008).

Protection of Natural Resources by Local Control and Stewardship

Social dimensions of green economy are often addressed in terms of green jobs, green consumerism, and the kinds of education/retraining, social safety-nets and social dialogue required to facilitate the transition, etc. While poverty eradication is increasingly accepted as a goal of green economy, it is often assumed that it will follow inevitably from low-carbon growth. A green economy can generate as much growth and employment as a brown economy, and outperforms the latter in the medium and long run, while yielding significantly more environmental and social benefits (UNEP, 2011).

In the UNEP vision of ecotourism, biodiversity is conserved as strengthened protected area systems that attract ecotourists. The sustainable use of biodiversity is achieved through tourism-generated income, jobs, and business networks. The equitable sharing of the benefits of biodiversity is realized by obtaining the "informal consent and full participation" of local communities and indigenous people. (UNEP, 2002) It embraces the principles of sustainable tourism, which contributes actively to the conservation of natural and cultural heritage most importantly with the involvement of the real stakeholders and as per their liking as well. In this context ecotourism aims to develop the feeling of ownership and stewardship among the local people by control over the resources.

Functions of Green Economy and Ecotourism

In the context of green economy, the specific functions of the ecosystems are valued economically. In short, the green economy sets the stage not only for trading in the conservation of commons such as water and biodiversity, that so far had escaped commoditization; but also the trading in ecosystem functions, like carbon sequestration to other ecosystem functions such as water purification and possibly even evapo-transpiration and pollination. This has tremendous implications for communities who have been custodians of these ecosystems, in some cases for generations. (Varghese, 2012) Thus its main function is to attach economic value to the natural capital in its tangible as well as intangible form. In return it would help the economy to grow and move towards sustainability.

On the other hand, the fundamental function of ecotourism is protection of natural and cultural resources through generation of revenue, education and local participation and capacity building (Ross and Wall, 1999). It has been accepted that protection of natural resources is the prime objective of ecotourism, which is possible through revenue generation by attracting more tourists and making them environmentally aware. In this process, involvement of the local people is a must so that they can get ownership and control over the resources, with the feeling of stewardship. This will bring sustainability at the local level. Each of the function of ecotourism is basic to its success, and together, they can lead to fulfilment of more specific objectives. These objectives are intertwined because success or failure to achieve one objective may influence the success or capacity to achieve others.

Economic Aspects of Green Economy and Ecotourism

The loss of biodiversity and ecosystem services has direct economic repercussions that are systematically underestimated. In this regard Green Economy works towards attaching economic value to natural capital. The concept of green economy advocates that healthy ecosystems act as the foundation of economic activity and thus a prerequisite for achieving a green economic transition. Given this, it is essential that the economic value of these services are recognized, demonstrated and captured in the accounts and decision-making of governments, the private sector and consumers. Ideally, changes in stocks of natural capital would be evaluated in monetary terms and incorporated into the national accounts, as is being pursued in the ongoing development of the System of Environmental and Economic Accounting (SEEA) by the UN Statistical Division, and the adjusted net national savings methods of the World Bank. In turn this would help the society pave the way for more targeted and cost-effective solutions. (UNEP, 2011)

Ecotourism lately has emerged as a development strategy aiming to protect natural environment and cultural diversity of a particular geographic region by attracting eco-tourists and in turn generating revenue for the local people which further enables and encourages them to preserve their natural resources, in fact develop their surroundings. These areas are developed with a notion to protect them through the financial returns of ecotourism activities. Indeed, ecotourism aims to promote conservation through a sustainable management regime under a 'protection through usage' approach with the objective of creating a greener tourism industry and an increase in the ecological and socio-cultural awareness and behaviour of tourists. (Mekong Communications, 2010) Ecotourism Captures $77 billion of the global market and experiencing double-digit gains that are likely to accelerate as concern about global warming rises. The United Nations World Travel Organization estimates that in 2007, ecotourism captured 7 per cent of the international market. According to UNWTO, from January to August 2010 the international tourist arrival worldwide was 642 million, 40 million more than same month in 2009 and one million more than in 2008. (UNWTO, 2012)

A survey by the United States Fish & Wildlife Service puts the annual economic value generated by bird watchers (or 'birders') and other wildlife watchers at around US$ 32 billion per year in the United States alone. This amount corresponds to the GDP of Costa Rica, which, coincidently, is a popular destination for US birders. In Scotland, the Royal Society for the Protection of Birds (RSPB) found out in 2011 that between GBP 5 to 8 million (US$ 8 to 12 million) is spent annually by tourists wishing to see White-tailed Eagles on the Isle of Mull alone. The equivalent of at least 110 full-time jobs—4 per cent of jobs in Scotland associated with wildlife tourism—is supported by this expenditure every year. Economic benefits delivered by White-tailed Eagles on the Isle of Mull have more than tripled since 2005. (UNEP, 2011)

These mammoth numbers are a clear indicator that Ecotourism is already a strong economic force and will become all the stronger with passing time. With its ever increasing economic importance and its direct impact upon/over biodiversity conservation it is all set to become a force to reckon with.

How Far we are from Green Economy?

Over the last quarter of a century, the world economy has quadrupled, benefiting hundreds of millions of people. In contrast, however, 60 per cent of the world's major ecosystem goods and services that underpin livelihoods have been degraded or used unsustainably. Indeed, this is because the economic growth of recent decades has been accomplished mainly through drawing down natural resources, without allowing stocks to regenerate, and through allowing widespread ecosystem degradation and loss. According to the Forest Resource Assessment Report by FAO; the world forest area has decreased from 4.17 billion hectares in 2000 to 4.03 billion hectares in 2010 due to Annual net forest loss of 5.2 million hectares of forest annually during 2000-2010 decade (FA0, 2010). Water is becoming scarce and water stress is projected to increase with water supply satisfying only 60 per cent of world demand in 20 years; agriculture saw increasing yields primarily due to the use of chemical fertilizers, which have reduced soil quality and failed to curb the growing trend of deforestation (UNEP, 2011).

Ecosystem and biodiversity are shrinking from all sides because of the expansion of agriculture, urbanization, forest depletion, pollution and consequent activities. (Maunasinghe and McNeely, 1995; Buckley, 2001) These are resulting into low levels of human and ecosystem wellbeing, which has been revealed by Prescott-Allen (2001) in the study of 180 countries. He found that two-third of the world's human population lives in countries with a poor Human Wellbeing Index (HWI) and a poor Ecosystem Wellbeing Index (EWI) and covers almost half (48%) of the planet's land and inland water surface. Low level of human and ecosystem wellbeing are the result of high level of environmental degradation, which is leading the world towards massive ecological imbalance. (Kumari, 2008)

Ecological Scarcities are, therefore, seriously affecting the entire gamut of economic sectors, which are the bedrock of human food supply (agriculture, freshwater, forestry) and a critical source of livelihoods for the poor. And ecological scarcity and social inequity are definitional signatures of an economy which is very far from being "GREEN". If this pace of forest degradation is continued then world can go towards irrecoverable ecological damage without any remedy.

How Ecotourism can Support Green Economy

As has been established and amply supported that both Green Economy and Ecotourism care about Natural Capital, Environmental Conservation and Equitable Income generation by the Local people, for the local people and through the local people. Though the impact of Ecotourism is not only confined to the local community area as the known and unknown effects of any Environmental and Ecological Preservation activity can be observed by a much wider geography than the residence of the local community. And Green Economy obviously is a global aggregate of individual communities meeting the needs of its citizens through their responsible actions, local production and exchange of goods and services.

Thus we can ascertain that both the concepts under discussion share a common Goal that is of "Sustainable Development" but they have their own approach and scope toward the achievement of the same.

Ecotourism is the Greener variant of Tourism which advocates conservation of biodiversity and act as a development strategy to build a self sustainable system to help protect and further enhance the ecosystem through the income generated by ecotourism activities.

If private or public investment is not required to go in for maintenance of the fragile ecological balance and yet the locals are destined to benefit out of the normally available natural resources then why would they not contribute their part in capacity building of the ecosystem by NOT exploiting it !! They would certainly protect their natural surroundings and would in turn make others behave the same way as well. In simple words, if the economic interests of local rural communities are taken care of and their long term well-being is provisioned for through ecotourism practices then people will ensure optimal utilization of the natural capital and they will also play their part in the self-renewal process of the extremely limited natural resources.

On the other hand, Green Economy propagates a concept that if the goal of sustainable development is to be achieved then it can be done so by generating economic sustainable wealth, facilitating jobs and using natural resources in an optimal way. It says that the present day's "Brown Economy" model has not substantially addressed social marginalization and resource depletion. In fact the world has drawn upon the natural resource reservoirs with no consideration for improved human well-being over the long term. No one has put a thought to the fact that their acts are exposing future generations to significant environmental risks or ecological scarcities.

This is so because people and communities have kept their economic interests, personal benefits prioritized over their social responsibilities with respect to biodiversity conservation. And why have they done so because their current requirements of meeting their financial needs have been more compelling.

Thus concept of Green Economy already identified the shortcoming in our policies for achieving sustainable development wherein we did not think of provisioning for economic needs of the people before working upon the preservation and sustainability issues.

Crux is that, Ecotourism is an economic activity which works directly for sustainable development and then further discusses about just distribution of benefits, wherein Green Economy first tries to provide for the equitable sharing of the benefits of biodiversity to the concerned people and intends to obtain the participants support for preservation.

So, when Ecotourism and Green Economy both target the same thing but only tread different path towards its achievement, then they certainly can be of help to each other. Its also proved that the latter is practically a better approach than the former then Ecotourism can definitely play second fiddle to Green Economy to help achieve their common cause of "Sustainable Development".

Ecotourism Development in Sikkim : A Case Study

As per the above discussion UNEP report identified 5 key sectors like Renewable Energy sector, Sustainable forms of Agriculture, Sustainable Urbanization, Solid Waste Management, and Ecotourism, etc., which can guide any economy to achieve goal of Green Economy (UNEP, 2011). Endowed with rich and varied depository of biodiversity, the Government of Sikkim has launched a unique and innovative programme called "State Green Mission" with the view to raising avenue plantation and beautification of all vacant and waste lands to further reinforce widespread recognition of Sikkim being a Green State (ENVIS, 2006). Among several environmental friendly activities, promotion of ecotourism has been considered of prime importance.

Sikkim : An Introduction

With around 607,000 inhabitants as of 2011, Sikkim is one of the smallest and the least populous States in India. It covers an area of 7,096 sq. km, extending approximately

114 km from North to South and 64 km from East to West. A miniature of rich and varied depository of biodiversity, Sikkim is shelved in the womb of the mighty Mt. Khangchendzonga and her beautiful ranges extending from west to north. North by the Khangchen Gyao, Gurudokmar and Dongkhyala peak forming cold desert plateau with abundance of rare and endangered faunal resources (*Wekipedia*, 2012). Its geo-physical location has been proved advantageous for the State, this is providing base for the State economy to move towards the Green Economy.

Relevance of Ecotourism for State's Economy

Sikkim's economy is largely dependent on agriculture and tourism, and as of 2011 the State has the fourth-smallest GDP among Indian States, although it is also among the fastest-growing. In recent years, the government of Sikkim has extensively promoted tourism. As a result, State revenue has increased 14 times since the mid-1990s. In Sikkim over 82 per cent of the geographical area is under forest land-use and more than 42.55 per cent area is under Protected Area Network, i.e., under Wildlife and Biosphere Reserve. The entire State is Eco-Green- Pollution Free. The Ministry of Tourism, Government of India, in its national conference held at Gangtok on 6th December, 2009 has declared Sikkim as "ECO-TOURISM DESTINATION". Sikkim is the most eco-friendly State not only in the country but also in the world and is managing its natural resources in most efficient manner (Government of Sikkim, 2012).

Ecotourism Development Policy

The Government of Sikkim addresses ecotourism as a priority sector and, therefore, the state government has developed 'Sikkim Ecotourism Policy 2011' with an aim to not only to ensure sustainable livelihood of the local people but to protect and manage forest, as well. In line with the previous Five-Year Plans, the current Eleventh Five Year Plan seeks to achieve sustainable economic development by focusing on horticulture and floriculture, education, hydroelectric generation, and tourism. The main thrust of the Eleventh Plan is to make Sikkim 'a premium ecotourism destination (Bhaduria, 2010).

Development of Ecotourism in Sikkim

Tourism was declared a focus area in the Ninth and Tenth Five Year Plans and will continue to be of equal significance in the eleventh plan. Tourist inflow increased by more than 100,000 between 2001 and 2005 to a total of 268,267 in 2005. The State invested an estimated 1.9 per cent of its total expenditure on tourism during the Tenth Plan and has proposed a 2.2 per cent outlay for the Eleventh Plan, indicating the State's high expectations toward the sector as a major source of revenue and employment (Bhaduria, 2010). As per the "State Green Mission" Sikkim has been taking rapid strides in the ecotourism sector and has now become a major ecotourism destination in the North-East region. According to the data released by the Ministry of Tourism for the total number of domestic and foreign tourist visits to North-Eastern States for the period 2007-09, Sikkim received the maximum number of foreign tourists (Bhaduria, 2010).

With above objective the development of ecotourism is benefiting its economy in following way.

Contribution towards the Conservation of Biodiversity

The Forest Department, which oversees both the forestry sector and biodiversity conservation, has worked hard to protect the valuable assets of Sikkim. Here are some of the initiatives taken and achievements to date, like the adoption of Forest, Environment, and Wildlife Policy in 1999, ban on grazing of all domestic and semi-domestic animals in reserved forest areas and adoption of Ecotourism concept at many sites. This resulted into increase in forest cover from 42.3 per cent in 1991 to 46.0 per cent in 2005 and 2006 (Department of Forest, 2012).

Sustains the Wellbeing of the Local People

UNESCO, TMI along with the State based NGO ECOSS, are working with an aim at promoting community participation in the area's tourism development. Over the last few years, new trekking trails like the Yuksa -Dzongri Corridor, Everest Singalila in Hee Bermiok, Yambong Singalila at Nambu and Areylungchok Dzongri in Labdang have been opened with the help and training provided by the Ecotourism and Conservation Society of Sikkim (ECOSS) along with UNESCO. This also helps generate employment opportunities and income-generating activities for local people through the development and management of tourist home-stays, as well as through training in the skills needed for local people to become cultural and nature guides.

Offering Diversity of Ecotourism Attraction for Tourists

Ecotourism can be successful when it gives diversity of attraction depending upon the natural and cultural resources of the destination. These different types of attraction can be identified by the local people, which can attract large number of tourists in turn it gives livelihood for them. This is being done in Sikkim, where the Khangchendzonga Conservation Committee (KCC) has come up with innovative ecotourism packages to beat the lean season. These includes packages like the "Monsoon Magic" Alpine Treks and the Subtropical Winter Treks. Dzongri, Singalila, Yambong, Areylungchok and Yumesamdong are some of the top alpine destinations. In an eight days "Monsoon Magic" package one can guarantee at least 100 different alpine flowers in full bloom.

Stress Local Participation and Ownership and Business Opportunities particularly for the Rural People

The Forest Department of Sikkim Established Joint Forest Management Committees (JFMCs) and Ecodevelopment Committees (EDCs) for decentralization through participatory forest management, Zoning of protected areas into conservation zones and zones for community use like Preparation of the region for ecotourism development, e.g., preparation of regulations for trekking, etc. This is developing sense of belongingness among the local people.

Stakeholders Participation in Ecotourism Development

Sikkim Biodiversity Action Plan (SBAP) contains identification and analysis of problems related with biodiversity conservation in this State, as well as enlists detailed action plans for the concerned departments of State and central, concern research institutes, NGOs and communities. The Green Mission of the State and its working guidelines have saved its natural capital, which is reflected in the Green Protection Index of Sikkim is 0.903 (first in India) as against 0.425 of India.

Thus, Sikkim is following collaborative approach for achieving the goal of Ecotourism. The Government has developed Ecotourism Policy and Code of Ethics for Ecotourists, State NGO ECOSS are facilitating environmental awareness programme and training for the local people with partnership of UNESCO. In addition, community based NGO, KCC is working towards encouraging the local people for participating in ecotourism development. Apart from environment friendly ecotourism development, the Government is also working towards strengthening its Panchayati Raj institutions, development of Green rural road, Clean energy Development programme which can help the State to play a vital role in the development of the State as well as help achieve its Green Mission Programme and ultimately the Goal of Green Economy-Sustainable Development.

Conclusion

The authors with all the discussions and literature reviews have been able to establish that beyond their common goal, Ecotourism and Green Economy both have a lot more similarity in terms of their objectives, driving forces, target audience, thrust areas and their areas of implementation. They have also been able to get the resolution to their principal concern of whether the ecotourism practices can be supportive to and useful for Green Economy.

Indeed Ecotourism Practices can add a great value to the Green Economy drive. It not only can support the green-drive but if these independent forces join hands and work in tandem then achievement of their Goal-Sustainable Development, can prove to be less tough a task or a lot easier than it might seem at present.

The green economy advocates the economic valuation of all natural capital resources whether it is in tangible form or in intangible form. In this respect ecotourism can be proved to act as complementary in supporting thus economically valued regions by further enhancing their valuation by promoting ecotourism activities there both for the region's tangible or intangible resources.

Future Scope

Once the bridge between these two concepts under discussion has been defined a more prominent question has crossed the minds of the authors as to—How Green Economy can be promoted through Ecotourism on an operational level. This aspect the authors wish to explore further with case study of on the ground implementation of such economies (be it at National, State or Local levels) practicing Ecotourism and is progressing towards a Green Economy.

The objective of the above question can only be fulfilled when ecotourism operations and its level of sustainability can be measured at operational level. This need the development of Ecotourism assessment framework, development of the model to maximize its level of sustainability and assessment of field level ecotourism sustainability- The authors are in the process of answering these questions by solving following research questions like:

How to practice ecotourism sustainability at operational level? What are the method to assess the level of ecotourism sustainability and how to maximize its level of sustainability?

After getting all answers there is need of international level of co-operation. As green economy activities are commercial in nature, providing positive returns on investment, external financing is sometimes required to complement public financing and catalyze private investment from both domestic and foreign sources. The need for external financing is particularly critical for green projects in infrastructure services sectors such as energy and waste management to make them commercially attractive to private investors.

NOTES

Balmey, R. K., (2001). Principles of Ecotourism, In D.B.Weaver, (Ed.), *The Encyclopedia of Ecotourism*. CABI Publishing, New York. 1-22.

Bhadauria, S.B.S. (2010). Development and Promotion of Eco-Tourism In Sikkim, 010 ENVIS Centre, Forest Environment & Wildlife Management Department, Government of Sikkim available on http://www.sikenvis.nic.in/Reports%20and%20Publications/100years/100%20Years%205.pdf

Buchsbaum, B. D., (2004). Ecotourism and Sustainable Development in Costa Rica' Major Paper Submitted to Virginia Polytechnic Institute and State University Master of Public and International Affairs, College of Architecture and Urban Studies, May 3.

Ceballos-Lascurain, H., (2001). Integrating biodiversity conservation planning into tourism sector.

Industry and Environment, July-December, 24(3-4):38-41.

Cater, E., (1997). Ecotourism: Dimensions of Sustainability. In: *Ecotourism, Forest Conservation and Community Development*, Proceeding of an International Seminar, 26-31 January 14-25.

FAO; (2010).Global Forest Resource Assessment 2010, FAO Forestry Paper 163. Food and Agriculture Organization of the United Nations Rome, 2010

Government of Sikkim, (2012), Tourism Mission 2015, available on www.sikkimtourism.travel/ Webforms/./Tourism_Mission_2015.pdf

Hetzer, N. D., (1965). Environment, Tourism, Culture. *LINKS* (July), Reprinted in Ecosphere, 1970 1(2):1-3

Kumar, B. and Kumar, P. (2001), Green economy: policy framework for sustainable development, *Current Science*, Vol. 100, No. 7, 10 April 2011.

Kumari, S., (2008). Identification of Potential Ecotourism Sites, Sustainability Assessment and Management in Geospatial Environment, Ph.D., Thesis, Department of Humanities and Social Sciences, Indian Institute of Technology, Kharagpur, India.

Mekong Communications, (2010). UNWTO Study on Best Practice in Ecotourism: A Compendium of Best Practices and Guidelines for Ecotourism in the Asia Pacific Region, 22 June 2010. available on http://www.mekongtourism.org/site-t3/news/latest-news/news/?tx_ttnews [tt_news]=203&tx_ttnews[backPid]=28&cHash=0e1cd9d3fb

Munasinghe, M., and McNeely, J. (1995). Key Concepts and Terminology of Sustainable Development. In: Munasinghe and Shearer, Defining and Measuring Sustainability: The Biogeophysical Foundations. 19-56.

Newsome, D.; Moore, S.A. and Dowling, R. K., (2002). *Natural Area Tourism: Ecology, Impacts and Management*, Channelview publications, Sydney. 1-340.

Page, S.J. and Dowling, R. K., (2002). *Themes in Tourism*, Pearson Education Limited, Harlow. 1-338.

Prescott-Allen, R., (2001). The Wellbeing of Nations, Island Press and International Development Research Centre. http://www.idrc.ca/en/ev-5474-201-1-DO_TOPIC.htm. 1-440 27th August, 2005.

TIES. (1990). What is Ecotourism?. Available on http://www.ecotourism.org/what-is-ecotourism

Tambe, S. and Rawat, G. S., (2006). *An Ecological Study of Pastoralism in the Khangchendzonga National Park, West Sikkim*. The Mountain Institute - India, Sikkim. 53 pp.

UNEP, (2010). Eleventh Special Session of the Governing Council/Global Ministerial Environment Forum. Bali, Indonesia, 24-26 February 2010.

UNWTO, (2010), *World Tourism Barometer*, Vol. 8, No. 3, October 2010.Available at: http:// dtxtq4w60xqpw.cloudfront.net/sites/all/files/pdf/unwto_barom10_3_en.pdf

UNEP, (2002). Sustainable Tourism Development Guidelines: UNEP Principles on the implementation of sustainable tourism and the CBD sustainable tourism development guidelines for vulnerable ecosystems. http://www.uneptie.org/PC/tourism/policy/ cbd_guidelines.htm. Accessed on 15th Jan 2007.

UN, (2011). Green Economy and Sustainable Development: Bringing Back the Social Dimension Nations, Geneva Source http://www.uncsd2012.org/rio20/index.php?page=view&nr=264& type=13&menu=27

UNEP, (2012a). Bird-Watching Can Help Eco-Tourism Fly High in a Green Economy http:// www.cms.int/news/PRESS/nwPR2012/pressreleases/wmbd_100512_e.pdf, published on - 10/05/2012

UNEP, (2012b). Green Economy Valuing Nature, Available on http://www.unep.org/ greeneconomy/Portals/88/VALUING%20NATURE.pdf

UNEP, (2011), Towards a Green Economy: Pathways to Sustainable Development and Poverty

Eradication - A Synthesis for Policy Makers, www.unep.org/greeneconomy

UNWTO (2012) Aims of Sustainable Tourism in the UNEP - UN Environment Programme and UNWTO- UN World Tourism Organisation Publication: *Making Tourism More Sustainable — A Guide for Policy Makers.*

Varghese, S. (2012). Green Economy: Commoditization of the Commons, Institute for Agriculture and Trade Policy. Available on http://www.iatp.org/documents/green-economy-commoditization-of-the-commons

WRI. World Resource Institute, (2000). Ecosystems and Human Well-being: A Framework for Assessment: A Report of the Conceptual Framework Working Group of the Millennium Ecosystem Assessment, Island Press, Washington D.C. http://www.who.int/globalchange/publications/millassess.pdf. Accessed on 12th September, 2004

WTO, World Tourism Organization, (1995). Agenda 21 for the Travel and Tourism Industry: Towards Environmentally Sustainable Development. World Travel and Tourism Council, WTO, and Earth Council, London, Madrid, San Jose, Costa Rica.

World Economic Forum, *Environmental, Sustainability Index*, Geneva, Switzerland, 2005.

18
Sustainable Renewable Energy
Transition towards Green Economy

Sarina Kalia and *Deepika Sharma*

Introduction

Perpetuation of disparities, a worsening of poverty, hunger, ill health and illiteracy and the continuing deterioration of the ecosystem on which we depend for our well-being is what shackles our development today. The prevailing economic growth is focused on increasing GDP above all other goals. The nations consider GDP as a measure of progress, but they forget that it comes with significant and potentially irreversible social, environmental and economic costs.

In a recent global assessment, approximately 60 per cent of the world's ecosystem services were found to be degraded or used unsustainably. The gap between the rich and poor and income inequality (measured by the gap between the highest and lowest income earners) rose in more than two-thirds of countries. The cumulative effect of the inverse relationship between economic growth and sustainability is alarming all of us. We need to recognize that only such development is sustainable, which respect ecological constraints. Therefore, there is an urgent need for transition from GDP based economy to "**GREEN ECONOMY**"one that simultaneously promotes sustainability and economic growth.

Green Economy : A Step towards Sustainable Development

A Green Economy is an alternative vision for growth and development, that can generate growth and improvements in people's lives in consistent with sustainable development. A green economy can be thought of as one which is low carbon, resource efficient, socially-inclusive, and that protects and enhances biodiversity and ecosystem services. The concept of the green economy envisages economic growth that is environmentally sustainable and socially inclusive. It is a system of economic activities related to the production, distribution and consumption of goods and services that result in improved human wellbeing over the long term, while not exposing future generations to significant environmental risks and ecological scarcities.

Transition to this form of green economy requires significant up-front investment in renewable energy production, efficient use of natural resources and innovative technologies. This will require a fundamental shift in thinking about growth and development, production of goods and services, and can be done by adopting national "green growth" or "low carbon" economic strategies. Renewable energy can contribute to current transition to a green economy. The global community and national governments are faced with four major challenges with respect to the energy sector:

- Concerns about energy security;
- Combating climate change;
- Reducing pollution and public-health hazards; and
- Addressing energy poverty.

Greening the energy sector, by substantially increasing investment in renewable energy, provides an opportunity to make a significant contribution to addressing these challenges. There is a need for an action-oriented approach in order to address the objectives of securing universal access to sustainable energy and improving energy efficiency in connection with renewable energy.

Renewable energy is derived from relatively replenishable resources. This included wind, solar, biomass, hydropower, tidal and geothermal. Renewable energy forms the basis of sustainable development. It aims to preserve quality of environment and the world natural resource base for future needs and thus preserves the earth's life support system.

From all the available renewable energy resources, bio-energy derived from biomass (plant, animal origin or human origin) is the key energy source. Various biomass resources such as crop residue, animal waste, municipal solid waste, human waste, bio-fuel crops can be utilized to produce bio-energy in the form of heat, electricity, gas or liquid fuel through several technologies like direct combustion, gasification, anaerobic fermentation, tranesterification.

This chapter is an attempt to analyses the potential of bio-electricity in Rajasthan by assessing the availability of crop residue and its role in sustainable development.

Bioenergy : A Sustainable Resource in Rajasthan

Rajasthan is the largest State of India with a geographical area of 3.42 lakh sq. km. The State is situated in the north-western part of India. It is also known as the "Desert State of India" as the Thar Desert encompasses about 70 per cent of the total landmass of Rajasthan.

There has been tremendous growth in the use of electricity across the State over the last 50 years due to rapid industrialization, agricultural growth and increase in population. Taking normal growth in various categories, i.e., Domestic (13.97%), Agriculture (7.70%), Commercial (8.2%) and Industrial consumption (6.17%), the expected power demand and energy requirement at the end of financial Twelfth Plan (March-2017) will be 10,012 MW and 63,072 MU respectively. For meeting peak power demand of 10,012 the State needs about 13,349 MW installed capacity from various sources. However, currently the State has access to 8,791 MW of power from various sources. Thus the State will experience a deficit of about 4580 MW by the end of Twelfth Plan. Thus attention has been focused not only to conserve the conventional sources of energy but also to explore and tapped the vast potential of renewable energy resources to meet the growing energy demand.

Rajasthan has meagre resources of coal and petroleum thus increased energy input from renewable sources is required for sustainable and qualitative growth. Rajasthan has immense potential for renewable energy development as the region is endowed with high insolation rate, good wind regime, sufficient biomass (including animal waste, crop residues, organic wastes, etc.) for energy generation. These sources of energy including wind, solar and bio-energy will provide more sustained, eco-friendly renewable energy and help the government to fill the large gap between the demand and supply of power in the State.

There is a great importance of bio-energy for the sustainable economic and regional growth of Rajasthan, because the State is having high potential for the development of bio-energy as a renewable and eco-friendly source. The potential is based on:

- Large surplus agriculture waste and forest waste provide a significant potential for generating **Bio-Electricity.**
- Rajasthan is having large livestock population which provides large quantity of cow dung to generate **Biogas.**
- Strong prospects of production of **Bio-Diesel** on wastelands through cultivation of Jatropha, Karanj and other non-edible oilseed crops.

Potential of Electricity Generation from Crop Residue in Rajasthan

Agriculture sector forms the main basis of the economy of Rajasthan. Almost 79 per cent of its population is engaged in agriculture and allied pursuits for their livelihood. Agriculture contributes about 55 per cent to the State total income. Rajasthan has a total area of 342 lakh hectares of which 222 lakh hectares (about 64%) is under cultivation. As per 2009-10 agriculture production statistics, the production of cereals and pulses has been assessed as 116 lakh tonnes and 7 lakh tonnes respectively. The production of oilseed has been found 44 lakh tonnes while the production of cotton and sugarcane has been estimated as 1.53 lakh tonnes and 3.4 lakh tonnes respectively.

Rajasthan occupy a major status in Indian agriculture. Mainly crops like wheat, maize, bajra, cotton, groundnut, sugarcane tobacco and oilseeds are grown in Rajasthan.

It is the largest producer of mustard and bajra in the country with 45 per cent and 42 per cent of Indian production respectively.

With such a large area under agriculture and increasing agricultural production has naturally resulted in the generation of large quantities of agriculture and agro industrial residue in Rajasthan. The major crop residues produced in Rajasthan are straws of various cereals, pulses, oilseed crops, mustard husk, bajra husk, sugarcane bagasse and trash, cotton stalk and so on. These crop residues are vastly underutilized and thus can be used as potential source of renewable energy. These can be utilized as sustainable alternative for power generation which would minimize the effect of global warming through carbon sequestration, reduce air pollution and have substantially lower environmental impacts than fossil fuels. Some of the important properties (moisture, ash content and calorific value) of various biomass material used for electricity generation are shown in Table 18.1.

Table 18.1: Heat Potential of Agricultural Waste

Crop Residue	*Moisture % by wt basis*	*Ash*	*Calorific Value Kcal/kg*
Rice Straw	10.6	20.1	3000
Cotton Stalk	12	13.5	3300
Maize Stalk	11.5	14.2	4700
Maize Cobs	8.6	13.8	3500
Wheat Straw	9.2	18.8	3800
Groundnut Straw	12.1	1.3	4200
Groundnut Shell	9.9	4.7	4300
Mustard Husk	16.5	7.2	3900
Bajra Stalk	11.2	17.5	3850

Source: Vimal O.P. and Tyagi P.D. (1984) *Energy from Biomass*, p. 39.

Successful utilization of these crop residues for electricity generation in biomass based power plants either using combustion or gasification technology strongly depends on secure fuel supply. In case of agricultural residue, an associated issue as far as availability goes, is that of the seasonality of the residue as these are produced after the harvesting season of crops. Figure 18.1 shows the seasonal availability of crops residue in India. Therefore, these crop residues have to be stored for long time (for one year) in order to provide continuous supply of biomass to the power plant. This results in the deterioration of the fuel quality. Thus one possible solution is to go for combination of biomass feedstock to avoid extended storage in order to achieve long term fuel availability and power plant sustainability.

The residue generated from these crops like **Mustard and Bajra husk, Cotton stalks, Groundnut shells, Sugarcane trash** are generally not used as a fodder for animals, thus large quantity of surplus biomass can be utilized for power generation. Harvesting season of the above five crops is such that the residue from these crops can be available for the whole year without any long term storage. As per Figure 18.1 mustard husk is available during April, May and June; Cotton stalk are made available during January, February and March; Groundnut shells are available at two times in a year during May, June and in October, November and December; and Bajra husk and sugarcane trash are made available during October, November and December. Thus if power plant use these combinations of crop residues as a feedstock for power generation they can minimize the biomass losses during long term storage and also the dependency on one type of fuel can be reduced.

Residue availability	Jan.	Feb.	Mar.	Apr.	May	June	Jul.	Aug.	Sept	Oct.	Nov.	Dec.
Maize stalk								■	■	■	■	■
Maize cobs							■	■	■	■	■	■
Cotton stalks	■	■	■									
Mustard husk				■	■	■						
Jute & mesta sticks							■	■	■	■	■	■
Rice husk				■	■	■	■	■	■	■	■	■
Ground nut shells				■	■	■				■	■	■
Arhar stalks	■	■	■							■	■	■

Fig. 18.1: Seasonal Availability of Crop Residue in India

Surplus Crop Residue in Rajasthan and Power Generation Potential

Estimation of Agriculture residue production is based on the crop production and crop to residue ratio of a particular crop. Crop to residue ratio for various crops has been shown in Table 18.2.

Table 18.2: Air Dried Crop to Residue Ratio (Per tonne of Crop Production)

Crops	*Residue*	*Crop to residue ratio (CRR)*
Rice	Husk	0.22
	Straw	1.5
Maize	Straw	1.5
	Cob	0.2
Bajra	Stalk	2.0
	cob and husk	0.63
Groundnut	Shell	0.3
	Stalk	1.3
Cotton	Stalk	3.0
Soyabean	Stalk	1.7
Wheat	Straw	1.3
Sugarcane	Bagasse	0.3
	Top and Trash	0.1
Rapeseed and Mustard	Husk and straw	2.0

Source : Maurli. S. and Saxena. Mohini (2007), Quantification of Agricultural Residues for Energy Generation, *Journal of JPHE*, India, Volume II.

1. **Surplus Crop Residue for Energy Production:** With reference to Table 18.2 the estimation of the total residue generated in a year from a particular crop is done by using the following formula:

Formula (a)

Crop residue production (tonnes/year) = crop production × crop to residue ratio (CRR)

The surplus residue available for energy production is generally much lower in comparison to the value of crop residue production. This is because of:

Collection Efficiency Factor: It is generally not possible to collect entire agro residue, some of it is lost during the collection.

Alternate Uses: Agro residue has several other uses such as fodder for animals, mulching, thatching, burning in the field, use as fuel for domestic purpose, in local industries, e.g., in brick kilns, pottery, food processing, etc., which is expressed as X. To estimate surplus crop residue following formula is used:

Formula (b)

Surplus agriculture residue available for energy production = Residue production × Collection efficiency − X

2. Estimation of annual power generation potential from surplus biomass production

For calculating the annual power generation potential the formula worked out by MNRE has been used:

Formula (c)

Annual power generation potential = Total surplus agricultural residue/ 365×24 hrs×1.2

Where, it is undertaken that 1.2 tonnes of biomass can produce 1 MW/hr of electricity (Rabindranath *et al.*, 1995),

In order to assess the potential for power generation from biomass (crop residue) in Rajasthan following steps have been taken:

(i) As discussed above five crops have been considered as best suitable crops for residual availability to generate electricity by using either combustion technology (large scale power generation) or gasification (small scale power generation) technology.

(ii) The crop yield and production data for these crops for the year 2005 to 2010 have been collected from agriculture department of Rajasthan. The average data of these five years has been calculated and incorporated for analyzing the potential of bioelectricity generation from crop residue in Rajasthan. The district-wise details have been given in Annexures 1, 2, 3, 4, 5.

(iii) To find out the district-wise annual biomass production from each crop, crop production value has been multiplied by crop to residue ratio (CRR) of that particular crop (as given in Table 18.2) For mustard (husk) the CRR is 2, for cotton (stalk) the CRR is 3, for groundnut (shell) the CRR is.3, for Bajra (husk) the CRR is 0.6 and for sugarcane trash the CRR is 0.1.

(iv) District-wise surplus biomass production from each crop has been calculated at 80 per cent collection efficiency and 50 per cent availability for electricity generation (given in Annexures 18.1, 18.2, 18.3, 18.4 and 18.5) as 50 per cent of the total collected residue is used as fuel for domestic purpose and in local industries such as brick kilns, pottery making, etc., burn in the field used for thatching for manuring the agricultural field and a very small amount as fodder for animals (these are non-fodder crop residue) and the rest 50 per cent is available for the bio-energy production. The information about the consumption pattern of crop residue has been gathered by interacting with farmers and by visiting local brick kilns and other industries.

(v) To assess the total surplus biomass availability from the five crops in each district, district-wise surplus biomass production values from each crop has been undertaken for further calculations and the whole analysis is shown in Table 18.4.

(vi) By using the formula (c) worked out by MNRE, district-wise power generation potential from total surplus crop residue of Mustard, Cotton, Groundnut, Bajra and Sugarcane in Rajasthan has been calculated and shown in Table 18.3. This has been illustrated graphically in the Figure 18.2.

(vii) Composite index analysis has been done for calculating the potential of crop residues of selected five crops in Rajasthan. The results has been given in Table 18.4 and shown by Fig. 18.3.

The formula used to calculate composite index is:

$$\text{C.I.} = \frac{\text{Gross Value}}{\text{N}}$$

C.I. = Composite Index
N = Total No. of observations

To determine the gross value, the following calculations have been done

- $\bar{X} = \frac{\sum X}{N} = \bar{X}$ Mean
- $\text{Sd} = \sqrt{\frac{\sum (X-\bar{X})^2}{N}}$ Sd = Standard Deviation

  $\sum (X-\bar{X})^2$ = Sum total of squares of deviation
- $\text{SD} = \frac{X-\bar{X}}{\text{Sd}}$ SD = Standardized Value

  $X - \bar{X}$ = Deviation of the values from the mean

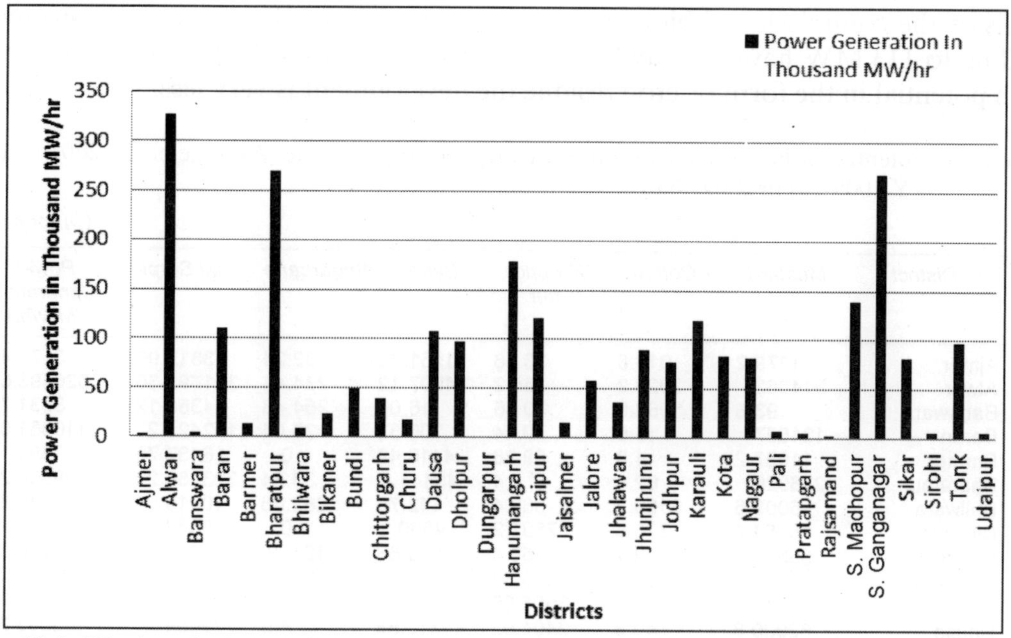

Fig. 18.2: District-wise Power Generation from Agro Residue in Rajasthan

The findings of the potential analysis for biomass power generation are summarized as follows:

- It has been estimated that annually about 30.5 lakh tonnes of crop residue from five major crops remain as surplus and this has the potential to generate 290 MW of electricity annually. The potential can be further increased by utilizing other crop residues for power generation.
- Districts Alwar, Bharatpur, Hanumangarh, Sri Ganganagar, Baran, Jaipur, Karauli, Sawai Madhopur have high potential for biomass electricity generation due to higher crop residue production in these districts.
- Districts like Banswara, Bhilwara, Dungarpur, Pali, Pratapgarh, Rajsamand, Sirohi and Udaipur have low potential for biomass electricity generation.
- Other districts have medium potential for biomass electricity generation.

The findings reveal that in Rajasthan there is enough potential to generate electricity from surplus biomass of Mustard, Cotton, Groundnut, Sugarcane and Bajra crops.

Development and Future Prospects of Bio-Electricity in Rajasthan

Rajasthan has an immense potential of generating electricity from surplus crop residue present in the State. The need is only to use all the potential resources in the best possible way Power Projects based on biomass (crop residue) with total aggregate capacity of 86.3 MW has been commissioned as on 28.02.2011. Table 18.5 shows the details of the commissioned biomass based power projects. Biomass power projects totalling to 113 MW have already been registered with RREC. Thus in spite having much potential in the form of crop residue the development is very less.

Table 18.3: Potential of Power Generation from Surplus Crop Residue (Average of FY 2005-06 to FY 2009-10) in Rajasthan

(*in tonnes*)

S. No.	District	Mustard	Cotton	Ground-nut	Bajra	Sugarcane	Total Surplus	Power Generation (MW/hr)
1.	Ajmer	1279.2	916.8	33.36	1551.36	32.24	3812.96	3177.47
2.	Alwar	314362	2023.2	265.2	74907.12	234.44	391792.36	326493.63
3.	Banswara	93.6	3963.6	30.36	16.08	254.48	4358.12	3631.77
4.	Baran	131597	15.6	79.44	706.32	23.04	132421.2	110351.00
5.	Barmer	11927.2	4.8	89.28	4416.48	0	16437.76	13698.13
6.	Bharatpur	276926	62.4	2.4	46441.2	33.84	323466.24	269555.20
7.	Bhilwara	6009.6	3199.2	380.4	143.76	282.08	10015.04	8345.87
8.	Bikaner	15754.4	376.8	7891.2	4530	0	28552.4	23793.67
9.	Bundi	56524.8	2.4	56.4	625.68	2191.48	59400.76	49500.63
10.	Chittorgarh	38572	3560.4	2479.32	1.2	2162.56	46775.48	38979.57
11.	Churu	15616	2.4	3505.56	2059.2	0	21183.16	17652.63
12.	Dausa	82900.8	40.8	2404.92	44312.88	4.6	129664	108053.33
13.	Dholpur	76272	2.4	11.64	40266	35.72	116587.76	97156.47
14.	Dungarpur	561.6	175.2	1.56	19.44	100.92	858.72	715.60
15.	Hanumangarh	122924	85232.4	582.96	5761.68	4.6	214505.64	178754.70
16.	Jaipur	81553.6	0	7032.36	57630.96	71.36	146288.28	121906.90
17.	Jaisalmer	17662.4	43.2	1027.56	267.6	2.32	19003.08	15835.90
18.	Jalore	64702.4	285.6	398.76	3915.12	0	69301.88	57751.57
19.	Jhalawar	43400	4.8	151.44	0.72	34.52	43591.48	36326.23
20.	Jhunjhunu	81588	12	273.72	26016.72	0	107890.44	89908.70
21.	Jodhpur	75488	5493.6	6432.72	4856.16	0	92270.48	76892.07
22.	Karauli	83703.2	61.2	181.44	59465.28	122.04	143533.16	119610.97
23.	Kota	101285	2.76	39.6	7.92	20.72	101355.8	84463.17
24.	Nagaur	46123.2	10388.4	2236.92	39573.84	4.6	98326.96	81939.13
25.	Pali	7853.6	1422	0	142.8	0	9418.4	7848.67
26.	Pratapgarh	5440.8	705.6	124.92	0.96	18.44	6290.72	5242.27
27.	Rajsamand	557.6	2706	177	43.2	544.52	4028.32	3356.93
28.	S. Madhopur	140585	0.84	655.08	25499.76	62.16	166802.64	139002.20
29.	S. Ganganagar	250734	62450.4	503.04	1079.28	6425.8	321192.12	267660.10
30.	Sikar	59344.8	67.2	3598.56	34737.84	27.64	97776.04	81480.03
31.	Sirohi	5512.8	504	624.12	770.64	0	7411.56	6176.30
32.	Tonk	106838	182.4	995.4	8595.36	77.68	116689.24	97241.03
33.	Udaipur	6142.4	313.2	268.32	7.44	1008.28	7739.64	6449.70
	TOTAL	2329835.2	184222	42534.96	488370	13780.1	3058741.84	2548951.53
	MEAN	70601.07	5582.47	1288.94	14799.09	417.58		
	SD	78625.3	17961.3	2092.96	21516.87	1188.01		

The major problems in use of crop residue as energy source relate to their thin spread over large area after crop harvest and their low bulk density and high moisture content. This creates problems in handling, transport and storage. Therefore successfully tested technology package are needed to promote use of crop residue based energy generation and these should include equipment and practices for all operations from harvesting and collection of crop residue to their supply in desired form to the power plants. In addition institutional barriers, policy barriers, technical barriers and information barriers could be crossed by making key policy decision, encouraging research and development on biomass energy sources and disseminating the information among potential investors.

Table 18.4: Composite Index Analysis of Potential of Agro Residue in Rajasthan (Average of FY 2005-06 to FY 2009-10)

S. No	*District*	*Mustard*	*Cotton*	*Groundnut*	*Bajra*	*Sugarcane*	*G.V.*	*C.I.*
1.	Ajmer	–0.88	–0.26	–0.60	–0.62	–0.32	–2.69	–0.54
2.	Alwar	3.10	–0.20	–0.49	2.79	–0.15	5.05	1.01
3.	Banswara	–0.90	–0.09	–0.60	–0.69	–0.14	–2.41	–0.48
4.	Baran	0.78	–0.07	–0.58	–0.65	–0.33	–0.86	–0.17
5.	Barmer	–0.75	–0.07	–0.57	–0.48	–0.35	–2.22	–0.44
6.	Bharatpur	2.62	–0.07	–0.61	1.47	–0.32	3.09	0.62
7.	Bhilwara	0.82	0.03	0.43	0.68	0.11	2.08	0.42
8.	Bikaner	–0.70	–0.07	3.15	–0.48	–0.35	1.56	0.31
9.	Bundi	–0.18	–0.07	–0.59	–0.66	1.49	0.00	0.00
10.	Chittorgarh	–0.41	–0.03	0.57	–0.69	1.47	0.92	0.18
11.	Churu	–0.70	–0.07	1.06	–0.59	–0.35	–0.65	–0.13
12.	Dausa	0.16	–0.07	0.53	1.37	–0.35	1.64	0.33
13.	Dholpur	0.07	–0.07	–8.61	1.18	–0.32	0.25	0.05
14.	Dungarpur	–0.89	–0.07	–0.62	–0.69	–0.27	–2.53	–0.51
15.	Hanumangar	0.67	1.01	–0.34	–0.42	–0.35	0.57	0.11
16.	Jaipur	0.14	–0.07	2.74	1.99	–0.29	4.51	0.90
17.	Jaisalmer	–0.67	–0.07	–0.12	–0.68	–0.35	–1.89	–0.38
18.	Jalore	–0.08	–0.07	–0.43	–051	–0.35	–1.43	–0.29
19.	Jhalawar	–0.35	–0.07	–0.54	–0.69	–0.32	–1.97	–039
20.	Jhunjhunu	0.14	–0.07	–0.49	0.52	–0.35	–0.25	–0.05
21.	Jodhpur	0.06	0.00	2.46	–0.46	–8.35	1.71	0.34
22.	Karauli	0.17	–0.07	–0.53	2.08	–0.25	1.39	0.28
23.	Kota	0.39	–0.07	–0.60	–0.69	–0.33	–1.30	–0.26
24.	Nagaur	–0.31	0.06	0.45	1.15	–0.35	1.01	0.20
25.	Pali	–0.80	–0.05	–0.62	–0.68	–0.35	–2.50	–0.50
26.	Pratapgarh	–0.83	–0.06	–0.56	–0.69	–0.34	–2.47	–0.49
27.	Rajsamand	–0.89	–0.04	–0.53	–0.69	0.11	–2.04	–0.41
28.	S. Madhopur	0.89	–0.07	–0.30	0.50	–0.30	0.71	0.14
29.	S. Ganganagar	2.29	0.72	–0.38	–0.64	5.06	7.06	1.41
30.	Sikar	–0.14	–0.07	1.10	0.93	–0.33	1.49	0.30
31.	Sirohi	–0.83	–0.06	–0.32	–0.65	–0.35	–2.21	–0.44
32.	Tonk	0.46	–0.07	–0.14	–0.29	–0.29	–0.32	–0.06
33.	Udaipur	–0.82	–0.07	–0.49	–0.69	0.50	–1.56	–0.31

Conclusion

To conclude power generation from biomass resource has immense scope and potential to serve for rural energy needs with sustainable environmental benefits. The power generation from crop residues will make the rural sector self-reliant which will lead to higher production and productivity. The use of local resources will enhance the employment opportunities and income in the rural areas and help in decentralizing the energy source. While there are several constraints to be overcome, it could be clearly seen that there are enormous opportunities to promote the utilization of biomass and

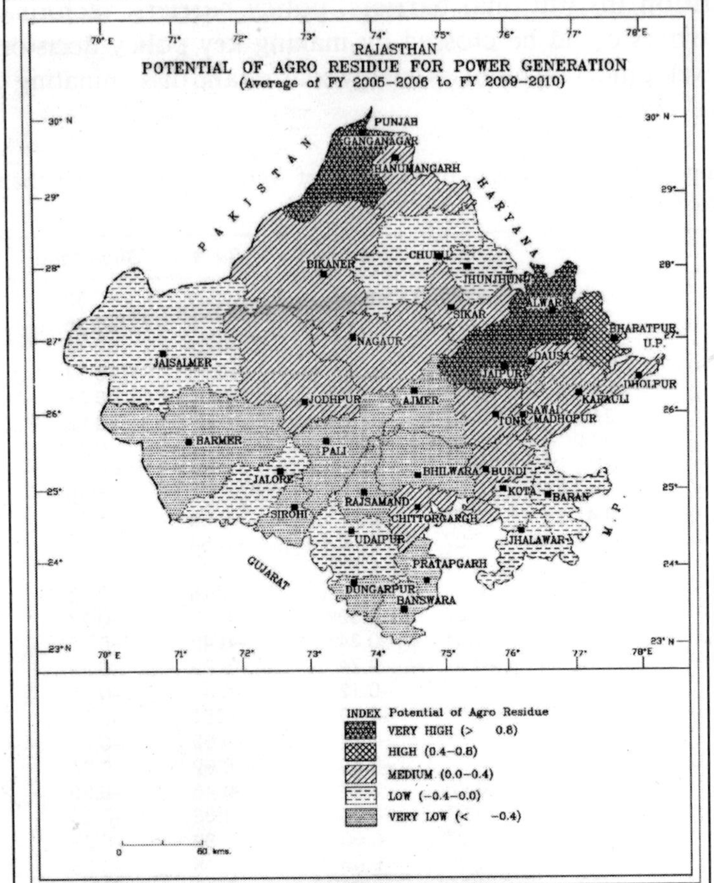

Fig. 18.3: Potential of Agro Residue for Power Generation

Table 18.5: Details of Commissioned Biomass Power Project (As on 28.02.2011)

Sr. No.	*Name of Power Producer*	*Capacity of Plant*	*Location of Plant*	*District*	*Reserved Area (Radius in Km)*
1.	Kalpataru Power Transmission Ltd.	7.8	Padampur	Sri Ganganagar	50
2.	Kalpataru Power Transmission Ltd.	8	Khotali-Uniara	Tonk	41
3.	Surya Chambal Power Ltd.	7.5	Rangpur-Ladpura	Kota	40
4.	Amrit Envi. Tech. (P) Ltd.	8	Kotputli	Jaipur	41
5.	Birla Corporation Ltd. (Captive)	15	Chanderia	Chittorgarh	60
6.	S.M. Environmental Technologies Pvt. Ltd. (Formerly Kalpataru Energy Venture Pvt. Ltd.)	8	Pachhar-Chhipa Barod	Baran	41
7.	Sambhav Energy Ltd.	20	Rampur	Sirohi	65
8.	Transtech Green Power Ltd.	12	Kachela Bagsari Sanchor	Jalore	51
	Total	86.3			

Source: RREC, Annual Report (2011)

improve the efficiency. With the available potential in the form of crop residue, biomass serves as a promising option of renewable energy which can take the State forward towards a clean and secure energy source.

Development of bio-energy in Rajasthan will help in resolving the power crisis in the State and open the path for sustainable development and boost the green economy in Rajasthan.

NOTES

Agarwal, A.K. and Agarwal, G.D. (1999), *Recent Technologies for the Conversion of Biomass into Energy, Fuel and Useful Chemicals.* pp. 15-19.

Mittal, N.K. *Non-Conventional Energy Systems*, 2007.

Biomass to energy: The science and technology of the IISc bioenergy systems. ABETS, Indian Institute of Science, 2003.

Kumar, M.; Patel, S.K. Energy values and estimation of power generation potentials of some non-woody biomass species, *Energy Sources*, Part A, 2008, 30, 765-773.

Shukla, P.R. (1997), *Biomass Energy in India*, Conference Proceedings - Biomass Energy: Key Issues and Priority Needs, International Energy Agency, Paris, France.

Somashekar, H. I.; Dasappa, S.; and Ravindranath, N. H., Rural bioenergy centres based on biomass gasifiers for decentralized power generation: case study of two villages in southern India. Energy Sustain. Dev. 2000, IV, 55-63.

Annexure 18.1
Biomass Production (tonnes/annum) from Mustard Crop
(Average of Year 2005-2010)

S. No.	*District*	*Area (hectare)*	*Production (tonnes)*	*Yield (kg/hec)*	*Crop to Residue ratio (CRR)*	*Biomass Production (tonnes)= Crop production (CRR)*	*Surplus Biomass Production (tonnes)*
1.	Ajmer	2857	1599	560	2	3198	1279.2
2.	Alwar	251375	392953	1563	2	785906	314362.4
3.	Banswara	89	117	1315	2	234	93.6
4.	Baran	114176	164496	1441	2	328992	131596.8
5.	Barmer	16854	14909	885	2	29818	11927.2
6.	Bharatpur	200126	346158	1730	2	692316	276926.4
7.	Bhilwara	9614	7512	781	2	15024	6009.6
8.	Bikaner	30279	19693	650	2	39386	15754.4
9.	Bundi	63718	70656	1109	2	141312	56524.8
10.	Chittorgarh	39415	48215	1223	2	96430	38572
11.	Churu	14546	19520	1342	2	39040	15616
12.	Dausa	78748	103626	1316	2	207252	82900.8
13.	Dholpur	65931	95340	1446	2	190680	76272
14.	Dungarpur	533	702	1317	2	1404	561.6
15.	Hanumangarh	89868	153655	1710	2	307310	122924
16.	Jaipur	99191	101942	1028	2	203884	81553.6
17.	Jaisalmer	34680	22078	637	2	44156	17662.4
18.	Jalore	62612	80878	1292	2	161756	64702.4
19.	Jhalawar	39401	54250	1377	2	108500	43400
20.	Jhunjhunu	78412	101985	1301	2	203970	81588
21.	Jodhpur	74197	94360	1272	2	188720	75488
22.	Karauli	75293	104629	1390	2	209258	83703.2
23.	Kota	86065	126606	1471	2	253212	101284.8
24.	Nagaur	46544	57654	1239	2	115308	46123.2
25.	Pali	10528	9817	932	2	19634	7853.6
26.	Pratapgarh	5562	6801	1223	2	13602	5440.8
27.	Rajsamand	970	697	719	2	1394	557.6
28.	S. Madhopur	158433	175731	1109	2	351462	140584.8
29.	S. Ganganagar	212656	313417	1474	2	626834	250733.6
30.	Sikar	59702	74181	1243	2	148362	59344.8
31.	Sirohi	7654	6891	900	2	13782	5512.8
32.	Tonk	173695	133548	769	2	267096	106838.4
33.	Udaipur	8615	7678	891	2	15356	6142.4

Annexure 18.2
Biomass Production (tonnes/annum) from Cotton Crop
(Average of Year 2005-2010)

S. No.	District	Area (hectare)	Production (tonnes)	Yield (kg/hec)	Crop to Residue ratio (CRR)	Biomass Production (tonnes)= Crop production (CRR)	Surplus Biomass Production (tonnes)
1.	Ajmer	6450	764	118	3	2292	916.8
2.	Alwar	4326	1686	390	3	5058	2023.2
3.	Banswara	11021	3303	300	3	9909	3963.6
4.	Baran	38	13	344	3	39	15.6
5.	Barmer	10	4	340	3	12	4.8
6.	Bharatpur	149	52	346	3	156	62.4
7.	Bhilwara	23335	2666	114	3	7998	3199.2
8.	Bikaner	5132	314	61	3	942	376.8
9.	Bundi	6	2	340	3	6	2.4
10.	Chittorgarh	8590	2967	345	3	8901	3560.4
11.	Churu	6	2	340	3	6	2.4
12.	Dausa	101	34	345	3	102	40.8
13.	Dholpur	6	2	340	3	6	2.4
14.	Dungarpur	423	146	3	3	438	175.2
15.	Hanumangarh	174821	71027	406	3	213081	85232.4
16.	Jaipur	0	0	0	3	0	0
17.	Jaisalmer	105	36	345	3	108	43.2
18.	Jalore	690	238	345	3	714	285.6
19.	Jhalawar	11	4	340	3	12	4.8
20.	Jhunjhunu	31	10	345	3	30	12
21.	Jodhpur	11022	4578	415	3	13734	5493.6
22.	Karauli	148	51	346	3	153	61.2
23.	Kota	7	2.3	340	3	6.9	2.76
24.	Nagaur	18093	8657	478	3	25971	10388.4
25.	Pali	4239	1185	280	3	3555	1422
26.	Pratapgarh	2260	588	260	3	1764	705.6
27.	Rajsamand	6529	2255	345	3	6765	2706
28.	S. Madhopur	2	0.7	340	3	2.1	0.84
29.	S. Ganganagar	163732	52042	318	3	156126	62450.4
30.	Sikar	161	56	345	3	168	67.2
31.	Sirohi	1790	420	235	3	1260	504
32.	Tonk	441	152	345	3	456	182.4
33.	Udaipur	756	261	345	3	783	313.2

Annexure 18.3

Biomass Production (tonnes/annum) from Groundnut Crop (Average of Year 2005-2010)

S. No.	District	Area (hectare)	Production (tonnes)	Yield (kg/hec)	Crop to Residue ratio (CRR)	Biomass Production (tonnes)= Crop production (CRR)	Surplus Biomass Production (tonnes)
1.	Ajmer	2390	278	116	0.3	83.4	33.36
2.	Alwar	1051	2210	2103	0.3	663	265.2
3.	Banswara	233	253	1086	0.3	75.9	30.36
4.	Baran	609	662	1087	0.3	198.6	79.44
5.	Barmer	684	744	1088	0.3	223.2	89.28
6.	Bharatpur	18	20	1111	0.3	6	2.4
7.	Bhilwara	8549	3170	371	0.3	951	380.4
8.	Bikaner	78547	65760	837	0.3	19728	7891.2
9.	Bundi	432	470	1088	0.3	141	56.4
10.	Chittorgarh	18675	20661	1106	0.3	6198.3	2479.32
11.	Churu	22664	29213	1289	0.3	8763.9	3505.56
12.	Dausa	15349	20041	1306	0.3	6012.3	2404.92
13.	Dholpur	89	97	1090	0.3	29.1	11.64
14.	Dungarpur	12	13	1083	0.3	3.9	1.56
15.	Hanumangarh	4468	4858	1087	0.3	1457.4	582.96
16.	Jaipur	38166	58603	1535	0.3	17580.9	7032.36
17.	Jaisalmer	9241	8563	926	0.3	2568.9	1027.56
18.	Jalore	2606	3323	1275	0.3	996.9	398.76
19.	Jhalawar	1321	1262	955	0.3	378.6	151.44
20.	Jhunjhunu	1555	2281	1467	0.3	684.3	273.72
21.	Jodhpur	43892	53606	1221	0.3	16081.8	6432.72
22.	Karauli	1297	1512	1166	0.3	453.6	181.44
23.	Kota	304	330	1086	0.3	99	39.6
24.	Nagaur	14251	18641	1308	0.3	5592.3	2236.92
25.	Pali	1587	0	0	0.3	0	0
26.	Pratapgarh	711	1041	1464	0.3	312.3	124.92
27.	Rajsamand	1802	1475	819	0.3	442.5	177
28.	S. Madhopur	6256	5459	873	0.3	1637.7	655.08
29.	S. Ganganagar	5007	4192	837	0.3	1257.6	503.04
30.	Sikar	22064	29988	1359	0.3	8996.4	3598.56
31.	Sirohi	3940	5201	1320	0.3	1560.3	624.12
32.	Tonk	14859	8295	558	0.3	2488.5	995.4
33.	Udaipur	3403	2236	657	0.3	670.8	268.32

Annexure 18.4

Biomass Production (tonnes/annum) from Bajra Crop

(Average of Year 2005-2010)

S. No.	District	Area (hectare)	Production (tonnes)	Yield (kg/hec)	Crop to Residue ratio (CRR)	Biomass Production (tonnes)= Crop production (CRR)	Surplus Biomass Production (tonnes)
1.	Ajmer	85338	6464	76	0.6	3878.40	1551.36
2.	Alwar	276962	312113	1127	0.6	187267.80	74907.12
3.	Banswara	169	67	396	0.6	40.20	16.08
4.	Baran	3593	2943	819	0.6	1765.80	706.32
5.	Barmer	956738	18402	19	0.6	11041.20	4416.48
6.	Bharatpur	12135	193505	1596	0.6	116103.00	46441.20
7.	Bhilwara	3167	599	189	0.6	359.40	143.76
8.	Bikaner	247150	18875	76	0.6	11325.00	4530.00
9.	Bundi	3182	2607	819	0.6	1564.20	625.68
10.	Chittorgarh	12	5	417	0.6	3.00	1.20
11.	Churu	407614	8580	21	0.6	5148.00	2059.20
12.	Dausa	138516	184637	1333	0.6	110782.20	44312.88
13.	Dholpur	86376	167775	1942	0.6	100665.00	40266.00
14.	Dungarpur	206	81	393	0.6	48.60	19.44
15.	Hanumangarh	53342	24007	450	0.6	14404.20	5761.68
16.	Jaipur	296379	240129	810	0.6	144077.40	57630.96
17.	Jaisalmer	161586	1115	—	0.6	669.00	267.60
18.	Jalore	328095	16313	50	0.6	9787.80	3915.12
19.	Jhalawar	6	3	500	0.6	1.80	0.72
20.	Jhunjhunu	263704	108403	411	0.6	65041.80	26016.72
21.	Jodhpur	574821	20234	35	0.6	12140.40	4856.16
22.	Karauli	133662	247772	1854	0.6	148663.20	59465.28
23.	Kota	83	33	398	0.6	19.80	7.92
24.	Nagaur	469992	164891	351	0.6	98934.60	39573.84
25.	Pali	79410	595	7	0.6	357.00	142.80
26.	Pratapgarh	10	4	400	0.6	2.40	0.96
27.	Rajsamand	458	180	393	0.6	108.00	43.20
28.	S. Madhopur	84792	106249	1253	0.6	63749.40	25499.76
29.	S. Ganganagar	4356	4497	1032	0.6	2698.20	1079.28
30.	Sikar	303948	144741	476	0.6	86844.60	34737.84
31.	Sirohi	15757	3211	204	0.6	1926.60	770.64
32.	Tonk	67765	35814	529	0.6	21488.40	8595.36
33.	Udaipur	78	31	397	0.6	18.60	7.44

Annexure 18.5

Biomass Production (Tonnes/annum) from Sugarcane Crop

(Average of year 2005-2010)

S. No.	District	Area (hectare)	Production (tonnes)	Yield (kg/hec)	Crop to Residue ratio (CRR)	Biomass Production (tonnes)= Crop production (CRR)	Surplus Biomass Production (tonnes)
1.	Ajmer	14	806	57571	0.1	80.60	32.24
2.	Alwar	101	5861	58030	0.1	586.10	234.44
3.	Banswara	140	6362	45443	0.1	636.20	254.48
4.	Baran	10	576	57600	0.1	57.60	23.04
5.	Barmer	0	0	0	0.1	0.00	0.00
6.	Bharatpur	16	846	52875	0.1	84.60	33.84
7.	Bhilwara	75	7052	94027	0.1	705.20	282.08
8.	Bikaner	0	0	0	0.1	0.00	0.00
9.	Bundi	1360	54787	40285	0.1	5478.70	2191.48
10.	Chittorgarh	989	54064	54665	0.1	5406.40	2162.56
11.	Churu	0	0	0	0.1	0.00	0.00
12.	Dausa	2	115	57500	0.1	11.50	4.60
13.	Dholpur	13	893	68692	0.1	89.30	35.72
14.	Dungarpur	68	2523	37103	0.1	252.30	100.92
15.	Hanumangarh	2	115	57500	0.1	11.50	4.60
16.	Jaipur	31	1784	57548	0.1	178.40	71.36
17.	Jaisalmer	1	58	—	0.1	5.80	2.32
18.	Jalore	0	0	0	0.1	0.00	0.00
19.	Jhalawar	15	863	57533	0.1	86.30	34.52
20.	Jhunjhunu	0	0	0	0.1	0.00	0.00
21.	Jodhpur	0	0	0	0.1	0.00	0.00
22.	Karauli	28	3051	108964	0.1	305.10	122.04
23.	Kota	9	518	57556	0.1	51.80	20.72
24.	Nagaur	2	115	57500	0.1	11.50	4.60
25.	Pali	0	0	0	0.1	0.00	0.00
26.	Pratapgarh	8	461	57625	0.1	46.10	18.44
27.	Rajsamand	301	13613	45226	0.1	1361.30	544.52
28.	S. Madhopur	27	1554	57556	0.1	155.40	62.16
29.	S. Ganganagar	2338	160645	68710	0.1	16064.50	6425.80
30.	Sikar	12	691	57583	0.1	69.10	27.64
31.	Sirohi	0	0	0	0.1	0.00	0.00
32.	Tonk	32	1942	60688	0.1	194.20	77.68
33.	Udaipur	391	25207	64468	0.1	2520.70	1008.28

19

Beach Tourism and Green Economy

An Environmental Performance of Hotel Industry of Coastal South Goa, India

Nandkumar Sawant

Introduction

In today's globalised world, tourism is the most promising driver of growth. While potential growth is enormous, the changes and challenges cannot be ignored. Global warming, pollution, degradation and various socio–cultural issues are greatly debated at various levels and forums. Goa has emerged as a tourist paradise, especially in the post-globalization era. Today it generates significant State's Domestic Product and Employment. Green economy may have diversed definition but from the context of the present study, it would be growth and development while protecting the earth's ecosystem as well improving human wellbeing and social equity. This also calls upon significant reduction of environmental risk and ecological scarcities. In the context of tourism, Greening of the tourism industry would focus on sustainable development through responsible and increased awareness of tourists demanding greening of the industry. Better information on the effects of their travelling to the surrounding natural and social environments.

Introduction to Goa and Tourism

Goa, the tropical paradise is blessed with nature's bounties and rich cultural heritage. Once a Portuguese colony, after liberation (1961), gained prominence as a world tourist destination. The serene beaches, swinging palms, dotted with white churches, colourful temples, world heritage sites, historical forts, colonial buildings, Indo–Portuguese culture and several cultural events throughout the year makcs it a perfect tourist destination but it is basically professing beach centric tourism, though it has now ventured into other forms of tourism namely, spiritual, adventure, health (Smita Kamat, 2010).

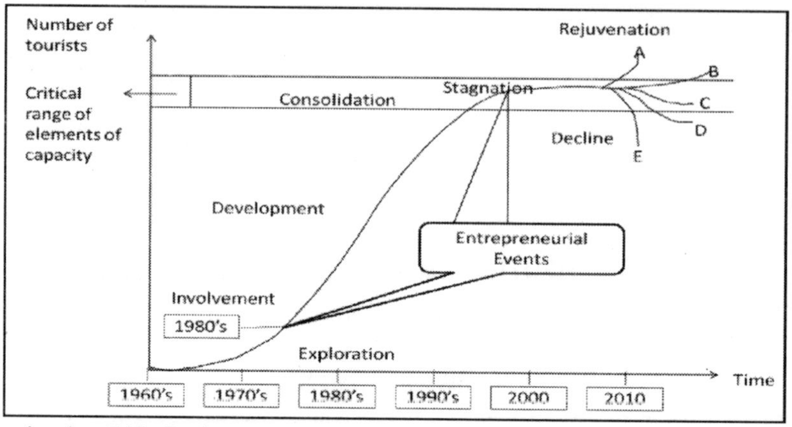

Fig. 19.1: Destination Life Cycle
Source: Butler, 1980

Considering the development of tourism, in early 1960's, and comparing with Butler's Destination Life Cycle, the Hippies explored the virgin beaches of Goa and made a tourist haven (Fig 19.1). Following the liberation of Goa in 1962, the socio–economic development of Goa were rather rapid as compared to other regions in India (Sanganal, 2010). Several sectors, i.e., tourism, fisheries and mining grew in rapid strides. As Goa brought about socio–economic development in 1970s and 80s and boost in tourist services. The tourist services like hotels and resorts, proper transport linkages, the commencement of the charter flights and later globalization and liberation, witnessed the influx thousands and lakhs of foreign and Indian tourists, however, the figures from 11.2 lakhs in 1996 increased to more than double, i.e., 24 lakhs within a decade (Table 19.1). This shows the multiple increase in the tourist arrivals. However, the State has increased it stagnation. Therefore, there is need to review the state of affair of the tourism.

Accommodation is an important component of the tourism industry. Boost in tourism is partly shouldered by the hotel industry. The location, type and tariff affect the rate of occupancy of the hotel. The voluminous tourist arrival shows skewed pattern spatially and seasonally. 95 per cent of the tourist arrivals are Beach tourists. Therefore, the four coastal talukas, i.e., Bardez, Tiswadi, Mormugao and Salcete support nearly are under tremendous pressure. Tourism was declared as the potential revenue earner and, therefore, in 1970s tourism services were given priority. The upcoming of first beach resort at Sinquerim (North Goa) in 1974 followed by at Dona Paula and Majorda (South Goa) in early 1980s complemented with the commencement of charter flights, saw upscaling to tourism industry in Goa.

The tourist infrastructure has increased significantly over the years. The members of hotels has increased from 310 in 1989 to 2,322 in 2006. Further Quantum leap to 2,444 in 2008 of which 83 belonged to star or Heritage category. During the same period, the hotels in South Goa increased from 93 to 212. Salcete has an array of hotels ranging from luxury to budget. Some of the most famous hotels are Leela Beach resort, Taj Exotica, Alila, Majorda Beach resort, Mahindra Holidays, Royal Orchids, etc. This puts a pressure on the resources is significant. The mushrooming of the hotels and related services have brought about changes in the land-use and subsequently demand for infrastructure and facilities.

Table 19.1: Goa : Tourist Arrivals

Years	*Number of Tourists*			*Annual Growth (%)*
	Domestic	*Foreign*	*Total*	
1996	888914	237216	1126130	1.7
1997	928925	261673	1190598	5.7
1998	953212	275047	1228259	3.2
1999	960114	284298	124412	1.3
2000	976804	291709	1268513	1.9
2001	1120242	260071	1380313	8.8
2002	1325296	271645	1596941	15.7
2003	1725140	314357	2039497	27.7
2004	2085729	363230	2448959	20.1
2005	1965343	336803	2302146	–6.0
2006	2098654	380414	2479068	7.7
2007	2208986	388457	2597443	4.8
2008	2020416	3511123	2371539	–8.7
2009 (P)	2127063	376640	2503703	5.6

Source: Tourism Department, Panjim, Goa.

Objective

Taking the case of coastal south Goa, the present study envisages the impact of increasing accommodation due to tourism and the consequences on the environment. Thus the objectives are:

(a) To envisage resource utility is the term of land and water use by sector.
(b) To assess the environmental and performance indicators for accommodation sector.
(c) Perception of the people towards hotels and environment.

Data Base and Methodology

The following research is a culmination of varied data. This could be synoptically shown by Fig. 19.2.

Secondary as well as primary data has been used for the present study. To assess the temporal change in land-use and land cover for selected destinations of coastal South Goa, i.e., Colva-Benaulim belt between 2005 and 2011, Google map has been used. Secondly, to assess the perception of locals in the context of hotel industry and environment, 214 localities, who have resided for more than for 15 years have been surveyed by using a questionnaire method. Lastly, To calculate the environmental index, eight indicators were taken and Likert Scale (1-5) was used, where in opinion–based questions on solid waste, environmental policies, environmental responsibilities, etc. questions were put forth. The scores were assigned 1-5 (very low to very high). For these purposes, 36 hotels were randomly chosen and categorized as low, middle and high.

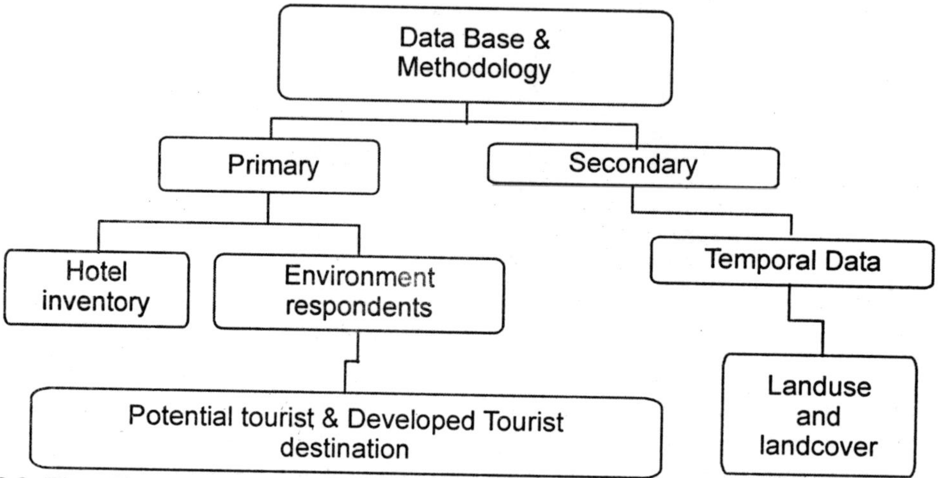

Fig. 19.2: Flow Chart of Database and Methodology

Discussion

The following discussion has been divided into three sections. The first section envisages how the hotel industry brings about a change in land-use and land cover. The second part focuses on the use of natural resources and the third perception of the local populace and the hotel industry towards environment.

(A) Natural Resources

The advocacy of tourism has serious implications on land-use and land cover. The traditional occupation of the local population was agriculture with orchard plantation, a few communities practicing fishing and toddy tapping As tourism sets in, the local population shun away from traditional occupations and subsequently the agricultural plots and plantation orchards are replaced by resorts and other tourist services (Sawant, 2010). The uprise of tourism brought transformation in occupational pattern and subsequently the impact was witnessed on the land-use patterns. As tourism demands accommodation, a number of hotels mushroomed along the coast. Subsequently, with increase in tourist arrivals, the coastal belt has become not only overcrowded but added various environmental issues like garbage and sewage management.

With accommodation sector demands increase of land conversions and changing landscapes. Taking the case of two coastal developed tourist destination, i.e., Colva and Benaulim, the popular beaches of south Goa are examined in the context of land-use/land cover change for 2003 and 2010.There has been significant decline in land under agriculture (25%) and also in vegetation cover. As the youth give up tradition occupations, compiled with demand for land construction. With a span of 6 years, the increase of built up, barren land is by 75 per cent (Table 19.2). This has been more towards the proximity to the beaches (Fig. 19.3).

Table 19.2: Benaulim-Colva Land-use 2005-2011 (Area in mtrs)

Land-use	Coverage 2005	Coverage 2011	% change
Agriculture	12036. 05	9047. 86	–24.82%
Built-up	5013.45	8754. 82	74.62%
Barren land	4263.97	7501. 47	75.92%
Vegetation	15250.84	13104.59	–14.07%
Seashore	2507.30	2150. 68	–14.22%
Scrub land	21344. 65	19093. 32	–10.54%

Source: Google Map, 2005 and 2011.

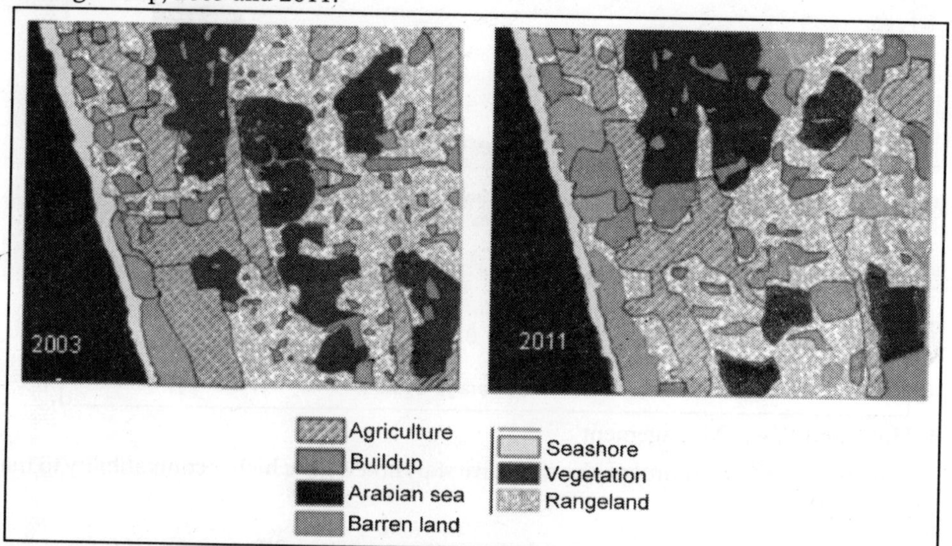

Fig. 19.3: Land-use and Land cover: Colva & Benaulim

Micro lab analysis shows that most of the hotels have come up on the agricultural lands and partly plantations. Agricultural lands, sand dunes and vegetation, mangroves and coconuts grows are experiencing a number of changes. These changes are result of tourism as well as changes in the social, political, legal institutions (Noronha L. *et al.*, 2002).

(A) Water Requirement

The water requirement ranges from 600 litres/room to 1000 litres/room per day. The water requirement of the luxury hotels is exceptionally high, i.e., 1500 litres per day/ room. This is basically star hotel have various tourist services like swimming pools, landscape, lawns and gardens, and even golf courses. Many real estate owners and hotel promoters sink bore wells to extract groundwater to maintain artificial greenery. Bore wells are not allowed with 200 m form High Tide Line (HTL). This has resulted in salt water Intrusion. For example, in 1990s most well on the Mobor Peninsula had turned saline and the Leela beach which had excavated a huge area for sweet water lagoon was converted to salt water lagoon (Claude Alvares, 2002). Similarly, Kenilworth Resort has to shut down wells put with 500 m zone for similar reasons.

Excessive extraction of groundwater has affected the groundwater and some of the coastal villages salt water ingression as the water table is lowering. The main pressure on aquifer in Goa is the resultant increase in local population, growth in tourism and related activities, lack of policy to regulate access and use, technology of extractive equipment and absence of protecting zoning of well heads (Noronha L., 2003).

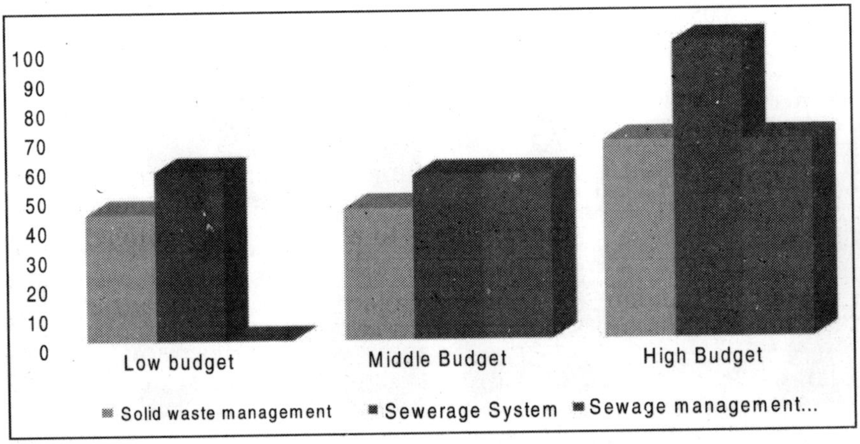

Fig 19.4: Hotel and Waste Management

Source: In the context of environment, the high (five star) hotels have higher compatibility to manage their waste.

(B) Responses of the Locals

Here an attempt has been made to put forth the opinions/ perception of the residents of the two tourist destinations. The first tourist destination is Developed coastal tourist belt, i.e., Colva, Benaulim and Varca; and second of the developing or potential tourist destination, i.e., Palolem and Agonda.

The perception of the residents of the developed beach destination related to upcoming of hotels, tourist arrivals and natural environment, shows that significant positive correlation (4.12) between construction of hotels and destruction of natural environment. It is strongly stated that there has been transformation of land.

Likewise the beaches have unpleasantly overcrowded (4.00). This has also in some way affected the socio-cultural life of the residents too. In short, people do agree that the uprise in the hotel industry has definitely overcrowded the coastal areas and destroyed the natural environment.

Table 19.3: Perception of Residents : Developed Tourism Destination

	Strongly Disagree	*Disagree*	*Neutral*	*Agree*	*Strongly Agree*	*Mean*
Construction of hotels & tourist facilities have destroyed the natural environment	2.56	7.69	3.85	37.18	44.87	4.12
High spending tourists have negatively affect our way of life.	4.49	16.67	7.05	37.82	28.85	3.68
Local residents have suffered from living in tourism destinations area	5.13	17.31	3.85	39.74	25.64	3.61
Tourism has resulted in traffic congestion, noise & pollution	3.21	13.46	7.05	31.41	36.54	3.76
Tourist has resulted in unpleasantly overcrowded beaches, hiking trails, parks and other outdoors places in our community	3.21	7.05	7.05	41.03	37.18	4.00

(C) Hotel Inventory

In the context of hotels *vis-à-vis* environmental conservation, the hoteliers across the categories believe that hotels do not add to environmental degradation, neither coastal environment of Goa in vanishing. Further, it is also felt that the government is not doing enough to address environmental related issues. Further as initiatives towards environmental protection, all hotels should maintain green zones and use solar energy. In addition, there are conscious efforts to conserve natural resources.

Most of the luxury hotels have sewage treatment plant to make used water reusable. Where the low and middle budget hotels have limited resources to manage their waste. In fact the low budget hotels do not manage the sewage. A low value of the index indicates greater stress by that of accommodation sector in the natural resources and consciousness towards environment. Using the average index, the luxury hotels secured the highest in the environmental domain *vis-à-vis* the other two categories of

Table 19.4: Environmental Index

Sr. No	*Environmental Indicators*	*Low Budget*	*Middle Budget*	*High Budget*
1.	Solid Waste Management	3	3	4
2.	Sewerage System Management	3	3	5
3.	Sewage Management	0	3	4
4.	composting	1	0	4
5.	Recycling	2	1	2
6.	Environmental policies	2	3	5
7.	Coordinator responsible for environmental issues	1	2	3
8.	Environmental reach out program?	0	2	4
9.	environmental domain Index	1.5	2.12	3.87

accommodation. Within the domain, the performance of the luxury hotels is the high in solid waste management, disposal management, maintain green zones and also participation of environmental initiatives.

Table 19.5: Perception of Hoteliers

Sr. No.	*Perception of Hoteliers*	*Low budget*	*Middle Budget*	*High Budget*	*Mean*
1.	Do you think that hotels should be allowed at the coast	4	3.89	4.67	4.19
2.	Do you think hotels add to environmental degradation	4.57	3.44	2.67	3.56
3.	Do you feel that star hotel have more wastage over small hotels	4.14	3.67	1.67	3.16
4.	Do you think star hotel manage sewage better than the small hotel	3.29	3.44	4.67	3.80
5.	Is the government doing enough to manage environmental related issues?	3.57	2.67	3.67	3.30
6.	Is the coastal environmental of Goa is vanishing because of the hotel	3.57	3.67	2.33	3.19
7.	Do you think that star hotels should maintain or demarcate green zones	3.57	3.78	4.67	4.01
8.	Should hotels start harvesting solar energy	3.86	4.56	5	4.47

Drawbacks can be converted into strengths with the light policy.

Low and middle budget hotels dispose sewerage and sewage through soak pits and tracks the low budget hotels just let it flow in the open fields or orchards. This has affected the groundwater quality.

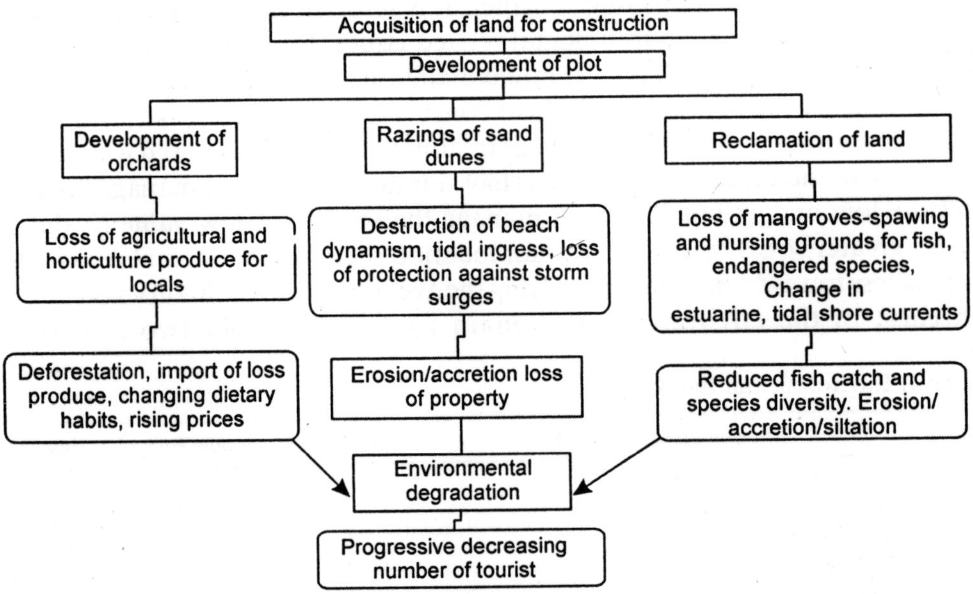

Fig. 19.5:

Source: National Institute of Oceanography of India 1996b.Comments on the coastal zone management plans of Goa. Report submitted to India, Ministry of Environment and Forest, New Delhi

The flowchart provides an insight degradation affect tourist influx, thereby questioning employability and sustainability.

Occupance of land along the coast and development for built-up result in loss of local agricultural productivity as well natural landscape and habitat. As consequences of erosion, deforestation, loss of property and other related factors, there is environmental degradation.

Suggestions

- Goa propagating beach centric tourism has undoubtedly reached the saturation and overcrowding, eventually pressuring the resources. The sustenance of the industry would call upon for specific measures.
- Promotion of Alternative tourism, which means push tourist towards interior portions of Goa. This would a step towards promoting ecotourism in the form of 'spice farms', Nature based trials, etc.
- Responsible tourist wherein both the tourist and the tourism industry/ accommodation industry proactively participate in judicious use of resources and put in efforts in conserving resources.
- Coastal regulation Zone defined by CRZ notification, limits development to beyond 200 m from the high–tide time in CRZ III and beyond 500 m in CRZ I. Tourism development in the courses under purview of the CRZ.
- There is a head to connect the coastal resource use and development. The approach would yield income as well as conserve natural resources to enhance sustainability.

NOTES

——(2011), *Economic Survey 2010-11*, Directorate of Planning, Statistics and Evaluation, Government of Goa, Panjim.

——(2011), *The Transition to a Green Economy: Benefits, Challenges and Risks from a Sustainable Development Perspective*, UN Conference on Sustainable Development.

lvares Claude (Ed) (2002), *Fish Curry and Rice*, Other Indian Bookstores, Mapusa-Goa (India), 128-157.

Butler, R.W. (1980), The concept of a Tourism Area Cycle of Evolution: Implication for Management Resources, *The Canadian Geographer*, 24(1) 5-6.

Chawla, Romila (2004), *Coastal Tourism and Development*, Sonali Publications, New Delhi.

D'Costa Lillian (2008), Can tourism in Goa Go Green? Article in news daily *Herald*, Goa 24th October 2008.

Kamat, Smitha (2010), Destination Life Cycle and Assessment - A Study of Goa Tourism Industry, *South Asian Journal of Tourism and Heritage*, 3(2) 140- 148.

Kazi, S. and K.S. Nairy (2003), Resource Use Pattern in Tourism Accommodation Sectors, Published in Coastal Tourism, Environment and Sustainable Local Development (ed) Naronhna *et al.* TERI, New Delhi, pp. 95-120.

National Institute of Oceonography, (1996), Comments on the Coastal Zone Management Plans of Goa, Report submitted to MOEF, Government of India, New Delhi.

Noronha L., N. Lourence, J.P. Lobo-Fareira *et al.*, (2203), Managing Coastal Tourism: Perspective from Indian and the European Union, TERI, Coastal Tourism Environmental and Sustainable Local Development (ed) 419-446.

Noronha, L.; Siqueria, A.; Sreekesh, S.; Queshy, L.; Kazi, S. (2202), Goa: Tourism, Migrations and Ecosystem Transformations, *Ambio* XXXI (4) 295–302.

Patil, Ashok and Choubey, Kailash (2010), Geographical Analyses of Tourism and Concerning Problems in Goa (India) *IRRJ*, II (19) Aug 2010, 4-7.

Raikar Avinash (2010), Growth, Problem and Prospects to Tourism in Goa, Articles in *Contemporary Tourism Development Issues and Challenges* (ed), Rajat Publications, New Delhi. pp. 10-30.

Sanguanal, V.S. (2010), Tourism and Environmental Degradation—Goa, Article in *Contemporary Tourism Development–Issues and Challenges*, Rajat Publication, New Delhi, 43-264.

Sawant, Gaikwad and Ghatage (2010), Coastal Tourism, Occupational Changes and Environmental Implications- Palolem- Goa. articles in *Contemporary Tourism Development Issues and Challenges* (ed), Rajat Publications, New Delhi.

Sawkar Kalidas *et al.*, (1998), *Tourism and the Environment, Issues of Concern in the Coastal Zone of Goa*, The Economic Institute of World Bank, Stock No. 37134, pp. 9-20.

20

Impact of Rubber Plantation on Environment

A Study on Some Selected Areas of Tripura, India

Sunil Kumar De

Introduction

Natural Rubber is the most widely used elastomer in the world. It is used as industrial raw material for almost 50,000 goods. The demand for natural rubber arises in all the countries and is highly correlated to the levels of industrialization and the standard of living. China is the world's largest consumer of natural rubber followed by USA and Japan. India is the fourth largest consumer of natural rubber in the world. Natural rubber consumption in the world has been growing at an average rate of 3 per cent. However, the production of natural rubber is limited to certain countries owing to the requirement of specific agro-climatic conditions. The supply of NR in the world market comes mainly from the South East Asian countries. Thailand is the world's largest producer of NR followed by Indonesia and Malaysia. Unlike other major producers of NR, India's NR production caters to the domestic market mostly (Bhowmik, 2006).

Kerala and the Kanya Kumari district of Tamil Nadu comprise the traditional zone of natural rubber production in India and the history of rubber plantation dates back to the early nineteenth century. Kerala occupies 84 per cent of the total area under rubber plantation in the country but contributes about 92 per cent of the total production (Bhowmik, 2006, *Rubber Statistics* 2005). Out of 615,000 ha of rubber plantation in the country, Kerala alone accounts for 502,000 ha. (Sinha, 2010).

Non-traditional areas found suitable for rubber cultivation, despite varying degrees of limitations, are hinterland of coastal Karnataka, Goa, Konkan region of Maharashtra, coastal Andhra Pradesh and Orissa, Siliguri area of West Bengal and the low elevation regions of the Northeastern States excluding Sikkim. Rubber is a versatile strategic raw material and its demand has been growing fast with rapid modernization. Traditional rubber belt in India has almost reached at saturation level and it was thought that the growth potential lies in non-traditional belt. Tripura has become the second largest rubber State in India despite agro-climatic limitations and the State has successfully (especially tribal community) seized the economic opportunity of producing rubbers. Apart from the economic benefit, it has found that the rubber plantation has many negative impacts on the physical environment, such as, changing rainfall pattern, increasing temperature, enhancing soil erosion and lot more. This chapter is aiming at finding out some of the negative impacts of rubber cultivation in the State of Tripura citing example from the Bankumari rubber plantation in the Bishalgarh subdivision.

Location of the Study Area

Tripura, is situated between 20°56′ north and 24°32′ north and 91°10′ east and 92°21′ east. It is almost completely surrounded by Bangladesh on three sides, having a tenuous link with the rest of the country through the Cachar district of Assam. There are many rivers in the State which rise from the six principal hill ranges and, after being fed by innumerable small *cherras*, flow through the valleys into the mighty rivers of East Bengal. Out of the total geographical area of 10,486 km^2 of the State of Tripura, 6,292 km^2 is covered by notified forestland and 2550 km^2 is under net sown area. 70 per cent of the total land of the State is characterized by hills and uplands. The total number of population according the 2011 census is 35.57 lakh out of which 11.12 lakh is ST category and 6.22 lakh is of SC category population.

For the present study a rubber garden, viz., Bankumari Rubber Plantation Centre (23°41′N and 91°19′E) from Bisalgarh sub-division of the State of Tripura has been selected, because rubber cultivation in Bisalgarh subdivision is practiced widely covering about 17 per cent of the total area. The present garden is maintained by TFDPC. Other gardens in the subdivision are controlled by TRPC, TTAADC and some private owners.

Bankumari Rubber Plantation Centre comprises an area of 70 h and located at an elevation of 40 m from mean sea level. This garden falls under North Charilam tahasil of Bishalgarh subdivision (Fig. 20.1).

Ideal Climate for Rubber Plantation

Commercial Rubber tree, *Hevea brasiliensis* grows wild in the Amazon basin and adjacent areas of Brazil. Important climatic requirements within the tropics for commercial cultivation of rubber generally are well-distributed high rainfall with no pronounced droughts; mild and short winter, high humidity and free from strong and prolonged seasonal winds.

Rubber is adaptable to a wide range of agro-climatic conditions. Deviation in respect of rainfall, temperature, humidity and wind affect growth and yield, but the adverse impact from variations to a limit can in most cases be moderated through management intervention. That it is so demonstrated by the generally good performance of rubber in Tripura and elsewhere in the North-East (Sinha, 2010).

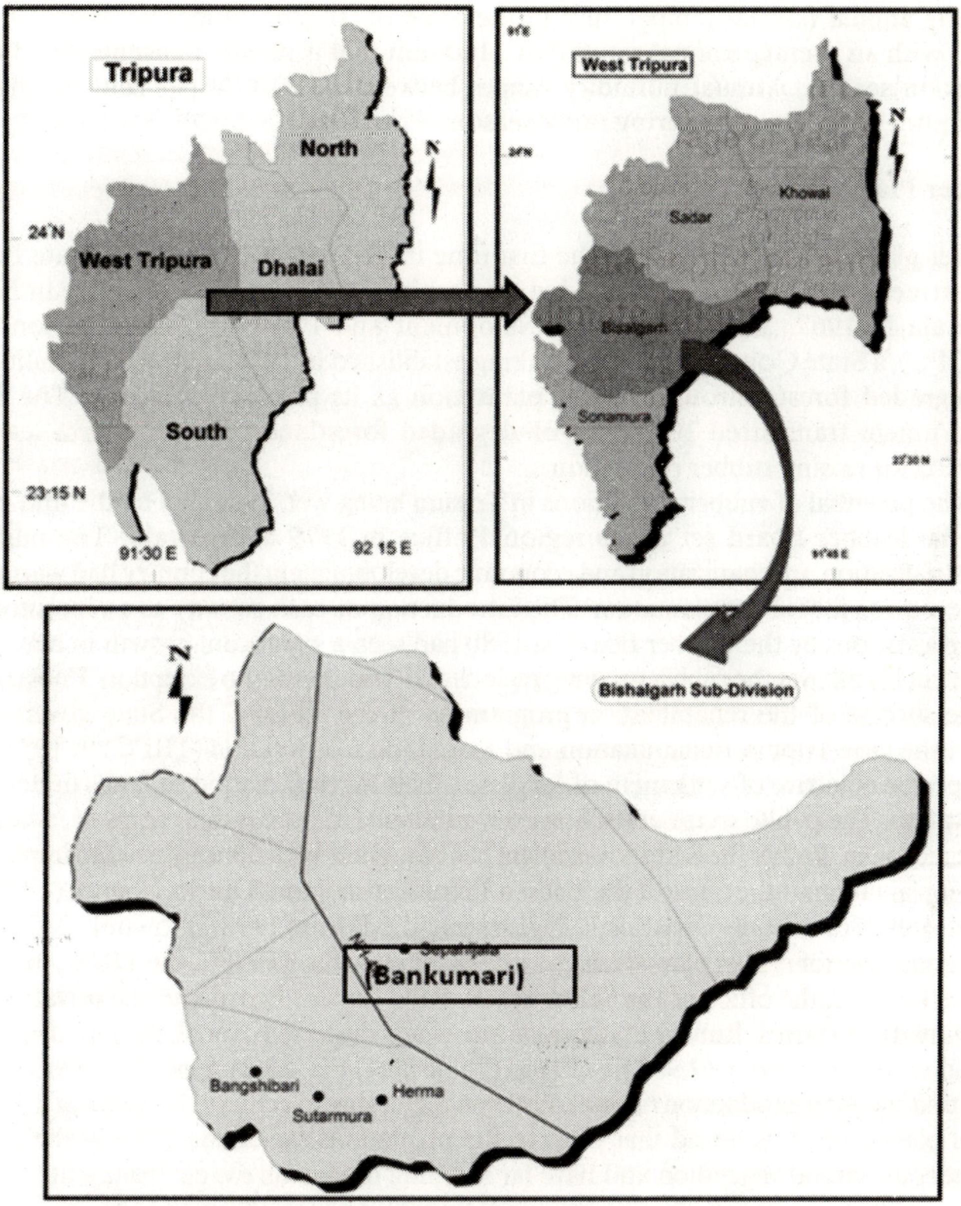

Fig. 20.1: Location Map of Tripura Bishalgarh Subdivision

The ideal climatic condition for the growth of rubber includes the evenly distributed annual rainfall of 200 to 300 cm, temperature of 21° to 35° C and strong seasonal winds for prolonged periods lead to stunted growth of young trees. Rubber is mostly grown below the altitude of 450 m from the MSL with gently undulating topography having slope between 5° and 15° along with well drained and of medium textured soils having a depth of more than 100 cm.

The annual normal temperature in the State of Tripura ranges between 10° and 35° C with an average annual rainfall of 2100 mm that is mostly concentrated during monsoon season. Annual humidity ranges between 50 and 74 per cent having the maximum of 85 per cent during rainy season.

Rubber Plantation in Tripura

Rubber plantation was raised for the first time in Tripura in 1963 by the State Forest Department. Rubber Board responded by opening its first (one-man) office in NE at Agartala in 1967. Tripura Forest Development and Plantation Corporation Ltd. (TFDPC), a State Government Undertaking established in 1976, adopted rehabilitation of degraded forests through rubber plantation as its primary objective. The State Government transferred 10,000 ha of degraded forestland on a long-term lease to TFDPC for raising rubber plantation.

The potential of rubber plantations in Tripura being well identified by the mid 1970s saw the Rubber Board set up its regional office in 1979 at Agartala. The ongoing industrialization, mechanization and economic development in the country had warranted for new areas for rubber plantation. The introduction of cash subsidy as an incentive for new plantations by the Rubber Board in 1980 had seen a significant growth in new areas under rubber all over the rubber growing regions and Tripura was no exception. Encouraged at the success of the rehabilitation programme of the TFDPC the State government established the Tripura Rehabilitation and Plantation Corporation (TRPC) in 1983 with the specific objective of settlement of landless tribals for reclamation of lands under *jhum* cultivation. The public sector units, however, continued to pursue their goals of increasing plantation area. Rather the State government has brought in two more agencies to implement rubber plantations—the office of the Tripura Tribal Autonomous District Council (TTADC) and the office of the Sub-Divisional Magistrates (SDM) from 1998 (Bhowmik, 2006). The major agencies for rubber plantation and extension are—the TFDPC, the TRPC, the office of the TTADC, the office of the SDM, the Jt. RPC Rubber Board and the private sector individuals and farms. Rubber Plantations are widely dispersed over all the four districts of the State covering about 49,960 ha of land (Table 20.1, Fig. 20.2). Since the inception the area under rubber production has steadily been increasing to achieve the target of 1,00,000 ha of plantation. It is found that most of the plantations are taking place at the cost of indigenous natural vegetation and have far reaching impact on the environment.

Impact of Rubber Plantation on Environment

Recent intensification of rubber plantation has had numerous negative environmental consequences: Erosion has accelerated and stream sediment loads have increased where repetitive cultivation is performed on steep slopes without appropriate conservation methods; permanent conversion of hill slopes and road building have increased the risk of landslides.

Table 20.1: Year-wise Growth of Rubber Plantation Area in Tripura

Year	*Area (ha)*	*Year*	*Area (ha)*
1979	1965.25	1995	17844.99
1980	2724.51	1996	19178.79
1981	3578.63	1997	20908.01
1982	4653.5	1998	22959.82
1983	5461.2	1999	24179.46
1984	6431.84	2000	25432.16
1985	7237.92	2001	26866.88
1986	8721.83	2002	27743.69
1987	10180.48	2003	29052.18
1988	11172.62	2004	30250.1
1989	12737.15	2005	31700.19
1990	13167.7	2006	32882.19
1991	13570.15	2007	34146.19
1992	14337.75	2008	35760.19
1993	15174.26	2011	49960
1994	16194.37		

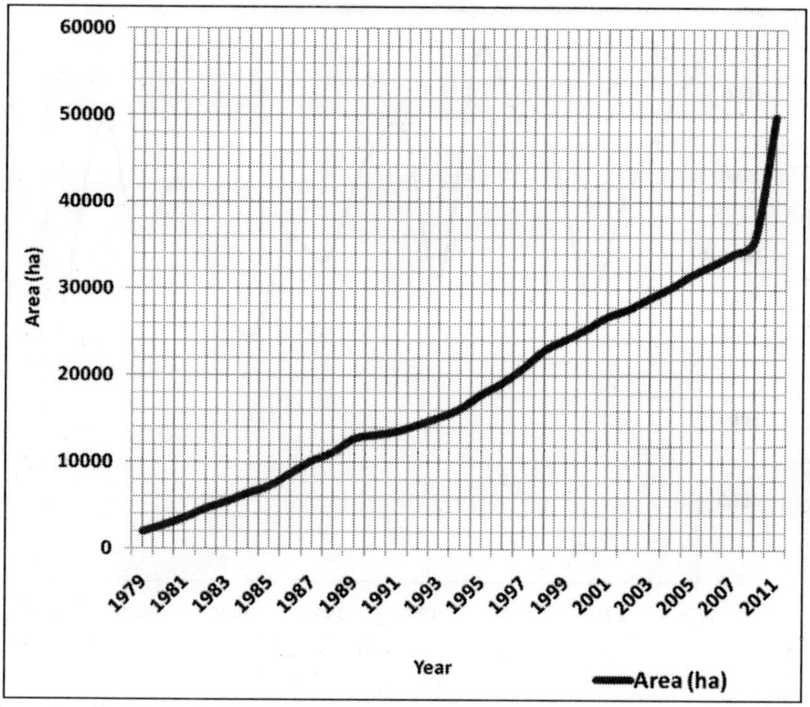

Fig. 20.2: Growth of rubber plantation area in Tripura

Since the beginning of rubber plantation the overall environmental scenario in the State of Tripura has started to change. The most serious impact has been noticed in climate and soil erosion.

Table 20.2: Trend of Annual Rainfall of Sadar Subdivision

Year	Rainfall (mm)
1975	3203.6
1977	2835.2
1979	2361.7
1981	1787.8
1983	2629.7
1985	2879
1987	3233.8
1989	1874.9
1991	2773.8
1993	2569
1995	1983.9
1997	2085
1999	1505.3
2001	1877.5
2003	2459.7
2005	2077.5
2007	2819.8
2009	1721.1

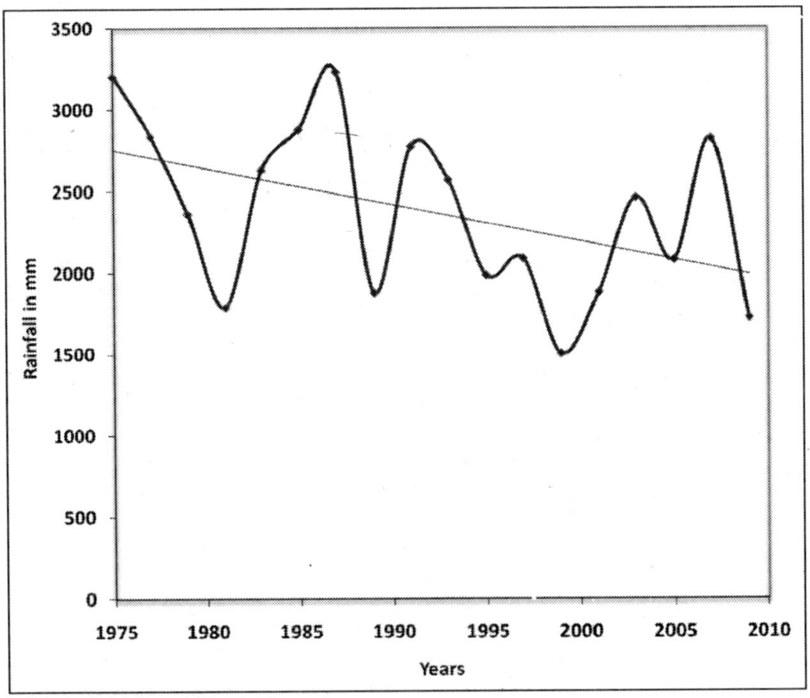

Fig. 20.3: Changing Trend of Rainfall of Sadar Subdivision

Climate

Belonging to the tropical monsoon type of climate, rubber is produced seasonally in the State. It absorbs huge amount of moisture from the environment during monsoon season because of which the rainfall pattern in the State has become erratic, the average temperature of the districts surrounding the plantations has increased in the past few years and the climate is gradually been transformed into semi-arid type.

Table 20.3: Trend of Annual Rainfall of Sadar Subdivision

Year	*Temp. (°C)*
1991	26.76
1992	26.68
1993	27.22
1994	26.45
1995	26.98
1996	27.25
1997	26.73
1998	23.69
1999	27.11
2000	25.98
2001	27.24
2002	26.32
2003	27.29
2004	28.41
2005	27.49
2006	27.93
2007	28.24
2008	27.17
2009	27.51

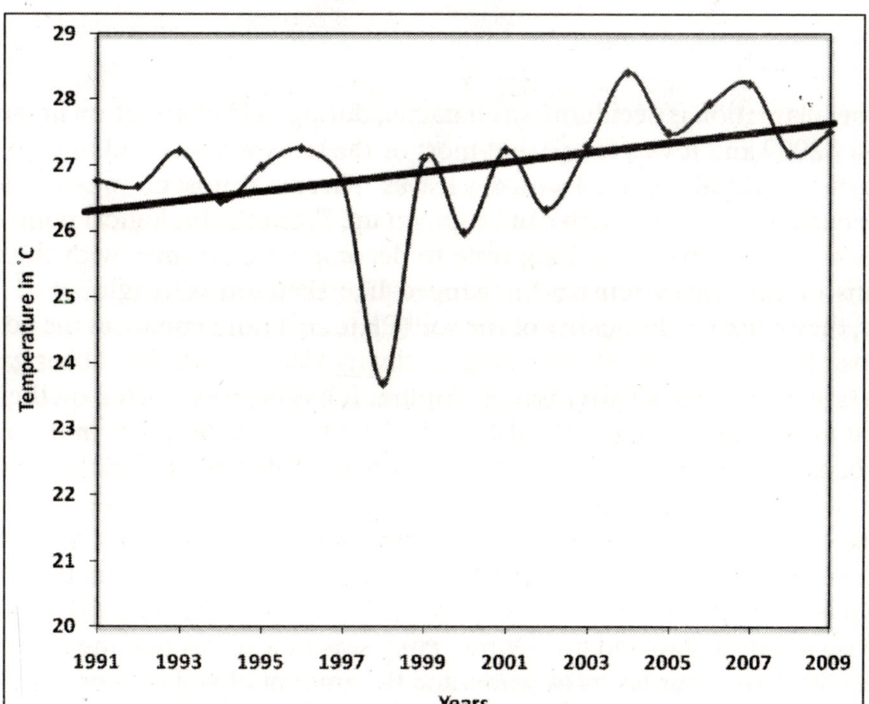

Fig. 20.4: Changing Trend of Temperature of Sadar Subdivision

The nature of changing rainfall and temperature of Sadar Subdivision has been analyzed on the basis of the collected data from the Agartala Meteorological Office. The overall trend of average annual rainfall clearly indicates a decreasing trend with annual fluctuations (Table 20.2, Fig. 20.3). The recurrence period of average annual rainfall change has almost become one year. Rubber is a rainfed crop and Tripura does fall in the zone of maximum rains. But the past few years have witnessed no uniformity in the pattern of rains in the State. At times there are massive downpour while longer periods pass without any trace of rain (Bhowmik, 2006). If human activity be considered as one of the major causes of climate change, the present scenario changing rainfall trend establishes the fact (Bhattacharyya,. The average annual temperature is also showing an increasing trend with an average increase of 1.5° C during the last 20 years. The recurrence period of fluctuating average annual temperature is almost one year. The lowest average annual temperature (23.69°C) was noticed in the year 1998 (Table 20.3, Fig. 20.4).

For mono crop plantation like rubber often natural forests are cleared and timber is often stalked and burned. This results the killing of vast majority of forest species including those which are living in the soil like micro-organisms. These organisms are directly exposed to the sun and rainfall; and cannot survive or withstand the fluctuating heat and moisture levels. Once rubber trees are planted, re-growth of any other vegetation or undergrowth is impossible until the seeds are completely depleted away from the soil or the canopy is closed (Jason, 2004).

Soil Loss

Since rubber plantation is deciduous in character, during field observation in the month of February 2009 and it was found that most of the trees were devoid of leaves while ground surface was fully covered with dry leaves. There is almost no vacant place from where the bushes or undergrowth can be grown up. From the biological point of view the cellulose of the leaves take long time to decompose or to mix with the soil but small veins of the leaves remain unchanged like skeleton (Greggio, *et al.*, 2008). Moreover, they enhance the acidity of the soil (Plate 20.1) and enhances the possibility of soil erosion. An example of such event is clearly visible from the rubber garden of Bankumari in Bishalgarh subdivision of Tripura. It has been estimated on the basis of USLE that an average amount of soil loss of 11.6 $t.h^{-1}.y^{-1}$ takes place in the area.

The roots of trees are laterally spreading instead of deeply penetrating downward into the soil. The soil holding capacity of such laterally spreading roots is very low that allows huge amount of soil erosion. As the secondary roots do not have any other sub-veins or smaller roots, they allow enormous amount of soil erosion from the bottom of the roots. It is found that the roots along most of the slopes are existing nakedly without any soil base [Plate 20.2 (a and b)]. During rainy season it seems that the top soil cover is covered with three/four layers of leaves and the amount of soil erosion is less. But in reality, huge amount of soil loss takes place along the slopes due to excessive amount of surface run-off and thereby the gradual reduction of tilla side slopes (De, 2010).

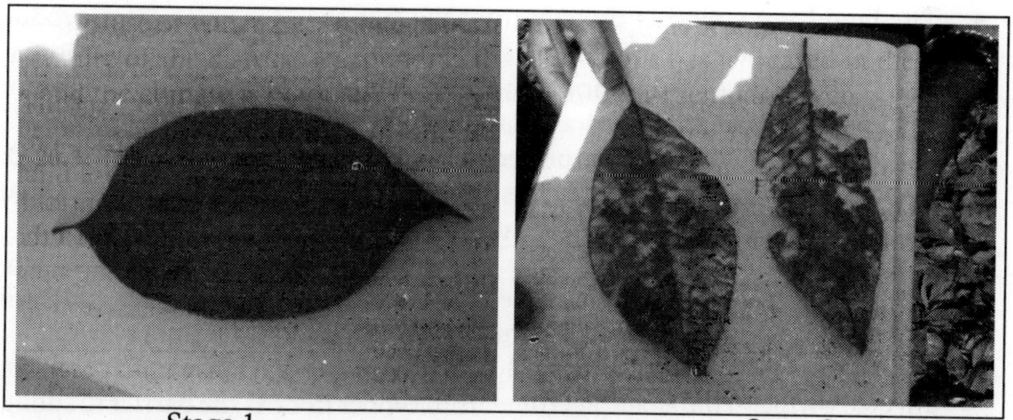

Stage 1 Stage 2

Plate 20.1: Decomposition of Cellulose leaving the skeletal remains of veins

Plates 20.2: (a and b). Lateral Root System Unable to hold Top Soil cover leading to Erosion

For creating land for rubber plantation, indigenous vegetation species are cleared along the tilla-side slopes. Contour terracing is then made artificially for the plantation of the seedlings. This deforestation activities as well as such unscientific contour terracing enhances the possibility of soil erosion.

It is noticed that the labourers are cleaning such dry leaves during winter season and gathering them in some places within the garden for burning. This practice is going on in some places along the lower part of the tilla slope but most places of the garden remain unchanged. As a result the contour strips those are made for plantation along the tilla-side slopes have totally been eroded away and their present existence is very difficult to notice especially in Bankumari rubber garden (Plate 20.3).

Other Impacts

Therefore, different types of environmental and technological problems are being encountered by the State, such as:

- The liquid sap collected directly from the tree for processing latex produces several toxic chemical effluents (Serum). Solid latex requires huge amount of energy to separate water from it after coagulation. Vulcanization of latex into rubber also produces effluences that are highly toxic to the environment.
- Some are of the opinion that the decomposition of rubber leaves in water takes longer time and impairs the process of aquaculture due to which the aquatic ecosystem falls under serious threat.
- The use of chemicals and smoke release a typical pungent smell which is harmful. The people can be relieved from this consequence if the processing units are constructed far away from the residential and plantation areas.
- Production in the agricultural plots adjacent to the rubber plantations is affected owing to the shade accruing from rubber trees as the barrier to sunlight affects productivity. The decrease in productivity has occurred by 10 per cent in those regions which are shadowed or has been intercepted of sunbeams. On the other hand, crop yields in low lying "lunga" lands adjacent to rubber plantation on "tilla" lands show increased yield owing to drainage of nutrients from the plantation areas with the help of rainwater. As a consequence, productivity of agricultural crops increases while production of latex decreases due to draining away of nutrients like nitrogen, potassium and sodium from the tilla lands.

Plate 20.3: Degraded Contour Strips due to Erosion

Conclusion

Although rubber plantation looks like a profitable one in the short run, but the overall environmental deterioration is manifold in terms of climatic change, destruction of indigenous plant and animal species that will put not only the State but also the whole region under severe threat in the long run (Liu, *et al.*, 2006).

Tree plantations are not forests. A plantation is a highly uniform agricultural system that replaces natural ecosystems and their rich biodiversity. The trees planted are geared to the production of a single raw material, whether it is timber, pulp, rubber, palm oil or others." (Gauntlett, 2008). Several studies have demonstrated that when an ecosystem is destroyed for a monoculture plantation, biodiversity suffers.

The forestry industry has argued for year that tree plantations should receive carbon credits for being carbon sinks, however, recent research reveals the opposite. Old-growth forests store carbon for centuries, whereas plantations and young forests are actually net emitters of carbon due to the disturbance of the soil and the degradation of the previous ecosystem (Carlson, 2012). A substantial increase in natural reserve areas could help to reduce the threats to biodiversity and carbon stocks.

Rubber plantation possess very high economic significance in the State of Tripura although many drawbacks. More than 50,000 people is engaged in the sector for their bread and butter. Thus, instead of many shortcomings of rubber plantation in the State some alternatives must be found for the survival of the people of the State.

NOTES

Bhowmik, Indraneel (2006), Status Report on Rubber Plantations in Tripura. Tripura Rubber Mission- Technical Bulletin 1, 2006.

Carlson, K. M., *et al.,* (2012), Committed carbon emissions, deforestation and community land conversion from oil palm plantation expansion in West Kalimantan, Indonesia. PNAS, 109: 19, 7559-7564

De, Sunil Kumar; Ray, S. amd Pal, T. (2009), Physico-economic sustainability of rubber plantation - A Case Study on Bishalgarh Subdivision of Tripura State. ed. by K.N. Jena, Globalized India and Developmental Issues. Abhijeet Publications, New Delhi; Ch. 21, pp. 244-255.

Gauntlett, Sandy (2008), Global Justice Ecolocy Project. Annual Report 2008. http://www.globaljusticeecology.org/files/annual%20report%202008%20%28compressed%29.pdf

Greggio, T. C.; Assis, L.C. and Nahas, E. (2008), Decomposition of the rubber tree (Heavea Brasiliensis) litter at two depths. *Chilen Journal of Agricultural Research*, 68: 128-135 (April-June, 2008)

Jason, W.C. (2004), *World Agriculture and the Environment: A Commodity* - by - guide tom impacts and practices. Island Press, 340-345.

Liu, W.; Hu, H.; Ma, Y and Li, H. (2006). Environmental and Socio-economic Impacts of Increasing Rubber Plantations in Menglun Township, Southwest China. *Mountain Research and Development*, 26:3; 245-253

Rubber Statistics (2005), Tripura Rubber Board, Tripura

Sinha, A.K. (2007), *Prospect of rubber plantation in NE region with special reference to the State of Tripura.* 'Saving Soils and Croplands', Guwahati.

21
Carbon Credit Accounting

M.L. Sharma and *Dhiraj Sharma*

Introduction

In 2005, Kerry Emanuel, a professor of atmospheric science at MIT, published a controversial paper in Nature linking global warming with the rising intensity of hurricanes (Emanuel, 2005). The paper relied on historical records showing the intensity of Atlantic storms had nearly doubled in 30 years. What caught people's attention, however, was not this alarming statistic, but rather that it was released just three weeks before Hurricane Katrina displaced one million people and left an estimated 1,836 dead.

Conditions are changing in our world. Some are feeling the impact now, from the heat wave and wildfires in Russia of the last two years, the devastating floods in Pakistan and Australia, tornadoes in the United States, mudslides in Brazil, and drought in China. Others are worrying about the impacts to come. The tea growers in Kenya's highlands who are seeing cases of malaria they didn't see only five years ago; the cocoa farmers in Ghana who think about how changes in rainfall will affect their sensitive crops; the rice farmers in Vietnam who are increasingly concerned about rising water levels.

One of the environmental threats our planet faces today is the potential for the long term changes in Earth's climate temperature pattern known as global climate change. Scientists estimate that as a result of global climate change, the Earth's average temperature could increase as much as six and one-half degrees Fahrenheit by the year 2100. While this may not sound like much of an increase, if the temperature increase approaches the six and one half degree mark, the earth will be a much different place than we know it today. To again an appreciation of how different the Earth could be, consider that during the last ice age, when our planet was on average only nine degree Fahrenheit cooler, the area that is New York city was 1000 feet of ice.

Emission of GHGs is one of the major causes behind global warming. Studies reveal that there is a strong correlation between per capita GHG emission and the economic prosperity of a country. It is true that irrational human activity triggered by industrialization is mainly responsible for most of the earth's warming and that is mostly by CO_2 (carbon dioxide) emissions.

In order to reduce to Green House Gas effect throughout nations of the world the Kyoto Protocol came into force on 15th February, 2005. The Annex I Countries (36 developed countries) are targeted to reduce the GHG emissions by at least 5 per cent below their 1990 baseline emission. India, along with China and Brazil has emerged as naturally one of the largest beneficiaries of the Kyoto Protocol. India is quite bullish of carbon trade market having enormous opportunities in various segments.

One carbon credit is equal to one metric tonnes of carbon dioxide or equivalent green house gases emissions like methane, nitrous oxide, Hydroflourocarbon, Sulfur hexafluoride, etc.

Majority of contribution to CO_2 emissions are commercial, manufacturing, mining and consumptions based. Hence, there is a need to take concrete steps to safeguard the environment. At international level, the steps initiated are (1) United Nations Framework Convention on Climate Change (UNFCCC); (2) Kyoto Protocol; and (3) Copenhagen Deal.

India, being a Non-Annex-1 country, is naturally one of largest beneficiaries of the Kyoto Protocol. Trading of carbon credits happens in the form of CERs or Certified Emissions Reductions. CERs are in the form of certificates, just like a stock. A CER is given by CDM (Clean Development Mechanism) Executive Board to project in developing countries to certify that they have reduced GHG emissions by one tonne of CO_2 per year. For example, if a project generates energy using wind power instead of burning coal, and in the process saves 25 tonnes of CO_2 per year, it can claim 25 CERs.

India now can earn huge benefits like advanced technological improvements and related foreign investments, contributes to greenhouse gas reduction by adopting alternative sources of energy. Indian companies can make huge profits and that too without losing anything. How? The answer is simply by selling carbon. Carbon credit trading is the newest concept which offers an opportunity to earn money on the bases of reduced carbon emission. Every industry is rewarded a carbon credit or "entitlement certificate" by the United Nations Framework Convention on climate Change (UNFCCC) to adopt a clean Development Mechanism (CDM) for their projects, which could be on agriculture, energy, manufacturing, fuel emissions, metal production, mining, etc. It is to motivate companies to emit less carbon. A carbon credit is a certificate which gives permission to the receiver to emit one metric tonne of carbon dioxide or other equivalent gases that ceases green house effect. The total emission is restricted and the market allocates monetary value for this shortfall through trading. So industries can buy and sell these credits based on the price set by the carbon exchange. The more credits an industry earns, the more money it can make by selling it off.

Object

The purpose of this chapter is to explore the impact of global warming on accounting policy. As a first step in this exploration it is useful to understand the environment in economic terms. It is difficult to turn a public resource into a private good. Market driven solutions viewed through the lens of economic theory are difficult to implement. The pace of politically motivated regulation regarding the environment is increasing. It is also very preliminary in the light of what the science literature suggests is necessary. As environmental impacts are privatized a whole new class of assets and liabilities will emerge. The specific purpose of this chapter is to examine how these items fit into the current financial reporting framework. Accounting policy that does this will ensure the relevance of accounting when it comes to evaluating corporate performance.

What are Carbon Credits?

Carbon credits are a key component of national and international emission trading schemes that have been implemented to mitigate global warming. Credits can be exchanged between businesses or bought and sold in international markets at the prevailing market price. Credits can be used to finance carbon reduction schemes between trading partners and around the world.

- The "currency" for this trade is called Carbon Emission Reduction (CER) commonly called as Carbon Credits.
- One unit of CER is equivalent to the reduction of one metric tonne of CO_2 or its equivalent.

Symbolically, 1 CER= 1 tonne of CO_2 (or equivalent gases) Carbon Credits have been given the recognition of an intangible commodity and can be traded on the commodities market. Trading of carbon credits happens in the form of CERs or Certified Emissions Reductions. CERs are in the form of certificates, just like a stock. A CER is given by the CDM Executive Board to projects in developing countries to certify that they have reduced green house gas emissions by one tonne of carbon dioxide per year. For example: If a project generates energy using wind power instead of burning coal, and in the process saves, say 25 tonnes of carbon dioxide per year, it can claim 25 CERs (One CER is equivalent to one tonne of carbon dioxide reduced).

A Brief Process to calculate Carbon Credit

Carbon credits will be allowed for less carbon emission in compare of baseline emission of a project.

A company can measure their carbon credit stock by following methods:

(1) Gain–Loss Method

In the gain–loss method, annual changes in carbon stocks are estimated by summing the differences between the gains and losses in carbon stock. Gains occur due to growth of biomass and forestation activities. It can be more expressive by following equation:

$$\Delta C = \Delta C_G - \Delta C_L$$

where:
ΔC = annual change in carbon stocks, tC/yr
ΔC_G = annual gain of carbon, tC/yr
ΔC_L = annual loss of carbon, tC/yr

(2) Stock-Difference Method

The stock-difference method can be used where carbon stocks are measured at 2 points in time to assess carbon stock changes. The following equation is applied:

$$\Delta C = \frac{(C_{t2} - C_{t1})}{(t_2 - t_1)}$$

where:
ΔC = annual change in carbon stocks, tC/yr
C_{t1} = carbon stocks at time t_1, tC
C_{t2} = carbon stocks t_2, tC

(3) Sequestration Technology

By this technology company can use improved technology in place of present technology. Carbon credits will be allowed for less carbon emission in comparison of baseline emission of a project.

Baseline emission can be calculated from the guideline provided in Greenhouse Gas Protocols for each specified industry. GHG Protocols provided a calculation tool for a specified industry. For the less carbon emission, various sequestration technologies can be used by an industry. The performance objective for a sequestration technology is not necessarily zero emission of CO_2 but rather a reduction compared with the baseline of current practice. To make sure that all carbon aspects are considered, care must be taken to ensure that there are no hiding emissions when making an alteration from the baseline. The fundamental question underlying an analysis of merit of a process or alteration of a process is as follows:

> "How much CO_2 is generated as a result of the operation (or change) of this process, and what is its ultimate result?"

Both inputs and outputs must be considered to obtain a total picture. When we speak of carbon sequestration in this chapter, we refer to all green house gas sequestration measured in carbon dioxide or carbon equivalence (CE). The carbon dioxide equivalence is also called the Global Warming Potential (GWP). A complete list of GWP values has been prepared by IPCC [1].

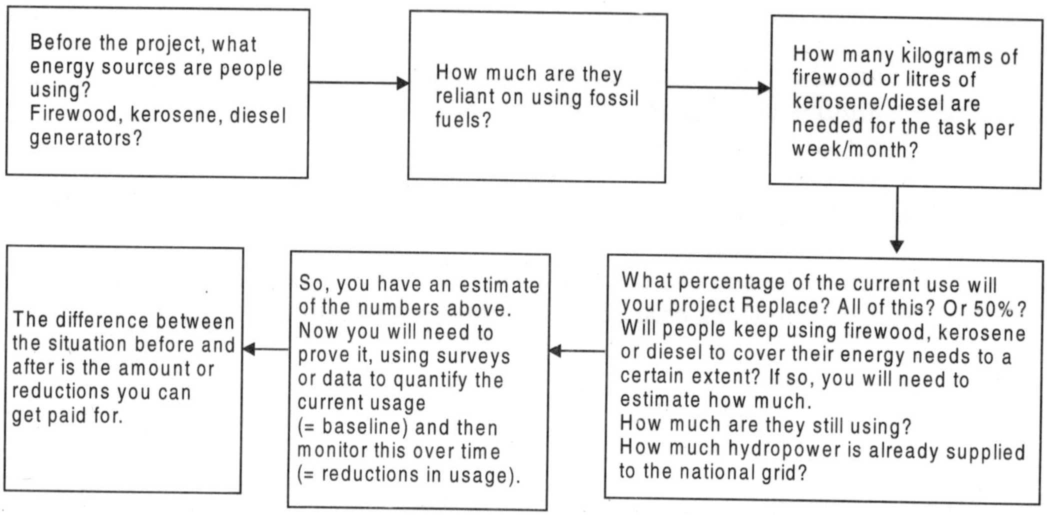

First of all we will calculate Carbon Emission Reduction.
Baseline emissions—Project emission

1. tonne Carbon Emission reduction = 1 CER

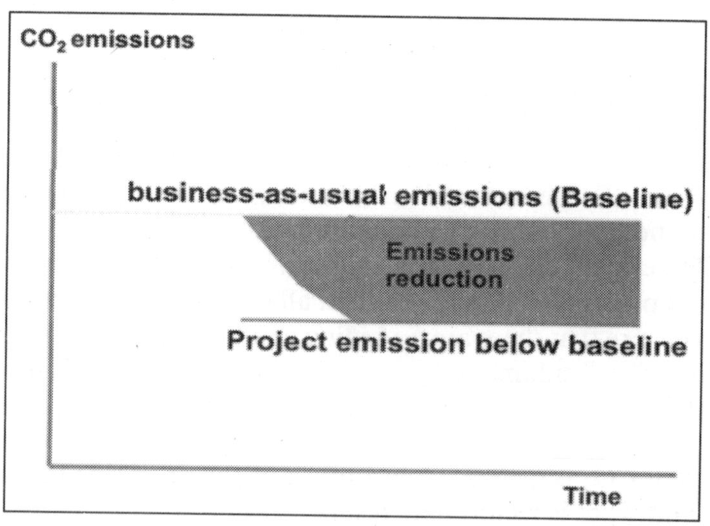

Administrative Process of CER.

1. Setting up a Clean Development Mechanism (CDM)
   ↓
2. Registration with CDM Executive Board of the UNFCCC
   ↓
3. Development of Project Design Document
   ↓
4. Approval from the Designated National Authority (*DNA)
   ↓
5. Validation by **DOE
   * Designated National Authority (DNA) which is an office, ministry or other official entity appointed by a Party to the Kyoto Protocol to review and give national approval to projects proposed under the CDM.
   ** A DOE (Designated operational Entity) is a company/organisation accredited by the CDM Executive Board that checks whether projects meet the CDM criteria.
   UNFCCC= United Nations Framework Convention on Climate Change
   Accounting Treatment and Taxability of Carbon Credit Accounting Treatment

Treated as Asset

1. Expenses incurred in the administrative process can be treated as contingent assets according to AS 29 because of Provisions, Contingent Liabilities and Contingent Assets, which defines a contingent asset as ***"a possible asset that arises from past events the existence of which will be confirmed only by the occurrence or non-occurrence of one or more uncertain future events not wholly within the control of the enterprise".*** This is because when the generating entity reduces carbon emissions by way of a CDM project, the generating entity becomes eligible to receive CERs from UNFCCC. However, whether CERs will actually arise and be received by the generating entity or it will not depend on a future uncertain event, i.e., certification of the same by UNFCCC.
2. When CER comes into existence by communication received by UNFCCC, it can be treated as "***Intangible Asset*** " according to AS 26 which provided following definition:
   "***An intangible asset is an identifiable non-monetary asset, without physical*** substance, held for use in the production or supply of goods or services, for ***rental to others, or for administrative purposes.***"

The other requirements of AS 26, such as the following, indicate that intangible assets include assets which are developed by an entity for sale:

"44. An intangible asset arising from development (or from the development phase of an internal project) should be recognised if, and only if, an enterprise can demonstrate all of the following:

(a) the technical feasibility of completing the intangible asset so that it will be available for use or sale;
(b) its intention to complete the intangible asset and use or sell it;
(c) its ability to use or sell the intangible assets;
(d) ......
(e) *the availability of adequate technical, financial and other resources to complete the development and use or sell the intangible asset."*

Treated as Inventory

Intangible assets held for the purpose of sale in the ordinary course of business are excluded from the scope of AS 26 (paragraph 2). Therefore, such assets are to be accounted for as per Accounting Standard (AS) 2, Valuation of Inventories. In this context, the definition of the term 'inventories' as given in AS 2 is noted below:

"Inventories are assets:

(a) held for sale in the ordinary course of business;
(b) in the process of production for such sale;
(c) in the form of materials or supplies to be consumed in the *production process or in the rendering of services."*

So, according to above provision, CER can be treated as inventory.

Treated as a Government Grant

If Carbon Emission Reduction Exchange will be developed under government supervision or government restriction, then carbon credits can be treated as government grant or aid. According to provisions of AS-12 government grant can be classified in two types, i.e., (1) Revenue Income Approach, and (2) Capital Income Approach.

Provisions are as follows available in AS 12

5.2 Those in support of the 'capital approach' argue as follows:

(i) Many government grants are in the nature of promoters' contribution, i.e., they are given with reference to the total investment in an undertaking or by way of contribution towards its total capital outlay and no repayment is ordinarily expected in the case of such grants. These should, therefore, be credited directly to shareholders' funds.
(ii) It is inappropriate to recognize government grants in the profit and loss statement, since they are not earned but represent an incentive provided by government without related costs.

5.3 Arguments in support of the 'income approach' are as follows:

(i) Government grants are rarely gratuitous. The enterprise earns them through compliance with their conditions and meeting the envisaged obligations. They should, therefore, be taken to income and matched with the associated costs which the grant is intended to compensate.

(ii) As income tax and other taxes are charges against income, it is logical to deal also with government grants, which are an extension of fiscal policies, in the profit and loss statement.

According to above provisions on the sale of Carbon credit it can be accounted as Government grant.

Taxability of Sale of Carbon Credit

1. If Carbon Credit accounted as intangible assets, than gain received from sale or transfer of such type of assets, will be taxable under the head of "Income from Capital Gain" of Income Tax Act. According to, Income Tax Act capital gain can be long term or short term regarding term of holding period of capital assets. It is suggested that holding period can be calculated from the validation date of CER.
2. If Carbon Credit accounted as inventory than sale proceeds will be taxable under the head of "Business and Profession" of Income Tax Act as receipt of normal course business u/s 28.
3. If Carbon Credit accounted as government grant, it will be taxable under the head of "Business and Profession" of Income Tax Act u/s 28.
4. The recent notification has been issued by National Capital Territory, Delhi regarding carbon credit transaction. The Notification explains that in substance CERs are tradable commodity. They have a market value, having a ready market, with willing buyers and sellers and are freely transferable as other marketable commodities. Thus carbon credits should be considered to be goods under the sales tax laws and any person/company/undertaking/entity engaged in the activity of sale or purchase of carbon credits is a "dealer" in terms of the definition of dealer as contained in Section 2(1) of the DVAT Act 2004.

The notification further explains that under section 2(1)(m) of the DVAT Act 2004, "goods" has been defined as–

> "goods" means every kind of movable property (other than newspapers, actionable claims, stocks, shares and securities) and includes (i) livestock, all material, commodities, grass or things attached to or forming part of the earth which are agreed to be served before sale or under a contract of sale; and (ii) property in goods (whether as goods or in some other form) involved in the

execution of a works contract, lease or hire-purchase or those to be used in the fitting out, improvement or repair of movable property".

5. Trading in CER is carried out either in spot market or in futures. Service Tax will be applicable on account of dealing in CERs on the exchange platform and in case of contracts resulting in delivery VAT will be applicable.
6. According to Kyoto Protocol, India is an Annexure II country who can sale carbon credit. Carbon Credit in India are sold to overseas buyers and hence there would be no VAT applicable on these goods.

NOTES

C.A. Sanjay K. Agarwal, Accounting and Taxation Aspect of Carbon Trading, October 2006 *The Chartered Accountant*, p. 509

Nidhi Bothra, Vinod Kothari & Company, *Carbon Credits—Unravelling Regulatory, Taxation and Accounting Issues*, p. 2

Exposure Draft: *Guidance Note on Accounting for Self-generated Certified Emission Reduction (CERs)*, June 23, 2009, Accounting Standard Board, Institute of Chartered Accountant of India, p. 6

David Disch, *Carbon Finance: A Guide for Sustainable Energy Enterprises and NGOs*, p. 7

Making of a Carbon Conscious Company: An Introduction to Carbon Accounting Advisory, 2009, p. 13

Ricardo Bayon, Amanda Hawn: *Voluntary Carbon Markets*, p. 1.

22

Climate Change and Green Power Generation in Rajasthan

Anand Chopra

Introduction

India is faced with the challenge of sustaining its rapid economic growth while dealing with the global threat of climate change. This threat emanates from accumulated green house gas emissions in the atmosphere, generated through long-term and intensive industrial growth and high consumption lifestyles in developed countries. While engaged with the international community to collectively and cooperatively deal with this threat, India should strongly push the national strategy to firstly, adapt to climate change; and secondly, to further enhance the ecological sustainability of India's development path.

Climate change has altered the distribution and quality of India's natural resources and adversely affect the livelihood of its people. With an economy closely tied to its natural resource base and climate-sensitive sectors such as agriculture, water and forestry, India is facing a major threat because of the projected changes in climate.

India's development path is based on its unique resource endowments, the overriding priority of economic and social development and poverty eradication, and its adherence to its civilization legacy that places a high value on the environment and the maintenance of ecological balance.

Clean Development Mechanism (CDM)

The CDM allows emission-reduction projects in developing countries to earn certified emission reduction (CER) credits, each equivalent to one tonne of CO_2. These CERs can be traded and sold, and used by industrialized countries (Developed Countries) to meet a part of their emission reduction targets under the Kyoto Protocol.

The mechanism stimulates sustainable development and emission reductions, while giving industrialized countries some flexibility in how they meet their emission reduction limitation targets.

India as a developing country has resource availability as well as investment opportunity. Hence, developed countries/NGOs and other societies have eye on India to invest in these resources and investment opportunities to meet out there emission reduction targets under Kyoto Protocol.

KPTL Vision and Action

Kalpataru Power Transmission Limited (KPTL) is a leading EPC Contractor for Power Transmission Lines (132 KV to 800 KV) for Power Utilities both in India and overseas. KPTL is a publicly listed company operating in various fields of Infrastructure and Contracting. KPTL has a world-class facility for designing, fabricating, galvanizing and testing of tower structures. It also has a full-fledged construction team for civil works, erection and stringing of power transmission lines.

In 2000, KPTL decided to gradually diversify its business activities by including Green Power Generation. Small-scale biomass based Power Projects were targeted by KPTL as a first step into that direction, through the assistance of Clean Development Mechanism (CDM) revenue on projects that might otherwise be too unattractive.

The 7.8 MW project in Ganganagar was the first of these projects started its operation in the year 2003 and is registered with the UNFCCC. Based on this first successful experience a final investment decision on the Tonk project was taken in March 2005 and its operation commenced in November 2006. This project is also registered with UNFCCC and reducing carbon emissions. This is a Gold Standard Project.

The electricity generated is sold to the State Grid.

Fig. 22.1: Uniara Power Plant 8.0 MW

Purpose of the Project

The purpose of the Rajasthan biomass projects is to contribute to an economically, environmentally and socially sustainable development in the region through the commercial operation of a power plant and thereby creating sustainable stakeholder value. The generated electricity replaces a mixture of coal and gas-based power generation which are emitting high range of pollution in the atmosphere.

Technology of Project Activity

The project involves the implementation of a biomass-based power generation plant using direct combustion boiler technology. The project is a case of utilizing an existing technology of combusting biomass fuels in a medium-pressure steam boiler.

The power plant is designed to operate with multiple agricultural biomass residues. The most important part of the feedstock comprises mustard crop residue, together with a small amount Cotton Crop Residue and other such Biomass only in case of shortages. A small quantity of wood chips is also consumed as a start-up auxiliary fuel.

The technology of biomass based medium steam pressure power generation itself is known and in use in India. However, the use of mustard crop residue as a fuel for power generation is still a pioneering effort by KPTL.

Contribution to Sustainable Development

The project contributes to the sustainable development of the region and country in the following ways:

1. Rajasthan had a peak deficit of 2.9 per cent at the end of the 2003, according to the Ministry of Power (MoP, 2003b). The biomass units contribute, though in a small measure, to bridging the gap between the supply and demand of power in the State.
2. The units are located at dispersed rural locations, which reduces the transmission and distribution (T&D) losses to some extent.
3. The project is in line with the policies of Ministry of Non Conventional Energy Sources (MNES) and Government of Rajasthan (GoR). It contributes to achievement of the Eleventh Plan target set by MNES. The objective of the Government of Rajasthan (GoR) is to raise the share of non-conventional Energy Sources. The project releases low carbon gases, thereby reducing carbon emissions, pollution and environmental risks.
4. Contrary to certain fossil fuel-fired plants, the project does not lead to an outflow of foreign exchange capital, since most capital equipment is locally produced and the biomass waste does not have to be imported. This is in accordance

with India's policy of self-reliance.

5. The plant is situated far from an urban centre, creating rural employment. Creation of employment opportunities in rural areas has long been recognised as a major element of sustainable development and to stem the large-scale migration from rural to urban areas. This concern has formed the cornerstone of most of the Government of India's (GoI) rural development programmes. To this extent, the project directly addresses a core national concern.
6. Besides creating income for farmers through the purchase of fuels, is also engaged in capacity building among local farmers and contribute to community development and human wellbeing.

Kalpataru Power Transmission Limited

Biomass Energy Division

Green Power Generation by Kalpataru

Kalpataru Power Transmission Limited (KPTL) situated in Gandhinagar, was established in 1981. It is a part of the KALPATARU GROUP, which has vast experience and expertise in Real Estate and Property Development, Industrial and Infrastructure Construction, Power Transmission, Distribution and Generation, Gas and Oil Pipeline laying, Manufacturing Plastic Films and Sheets, Trading in Office Stationery and International Trading, etc.

KPTL is one of India's leading companies in the Infrastructure Sector.

Introduction

Water and Electricity, being the basic need of every human being, have become vital and of core importance in the present scenario. Rajasthan has been facing shortage of Electricity (Power), due to lesser generation than demand through conventional sources, i.e., Thermal, Hydel, Gas, etc. This is because of geographical, climatic and other similar conditions and scarcity/no availability of conventional fuels in the State of Rajasthan.

Under the Government of Rajasthan Policy for Promoting Non-Conventional Energy Sources dated 11/03/1999, Kalpataru Power Transmission Limited has diversified into Power Generation by non-conventional fuels using Biomass like agro-forest waste, (Juli Flora, Firewood), agriculture waste (Mustard Crop Residue, Cotton Sticks), etc., in the State of Rajasthan.

In August 2003 KPTL created History in Power Generation Sector of Rajasthan and established first 7.8 MW Biomass Based Power Plant at Chak 27 BB, Tehsil Padampur, District Sri Ganganagar. This Plant is the first Biomass based Power Plant of Rajasthan and also first of its kind in the world using Mustard Crop Residue as fuel and, therefore, is a unique project.

Another 8 MW Biomass based Power Plant has been successfully started on 10th November 2006 at Village Khatoli, Tehsil Uniara, District Tonk, Rajasthan. Both the plants are running at their optimum level and each plant are generating approximately 5.5 crore Units per annum at a PLF of 84 per cent on Working Days and 92 per cent on running days.

Biomass Based Power Plant

Policy for Promoting Non-Conventional Energy Sources: In the State of Rajasthan Government has issued policy for promoting non-conventional energy sources for power generation. At present the policy 2004 is active.

Biomass as Fuel and its Procurement

Biomass is a promising fuel. It should have a good GCV and reliable availability. The Biomass of such kind available in Rajasthan is like mustard crop residue, cotton sticks, Gwar residue, groundnut residue, soyabeen residue, julie flora, and fire wood.

Location of the Plant

The ideal location for Biomass based power plant should be:

1. The area where Biomass availability is regular and abundant;
2. Water availability; and
3. Near to 132 KV Grid sub station.

Sale of Electricity

For Electricity sale, the Power Purchase Agreements with Rajasthan Vidhyut Parsaran Nigam Limited (RVPNL) and three Discoms JVVNL, AVVNL and JD.VVNL is to be entered.

The Power can also be sold to third party, consumer having capacity more than one MW, for which power and wheeling agreement is to be entered with RVPN.

Socio-Economic and Environmental Benefits

Biomass Power Plants are pioneering efforts in Rajasthan and bring various social, economical and environmental benefits such as:

1. Bulk utilization as fuel of an agro-waste, so far of no use.
2. Realizing fair/reasonable prices for crop residue.
3. Greatly increasing income of farmers by disposal of agro-wastes.
4. Reducing in use of conventional/fossil fuels (coal).

5. Neutralizing greenhouse effects.
6. Helping in reducing T & D losses of the State, being generation dispersed mostly in rural areas.
7. Increasing availability of qualitative power in the region.
8 Providing Rural employment directly and indirectly.

Index